T0292703

CAMBRIDGE LIBRARY COLLECTION

Books of enduring scholarly value

Botany and Horticulture

Until the nineteenth century, the investigation of natural phenomena, plants and animals was considered either the preserve of elite scholars or a pastime for the leisured upper classes. As increasing academic rigour and systematisation was brought to the study of 'natural history', its subdisciplines were adopted into university curricula, and learned societies (such as the Royal Horticultural Society, founded in 1804) were established to support research in these areas. A related development was strong enthusiasm for exotic garden plants, which resulted in plant collecting expeditions to every corner of the globe, sometimes with tragic consequences. This series includes accounts of some of those expeditions, detailed reference works on the flora of different regions, and practical advice for amateur and professional gardeners.

Planting and Ornamental Gardening

William Marshall (1745–1818), an experienced farmer and land agent, published this work anonymously in 1785. (His later, two–volume *Planting and Rural Ornament* is also reissued in this series.) His intention here is 'to bring into one point of view, and arrange in compendious form, the Art of Planting and Laying–Out Plantations', which had been treated by previous authorities as two distinct subjects. The book begins with instructions on propagation, planting out and transplanting, followed by an outline of the Linnaean system and an extensive alphabetical plant list (by Latin names). The second part consists of advice on the use of plants in the wider landscape: stands of timber for cutting, hedges, woodlands, and the landscaping of 'grounds'. This includes a history of modern gardening, and discussions of 'factitious accompaniments', including the hunting–box and the ornamented cottage, though not the temple, 'a great act of folly' according to Marshall's criteria.

Cambridge University Press has long been a pioneer in the reissuing of out-of-print titles from its own backlist, producing digital reprints of books that are still sought after by scholars and students but could not be reprinted economically using traditional technology. The Cambridge Library Collection extends this activity to a wider range of books which are still of importance to researchers and professionals, either for the source material they contain, or as landmarks in the history of their academic discipline.

Drawing from the world-renowned collections in the Cambridge University Library and other partner libraries, and guided by the advice of experts in each subject area, Cambridge University Press is using state-of-the-art scanning machines in its own Printing House to capture the content of each book selected for inclusion. The files are processed to give a consistently clear, crisp image, and the books finished to the high quality standard for which the Press is recognised around the world. The latest print-on-demand technology ensures that the books will remain available indefinitely, and that orders for single or multiple copies can quickly be supplied.

The Cambridge Library Collection brings back to life books of enduring scholarly value (including out-of-copyright works originally issued by other publishers) across a wide range of disciplines in the humanities and social sciences and in science and technology.

Planting and Ornamental Gardening

A Practical Treatise

WILLIAM MARSHALL

CAMBRIDGE
UNIVERSITY PRESS

CAMBRIDGE
UNIVERSITY PRESS

University Printing House, Cambridge, CB2 8BS, United Kingdom

Cambridge University Press is part of the University of Cambridge.
It furthers the University's mission by disseminating knowledge in the pursuit of
education, learning and research at the highest international levels of excellence.

www.cambridge.org
Information on this title: www.cambridge.org/9781108075930

© in this compilation Cambridge University Press 2017

This edition first published 1785
This digitally printed version 2017

ISBN 978-1-108-07593-0 Paperback

PLANTING

AND

ORNAMENTAL GARDENING;

A

PRACTICAL TREATISE.

PLANTING

AND

ORNAMENTAL GARDENING;

A

PRACTICAL TREATISE.

LONDON:

Printed for J. DODSLEY, Pall-Mall.

MDCCLXXXV.

ADVERTISEMENT.

THE Intention of this Publication is to bring into one point of view, and arrange in a compendious form, the Art of Planting and Laying-out Plantations: an art which, though in itself a unity, has hitherto been treated of as two diſtinct ſubjects. Books upon Planting we have many; and thoſe upon ornamental Gardening are not leſs numerous; but a Practical Treatiſe comprehending the entire ſubject of conducting rural improvements upon the principles of modern taſte, has not hitherto appeared in public. This circumſtance, however, is the leſs to be wondered at, as the man of buſineſs and the man of taſte are rarely united in the ſame perſon. There are many Nurſerymen who are intimately acquainted with the various methods of propagating trees and ſhrubs; and many gentlemen whoſe natural taſte, reading, and
obſervation,

obfervation enable them to form juft ideas of rural embellifhment; but where fhall we find the Nurferyman who is capable of ftriking out the great defign, or the Gentleman equal to the management of every tree and fhrub he may wifh to affemble in his collection? To proceed one ftep farther, where is the Gentleman, or the Nurferyman, who is fufficiently converfant in the after treatment of Woodlands, Hedges, and the more ufeful Plantations? In fine, where fhall we look for the man who in the fame perfon unites the Nurferyman, the Land-Steward, the Ornamentalift, and the Author? We know no fuch man: the reader therefore muft not be difappointed when he finds that, in treating of exotic trees and fhrubs, the works of preceding writers have been made ufe of.

Cook is our firft writer on Planting; neverthelefs Evelyn has been ftyled the Father of Planting in England. It is probable that, in the early part of life, Evelyn was a practical planter upon his eftate at Wotton in Surrey; but his book was written in the wane of life, at Greenwich, during a long and painful fit

of

of the gout. His *Sylva* contains many practical rules, valuable, no doubt, in his day,
but now fuperfeded by modern practice; and
may be faid to lie buried in a farrago of traditional tales and learned digreffions fuited to
the age he lived in *. MILLER at length arofe
among a group of minor planters; and after
him the indefatigable HANBURY, whofe immenfe labours are in a manner loft to the
Public.

Cook and Evelyn treated profeffedly of Fo-
REST-TREES, Miller and Hanbury include
ORNAMENTALS; but their works, which are
voluminous and expenfive, alfo include kitchen-
gardening, flower-gardening, the management
of green-houfes, ftoves, &c. &c. the propagation of trees and fhrubs, adapted to the
open air of this climate, forming only a fmall
portion of their refpective publications.

Miller and Hanbury, however, are the only
writers who could afford us the required affiftance; and we were led to a choice of the
latter, as our chief authority, by three prin-

* The firft Edition was printed in 1664, having been previoufly read before the Royal Society in 1662.

cipal

cipal motives :—Hanbury wrote fince Miller,
and having made ample ufe of Mr. M.'s book,
his work contains in effect the experience of
both writers : Miller is in the hands of moft
gentlemen ; Hanbury is known to few; his
book, either through a want of method, a
want of language, or through an ill-judged
plan of publifhing on his own account, has
never *fold :* and laftly, Miller's botanical arrangement is become obfolete; Hanbury's is
agreeable to the Linnean fyftem.

Since Mr. Hanbury's death, the public have
been favored with a new and fumptuous edition
of Evelyn's Sylva; with notes by Dr. Hunter
of York, confifting of botanical defcriptions,
and the modern propagation of fuch trees as
Evelyn has treated of. Thefe notes, however,
contain little new information ; the defcriptions
being principally copied from Miller, and the
practical directions from Hanbury.

Left unacknowledged affiftance, or affiftance
acknowledged indirectly fhould be laid to our
charge, it is thought proper in this place to
particularize the feveral parts of this publication which are *written* from thofe which are

ADVERTISEMENT. ix

The INTRODUCTORY DISCOURSES, containing the Elements of Planting, and the Outline of the Linnean Syſtem, are, as rudiments, entirely new; excepting the quotations from Linneus's work, which quotations are extracted from the Lichfield Tranſlation of The *Syſtema Vegetabilium* of that great man.

The ALPHABET OF PLANTS, ſo far as it relates to TIMBER-TREES, and other NATIVE PLANTS, as well as to ſome of the more USEFUL EXOTICS, is either wholly our own, or contains ſuch additions as have reſulted from our own obſervation and experience : ſo far as it relates to ORNAMENTAL EXOTICS, it is entirely HANBURY's ; excepting the quotations which are marked, and excepting the GENERAL ARRANGEMENT, which is entirely new. HANBURY has not leſs than ſix diſtinct claſſes for the plants here treated of, namely, deciduous Foreſt-Trees, Aquatics, evergreen Foreſt-Trees, deciduous Trees proper for ornament and ſhade, evergreen Trees proper for ornament and ſhade, and hardy climbing Plants. The firſt three claſſes are without any ſubordinate arrangement ; in the laſt three the plants are arranged alphabetically, agreeably

to

to their genera. This want of fimplicity in
the arrangement renders the work extremely
heavy and irkfome to refer to; and is productive of much unneceffary repetition, or of
tirefome references from one part of his un-
wieldy work to another. His botanical fyno-
nyms we have wholly thrown afide, as being
burdenfome, yet uninftructive; and in their
place we have annexed to each Species the tri-
vial or fpecific name of LINNEUS, which in
one word identifies the plant with a greater
degree of certainty than a volume of Synony-
ma. Other retrenchments, and a multiplicity of corrections have taken place : however,
where practical knowledge appears to arife
incidentally out of our author's own experience, we have cautioufly given it in his own
words : likewife, where interefting information
lies entangled in a fingularity of manner, from
which it could not well be extricated, we have
marked the paffages containing it, as literal
quotations ;---to diftinguifh them from others,
which, having been written in a manner more
properly didactic, or brought to that form by
retrenchment or correction, we confider as being

ing more fully entitled to the places we have affigned them.

The articles TIMBER, HEDGES, and WOOD-LANDS, are altogether new *, being drawn from a confiderable fhare of experience, and an extended obfervation.

The article GROUNDS is likewife new, if any thing new can be offered on a fubject upon which fo much has been already written. Tafte, however, is a fubject upon which all men will think and write differently, even though their fources of information may have been the fame. WHEATLEY, MASON, and NATURE, with fome EXPERIENCE, and much OBSERVATION, are the principal fources from which this part of our work was drawn : if we add that it was planned, and in part writ-ten, among the magnificent fcenes of nature in Monmouthfhire, Herefordfhire, and Gloucef-terfhire, where the rich and the romantic are happily blended, in a manner unparalleled in any other part of the Ifland, we flatter our-felves no one will be diffatisfied with the *origin* of the *production*, let the Public fpeak.

---

* Excepting fuch extracts and quotations as are marked, and have their refpective authorities fubjoined.

CONTENTS.

GROUNDS

C O N T E N T S.

ERRATUM.

Page 14. l. 3. *after* them, *read*—the part placed in contact with the foil fends forth roots, whilft that expofed to the open air fends forth branches.

INTRODUCTION

TO

PLANTING

AND

ORNAMENTAL GARDENING.

GENERAL VIEW.

THE earth produces an almoſt infinite variety of Plants, poſſeſſing various properties, and different degrees of ſtrength and ſtature. In the vegetable as in the animal world, the ſtronger ſubdue the weaker : the herbaceous tribes bow to the ſhrub, and this to the more robuſt foreſt-tree; and in an unpeopled country a ſtate of woodineſs prevails. The interior parts of America are at this day a foreſt : the Continent of Europe too has ſtill its foreſt; and England once was famous for her's.

As inhabitants increaſe, woodineſſes give way to huſbandry and the arts; not merely as incumbrances, but as affording uſeful materials. Population ſtill increaſing, the *foreſt* breaks into *woods*.

Commerce

Commerce and luxury advancing, the canoe becomes a ſhip, and the cottage a manſion; at length even the woods dwindle away, and *plantations*, or an *import of foreign timber*, become neceſſary to ſupply the want.

England has experienced, more or leſs, every ſtage of this decline. Its preſent ſtate, in reſpect to timber, we conceive to be this : A few broken foreſts and many extenſive woodlands ſtill remaining; a great number of plantations of different growths, and a vaſt ſupply of foreign timber of various kinds. Indeed, we are of opinion, that had it not been for this foreign ſupply, ſcarcely a timber-tree, at this day, would have been left ſtanding upon the iſland.

Our exiſtence, as a nation, depends upon a full and *certain* ſupply of ſhipping; and this, we may venture to ſay, upon an *internal* ſupply of ſhip-timber. That there is no want of oak-timber at preſent in this iſland is, we believe, a fact; but that the article of *ſhip*-timber is growing ſcarce, as we ſhall explain more fully in its proper place, is, we believe, alſo a fact which cannot be controverted. This is an important matter, which demands the firſt attention of Government, and is not unworthy the notice of every landed individual.

Mankind, however, do not view the face of nature in the light of ſelf-preſervation only; the great

Author

Author of creation has wonderfully adapted our fenfes to the enjoyment of its delights; the eye is gratified by tints of verdure, and the ear by the mufic of the woods and the mellownefs of echo— and both by the voice and majefty of a foreft roufed by the breath of Nature. Our plan therefore has two objects, UTILITY and ORNAMENT; they are nearly allied, however, as labour and recreation, or as the ufe and the ornament of drefs.

But before we give directions for raifing a wood, or ornamenting the face of a country, we muft firft treat feparately of each individual tree and fhrub adapted to our purpofe; and, preparatory to this, give a comprehenfive view of the operations incident to

| PROPAGATING, | PLANTING-OUT, and |
| TRAINING-UP, | TRANSPLANTING. |

Trees and Shrubs in general.

PROPAGATING TREES AND SHRUBS.

TREES and Shrubs are propagated

From SEEDS,	by LAYERING,
—— SUCKERS,	— BUDDING,
—— CUTTINGS,	— GRAFTING.

But before the young planter put his foot upon the fpade, we beg leave to caution him in the ftrongeft terms againft a WANT OF SPIRIT. A flo-

venly planter ranks among the moſt extravagant
order of ſlovens : the labour, the plants, and the
ground are thrown away ; beſides the conſequent
diſgrace, not only to the individual himſelf, but to
the profeſſion in general. Anxious and intereſted
as we are in the cauſe of planting, we would rather want pupils than have them paſs through our
hands *unfiniſhed:* we therefore rejeſt all ſuch as
have not induſtry, ſpirit, and perſeverance, to go
through with what they undertake ; and we recommend to ſuch as are poſſeſſed of theſe valuable
qualifications, *to begin upon a ſmall ſcale,* and to let
their ſeminary, their nurſery, and their plantations
increaſe with their experience.

Whilſt, however, we caution againſt entering
immaturely upon the buſineſs of planting, we cannot refrain from mentioning the PLEASURES which
reſult from it. How rational, and to a contemplative mind how delightful, to obſerve the operations
of Nature ;—to trace her in every ſtage, from the
ſeed to the perfeſted plant ; and, from beneath the
leaf-ſtalk of this, through the flower-bud, the
flower, and the ſeed-veſſel, to the ſeed again ! Man
muſt be employed ; and how more agreeably than
in converſing with Nature, and in ſeeing the works
of his own hands, aſſiſted by her, riſing into perfection.

Nor

Nor do we mean to hold out pleafure alone as an inducement to planting;—its PROFITS are great, when properly executed, and this idea adds folidity to the enjoyment. Pleafure alone may fatiate; but profit and pleafure united feldom fail of producing a lafting gratification.

There is another incitement to planting, which alone has been generally held out as a fufficient inducement. We are forry to confefs, however, that we know too much of mankind, to believe that PATRIOTISM, unaided by perfonal intereft, will ever produce a fupply of fhip-timber to this or any other nation. Far be it from us, however fafhionable it may be, to fpeak irreverently of patriotifm; we confider it as the nobleft attribute of the human mind. Young men, to whom we more particularly addrefs ourfeives, are feldom without fome fhare of it; and we flatter ourfelves that this virtuous principle, affifted by the pleafure, the profit, and the POPULARITY which attends planting,—ornamental plantations more particularly,—will induce the young men of the prefent age to ftudy and practife it; not more for themfelves than for future generations.

PROPAGATING FROM SEED.—There are four ways of raifing from feed the trees and fhrubs adapted to our purpofe:

In

In Beds of natural Soil,
In Beds of Compoſt,
In Pots,—and ſome few
In Stoves, or under Glaſſes.

It will be expeĉted, perhaps, before we begin
to treat of the different methods of ſowing, that we
give ſome direĉtions for GATHERING and preſerving ſeeds. Little, however, can be ſaid upon the
ſubjeĉt under this general head ; different ſpecies
requiring a difference in management. We may,
nevertheleſs, venture to ſay, that all ſeeds ought to
be fully matured upon their native plants ; and we
may further add, that ſuch as drop ſpontaneouſly
from the ſeed-veſſel, or are ſhed by a moderate
wind, or other gentle agitation, are preferable to
thoſe which are torn from the tree immaturely.
The ſeeds of ſcarce or valuable plants may be gathered thus : As ſoon as they begin to fall of themſelves, ſpread a cloth under the plant, and agitate it
moderately, until all that are ripe have fallen ;—
and repeat this whenever a ſecond and a third
ſpontaneous fall takes place.

The art of PRESERVING ſeeds reſts chiefly upon
that of *curing* them immediately after gathering.
If graſs were put into the ſtack immediately after
mowing, or corn threſhed out at harveſt and laid in
heaps, it would preſently heat and become entirely
ſpoiled. So it is with the ſeeds of trees and ſhrubs :
therefore

therefore they ought, as foon as they are gathered, to be fpread thin in an airy place, and be turned as often as a clofe attention fees neceffary. When the fuperfluous moifture has evaporated, they may be collected into bulk; remembering, however, to run them every now and then down a fkreen, or fhake them in a fieve, that their brightnefs and fweetnefs may be preferved. Some of the larger feeds, acorns efpecially, are difficult to cure, and require a very ftrict attention. It muft alfo be remembered, that mice and other vermin are dangerous enemies to feeds. Thofe which are particularly valuable, may be hung up in bags to the ceiling of a dry room.

In procuring feeds from the SHOPS, or from ABROAD, fome caution is neceffary. A reputable feedfman, and a correfpondent who is himfelf a judge of the quality of feeds, are the beft general guards againft impofition and difappointment.

There are feveral ways of TRYING THE QUALITY of feeds. The heavier kinds may be proved in water; fuch as fwim are at leaft doubtful. The lighter forts may be tried by biting them: if they break abruptly between the teeth, they are generally good; but if they be tough and leathery, they are moftly the contrary. If when crufhed, or feparated by a knife or fciffars, they appear firm, white, and farinaceous, they may generally be efteemed good; but if on the contrary they be fpungy and difco-

B 4 loured,

loured, they are generally of a bad quality. But the moſt certain mode of trial, and that which in caſes of ſuſpicion ought never to be omitted, is to *force* a few of them in a garden-pot, placed in an artificial heat, or other warm ſituation. Put in ſome certain number, taken promiſcuouſly from the parcel, and, from the proportional number that vegetate, a tolerably juſt idea may be formed of the quality of the whole. Without this precaution a ſeaſon may be loſt, and the uſe of the land, together with the labour, be thrown away.

All the natives, and many exotics, may be raiſed in BEDS OF NATURAL MOULD. The ſoil ſhould be rich, and ſufficiently deep to admit of being trenched or double dug two ſpit deep. If it will not bear one ſpit and a half, namely about fourteen inches, it is improper for ſeed-beds, and ſhould either be wholly rejected, or (if the ſubſtratum is not of too hungry and poiſonous a nature) be trenched a ſpit and a half deep, and the crude mould meliorated by manure and repeated diggings. Autumn is the beſt time to bring up the ſubſtratum, letting it lie in rough trenches all winter to take the froſt. In the ſpring put on a quantity of dung, in proportion to the poverty of the ſoil; turning it in ſuperficially, and mixing it well with the ſoil to be improved. Repeat this ſingle digging, through the ſummer, as often as convenient, or as often as the weeds, which

never

never fail to rife in great abundance from a fubftratum expofed to the fun and air, require it. In
autumn turn up the foil from the bottom, and mix
the whole well together. The longer the foil and
fubftratum lie in the ftate of inverfion, the better
tempered the frefh mould will become, and the
mellower will be the old cultivated foil. In a manner fimilar to this, all foils which are not naturally
rich ought to be treated. No department of planting calls more loudly for a fpirited management
than the feminary, which, if not rich and deep by
nature, ought to be made fo by art, at almoft any
expence.

In large undertakings a feparate feminary may be
neceffary ; but, in general, a portion of the kitchen
garden is better adapted to the purpofe. There are
indeed two very great advantages in mixing the
feminary with the kitchen-garden : the feed-beds
are always under the eye, and are more likely to
be defended from weeds and vermin there, than in
a detached feminary vifited only now and then ;
and, when the ground has borne a crop of feedling
plants, it may be applied to the purpofe of culinary
herbs ; whilft that which has been long under crops
of thefe may be changed to feed-beds. In whatever fituation they are placed, they muft be carefully fenced againft hares and rabbits, or the labour
of a whole feafon may be cut off in a few nights : in
this light alfo the kitchen-garden has a preference.

It

It would be idle to give particular directions for laying out a SEMINARY, or to say under this general head where this or that seed should be sown. Suffice it therefore to mention here, that SEED-BEDS are generally made from four to four feet and a half wide, with intervals of one foot and a half to two feet. These dimensions render them convenient to be weeded, without the plants being trodden or kneeled upon.

The METHOD OF SOWING is various: By DIBBLING, by DRILLING, and by BROADCAST, which last is the most prevailing method. Seeds sown in the promiscuous broadcast manner are covered either with the rake, or with the spade (or sieve). COVERING WITH THE SPADE (or sieve) is the common practice, and is thus performed : The surface being made light and fine by a recent digging and raking, and the beds formed (operations which every gardener and gardener's man are acquainted with), a thin coat of mould is raked off the beds into the intervals, in proportion to the depth the seeds require to be buried, and according to the nature of the soil, taken jointly. In a light sandy soil, the seeds require to be buried deeper than they do in a strong loam; and whilst an acorn may be covered from one to three inches, the seeds of Larch will not bear more than from a quarter to three-fourths of an inch. The new surface being rendered perfectly fine and level, the seeds are

sown,

fown, and in fome cafes preffed gently into the
mould, by patting it with the back of the fpade.
The earth which was raked off into the interval (or
taken off with a fpade and placed in little hillocks
in it) is now returned ; either by cafting it on with
the fpade, with a kind of fleight which nothing but
practice can give, or by fifting it on through a
fieve (an operation more eafy to the inexpert, and
in many cafes preferable) as even and regularly as
poffible. The intervals cleared, the beds neated
up, and if the foil be light, or the feed requires
it, their furfaces patted with the back of the fpade,
fo as to give them a kind of polifhed firmnefs, the
bufinefs is finifhed. DRILLING is performed two
ways : By drawing open drills with hoes in the
common manner, or by taking off the furface of
the beds, drawing lines upon the new furface, laying or fcattering the feeds along thefe lines, and
covering them with the fpade or fieve, as above
directed for broadcaft fowing.

The next bufinefs of the feminary is to DEFEND
the feed and feedling from BIRDS, VERMIN, the
WEATHER, and WEEDS. Nets are the beft guard
againft birds, and traps againft vermin. As a defence againft the fcorching heat of the fun, the
beds fhould be hooped, and matts occafionally
fpread over them, in the manner of a tilt or awn-
ing ; but when the fcorching abates the matts fhould
be taken off, to give the plants the benefits of the
atmofphere ;

atmofphere; and in dry weather the beds fhould be kept conftantly watered. The awnings are equally fafe-guards againft fpring-frofts, than which nothing is more injurious to feedling plants. In refpect to WEEDS, there is a general rule, which ought not to be departed from ; that is, not to fuffer them to get too ftrong before they be drawn ; for, if they be permitted to form large roots, they not only encum r and rob the ground, but in drawing them many feeds or tender feedlings will be drawn out with them. To prevent the young plants from be_ing DRAWN OUT OF THE GROUND BY WINTER-FROSTS, which they are very liable to, efpecially by a continuance of froft and thaw alternately, coal-afhes may be fifted over them. If this evil has already taken place, and the roots appear expofed aboveground, fome fine mould fhould firft be fifted on to cover the roots, and then the afhes fifted over the mould. If the plants be BEATEN OUT OF THE GROUND BY HEAVY RAINS, the remedy is fimilar.

The length of time between the fowing of the feed and the appearance of the plant is very uncertain : much depends upon the feafon, and ftill more upon the nature of the plant itfelf. Some feeds lie in the ground a whole year before they vegetate, and fome two or three years (as will be mentioned under their refpective fpecies). During this time the beds fhould be kept free from weeds and mofs; and, in cafe of a long continuance of dry weather,

fhould

fhould be well watered. After very heavy rains, which are liable to run the furface to a batter and wafh away part of the foil, it is well to rake the beds flightly, and fift over them a little frefh mould : this prevents the furface from baking, and at once gives a fupply of air and nourifhment to the embryo plants.

BEDS OF COMPOST are made by mixing drift-fand, or other materials, with the natural foil of the feminary, or with virgin mould taken from a rich meadow or old pafture-ground. But the particular ingredients of a compoft depend upon the nature of the plant to be raifed; and the reader is referred to the refpective Species in the ALPHABET OF PLANTS for further information on this head.

The mode of raifing plants in POTS and BOXES alfo depends greatly upon the particular plant to be raifed. The chief intent of this method is to guard the embryo and feedling plants from the extremes of heat and cold. The pots are filled with compoft fuited to the plant. For examples, fee the articles ANNONA, ARALIA, AZALIA, MELIA, PISTACIA, &c. &c.

PROPAGATING FROM CUTTINGS.—It is not from feeds alone that plants may be increafed; fo great a fimplicity prevails in the fyftem of vegetation, that numerous tribes may be propagated from twigs or truncheons cut out of the woody

parts

parts of the plants themfelves, and ftuck naked into the ground without either root or branch upon them : the part in contact with the foil fends forth roots, which, expofed to the open air, fend forth branches!

But altho' moft of the aquatics, and many other genera of trees and fhrubs, may be raifed from CUTTINGS planted in common earth and in the open air, there are others which require more care and greater helps. Some require a warm, others a cool border : fome muft be rooted in pots, others in ftoves, or in the green-houfe. Again, fome fhould be taken from the older branches, others from younger fhoots: fome require to be planted in autumn, others in the fpring. Thefe and other peculiarities of treatment will be fpecified, when we come to treat feparately of each individual.

PROPAGATING FROM SUCKERS.—There is a great fimilarity between the branches and the roots of plants. If the fibres of fome fpecies become expofed to the air, they quit their function of fupplying the parent plant with nourifhment, and, taking upon them the nature of feedlings, put forth leaves and branches. Thefe rootling plants are called SUCKERS; and if they be flipt off from the parent root, and planted in a foil and fituation fuited to their refpective natures, they will grow up in the manner of feedling plants.

Various

Various opinions are held refpecting the propriety of raifing trees and fhrubs from fuckers: Evelyn and Miller are againft the practice; faying, that plants raifed from fuckers are more apt to fend up fuckers (which are troublefome intruders, efpecially in ornamental grounds) than thofe of the fame fpecies which have been raifed from feeds. Hanbury, however, is of a contrary opinion: he fays, " What might incline people to this notion was, that they have obferved trees raifed from feeds very long before they produced fuckers: but they fhould confider, that no tree or plant will produce fuckers till it is of a fuitable fize or ftrength for the purpofe, any more than animals can produce young before they are of proper age; and let them plant a feedling that is grown ftrong, a layer of the fame ftrength, and one which has been raifed from a fucker, exactly of the fame fize, and with the fame number of fibres to the root, and they will find that the feedling or the layer will not be behindhand with the other in producing fuckers, if they have all a like foil and fituation; for it is peculiar to them to fport under the foil in this manner; and Nature will ever act agreeably to herfelf, if not ftopped in her progrefs by art." Neverthelefs, in fpeaking of particular plants, we find him holding forth a different language.

PROPAGATING BY LAYERING.—As the roots of some plants when expofed to the air fend forth fhoots and branches, fo the branches of others, when placed in contact with the earth, fend out fibres and roots, which being fevered from the parent plant, a feparate tree is produced.

LAYERING being an operation by which a great majority of trees and fhrubs may be propagated, and by which the many beautiful variegations are principally preferved, we fhall here give fome general directions for performing it ; referving, however, the minutiæ peculiar to each fpecies until we come to treat of the individual fpecies feparately.

LAYERS are bent either from the *ftools* of trees and fhrubs headed down to a few inches above the furface of the ground, or from *boughs*, plafhed fo as to bend their tops to the ground ; or from *trees* brought into a ftooping pofture, by excavating the foil on one fide of them, until their heads are lowered into a fimilar fituation.

STOOLS afford the fimpleft, and are the moft common, fupply of layers. Where a great number of layers are wanted, plants fhould be raifed for the purpofe, and planted in fome well-fenced ground, or in fome vacant part of the feminary or nurfery ; and, when of a proper age and fize, headed down to the height of about eight inches for ftools. In many cafes, trees ftanding in grounds or woods may be

be cut down, and give a fufficient fupply. In whatever fituation they are, the earth round them muft
be double-dug as deep as the foil will allow, and be
treated in a manner fimilar to that of a feed-bed.
The METHOD OF LAYERING is this : Dig a ring-
trench round the ftool, of a depth fuitable to the nature of the plant; and having pitched upon the
fhoots to be layered, bend them to the bottom of
the trench (either with or without plafhing, as may
be found moft convenient), and there *peg* them
faft; or, putting fome mould upon them, *tread*
them hard enough to prevent their fpringing up
again ;—fill in the mould ;—place the top of the
layer in an upright pofture, treading the mould hard
behind it; and cut it carefully off above the firft,
fecond, or third eye.

In this fimple way a numerous tribe of plants
may be layered : there are many, however, which
require a more complex treatment. Some will
fucceed by having a *chip* taken off the under fide of
the lower bend of the layer, which gives the fibres
an opportunity of breaking out with greater freedom : others by having a *cleft* made in that part by
thrufting an awl or bodkin through it, keeping the
cleft open by a chip or wooden peg; or by making
a longitudinal *flit in the bark* only : others fucceed
better by *twifting* the part : and others again by
pricking it, and binding a *wire* round it. But when
SIMPLE LAYERING will not fucceed, the moft pre-

C   vailing,

vailing, and in general the moſt certain method is
that of TONGUE-LAYERING, which is thus performed: The excavation being made, and the layer choſen
and trimmed, aſcertain where the lower bend of it
will fall, by taking it in the left hand and bending
it down to the bottom of the trench ; then placing
the thumb of the right hand firmly againſt the part
oppoſite which the tongue falls, inſert the edge of
the knife as with an intent to cut the layer off ſhort
in that place; but having cut about half way thro'
it, turn the edge of the knife abruptly upwards,
drawing it along the pith half an inch, or an inch,
according to the ſize of the layer. The whole ſtool
being treated in this manner, proceed to peg the
layers cloſe to the bottom of the trench, bedding
the cleft or mouth of each in fine mould for the
fibres to ſtrike into. (If the mould and the ſeaſon
be very dry, it is well to moiſten ſome fine
mould with ſoft water, making it into a paſte, and
wrap the wounded part in a handful of this prepared earth.) This done, level in the mould,
draw the point of the layer upright, and ſhorten it
as above directed ; being careful to diſturb the
wounded part as little as poſſible. It is a practice
with ſome to trim the ſtools entirely after layering :
we would rather recommend, however, to trim off
ſuch ſhoots only as are too old, or are defective,
leaving ſuch as are too young to increaſe in growth ;

by

by which means an annual, inftead of a biennial, fucceffion of layers will be had.

The TIME OF LAYERING is generally autumn; fpring is favourable to fome plants, and midfummer to others; but trees and fhrubs in general may be layered at almoft any time of the year.

The length of time requifite for ROOTING a layer depends upon the nature of the plant : twelve months is generally confidered as a fufficient time, during which the layers fhould be kept clear from weeds; and when the rooted plants are taken off, the ftumps from which they were fevered fhould be cut off clofe to the ftools, in order that they may fend forth a future fupply of fhoots.

BUDDING and GRAFTING are operations more particularly applicable to fruit-trees, and belong to the *kitchen-gardener* rather than to the *planter*. They are operations difficult to defcribe upon paper; and are known to every nurferyman and gardener. The great art in *grafting* lies in uniting the graft clofely and firmly with the ftock; and, in *budding*, not to leave too much wood, nor yet to pare it off too clofe to the eye.

TRAINING-

TRAINING-UP TREES AND SHRUBS.

TREES and fhrubs may be trained up from the feed-bed, &c. until they be fit to be planted out to ftand, either in nurferies fet apart for the purpofe, or in YOUNG PLANTATIONS; which laft are frequently the moft eligible nurferies, as will be explained hereafter. A SEPARATE NURSERY however is neverthelefs neceffary; and, in this place, it will be proper to give fome general ideas of the foil, fituation, and bufinefs of a nurfery-ground.

The SOIL of the nurfery, like that of the feminary, fhould be rich and deep, and like that alfo fhould be prepared, by double-diggings and fuitable meliorations : if not deep and rich by nature, it muft be made fo by art, or be wholly rejected as unfit for the purpofes of a nurfery-ground. For if the roots of the tender plants have not a foil they affect, or a fufficient room to ftrike in, there will be little hopes of their furnifhing themfelves with that ample ftock of fibres which is neceffary to a good plant, and with which to fupply them is the principal ufe of the nurfery.

The SITUATION of the nurfery is frequently determined by the foil, and frequently by local conveniencies : the nearer it is to the garden or feminary, the more attendance will probably be given it; but the nearer it lies to the fcene of planting, the lefs carriage will be requifite. In whatever
fituation

fituation the nurfery be placed, it muft, like the feminary, be effectually fenced againft hares and
rabbits.

The BUSINESS OF THE NURSERY confifts principally in

PREPARING THE SOIL,	PRUNING,
TRIMMING THE PLANTS,	THINNING,
PUTTING THEM IN,	TAKING-UP, and
KEEPING THEM CLEAN,	PACKING FOR CARRIAGE.

The PREPARATION OF THE SOIL has already been
mentioned : too much pains cannot be taken in this
department ; it is the foundation upon which the
fuccefs of the whole bufinefs greatly depends.

In TRIMMING feedlings, layers and fuckers, for
the nurfery, the ramifications of the roots fhould
not be left too long and fprawling ; but *(in this
cafe)* fhould be trimmed off pretty clofe, fo as to
form a fnug globular root : by this means the new
fibres will be formed immediately round the root
of the plant, and may of courfe be eafily removed
with it, and without difturbing the earth interwoven
amongft them. The tops fhould in moft cafes
be trimmed quite clofe up to the leader, or (if auk-
ward or defective) be cut off a little above the
root.

In PUTTING-IN feedlings, various methods are
practifed : by the *dibble* ; by the *fcoop* ; by a fingle

C 3   chop

chop with the *fpade*, or by two chops, one acrofs the other : by *fquare holes* made by four chops of the fpade, bringing up the mould with the laft : or by *bedding* ; a method chiefly made ufe of for quick-fets. If the foil be well-prepared, and the plants properly trimmed, the chief art in putting them in lies in not cramping the fibres of the roots; but, on the contrary, in letting them lie free and eafy among the mould : and the particular mode, or in-ftrument to be made ufe of depends much upon the fize of the plants to be put in. This alfo de-termines, in a great meafure, the proper *diftance* between the rows and between plant and plant. Strong fuckers or layers require larger holes and a greater diftance than weak feedling plants. The propofed method of cleaning too is a guide to the diftance : the plow cannot work in fo narrow a com-pafs as the fpade. The natural tendency of the plant itfelf muft alfo be confidered; fo that few general directions can be given under this head. If we fay from fix to twenty-four inches in the rows, with intervals from one to four feet wide, we fhall comprehend the whole variation of diftances.

CLEANING THE NURSERY is a bufinefs which muft not, of all others, be neglected : all plants are enemies to each other. If grafs and weeds are fuf-fered to prey upon the foil, the young plants will be deprived of their proper nourifhment and moif-ture : in fhort, it is neceffary that the nurfery fhould

be

be kept equally clean as the feminary, and this as
clean as the kitchen-garden: it would be more
pardonable to fuffer the plants to be fmothered in
the feed-beds than in the nurfery quarters; for in
that cafe only a fmall part of the expence would be
thrown away. Nor is merely keeping the weeds
under, the only care in a nurfery : the intervals muft
be kept ftirred, in order to give air and freedom
to the fibres. This may be done either with the
fpade, which is called turning-in; or, if the intervals be wide enough and the nurfery extenfive,
with the plow, which is attended with much lefs
expence.

The next bufinefs of the nurfery is PRUNING :
this is neceffary to prevent the plants from crouding each other, and to give them ftem. Shrubs
which do not require a ftem fhould not be placed
in nurfery-rows, but in the quincunx manner, that
they may have an equal room to fpread on every
fide ; but foreft-trees, and trees in general, require
fome length of ftem ; and in giving them this, the
leading fhoot is more particularly to be attended
to. If the head be double, one of the fhoots muft
be taken clofe off : if it be maimed, or other ways
defective, it may be well to cut the plant down to
the ground and train a frefh fhoot; or if the head
be taken off fmooth, immediately above a ftrong
fide-fhoot, this will out-grow the crookednefs, and
in a few years become a ftraight plant.

The time of the plants remaining in the nurfery
is determined by a variety of circumftances, and a
feafonable THINNING frequently becomes neceffary.
In this part of the bufinefs there are general rules
to go by : the fhrubby fpreading tribes fhould be
thinned whenever their branches begin to interfere;
and the ftem-plants, whenever their roots get into
a fimilar ftate. If either the tops of the one, or
the roots of the other, be fuffered to remain in a
ftate of interference and warfare with each other,
the beauty of the fhrubs will foon be deftroyed,
and the thriftinefs of the trees will be checked. If
the plants be wanted for planting out, it is fortunate ; if not, every alternate plant fhould be moved
to a vacant ground prepared for the purpofe. If
fuch as ftand in rows be removed alternately into
the intervals, and fet in the quincunx form, a temporary relief will be gained at a fmall expence.

Planters in general are not aware of the caution
neceffary in TAKING-UP plants, for the purpofe of
planting them out to ftand. *In this cafe*, every root
and every fibre ought as much as poffible to be preferved. No violence fhould therefore be ufed in
this operation. The beft way is to dig a trench by
the fide of the plant or plants to be taken up; and
having undermined the roots let the plant fall of
itfelf, or with a very little affiftance, into the
trench : if any licentious root or roots ftill have
hold, cut them off with fome fharp inftrument, fo

as

as to jar the main root as little as may be. If the root was properly trimmed before planting, it will now turn out a globular bundle of earth and fibres, the beſt characteriſtic of a well-rooted plant.

When the nurſery lies at a diſtance from the plantations, or when the plants are to be ſent to ſome diſtant place, much depends upon PACKING them up judiciouſly. Valuable plants may be packed in pots or baſkets;—ſtraw may, however, in general be uſed, and will equally preſerve them from froſt in winter and the drought of autumn or ſpring; eſpecially if, in the latter caſe, the ſtraw be occaſionally moiſtened with ſoft water. Large plants ſhould be packed ſingly with as much earth about the roots as conveniency will allow. If a piece of matt be put over the ſtraw, it will ſave ſome trouble in cording, and be more effectual than ſtraw alone.

PLANTING-OUT TREES AND SHRUBS.

ALL that we propoſe in this place is to convey to our readers ſome general ideas of

PREPARING THE SOIL,
TRIMMING AND SORT-
  ING THE PLANTS,
PLANTING THEM OUT,
ATTENDING THEM AF-
  TER PLANTING,

CLEANING THE PLAN-
  TATION,
PRUNING THE PLANTS,
  AND
THINNING THEM;—

in

in order to avoid ufelefs repetitions, when we come to fpeak feparately of each individual fpe- cies; and to enable fuch of our readers as are totally unacquainted with the fubject, to follow us thro' the ALPHABET OF PLANTS with a degree of eafe and fatisfaction to themfelves, which, with- out thefe previous inftructions, they would not be able to do.

The PREPARATION OF THE GROUND depends in fome meafure upon the fize of the plants. To fpeak generally upon the fubject,—For plants under four feet high the foil ought to receive a double- digging, or a fummer's fallow under the plow, or a crop of turnips well-hoed; but, for larger plants, feparate holes dug in the unbroken ground are frequently made ufe of; though we cannot by any means recommend the practice. Trees and fhrubs never thrive better than when they are plant- ed upon *made ground*; for here the fibres rove at large, and the nearer the foil of a plantation is brought to the ftate of made ground; that is to fay, the more it is broken, and the deeper it is dug, the greater probability there will be of fuccefs. Plants put in a hole may thrive very well whilft the fibres have loofe mould to work in; but, whenever they reach the firm unbroken fides of the hole, they will, except the foil be of a very rich loamy nature indeed, receive a check which they will not

overcome

overcome for many years. The fize of the holes, whether in broken or unbroken ground, muft be in proportion to the roots of the plants to be put in. For large nurfery plants the holes in unbroken ground fhould not be lefs than two feet deep ; and, for plants from four to eight feet high, the holes ought to be made from two to three feet diameter : the different ftrata fhould be kept feparate ; laying the fod on one fide of the hole, the corn mould or foil on another, and the fubftratum on a third ; and in this ftate they fhould lie fome weeks before the time of planting.

This, namely the TIME OF PLANTING, varies with the fpecies of plant, and with the nature of the foil. Plants in general may be fet out either in the autumn or in the fpring. In a bleak fituation the latter is generally preferable ; provided the planting be not done too late. The latter end of February, and all March, is a very proper feafon for moft plants : but where the fcene of planting is extenfive, every fit of open weather, during the fix winter months, may be embraced. Some plants, however, are partial to particular feafons : thefe peculiarities will be mentioned in their proper places.

It has been already intimated, that when trees and fhrubs are planted out finally, their *roots* fhould be left UNTRIMMED. It is ufual, and may be proper, to take off the bruifed and maimed

parts ;

parts; but even this fhould be done with caution.
Their *tops*, however, require a different treatment.
Foreft trees and other ftem-plants may in general
be trimmed clofe, by which means the roots will
be able to fend up a fufficient fupply of nourifhment and moifture the firft year, and thereby preferve the life of the plant : whereas, on the contrary, if a number of fide-fhoots be left on, the
quantity of leaves and fhoots becomes fo great, that
the plant probably is *ftarved* for want of that neceffary fupply. This renders the fuccefs of
fhrubby plants uncertain ; and is an argument
againft their paffing through the nurfery ; and, of
courfe, in favour of their being moved (when practicable) from the feminary into the place in which they
are intended to remain. A well-rooted plant, however, if planted in a good mould and a moift feafon, will fupport a confiderable top ; and there is a
general rule for trimming of plants : Leave them
tops proportioned to their roots ; for no doubt the
larger the top, provided the root can fupport it,
the quicker progrefs the plant will make : neverthelefs it is well to be on the fafer fide ; a fure
though flow progrefs is preferable to a dead plant,
which is always a reflection upon the planter, and
an unfightly incumbrance in the plantation. A
judicious planter, whilft he trims his plants, will at
the fame time SORT them : inftead of throwing
them out of his hand into one heap promifcuoufly,

he

he will lay the weak ill-rooted plants in one place, the middle fort in another, and the ftrong well-rooted ones in a third; in order that, when they are planted out, each plant may have a fair and equal chance of rifing; which, without this precaution, cannot be the cafe.

We now come to the operation of PLANTING; which is guided in fome meafure by the fpecies of plantation. If the plants be large and the plantation chiefly ornamental, they ought to be planted out promifcuoufly in the fituation in which they are intended to remain; but if the plants be fmall, and the plantation chiefly ufeful, nurfery-rows ought generally to be preferred. For in this manner the tender plants give warmth to each other; the tranfition is lefs violent than when they are planted out immediately from the nurfery or feed-bed fingly, and at a diftance from each other: the ground is more eafily kept clean than where the plants ftand in the random manner; befides, the intervals may, whilft the plants are young, be cropped with advantage: whilft the remainder of the intended plantation may be kept in an entire ftate of cultivation until the plants acquire a confiderable fize; or, if the whole ground be ftocked in this nurfery manner, the fuperfluous plants may in almoft any country be fold to great profit. We do not recommend planting thefe nurfery-plantations too thick: four feet between the rows and two feet between

the

the plants are convenient diftances; or if the intervals be fet out exactly a quarter of a rod wide,
namely four feet one inch and an half, and the
plants be put in at twenty-four inches and three
quarters apart, the calculation of how many plants
will be required for an acre or any other given portion of ground, or, on the contrary, how much
ground will be neceffary for a given number of
plants, will be made eafy and certain. The method
of putting in the plants in thefe nurfery-rows is
this: The ground being brought to a proper ftate of
cultivation, as directed above, the plants trimmed
and forted, and the rows fet out, a line is laid along
to make the holes by. To afcertain precifely the
center of each hole, a mark is made in the line (or
land-chain, which is not liable to be varied in its
length by the weather), and a ftick or other guide
placed where the center of each hole falls. The
workmen begin to make the holes by chopping a
ring round each ftake, with the fpade, of a diameter proportioned to the fize of the plants, and of a
depth equal to that of the cultivated mould. A
row of holes being finifhed, the plants, in this
cafe, may be immediately put in, which is done in
this manner: One man, or boy, holds the plant
upright with its ftem in the center of the hole, at
the fame time looking along the row to fee that it
ftand in its proper line, whilft another fills in the
mould; firft fpreading the roots and fibres level in

the

the bottom of the hole; being careful not to fuffer any of them to lie in a cramping folded ftate;
but opening them wide, and fpreading them
abroad in the manner of a bird's foot. Whilft the
fpade-man is bedding the roots in the fineft of the
mould, the perfon who fteadies the plant fhould
move it gently up and down, in order to work
in the mould more effectually among the fibres;
which done, they fhould be preffed down gently
together with the foot; and the treading, if the
foil be light, fhould be repeated two or three times,
until the hole be filled up level, and the plant
firmly fixed at the fame depth at which it ftood in
the place from whence it was taken : if on trial the
hole be found too fhallow, it muft be deepened; if
too deep, fome of the rougheft of the mould muft
be thrown to the bottom, until the roots be brought
to their own natural level. The row being finifhed, the planter walks back along it, and adjufts
fuch plants as lean or ftand out of the line, whilft
his helper diftributes the plants of the next row.
In a fimilar manner the plants are put in when the
holes are made in whole ground. The fods are generally thrown to the bottom of the hole; and if
thefe be not fufficient to raife the plant high
enough, fome of the fubftratum is mixed with
them; or if this be of a very bad quality, fome of
the top foil is dug from the intervals and thrown
into the hole. The roots are bedded in the beft of
the

the mould, and the hole rounded up, either with the fubftratum or with the foil of the interval, fo as to form a hillock or fwell round the ftem of the plant, in order to allow for the fettling of the broken mould.

Plantations require a clofe ATTENDANCE AFTER PLANTING; efpecially in windy weather. Large plants are generally ftaked ; but this is a practice we do not recommend, except for large tranfplant-ed trees or fhrubs : but of thefe in the next chap-ter. Plants, even of fix or eight feet high, if well-rooted and *firmly* planted, will withftand a great deal of weather. The plantation however fhould frequently be gone over, and fuch plants as have loft their upright pofture, or are loofe at the roots, fhould be righted and *rammed* ; efpecially if the foil be of a light open texture : indeed, in fuch a foil, it is prudent to ram them at the time of planting, which not only prevents their being mif-placed by the winds, but alfo prevents the drought from reaching the roots fo foon as when the mould is left light and porous. In this cafe, however, it muft be remembered, that when the plants have got foot-hold, the mould which has been rammed fhould be loofened with the fpade, in order to ad-mit a full fupply of air to the roots, without which no plant can flourifh. If a continuance of drought fet in after planting, it will be prudent to WATER the plants ; not partially, by pouring a fmall quan-

tity

tity of water againſt the ſtem of each; but in large
quantities poured into a ring made near the outſide
of the hole; ſo that the whole maſs of broken
earth may be thoroughly moiſtened, without waſh-
ing off the finer mould from the fibres. A ſuper-
ficial watering tantalizes the plants, and leads the
fibres towards the ſurface for nouriſhment: the
moiſture thus partially given ſoon evaporates, and
the diſappointed fibres become expoſed to the
parchings of the ſun and wind.

Plantations in rows are beſt CLEANED with the
plow. In the ſpring gather two furrows, or if room
four furrows, into the middle of the interval: in
ſummer ſplit theſe interval-ridges, throwing the
mould to the roots of the plants, to ſave them from
the drought: in autumn gather them again into
the interval; and in winter again return them to
the rows to keep out the froſt. If the ſoil be good,
and dung can be had, a row of potatoes, cabbages,
&c. may be planted in each interval, or turnips
ſown over the whole: in either caſe, the inter-ſpaces
of the rows ſhould be kept clean hoed or hand-
weeded. To be brief, a plantation of this kind
ſhould be conſidered, whilſt young, as a *kitchen-
garden*, and ought to be treated accordingly.

As the plants increaſe in bulk and ſtature, they
will require PRUNING. Much depends upon doing
this judiciouſly. If it has been neglected too long,
care muſt be had not to do too much at once. The

D   leader

leader is the principal and firſt objeċt; the ſide-
branches may be afterwards taken off gradually, ſo
as not to wound the plant too much, nor let too
much air at once into the plantation. The time of
pruning is generally conſidered to be in autumn or
winter, when the ſap is down and the leaves off;
but for plants which are not liable to *bleed*, we rather recommend midſummer, as ſhoots taken off at
that time are not ſo apt to be followed by freſh
ſhoots as thoſe taken off in winter. If the ſhoots be
young and ſlender, it is better to *rub* them off than
to cut them off clean with a ſharp inſtrument:
boughs and ſtrong ſhoots however require an inſtrument, and from young trees they ſhould be taken
off as ſmooth and cloſe to the ſtem as poſſible. If a
ſtump be left, it will be ſome years before it be
grown over, and a flaw, if not a decayed place,
will be the conſequence; but if a ſhoot, or even a
confiderable bough of a young growing tree, be
taken off level with the bark of the ſtem, the
wound will ſkin over the firſt year, and in a year or
two more no traces of it will be left. A large
bough of an aged full-grown tree requires a different
treatment, which will be given under the article
HEDGE-ROW TIMBER.

Great judgment is required in THINNING plantations. The ſame rule holds good in nurſery-plan-
tations as in the nurſery itſelf: and the ſame general rule (liable no doubt to many exceptions) may

be

be extended to woodlands and ornamental plantations. But of thefe hereafter : fuffice it to repeat
in this part of our Work, that whenever the roots
of plants begin to interfere with each other, their
growth from that time is more or lefs checked ; and
whenever their branches are permitted to clafh,
from that time their beauty and elegance are more
or lefs injured.

TRANSPLANTING TREES AND SHRUBS.

BY this is meant the removal of trees and fhrubs,
which, having formerly been planted out, have
acquired fome confiderable fize. We do not mean
to recommend the practice in general terms ; but
for thinning a plantation, for removing obftructions, or hiding deformities, or for the purpofe of
raifing ornamental clumps or fingle trees expeditioufly, it may frequently be ufeful, and is univerfally practifed, though feldom with uniform fuccefs. This is indeed the moft difficult part of planting, and requires confiderable fkill—with great
care and attention in applying it.

It is in vain to attempt the removal of a tap-
rooted plant (as the oak), which has not previoufly
been *tapped* ; that is, its tap-root taken off ; and
not lefs arduous to make a weak-rooted plant of
almoft any fpecies (the aquatics excepted) fucceed

D 2 with

with a large top upon it : much therefore depends upon taking-up and trimming trees and fhrubs for tranfplantation.

Before a tap-rooted plant, which has never been removed from its place of femination, can be taken up with propriety, it muft be tapped in this manner : Dig a trench or hole by the fide of the plant, large enough to make room to undermine it, in fuch a manner as to be able to fever the tap; which done, fill in the mould, and let the plant remain in this ftate one, two, or three years, according to its fize and age. By this time the horizontal roots will have furnifhed themfelves with ftrength and fibres ; efpecially thofe which were lopped in the excavation; and the plant may be taken up and removed in the fame manner as if it had been tapped and tranfplanted whilft a feedling, though not with equal fafety ; for plants which have never been removed have long branching roots, and the *fibres* lie at a diftance from the body of the plant; whilft thofe which have been taken up and have had their roots trimmed when young are provided with fibres, which, being lefs remote from the ftem, may be taken up with the plant, and conveyed with it to its new fituation. This naturally leads to what may perhaps be called a *refinement* in taking up large plants for tranfplantation; namely, lopping the whole, or a part of the horizontal roots, twelve months or a longer time before the plant be taken

up ;

up; leaving the downward roots, and (if neceffary) part of the horizontal ones, to fupport the plant until the time of removal *. It would be needlefs to add, that in taking up plants in general, the greater length of root and the greater number of fibres there is taken up, the more probable will be the fuccefs. It is alfo a circumftance well underftood, that too much earth cannot be retained among the fibres.

The plant being thrown down, and the roots difentangled, it is proper, before it be removed from its place, to trim the top, in order that the carriage may be lightened. In doing this, a confiderable fhare of judgment is requifite : to head it down in the pollard manner is very unfightly ; and to trim it up to a mere May-pole, or fo as to leave only a fmall broom-like head at the top, is equally deftructive of its beauty. The moft rational, the moft *natural*, and, at the fame time, the moft elegant, manner of doing this, is to *trim the boughs* in fuch a manner as to form the head of the plant into a conoid, in refemblance of the natural head of the Lombardy poplar, and of a fize proportioned to the *ability of the root*. Whoever was the inventor of this method of trimming the heads of trees, deferves infinite credit : it only wants to be known in order to be approved ; and we are happy to fee it growing into univerfal practice.

* In this cafe the head ought at the fame time to be trimmed, and the plant, if expofed, fupported.

D 3 The

The mode of carriage refts wholly with the fize of the plant : if fmall, it is beft carried by hand, either upon the fhoulder, or upon hand-fpikes :— if larger, two fledges, one for the root, the other for the head, may be ufed :—if very large, and the ball of earth heavy, a pair of high timber-wheels, or a timber-carriage, will be found neceffary.

The hole muft be made wide enough to admit the root of the plant, with a fpace of a foot at leaft all round it, for the purpofe of filling in the mould with propriety ; fo that if the tree was taken up with a root of fix feet diameter, the hole muft be made of the diameter of eight or nine feet, and of a depth fufficient to admit of the tree's being feated (when the mould is fettled) at its natural depth, as alfo to receive the fods, and other rough unbroken mould, at leaft a foot thick underneath its root.

The method of planting depends upon the ftate of the root, and the temperature of the mould and the feafon. If the root be well furnifhed with fibres and mould, and the foil be moift from fitua-tion, or moiftened by the wetnefs of the feafon, no artificial preparation is neceffary. The bottom of the hole being raifed to a proper height, and the tree fet upright in the center of it, the mould may be filled in ; being careful to trim it well in amongft the roots, and to bed the fibres fmoothly amongft it ; treading every layer firmly, and, with a carpen-ter's rammer, filling every crevice and vacancy

among

among the roots, fo that no foft place nor hollownefs remain; and proceed in this manner until
the hole be filled, and a hillock raifed round the
plant to allow for its fettling. But if the roots be
naked of mould and thin of fibres, and the foil, the
fituation, and the feafon be dry, or if any other
doubt or fcruple, with refpect to a probability of
fuccefs, arife, we recommend in the ftrongeft terms
the following method : The requifite depth of the
hole being afcertained, and its bottom raifed to a
proper height, add a confiderable quantity of the
fineft mould, pour upon it water fufficient to moif-
ten it thoroughly, and work them up together into
a mortar-like confiftency : having fpaded out a principal part of this pafte, and laid it in heaps by the
fide of the hole, fpread the remainder thin over the
bottom, and fet the plant upon it : with the other
part of the pafte cover up the wounded parts, and
the fibres; and this done, fill in the remainder of
the mould, as above directed. If the tranfplantation be done in autumn, it will require nothing farther at that time ; but, if made in the fpring, waterings will almoft immediately be wanted. Therefore, at once, draw a ring near the outfide of the
hole, and in the bottom of its channel make fix,
eight, or ten holes, (by means of a fpike and beetle)
at equal diftances, and of a depth equal to that of
the roots of the plant. Thefe holes will not only
convey water, but air alfo, to the immediate region

in

in which they are both indifpenfibly neceffary to the health of the plant. We have been the fuller in our inftruĉions relative to tranfplanting, as being a procefs little underftood by profeffional men. Every nurfery-man, and almoft every kitchen-gardener, can raife, train, and plant out feedling and nurfery plants; but the removal of trees feldom occurs in their practice; and we have met with very few men indeed who are equal to the tafk. The foregoing rules are the refult of experience.

Having thus attempted to give our readers a general knowledge of what may be called the manual operations of planting, we will next enter into particulars, and endeavour to complete his fkill in the bufinefs of the nurfery, by treating feparately of each refpective tree and fhrub adapted to the purpofe of ufeful and ornamental planting.

AN

A N

ALPHABET of PLANTS,

PROPER FOR

USEFUL and ORNAMENTAL PLANTATIONS;

ARRANGED UNDER THE

GENERIC NAMES of LINNEUS:

WITH A

DESCRIPTION,

AND THE

PROPAGATION,

Of EACH SPECIES.

INTRODUCTION.

FOR a Book of Reference the DICTIONARY form is the moft convenient. This part of our Work is entirely of that nature ; we have therefore adopted an alphabetical arrangement. Our reafons for making ufe of the Linnean names as the ground-work of this arrangement, are manifold : a great number of the plants here treated of have no Englifh generic name belonging to them ; yet it was neceffary to that concifenefs and fimplicity which is the bafis of our plan, to arrange

them

them agreeably to their refpective genera; becaufe, in general, the individuals of the fame genus have fimilar appearances and fimilar propenfities, which being placed together in one point of view, their defcription and mode of culture are rendered infinitely more eafy and compendious than they could poffibly be, if treated of feparately under diftinct and detached fpecies. Befides, even many of the fpecies now common in our ornamental grounds and fhrubberies have not yet had any Englifh name given to them; and there are many more, whofe Englifh names are local and unfettled; whereas the Linnean names are the fame every where, and are known to the whole world *.

We do not mean to enter into the difpute about the Sexual Syftem of Linneus: it is enough for our purpofe that it is at prefent the prevailing fyftem; and that, being founded in nature, its principles can never be overturned : we are neverthelefs fo far from thinking it a *perfect* fyftem, that we believe it capable of very great improvement : at prefent, however, it is our bufinefs to take it as we find it; and for the ufe of fuch of our readers as are unacquainted with its principles, it is proper that we fhould here give its outline.

Every PERFECT FLOWER has four principal parts, which in general are obvious to the naked eye;

* An alphabet of Englifh names will be given in the general Index to the work.

namely,

namely, the CALYX, or outer guard; the COROLLA, or coloured leaves; the STAMINA; and the PISTIL-LUM. The calyx is evident in the mofs-rofe, being thofe elegant rough leaves which inclofe the blufh-ing beauties of the flower: it is alfo confpicuous in the primrofe, being that angular tube out of which the more delicate parts of the flower iffue. The corolla of the primrofe is that yellow ornament which by unbotanical obfervers is itfelf confidered as the flower. The ftamina are confpicuous in moft flowers, and are diftinguifhed by the farina or duft with which they are covered. In the primrofe they proceed from the infide of the tube of the corolla; and, when matured, form themfelves into a circle round the top of the tube. The piftillum, in the primrofe, is that delicate white pillar, which, rifing from the bottom of the tube of the corolla, fhews its flatted top in the middle of the ftamina, and in the center of the flower. According to Lin-neus, the calyx is an expanfion of the outer bark; the corolla, of the inner bark; the ftamina, of the wood; and the piftillum, of the pith of the plant; and according to his Sexual Syftem, the ftamen is the male, and the piftillum the female, part of generation.

Whether this laft is or is not a fa\ć in nature, has been the fubje\ć of much difpute. But, to the FLORAL SYSTEM, it is, we are humbly of opinion, a matter of no great import. The parts themfelves,

and

and not their functions, are the bafis of the Linnean Syftem; and we are clear in our opinion, that if that great man had confidered his Syftem as being, what it in reality is, merely FLORAL, without having unfortunately clogged it with the idea of SEXUAL, he would have faved himfelf a hoft of enemies, and would beyond a doubt have rendered his Syftem infinitely more fimple and fcientific, and confequently more ufeful, than it really is. But it is now too late to regret : his Syftem is eftablifhed; and himfelf no more. Having however faid thus much, it would be unpardonable in us not to add, that whether we confider his genius, his perfeverance, or the Syftem he has formed, notwithftanding its imperfections, he died one of the greateft characters the world has known.

The VEGETABLE KINGDOM is divided by LINNEUS into twenty-four CLASSES : thefe Claffes are fubdivided into ORDERS; the Orders into GENERA; the Genera into SPECIES; and the Species into VARIETIES.

His principle of Claffification is feen in the following

" KEY

" KEY of the SEXUAL SYSTEM.

MARRIAGES of PLANTS.
Florefcence.
PUBLIC MARRIAGES.
Flowers vifible to every one.
In one Bed.
Hufband and wife have the fame bed.
All the flowers hermaphrodite : ftamens and piftils in the fame flower.
Without Affinity.
Hufbands not related to each other.
Stamens not joined together in any part.
With Equality.
All the males of equal rank.
Stamens have no determinate proportion of length.

1. ONE MALE.	7. SEVEN MALES.
2. TWO MALES.	8. EIGHT MALES.
3. THREE MALES.	9. NINE MALES.
4. FOUR MALES.	10. TEN MALES.
5. FIVE MALES.	11. TWELVE MALES.
6. SIX MALES.	12. TWENTY MALES.
	13. MANY MALES.

With Subordination.
Some males above others.
Two ftamens are always lower than the others.
14. TWO POWERS. | 15. FOUR POWERS.
With Affinity.
Hufbands related to each other.
Stamens cohere with each other, or with the piftil.

16. ONE BROTHERHOOD.	19. CONFEDE-
17. TWO BROTHERHOODS.	RATE MALES.
18. MANY BROTHERHOODS.	20. FEMININE MALES.

In Two Beds.
Hufband and wife have feparate beds.
Male flowers and female flowers in the fame fpecies.

21. ONE HOUSE.	23. POLYGAMIES.
22. TWO HOUSES.	

Clandestine Marriages.
Flowers fcarce vifible to the naked eye.
24. CLANDESTINE MARRIAGES."

His CLASSES are :

I. ONE MALE *(Monandria.)*
One hufband in marriage.
One ftamen in an hermaphrodite flower.

II. TWO MALES. *(Diandria.)*
Two hufbands in the fame marriage.
Two ftamens in an hermaphrodite flower.

III. THREE MALES. *(Triandria.)*
Three hufbands in the fame marriage.
Three ftamens in an hermaphrodite flower.

IV. FOUR MALES. *(Tetrandria.)*
Four hufbands in the fame marriage.
Four ftamens in the fame flower with the fruit.
(If the two neareft ftamens are fhorter, it is referred to Clafs 14.)

V. FIVE MALES. *(Pentandria.)*
Five hufbands in the fame marriage.
Five ftamens in an hermaphrodite flower.

VI. SIX MALES. *(Hexandria.)*
Six hufbands in the fame marriage.
Six ftamens in an hermaphrodite flower.
(If the two oppofite ftamens are fhorter, it belongs to Clafs 15.)

VII. SEVEN MALES. *(Heptandria.)*
Seven hufbands in the fame marriage.
Seven ftamens in the fame flower with the piftil.

VIII. EIGHT MALES. *(Octandria.)*
Eight hufbands in the fame marriage.
Eight ftamens in the fame flower with the piftil.

IX. NINE MALES. *(Enneandria.)*
Nine hufbands in the fame marriage.
Nine ftamens in an hermaphrodite flower.

X. TEN MALES. *(Decandria.)*
Ten hufbands in the fame marriage.
Ten ftamens in an hermaphrodite flower.

XI. TWELVE MALES. *(Dodecandria.)*
Twelve hufbands in the fame marriage.
Twelve ftamens to nineteen in an hermaphrodite flower.

XII.

XII. TWENTY MALES. (*Icofandria.*)
Generally twenty hufbands, often more.
Stamens inferted on the calyx (not on the receptacle *)
in an hermaphrodite flower.

XIII. MANY MALES. (*Polyandria.*)
Twenty males or more in the fame marriage.
Stamens inferted on the receptacle, *from* 20 *to* 1000 *in the fame flower with the piftil.*

XIV. TWO POWERS. (*Didynamia.*)
Four hufbands, two taller than the other two.
Four ftamens: of which the two neareft are largeft.

XV. FOUR POWERS. (*Tetradynamia.*)
Six hufbands, of which four are taller.
Six ftamens: of which four are longer, and the two op-pofite ones fhorter.

XVI. ONE BROTHERHOOD. (*Monadelphia.*)
Hufbands, like brothers, arife from one bafe.
Stamens are united by their filaments † into one body.

XVII. TWO BROTHERHOODS. (*Diadelphia.*)
Hufbands arife from two bafes, as if from two mothers.
Stamens are united by their filaments into two bodies.

XVIII. MANY BROTHERHOODS. (*Polyadelphia.*)
Hufbands arife from more than two mothers.
Stamens are united by their filaments into three or more bodies.

XIX. CONFEDERATE MALES. (*Syngenefia.*)
Hufbands joined together at the top.
Stamens are connected by the anthers ‡ forming a cylinder (feldom by the filaments).

XX. FEMININE MALES. (*Gynandria.*)
Hufbands and wives growing together.
Stamens are inferted in the piftils (not on the receptacle).

XXI. ONE HOUSE. (*Monoecia.*)
Hufbands live with their wives in the fame houfe, but, have different beds.
Male flowers and female flowers are on the fame plant.

---

* " The bafe by which the parts of the fructification are connected."

† The *thread* or body of the ftamen.

‡ The *tips* or heads of the ftamen.

XXII.

XXII. TWO HOUSES. (*Dioecia.*)
    Hufbands and wives have different houfes.
    Male flowers and female flowers are on different plants.
XXIII. POLYGAMIES. (*Polygamia.*)
    Hufbands live with wives and concubines.
    Hermaphrodite flowers, and male ones, or female ones in
        the fame fpecies.
XXIV. CLANDESTINE MARRIAGES. (*Cryptogamia.*)
    Nuptials are celebrated privately.
    Flowers concealed within the fruit, or in fome irregular
        manner."

His ORDERS are diftinguifhed by different
parts of the flowers, according to the Claffes.
Thofe of the firft thirteen Claffes are taken from
the number of females or piftils, (reckoning
" from the bafe of the ftyle * ; but if there is no
" ftyle, the calculation is made from the number
" of ftigmas" †); as ONE FEMALE (*Monogynia*),
TWO FEMALES (*Digynia*), THREE FEMALES, (*Trigynia*), &c. Thofe of the fixteenth, feventeenth,
eighteenth, twentieth, twenty-firft, and twenty-
fecond Claffes are taken from the number of males,
or ftamens. Thofe of the fourteenth are diftinguifhed by SEEDS NAKED (*Gymnofpermia*), and
SEEDS CLOATHED (*Angiofpermia*). Thofe of the
fifteenth, by the formation of the feed-veffel, or
pod; as, WITH SILICLE (*Siliculofa*), and, WITH
SILIQUE (*Siliquofa*). Thofe of the twenty-third are
ONE HOUSE (*Monoecia*); TWO HOUSES (*Dioecia*);
and THREE HOUSES (*Trioecia*). Thofe of the twen-

    * The *fhaft* or body of the piftil.
    † The *fummits* or heads of the piftil.

ty-fourth

ty-fourth are FERNS, MOSSES, FLAGS, and FUN-
GUSSES. Thofe of the nineteenth Clafs (confifting
chiefly of plants with compound difcous flowers,
as the thiftle, dandelion, &c.) are, EQUAL POLY-
GAMY (*Polygamia Equalis*); SUPERFLUOUS POLY-
GAMY (*Polygamia Superflua*); FRUSTRANEOUS POLY-
GAMY (*Polygamia Fruftranea*); NECESSARY POLY-
GAMY (*Polygamia Neceffaria*); SEPARATE POLY-
GAMY (*Polygamia Segregata*); MONOGAMY (*Mono-
gamia*). The following is Linneus's account (*li-
terally* as it ftands in the Litchfield tranflation) of
the Orders laft mentioned.

EQUAL POLYGAMY confifts of many marriages with promifcuous intercourfe.
> That is, *of many florets furnifhed with ftamens and piftils.*
> *The flowers of thefe are vulgarly called* Flofculous.

SPURIOUS POLYGAMY, where the beds of the married
occupy the difk, and thofe of the concubines the circumference.
> That is, *the hermaphrodite florets occupy the difk, and the
> female florets without ftamens furround the border, and that
> in three manners :*

(*a*) SUPERFLUOUS POLYGAMY, when the married females are fertile, thence the concubines fuperfluous.
> That is, *when the hermaphrodite flowers of the difk
> are furnifhed with ftigmas, and produce feeds; and the
> female flowers alfo, which conftitute the circumference,
> produce feeds likewife.*

(*b*) FRUSTRANEOUS POLYGAMY, when the married
females are fertile, and the concubines barren.
> That is, *when the hermaphrodite flowers of the difk
> are furnifhed with a ftigma, and produce feeds; but the
> florets which conftitute the circumference having no ftig-
> ma, produce no feeds.*

(c) NECESSARY POLYGAMY, when the married females are barren, and the concubines fertile.

That is, when the hermaphrodite flowers, from the defect of the stigma of the pistil, produce no seed; but the female flowers in the circumference produce perfect seeds.

(e) SEPARATE POLYGAMY, when many beds are so united that they constitute one common bed.

That is, when many flower-bearing calyxes are contained in one common calyx, so as to constitute one flower.

His GENERA are taken from the construction of the parts of fructification. All plants, whether herbs, shrubs, or trees, whose flowers and seeds correspond as to figure and disposition, are of the same GENUS.

His SPECIES are distinguished by the leaves, and other more permanent parts of the plant. Or, speaking with greater precision, the SPECIES is determined by the nature and property of the seed: for, let the exterior of a plant, or tribe of plants, be what it may, if the seed do not produce near resemblances of the parent stock, but plants whose appearances or properties are different from it (as in the case of apples, pears, &c.); such plants are not considered as forming a distinct SPECIES, but are properly deemed VARIETIES (of some *natural species*), arising from cultivation, or some fortuitous circumstance.

ALPHA-

ALPHABET

OF

PLANTS.

ACER.

LINNEAN Claſs and Order, *Polygamia Monoecia:* Male flowers containing eight ſtamens, and hermaphrodite flowers containing eight ſtamens and one piſtil, upon the ſame plant. There are Eleven SPECIES: Ten of which are natives of, or have been introduced into, this country.

1. ACER *Pſeudo-platanus:* The SYCAMORE; *a tall deciduous tree;* native of the continent of Europe, but doubtful whether or not of this iſland.

2. ACER *Campeſtre:* The COMMON MAPLE; *a low deciduous tree;* common in our woods and hedges.

3. ACER *Negundo:* the ASH-LEAVED MAPLE; *a deciduous tree;* native of Virginia and Carolina.

4. ACER *Platanoides:* the NORWAY MAPLE; *a deciduous tree;* native of Norway and the north of Europe.

5. ACER *Monſpeſulanum:* the MONTPELIER MAPLE; *a low deciduous tree;* growing common about Montpelier.

6. ACER *Creticum:* the CRETAN MAPLE; *a low deciduous tree;* native of the Eaſt.

7. ACER *Rubrum:* the SCARLET MAPLE; *a deciduous tree;* native of Virginia and Penſylvania.

8. ACER *Saccharinum:* the SUGAR MAPLE; *a deciduous tree;* native of Penſylvania.

9. ACER *Tartaricum:* the TARTARIAN MAPLE; *a low deciduous tree;* native of Tartary.

10. ACER *Pennſylvaticum:* the PENNSYLVANIAN MAPLE; or the MOUNTAIN MAPLE; *a tall deciduous ſhrub;* native of Pennſylvania.

1. The

1. The SYCAMORE. This tree grows to a great heignt and ample fize, throwing out a wide-fpreading top. Its leaves are vine-fhaped; and, on their firft appearance, are of a plea-fant green; but their beauty foon goes off, being liable to be per-forated and disfigured by infects during the fummer months, which-reduces the value of the Sycamore as an *ornamental*: it has however, long been confidered as a *timber-tree* in this country, having been much ufed by the turners for wooden bowls, difhes, trenchers, &c.; but, fince the cuftom of ufing earthen ware has become fo prevalent, its value for this purpofe is greatly decreaf-ed. Neverthelefs, near the fea-coaft it may be planted with ad-vantage, as it is known to withftand the attacks of the fea-air with peculiar hardinefs. HANBURY fays, The Sycamore being wounded exudes a great quantity of liquor, of which is made good wine. There are two *Varieties* of the Sycamore: one with broad leaves and large keys; the other with variegated leaves.

The PROPAGATION of the Sycamore is very eafy. In the autumn, when the keys are ripe, they may be gathered, and in a few days after fown, about an inch and an-half deep, in beds of common mould. In the fpring the plants will appear, and make a fhoot about a foot and a half by the autumn following, if the ground of the feminary be tolerably good, and they are kept clean from weeds. The fpring after they come up, they fhould be planted in the nurfery, in rows two feet and a half afunder, and their diftance in the rows muft be one foot and an half. Here they may remain till they are big enough to plant out finally, with no farther trouble than taking off unfightly fide-branches, and fuch as have a tendency to make the tree fork-ed, except digging between the rows, which muft always be done every winter. This tree will grow upon almoft any foil.

2. The COMMON MAPLE is too well known to need a de-fcription. It is of much humbler growth than the Sycamore; and is by no means ornamental; nor is its timber of a good quality, being peculiarly brittle: The texture however is clofe and firm, and it is in good efteem amongft the turners. In the vale of Glocefter, where oak timber is fcarce, Maple is ufed for gate-ftuff and other purpofes of hufbandry; and fometimes fkrews for cyder-preffes are made of this wood. But the principal value of the Maple, is for *underwood*: it is of quick growth, and affords good fuel.

The

The method of PROPAGATION is the fame as that of the Sycamore ; and, like it, the Maple will grow in almoft any foil and fituation.

3. The ASH-LEAVED MAPLE grows to a large timber-tree : its leaves are of a pale green, and well adapted to give variety of tint ; but HANBURY fays, this tree is not proper to be planted in expofed fituations, the branches being fubject to be fplit off by the winds. Its ufes are fimilar to thofe of the Sycamore.

It may be PROPAGATED from the keys, which are perfected in this country; or by layering ; or from cuttings, planted in a moift fituation, in autumn.

4. The NORWAY MAPLE. This alfo grows to a large timber-tree. Its leaves are of a fhining green colour, and are as large or larger than thofe of the Sycamore; their edges are acutely and more beautifully indented ; they are not fo liable to be eaten by infects in the fummer ; and " in the autumn they die to a golden yellow colour, which caufes a delightful effect at that feafon, when the different tints of the decaying vegetable world are difplayed." The flowers are alfo beautiful ; they come out early in the fpring, are of a fine yellow colour, and fhew themfelves to advantage before the leaves come out. They are frequently fucceeded by keys, which fometimes arrive at maturity in this climate. There is a Variety with ftriped leaves.

The Norway Maple may be PROPAGATED from feed, as the Sycamore ; it may alfo be raifed by layers, and cuttings, planted in a moift foil.

5. MONTPELIER MAPLE grows to about twenty feet high, and is a very beautiful tree. The leaves are compofed of three lobes, are of a fhining green, a thickifh fubftance, and retain their verdure later in the year than moft of the other forts. The flowers come out in the fpring, but have very little beauty ; their blow is foon over, and fometimes they are fucceeded by feeds, which come to perfection in our gardens.

6. CRETAN MAPLE. This grows to about the height of the former. The leaves are downy, compofed of three lobes, and grow oppofite to each other on long downy footftalks. The flowers come out in the fpring, are inconfiderable to the florift, and are very feldom fucceeded by good feeds in England.

7. SCARLET-FLOWERING MAPLE. Of this there are two
forts; called, 1. *Virginian fcarlet-flowering Maple* ; and, 2. *Sir
Charles Wager's Maple*. Both of thefe are propagated for the
fake of the flowers, which are of a fcarlet colour, and come out
early in the fpring. The leaves are compofed each of five fharp-
pointed lobes, which are flightly indented or ferrated. They are
fmooth, of a pale green on their upper furface, glaucous * underneath ; and they grow on long, fimple, taper, reddifh footftalks.
The flowers come out in clufters from the fide of the branches.
They appear in April, and the feeds ripen in June. The fort
called *Sir Charles Wager*'s produces larger clufters of flowers
than the others ; on which account it is in moft efteem.

8. SUGAR MAPLE is a large-growing tree ; will arrive at the
height of forty feet, and has broad thin leaves, divided into five
principal-parts ; which are again indented or cut at the edges
into feveral acute fegments. Their furface is fmooth, of a light
green colour, whitifh underneath ; and they grow on pretty long
footftalks. The flowers come out in the fpring, about the time
of the Norway Maple ; and they are fucceeded by long keys,
which fometimes ripen in England. In America, the inhabitants tap this tree in the fpring, boil the liquor, and the fœces
afford a ufeful fugar. The Sycamore, the Afh-leaved and the
Norway Maples alfo abound with a faccharine juice, from which
there is no doubt but a ufeful fugar might be prepared.

9. TARTARIAN MAPLE will grow to upwards of twenty feet
high. The leaves are heart-fhaped, undivided, and their edges
are unequally ferrated. The flowers come out from the wings of
the leaves, in longifh bunches ; they appear early in the fpring ;
and fometimes are fucceeded by ripe feeds in our gardens.

10. MOUNTAIN MAPLE. The *ftalks* of this fhrub are flender,
covered with a whitifh bark, fend forth feveral red branches, and
grow about fifteen feet high. The leaves are three-lebed, pointed, and are unequally and fharply ferrated. The flowers come
out in longifh bunches, in the fpring : They are of a greenifh
yellow colour ; and are fucceeded by feeds which (like thofe of
the Norway Maple) generally fall off before they are ripe.

Thefe forts are all PROPAGATED, 1. by the feeds ; but as they

* *Glaucous*, of a fea green colour.

do

do not always ripen in this country, the beſt way will be to pro-cure them from the places where they naturally grow. A cool ſhady part of the ſeminary ſhould be appropriated for the purpoſe; the mould ſhould be made fine; beds ſhould be marked out four feet wide, and in length proportionable to the quantity; and in theſe the ſeeds ſhould be regularly ſown, ſifting over them about half an inch of the fineſt mould. When the plants come up, they muſt be kept clean from weeds, and fre-quently watered; and this work muſt be duly attended to all ſummer. The ſpring following, the ſtrongeſt may be drawn out, and planted in the nurſery, in rows two feet aſunder, and at the diſtance of a foot from each other in the rows; leaving the others in the ſeminary to gain ſtrength. The ſpring following they alſo muſt receive the ſame culture; and in the nurſery they may remain, with no other trouble than keeping the ground clean from weeds in the ſummer, digging between the rows in the win-ter, and taking off all ſtrong and irregular ſide-ſhoots, till they are planted out. Trees raiſed from ſeeds will grow faſter, and arrive at greater height, than thoſe raiſed from layers; but they will not produce ſuch quantities of flowers; which makes the latter method more eligible for thoſe who want theſe plants for a low ſhrubbery. 2. By layers all the ſpecies of this genus are to be propagated; though it is never practiſed for the Common Maple and the Syca-more. The young ſhoots may be at any time laid down in the autumn, winter, or early in the ſpring. By the autumn follow-ing, they will have ſtruck root, and become good plants; when the ſtrongeſt may be ſet out in the places where they are to remain; whilſt the weakeſt may be planted in the nur-ſery, like the ſeedlings, for a year or two, to gain ſtrength. 3. By cuttings alſo theſe trees are to be propagated: But this method is chiefly practiſed on the Aſh-leaved and Norway Maples, which more readily take root this way. The cuttings ſhould be the bottom parts of the laſt year's ſhoots: They ſhould be taken off early in October, and planted in rows in a moiſt ſhady place. The ſpring and ſummer following, they muſt be duly watered as often as dry weather makes it neceſſary, and be kept clean from weeds. By the autumn they will be fit to remove into the nur-ſery; though if the cuttings are not planted too cloſe, they may remain in their ſituation for a year or two longer, and then be ſet out finally, without the trouble of being previouſly planted in the

nurſery,

nurfery. 4. By budding, grafting, and inarching likewife Maples are to be propagated : But the other methods being more eligible, thefe are never practifed, except for the variegated forts and the large broad-leaved kind. The latter is to be continued no other- wife than by budding it on ftocks of the common Sycamore; for the feeds, though fo large themfelves, when fown afford you only the common Sycamore in return.

Seeds of the variegated kinds, however, when fown will pro- duce variegated plants in return ; which renders the propaga- tion of thefe forts very expeditious, where plenty of feeds may be had. Where thefe are not to be obtained, in order to propagate thefe varieties by budding, let fome plants of the common Sycamore, one year old, be taken out of the feminary, and fet in the nur- fery in rows a yard afunder, and the plants about a foot and a half diftance from each other in the rows : Let the ground be kept clean from weeds all fummer, and be dug, or, as the gar- deners call it, *turned in,* in the winter ; and the fummer follow- ing the ftocks will be of a proper fize to receive the buds, which fhould be taken from the moft beautifully-ftriped branches. The beft time for this work is Auguft ; becaufe if it is done earlier, the buds will fhoot the fame fummer ; and when this happens, a hard winter will infallibly kill them. Having, therefore, budded your ftocks the middle or latter end of Auguft, with the eyes or buds fronting the north, early in October take off the bafs matting, which before this time will have confined the bark and pinched the bud, but not fo as to hurt it much. Then cut off the ftock juft above the bud, and dig the ground between the rows. The fum- mer following, keep the ground clean from weeds ; cut off all natural fide-buds from the ftock as they come out ; and by autumn, if the land is good, your buds will have fhot forth, and formed themfelves into trees five or fix feet high. They may be then removed into the places where they are defigned to remain ; or a few of them only may be drawn out, leaving the others to be trained up for larger ftandards, to ferve for planting out in open places, or fuch other purpofes as fhall be wanting.

The Striped Norway Maple fhould be budded on ftocks of its own kind ; for on thefe they take beft, and both kinds are not very liable to run away from their colours. Variegated plants in general muft be planted in poor, hungry, gravelly, or fandy foils, to feed the difeafe which occafions thefe beautiful ftripes,

and

and caufe it to be more powerful. But thefe trees fhew their ftripes in greater perfection in a good foil: The plant, though in ficknefs, has the appearance of health; the fhoots are vigorous and ftrong; the leaves are large, lefs liable to be hurt by infects; and the ftripes appear more perfect, natural, and delightful, than thofe on ftunted trees growing on a poor foil.

Æ S C U L U S.

LINNEAN Clafs and Order, *Heptandria Monogynia:* Each flower contains feven *or* eight males and one female : There are only two SPECIES:

1. ÆSCULUS *Hippo-caftanum:* The HORSE-CHESNUT ; *a deciduous tree*; native of Afia.

2. ÆSCULUS *Pavia:* The SCARLET ESCULUS, or SCARLET-FLOWERING HORSE-CHESNUT ; *a tall deciduous fhrub*; native of Carolina, the Brazils, and feveral parts of the Eaft.

1. HORSE-CHESNUT. This is a large noble-looking tree; growing to feventy or eighty feet high, and throwing out its branches to a confiderable width; yet forming a clofe thickfet head; which, if left to nature, takes a moft beautifully-ftriking parabolic form. Its leaves are large, palmated, and of a dark-green colour: they appear very early in the fpring; their buds fometimes beginning to fwell fo early as Chriftmas, and anticipate the pleafures of the coming fpring. Its flowers are fingularly beautiful, ftanding in large fpikes thick among the leaves. This tree is peculiar in a quick formation of its fhoots, which are frequently perfected in lefs than three weeks from the time of foliation; "in which time," fays MILLER, "I have meafured fhoots a foot and a half long with their leaves fully expanded." For fingle trees the Horfe-Chefnut ftands, amongft the *ornamental* tribe, next to the oak; and in the fpring of the year, when its flowers are out, we know no tree equal to it in

---

* The Æfculus is one of the defective genera of Linneus. The *Pavie* having eight males in each flower, belongs properly to the eighth Clafs.

grandeur,

grandeur. It is improper however to be planted near gardens or kept walks, as it fheds its leaves early in autumn, and, being large and numerous, they create a difagreeable litter. The *ufes* of the Efculus are few: its timber is of an inferior kind, and its fruit of no great eftimation: deer are faid to affect it much; and MILLER fays, "in Turkey the nuts of this tree are ground and mixed with the provender for their horfes, efpecially thofe which are troubled with coughs or are broken-winded, in both which diforders they are accounted very good." HANBURY tells us, that fwine will fatten upon them; but does not fay how they are to be prepared. We have known them offered to hogs raw, alfo boiled, as likewife baked in an oven, but without fuccefs.

The Horfe-Chefnut is PROPAGATED from the nuts: In autumn, therefore, when they fall, a fufficient quantity fhould be gathered. Thefe fhould be fown foon afterwards in drills, about two inches afunder. If the nuts are kept till fpring, many of them will be faulty; but where the feminary-ground cannot be got ready before, and they are kept fo long, it may be proper to put them in water, to try their goodnefs: The good nuts will fink, whilft thofe which are faulty will fwim; fo that by proving them this way you may be fure of good nuts, and have more promifing hopes of a crop. In the fpring the plants will come up; and when they have ftood one year they may be taken up, their tap-roots fhortened, and afterwards planted in the nurfery. When they are of fufficient fize to be planted out finally, they muft be taken out of the nurfery with care, the great fide-fhoots and the bruifed parts of the roots fhould be taken off, and then planted in large holes level with the furface of the ground, at the top of their roots; the fibres being all fpread and lapped in the fine mould, and the turf alfo worked to the bottom. A ftake fhould be placed to keep them fafe from the winds, and they muft be fenced from the cattle till they are of a fufficient fize to defend themfelves. The beft feafon for all this work is October. After the trees are planted, neither knife nor hatchet fhould come near them; but they fhould be left to Nature to form their beautiful parabolic heads, and affume their utmoft beauty.

The Horfe-Chefnut, like moft other trees, delights moft in good fat land; but it will grow exceedingly well on clayey and marley grounds.

MILLER

Miller says, " when thefe trees are tranfplanted their roots fhould be preferved as entire as poffible, for they do not fucceed well when torn or cut ; nor fhould any of the branches be fhortened, for there is fcarce any tree which will not bear amputation better than this ; fo that when any branches are by accident broken, they fhould be cut off clofe to the ftem, that the wound may heal over."

2. The Scarlet Esculus grows to about fifteen or fixteen feet high ; and there is a delicacy in this tree that makes it defirable. The bark of the young fhoots is quite fmooth, and the growing fhoots in fummer are of a reddifh hue. The leaves are palmated, being pretty much like thofe of the Horfe-Chefnut, only much fmaller, and the indentures at the edges are deeper and more acute. The lobes of which they are compofed are fpear-fhaped ; they are five in number, are united at their bafe, and ftand on a long red footftalk. The leaves grow oppofite by pairs on the branches, which are fpread abroad on every fide. The flowers come out from the ends of the branches. The firft appearance of the buds is in May ; though they will not be in full blow till the middle of June. They are of a bright red colour, and confequently have a pleafing effect among the vaft tribe of yellow-flowering forts which fhew themfelves in bloom at that feafon. They continue in fucceffion for upwards of fix weeks ; and fometimes are fucceeded by ripe feeds in our gardens.

There are two ways of propagating this tree; 1. By budding it upon the young plants of the Horfe-Chefnut. Thefe ftocks fhould be raifed as was directed in that article. They fhould be planted in the nurfery way, a foot afunder, and two feet diftant in the rows, which fhould be kept clean of weeds, and muft be dug between every winter till the operation is to be performed. After they have ftood in the nurfery-ground about two years, and have made at leaft one good fummer's fhoot, the fummer following is the time for the operation. Then, having your cuttings ready foon after midfummer, the evenings and cloudy weather fhould be made choice of for the work. Whoever has a great number of trees to inoculate, muft regard no weather ; but keep working on, to get his bufinefs over before the feafon ends ; and indeed, a good hand will be always pretty fure of fuccefs, be the weather what it will. If the ftocks were healthy, the fummer
following

following they will make pretty good shoots; and in a year or two after that will flower. This is one method of propagating this tree; and those plants that are propagated this way will grow to a larger size than those raised immediately from seeds. 2. This tree also may be propagated by seeds; which will sometimes ripen with us, and may be obtained out of our own gardens. The manner of raising them this way is as follows: Let a warm border be prepared; and if it is not naturally sandy, let drift sand be mixed with the soil; and in this border let the seeds be sown in the month of March, about half an inch deep. After this, constant weeding must be observed; and when the plants are come up, if they could be shaded in the heat of the day, it would be much better. These, with now and then a gentle watering in a dry season, will be all the precautions they will require the first summer. The winter following, if the situation is not extremely well sheltered, protection must be given them from the hard black frosts, which will otherwise often destroy them: So that it will be the safest way to have the bed hooped, to cover them with mats in such weather, if the situation is not well defended: if it is, this trouble may be saved; for, even when young, they are tolerably hardy. In about two or three years they may be removed into the nursery, or planted where they are to remain, and they will flower in three or four years after. The usual nursery care must be taken of them when planted in that way; and the best time for planting them there, or where they are to remain, is October; though they will grow exceeding well if removed in any of the winter months; but, if planted late in the spring, they will require more watering, as the ground will not be so regularly settled to the roots, as if they had been planted earlier.

A M O R P H A.

Linnean Class and Order, *Diadelphia Decandria*: Each flower contains ten males and one female; the males being connected at the base in two divisions: There is only one known Species:

AMORPHA

AMORPHA *Fruticofa*: BASTARD INDIGO ; *a deciduous fhrub* ; native of Carolina.

THE AMORPHA has its beauties ; but it has other ill effects to detract from its value. The leaves are late in the fpring before their foliage is fully difplayed. The ends of their branches are generally deftroyed by the froft ; or, if they recover it, they have the appearance of being dead ; whilft other plants teftify their effects of the reviving months. But notwithftanding thefe defects, this tree has fome other good properties that in part make amends for them. The leaves, when out, which will not be before the middle of May, are admired by all : They are of a pleafant green colour ; are very large, beautifully pinnated, the folioles being arranged along the ftalk by pairs, and terminate by an odd one. The flowers are of a purple colour, and fhew themfelves in perfection with us the beginning of July. They grow in fpikes, feven or eight inches long, at the ends of the branches, and are of a fingular ftructure. In order to make this tree have its beft effect, it fhould be planted among others of its own growth, in a well-fheltered fituation ; by which means the ends will not be fo liable to be deftroyed by the winter's frofts ; the branches will not fuffer by the violence of the winds ; and as it is fubject to put out many branches near the root, thefe indelicacies and imperfections will be concealed ; whilft the tree will fhew itfelf to the utmoft advantage when in blow, by elevating its purple fpiked flowers amongft the others in a pleafing view.

This tree may be PROPAGATED two ways ; firft, by feeds, which muft be procured from America, where the plant is a native ; for they do not ripen with us in England. We generally receive the feeds from thence in February : and they fhould be committed to the ground as foon after as poffible. They will grow in almoft any foil that is tolerably good ; though the more fandy it be, it will be the better. After they are come up, they fhould have the ufual care of feedlings for a year or two, and then be planted, either where they are intended to remain, or elfe in the nurfery, where they will in a year or two make ftrong plants. This tree may be alfo propagated by layers ; and this operation I would have performed the latter end of fummer, whilft the fap is in motion ; for if it is deferred until winter, the

branches

branches are then so exceedingly brittle, that it will be with difficulty they are brought down, without breaking, a proper depth
into the earth: Let the utmost care be taken, or many of the
young branches that would have made layers will be lost. In
summer, then, let the branches be brought down while they are
pliable ; and by the autumn twelve-months after they will have
taken root, and be fit to remove.

A M Y G D A L U S.

LINNEAN Class and Order, *Icosandria Monogynia:* Each
flower contains about twenty males and one female : There are
four SPECIES ; three of which are more particularly to our purpose.

1. AMYGDALUS *Communis:* The COMMON ALMOND; *a tall
deciduous shrub* ; native of Africa.

2. AMYGDALUS *Nana:* The DWARF ALMOND ; *a deciduous
shrub* ; native of Asia Minor.

3. AMYGDALUS *Persica:* The PEACH ; *a deciduous tree* ; of
what country is uncertain.

1. THE COMMON ALMOND will grow to near twenty feet
high ; and whether planted singly in an open place, or mixed
with others in clumps, shrubbery-quarters, &c. shews itself one
of the finest flowering trees in nature. Those who never yet
saw it, may easily conceive what a noble appearance this tree
must make, when covered all over with a bloom of a delicate
red, which will be in March ; a time when very few trees are
ornamented either with leaves or flowers. No ornamental plantation, therefore, of what sort or kind soever, should be without
almond-trees. Neither are the beauties of the flowers the only
thing desirable in this tree : The fruit would render it worthy of
planting, were there no other motive. It ripens well, and its
goodness i not unknown to us.

The White-flowering Almond, well known in our nurseries, is
a *variety* of this species, and is cultivated for the sake of the
                                                            flowers

AMY

63

flowers and the fruit, though the flowers are inferior to the others. Neither is this tree so proper to plant singly in open places, or near windows, for the show of its flowers; for although they come out early, yet the whole bloom is subject to be taken off in one night's nipping weather, which frequently happens at this season. Its station, therefore, should be in shrubbery-quarters, in well-sheltered places; and in such it will flower exceeding well, and shew its white blossoms to great advantage. When it is designed for fruit, it should be set against a south wall, in a well-sheltered place, otherwise there will be little hopes of success.

2. The Dwarf Almond. Of this tree there are two sorts, the single and the double. Both grow to about four or five feet high, and are in the first esteem as flowering-shrubs. The single sort has its beauties; but the double kind is matchless. In both, the flowers are arranged the whole length of the last year's shoots; their colour is a delicate red; and they shew themselves early in the spring, which still enhances their value.

3. The Peach-tree has hitherto been planted against walls for the sake of the fruit; but, says Hanbury, as I hardly ever knew a person who was not struck with the beauty of the flowers when in full blow against a wall, why should it not have a share in wilderness-quarters and shrubberies, amongst the sorts of almonds, &c.? It may be kept down, or permitted to grow to the height of the owner's fancy; and the flowers are inferior to none of the other sorts. Add to this, they frequently, in well-sheltered places, produce fruit which will be exceedingly well-flavoured; and thus the owner may enjoy the benefit of a double treat. The above observations respect the single peach; with regard to the double-flowered, it is generally propagated for ornamental plantations, and is universally acknowledged to be one of the finest flowering-trees yet known. Against a wall, however, these trees are always the fairest; and if they have this advantage, they are succeeded by very good fruit.

All these sorts are propagated by inoculating them into plum-stocks, in August. The stocks should be first planted in the nursery, when of the size of a straw, and the first or second summer after they will be ready to receive the bud. The usual

method

method of inoculation muſt be obſerved, and there is no danger
of ſuccefs ; though it may be proper to obſerve, that the double-
bloſſomed peach ſhould always be worked into the ſtocks of the
muſſel-plum. The two ſorts of Dwarf Almond may alſo be
propagated by layers, or from the ſuckers, which they ſometimes
ſend forth in great plenty.

A N A G Y R I S.

LINNEAN Claſs and Order, *Decandria Monogynia:* Each
flower contains ten males and one female. There is only one
SPECIES :

ANAGYRIS *Fœtida :* The FETID ANAGYRIS, or Stinking
Bean Trefoil ; *a deciduous ſhrub;* native of Italy, Sicily, and
Spain.

THE ANAGYRIS is a ſhrub of about ten feet growth. The
leaves are different in the different varieties : In one ſort they
are oval, and moderately broad ; in the other they are oblong
and narrow ; but all of them are hoary. The flowers are produced from the ſides of the branches, in May, like thoſe of the
Laburnum : They are numerous, of a bright yellow colour, but
ſeldom ſucceeded by good ſeeds in theſe parts.

The beſt method of PROPAGATING theſe plants is, 1ſt, by
the ſeeds, which ſhould be procured from the countries where
they ripen well. Sow them in a border of good rich earth, in
a well-ſheltered place, and ſift over them about half an inch of
fine mould. March is a very good month for this buſineſs ; and
when the plants appear, if the weather proves dry, frequently
give them water ; keep them clean of weeds all ſummer, and at
the approach of winter prick round the beds ſome furze-buſhes
very cloſe : Theſe will break the keen edges of the black winds ;
for common froſts theſe plants bear moderately well. In the
ſpring let them be ſet out in the nurſery-ground, at a foot diſtance from each other. Here let them ſtand a year or two, and
they will be of a proper ſize to be finally planted out. 2. Theſe
plants may alſo be propagated by layers. For this purpoſe, a

few

few plants fhould be fet for ftools. Let them grow one fummer, to get good hold of the ground, and then head them down. The fummer following they will make ftrong fhoots, which in the autumn fhould be layered. They will readily ftrike root, and by the autumn following will be good plants. The weakeft of thefe may be fet out in the nurfery-ground for a feafon or two; but the ftrongeft may be immediately planted out.

A N D R O M E D A.

LINNEAN Clafs and Order, *Decandria Monogynia*: Each flower contains ten males and one female. There are fifteen SPECIES; three only of which are yet enured to this climate:

1. ANDROMEDA *Paniculata*: The VIRGINIAN ANDROMEDA; *a deciduous fhrub*; native of Virginia.

2. ANDROMEDA *Calyculata*: The CANADIAN ANDROMEDA; *a low deciduous fhrub*; native of Canada, Siberia and Ingria.

3. ANDROMEDA *Mariana*: The MARYLAND ANDROMEDA; *a very low deciduous fhrub*; native of Maryland and other parts of North America.

1. The VIRGINIAN ANDROMEDA is a branching fhrub, about four feet high. The leaves are oblong, pointed, plane, and are placed alternately on the branches. The flowers come out in panicles from the ends of the branches: They are of a pale yellow colour, and come out in July, but are rarely fucceeded by good feeds in England.

2. CANADA ANDROMEDA is a low branching fhrub, hardly a foot and a half high. The leaves are oval, fpear-fhaped, obtufe, reclined on their borders, and poffeffed of numerous fmall punctures. The flowers grow in fhort leafy fpikes, from the ends of the branches: Their colour is white, they appear in July, and are feldom fucceeded by good feeds in this country.

There is a *variety* of this fpecies, with oval obtufe leaves, of a thick fubftance, and which, in mild feafons, continue on the plants all winter.

F 3. MARYLAND

3. MARYLAND ANDROMEDA. This is a fhrub, about two feet high, fending forth feveral ligneous ftalks from the root. The leaves are oval, entire, of a pale green colour, and grow alternately on fhort footftalks. The flowers come out in fmall bunches from the points of the ftalk : They are of a greenifh colour, come out in June and July, and are fometimes fucceeded by five-cornered capfules, full of feeds ; which, neverthelefs, feldom ripen in England.

PROPAGATION : Thefe plants fucceed beft upon boggy and moift grounds. You muft procure the feeds from the places where they grow naturally ; a year before which a boggy or the moifteft part of your garden fhould be dug, and the roots of all weeds cleared off. As the weeds begin to rife, fo conftantly fhould the ground be again dug, and fea or drift fand fhould be plentifully mixed with the natural foil By this management till the feeds arrive, the ground being made tolerably fine, the feeds fhould be fown very fhallow in the moift or boggy land ; or if the land fhould be fo boggy that it cannot be eafily worked, fo as to be proper for the reception of the feeds, then let a fufficient quantity of foil from a frefh pafture, mixed with drift fand, be laid over the bog, and let the feeds be fown therein. The bog will in time abforb this foil, but the feeds will come up ; and this is the moft effectual method of procuring plants of this kind from feeds. The firft year after they come up they fhould be fhaded in very hot weather ; and after that they will require little or no care. Another method of encreafing thefe fhrubs is by layers, or fuckers ; fo that whoever has not the conveniency of procuring the feeds from abroad, fhould get a plant or two of the forts he moft likes : Thefe he fhould plant in a boggy fituation ; and in a very little time he will have encreafe enough ; for they throw out fuckers in prodigious plenty, and, if they like the fituation, to a great diftance. Thefe may be taken off, and planted where they are to remain.

A N N O N A.

LINNEAN Clafs and Order, *Polyandria Polygynia* : Each flower contains many males and many females : There are nine fpecies ;

one

one only of which is fufficiently hardy for the open air of this climate.

ANNONA *Triloba* : The PAPAW or CUSTARD-APPLE ; *a tall fhrub* ; native of the Weft-Indies.

The PAPAW grows to about fixteen or eighteen feet high. The leaves are large, and fhaped like a fpear, and they fall off pretty early in the autumn. The flowers, which will fhew themfelves in the beginning of May, are of a kind of chocolate colour tinged with purple, and grow two or three on a footftalk. The fruit is large, and never ripens in England ; but in the countries where it grows naturally, it is eaten by the meaneft of the inhabitants. The difference of its fhape from that of a pear is, that its wideft part is neareft the footftalk ; and it contains a number of large feeds lying in a row. It is a native of Maryland, Carolina, Virginia, and the Bahama-Iflands ; and from thence we have the feeds brought, by which numbers of plants are annually raifed.

The manner of RAISING them is thus : Let a bed be prepared in a moiftifh part, that is exceedingly well fheltered, and naturally fandy, or inclined thereto. If the foil is oppofite to this, let a fourth part of drift fand be mixed with the mould ; and having obtained the feeds from abroad, fow them in this bed about half an inch deep, letting the feeds be at fome diftance from each other. It is probable they will come up in the fpring, though they fometimes remain till the fecond, nay the third fpring before they make their appearance. When this happens, the beds muft be weeded all the time, and the mould at the furface gently loofened, if it fhould be inclined to cruft over. After the plants are come up in the fpring, no other than the ufual care of feedlings need be taken, until the autumn, when the beds muft be hooped over, to be covered with mats at the approach of any froft ; and the gardener muft conftantly obferve the weather, whether the air hath the leaft tendency to it, that he may cover the bed over ; for one night's hard froft, while they are fo very young, would deftroy them all. With this careful eye he muft conftantly watch over thefe plants all winter. He muft double his covering as the froft encreafes, and muft always uncover them again in mild and open weather. The fecond winter, the fame care muft be obferved, though fo ftrict an eye will not be neceffary ; for although they will be fubject to

be

be deftroyed by hard frofts, yet if a gentle froft fhould catch them unawares to the gardener in the night, there will not be much danger of their fuffering ; for they will be got tolerably ftrong by the fecond fummer's fhoot : They will, neverthelefs, be too tender to ftand the brunt of a winter's froft for a year or two after that ; and confequently muft have a proportional fhare of this attention every year during thefe months. By this time the plants will have grown to be tolerably ftrong, and may be taken up and planted where they are to remain ; though their fituation fhould be well defended ; for a fevere froft in an ex-pofed place would ftill overpower them ; though, after they have grown to be of larger fize, they are hardy enough.

If a perfon has the conveniency of a green-houfe, or fome fuch room, he may fow his feeds in boxes or pots filled with maiden earth, from a rich pafture, mixed with drift fand. Thefe boxes or pots fhould be afterwards plunged into the natural mould, in a fhady part of the garden ; and the autumn after the plants are come up, they may be removed into the green-houfe, where they will be naturally protected from the injuries of weather. This protection may be afforded them every winter, till they are ftrong enough to defend themfelves, when they may be turned out of the boxes or pots, mould and all, into the places where they are defigned to remain.

A R A L I A.

LINNEAN Clafs and Order, *Pentandria Pentagynia* : Each flower contains five males and five females : There are five SPECIES , only one of which is adapted to our purpofe :

ARALIA *Spinofa :* The ANGELICA TREE, or PRICKLY ANGE-LICA : *a deciduous fhrub* ; native of Virginia.

The PRICKLY ANGELICA : The height to which this tree will grow, if the foil and fituation wholly agree with it, is about twelve feet ; and the ftem, which is of a dark brown colour, is defended by fharp fpines, which fall off ; nay, the very leaves, which are branching, and compofed of many wings, and are of a pleafant green colour, have thefe defenders, which are both crooked

ed and ftrong, and ftand as guards to them till the leaves fall off in the autumn. The flowers are produced in large umbels from the ends of the branches : They are of a greenifh yellow colour; and their general characters indicate their ftructure. They make their appearance the end of July or beginning of Auguft; but are not fucceeded by ripe feeds in our gardens.

PROPAGATION : This tree will what gardeners call *fpawn* ; *i. e.* after digging among the roots young plants will arife, the broken roots fending forth frefh ftems; nay, if the roots are planted in a warm border, and fhaded in hot weather, they will grow ; but if they are planted in pots, and affifted by a moderate warmth of dung, or tanners bark, they will be pretty fure of fuccefs ; fo that the propagation of this tree is very eafy. But the general method of propagating it, and by which the beft plants may be had, is from feeds, which muft be procured from America, for they do not ripen in England ; and, after having obtained them, they muft be managed in the following manner : The time that we generally receive them is in the fpring; fo that againft their coming we muft be furnifhed with a fufficient number of large pots. Thefe, when the feeds are come, muft be filled with fine mould, which, if taken from a rich border, will do very well. The feeds muft be fown in thefe pots as foon as poffible after their arrival, hardly half an inch deep, and then the pots fhould be plunged in a warm place their whole depth in the foil. Care muft be taken to break the mould in the pots, and water them as often as it has a tendency to cruft over; and if they are fhaded in hot weather, the plants will frequently come up the firft fummer. But as this does not often happen, if the young plants do not appear by Midfummer, the pots fhould be taken and plunged in a fhady place ; nay, if they fhould, there will be ftill more occafion for this being done ; for they will flourifh after that better in the fhade ; and the defign of plunging them in a warm place at firft was only with a view of fetting the powers of vegetation at work, that, having natural heat, artificial fhade alfo may be given them, and water likewife, the three grand neceffaries for the purpofe. The pots, whether the plants are come up in them or not, fhould be removed into fhelter in October, either into a greenhoufe, fome room, or under an hotbed-frame ; and in the fpring, when all danger of froft is over, they fhould be plunged

F 3   into

into the natural ground their own depth in a shady place. Those
that were already come up will have shot strong by the autumn
following ; and if none of them have appeared, they will come
up this spring ; and whether they are young seedlings, or small
plants of a former summer's growth, they must be constantly
kept clean of weeds, and duly watered in the time of drought ;
and this care must be observed until the autumn. In October
they must be again removed into shelter, either into a greenhouse,
&c. as before, or fixed in a warm place, and hooped, that they
may be covered with mats in frosty weather. In the latter end
of March following, they should be planted in the nursery way,
to gain strength before they are finally planted out. The ground
for this purpose, besides the natural shelter, should have a reed-
hedge, or something of the like nature, the more effectually to
prevent the piercing winds from destroying the young plants. In
this snug place the plants may be set in rows : in each of which
rows furze-bushes should be stuck the whole length ; and all
these together will ensure their safety. But here one caution is
to be observed ; not to stick the furze so thick, but that the
plants may enjoy the free air in mild weather, and not to take
them away too early in the spring, lest, being kept warm the
whole winter, and being deprived of their protection, a cutting
frost should happen, as it sometimes does even in April, and destroy them. Weeding and watering in dry weather must be
their summer's care. They may be stuck again with furze-
bushes in the winter ; though it will not be necessary to do it in
so close a manner ; and with this care, still diminishing in proportion the number of furze-bushes, they may continue for three
or four years, when they may be planted out into the warmest
parts of the plantation. With this management these plants will
be inured to bear our winters, in well-sheltered places.

The spines which grow on the branches and the leaves admo-
nish us, for our own safety, not to plant this tree too near the
sides of frequented walks ; and the consideration of the nature of
the tree, which is rather tender at the best, directs us (if we have
a mind to retain the sort) to plant it in a warm and well-sheltered situation ; where the piercing frosts, come from what point
they will, will lose their edge ; for without this, they will be
too tender to stand the test of a severe winter ; though it has
often happened, that after the main stem of the plant has been

deſtroyed,

deſtroyed, it has ſhot out again from the root, and the plant by
that means been both encreaſed and preſerved.

A R B U T U S.

LINNEAN Claſs and Order, *Decandria Monogynia :* Each
flower contains ten males and one female. There are ten
SPECIES ; two of which are proper for our purpoſe.

1. ARBUTUS *Unedo :* The ARBUTUS, or the COMMON STRAW-
BERRY TREE ; *an evergreen tree* ; native of Ireland and many
parts of Europe.

2. ARBUTUS *Andrachne :* The ANDRACHNE; or the ORIEN-
TAL ARBUTUS ; or the ORIENTAL STRAWBERRY-TREE ; *an
evergreen tree* ; native of the Eaſt.

1. The ARBUTUS. Of this Species there are four *Varieties :*
namely,

THE OBLONG-FRUITED,	THE RED-FLOWERED, AND
THE ROUND-FRUITED,	THE DOUBLE-BLOSSOMED.

One deſcription is nearly common to them all : And their in-
conſiderable variation is almoſt ſufficiently ſhewn in their reſpective appellations.

The oblong-fruited Arbutus will grow to be a middling-ſized
tree in ſome countries ; for we read of the large uſes its wood has
been applied to ; ſuch as, *Arbuteæ crates*, &c. Arbutean har-
rows, &c. With us it may be kept down to any ſize. The
main ſtems are covered with a light-brown bark, rough, and falling. The younger branches are of a kind of purple colour,
whilſt the laſt year's ſhoots are of a fine red, and a little hairy.
The leaves grow alternately on the branches, and are of an ob_
long oval figure. They ſtand on ſhort footſtalks, and the oldeſt
leaves make a contraſt with the younger by having their footſtalk
and mid-rib of a fine ſcarlet colour. They are ſmooth, and beautifully ſerrated. Their upper ſurface (as in moſt trees) is of a ſtronger
green than their under ; and the young twigs are garniſhed with
them in plenty. Theſe are beauties in common to moſt trees, in
ſome degree or other ; but every thing elſe almoſt of this tree
that preſents itſelf to conſideration is ſingular : The time of its
flowering will be in November and December ; when it is rather
ſingular to ſee a tree in the open ground in full blow ; and the

fruit

fruit ripens by that time twelvemonth after. The manner and
nature of the fruit, which look like very large red ftrawberries,
give it alfo a fingular and delightful look ; and this is heightened
as they appear all over the tree among the flowers ; for that is
the time of its being ripe, when the flowers for the fucceeding
crop are fully out. The flowers themfelves make no great
figure ; they are of a kind of whitifh-yellow colour ; and are
fucceeded by the abovementioned Strawberry fruit, which will
require a revolution of twelve months, before they perfectly arrive at their maturity and colour. The flowers of the firft fort
are larger than thofe of the fecond ; and the fruit is oval, and
much larger than our Common Scarlet Strawberry.

The *round-fruited* fort has its pitcher-fhaped flowers, which are
fucceeded by round fcarlet fruit, as wide as they are long ; and
this is all the difference between thefe forts.

The *Strawberry-tree with red flowers* differs in no refpect from
the common fort, only the flowers are red, and thefe conftitute
a variety from the other forts of flowers ; but the contraft is not
fo great between their fruit and them, as of the other forts,
their colour approaching too near to a famenefs.

The *Double-bloffomed Strawberry-tree* differs in no refpect, only
that the flowers are double ; but this difference is fo inconfiderable, that it will not be feen without looking into the flower ; and
even then the doublenefs will appear fo trifling as fcarcely to merit notice ; fo that a plant or two, to have it faid that the collection is not without it, will be fufficient. Neither ought any
more to be admitted ; for they will not produce the fame plenty
of fruit, which conftitutes the greateft beauty of thefe trees, as
the fingle forts.

The method of PROPAGATING the *Varieties* of the Arbutus
is by layers and cuttings : the *Species* itfelf may be raifed from
feed. By layers they will all grow : The operation muft be
performed on the youngeft twigs ; and in fome foils they will
ftrike root pretty freely, whilft in others they can hardly be made
to grow at all : But before they have lain two fummers, you
may fcarcely venture to look for any. When the roots are
ftruck, the layers fhould be carefully taken off in the fpring,
and planted in feparate pots ; and after well watering them, they
fhould be plunged up to the rims in an hotbed, and this will fet
them forward ; for without this affiftance, many of the layers
will

will be loft; fince they are difficult plants to make grow. After the hotbed has forced the feeds into a ftate of vegetation, the pots may be taken out, and plunged up to the rims in fome natural mould, to keep them cool and moift; and here they may ftand for two or three years, or longer, if the pots are large enough, without ever removing or fheltering in winter; for they are hardy enough to refift our fevereft cold. When they are to be finally fet out, all the mould may be turned out of the pots hanging to the roots; and having proper holes made ready, they may be planted in them, and the plant will be ignorant of its new fituation.

Thefe plants may be encreafed by cuttings, which muft be planted in pots, and have the benefit of a good bark-bed; in which being conftantly fhaded and duly watered, many of them will grow. As the plants raifed this way will be rather tender by being forced in the bark-bed, it will be neceffary to remove them into the greenhoufe, or to place them under an hotbed-frame during the firft winter; and after that, the pots may be fet up to the rims in the ground, and, like the layers, the plants may be turned out at a convenient time into the places where they are to remain.

Next we proceed to the beft way of raifing the *Common Arbutus*; and that is from feeds. Let thefe be taken from the oblong or round-fruited fort. The feeds, which will be ripe fome time in November or the beginning of December, for they will not be ripe at the fame time in all places, muft be then gathered; and as they fhould not be fowed until the fpring, it will be proper to put them into a pot or jar, mixing with them a quantity of drift-fand; and this will preferve them found and good. The beginning of March is the beft time for fowing the feeds; and the beft foil for them is maiden earth, taken from a rich pafture at leaft a year before, with the fward; and this, by conftant turning, being well rotted and mixed, will be ready to receive them. Having filled a different quantity of pots with this fine mould, let the feeds be fown, and but juft covered, fcarcely a quarter of an inch deep. A dry day fhould be chofen for the bufinefs; and no watering by the hand fhould be given them, as it will endanger the fetting the mould hard in the pots. Leave them abroad until fome rain falls, which at that time may be hourly expected; and after that, having an hotbed ready,

plunge

plunge the pots therein. In lefs than fix wecks you may expect
your plants to appear; when much air fhould be afforded them,
and frequent waterings, in fmall quantities, gently fprinkled
over them. After this, they may be hardened to the air by degrees, and the pots fet.up to the rims in the natural mould, in a
fhady place. In October they fhould be removed into the greenhoufe, or fome fhelter, in frofty weather; though they fhould
always be fet abroad in mild open weather. In the fpring they
may be fhook out, and planted in feparate pots; and they
fhould have the advantage alfo of a hotbed to fet them a-growing: their future management may be the fame as was directed
for the layers. When thefe trees are to be planted out, very
little regard need be paid to the foil or fituation; for they will
grow almoft any where, and refift our fevereft northern blafts.
One thing, however, the gardener muft conftantly obferve, in
order to continue his trees in their beauty; viz. as often as a
heavy fnow falls, fo conftantly fhould he go and fhake the
boughs; for it will lodge amongft the leaves and branches, in
fuch great quantity, as to weigh down and fplit the largeft
branches; the deformity of which afterwards may be eafily conceived. Befides, many years muft expire before the tree will,
if ever it fhould, grow to its former beauty; to preferve this,
therefore, makes the narrowly watching thefe trees in fnowy
weather highly neceffary.

2. The ANDRACHNE will grow to a larger fize than the Arbutus.
The leaves are fmooth, and nearly of the fame figure as the
preceding fort; though they are larger, and have their edges
undivided. The flowers grow like the other forts; are of the
fame colour; and they are fucceeded by large, oval, fcarlet fruit.
It is called the Oriental Strawberry-Tree, becaufe this fort grows
plentifully in many parts of the Eaft, and is ufeful to the inhabitants for many purpofes in life.

The *Andrachne* may be PROPAGATED in the fame manner as
the *Arbutus*.

A R T E M I S I A.

LINNEAN Clafs and Order, *Syngenefia Polygamia Superflua*:
Hermaphrodite florets containing five males and one female, and
female

female florets containing one piftil, in the fame difcous flower: There are twenty-five Species, which are principally herbaceous; one only being intitled to a place amongft the tribe of ornamentals.

Artemisia *Arborefcens* : The Tree-Wormwood: *a non-deciduous fhrub* : native of Italy and the Eaft.

The Tree-Wormwood rifes with an upright ftalk to the height of about fix feet. The leaves are its chief excellence; and of thefe there are two or three forts : One fort is very much divided, or cut into feveral narrow fegments; thofe of the other are broader. They are very hoary; and as they continue on the branches all winter, they have a fingular and an agreeable effect among the evergreens at that feafon. The flowers are fmall, and have very little beauty; they are collected into roundifh heads, and I never perceived them to be followed by good feeds.

This plant is eafily propagated by cuttings. Plant them in May, June, July or Auguft, in a fhady place, and they will readily grow, efpecially if they are watered a few times at the firft planting. In the autumn thefe cuttings, which will then have become good plants, fhould be each fet in a feparate fmall pot, and placed under a hotbed frame, or in the greenhoufe, to be preferved all winter. In the fpring they may be turned out into the places where they are defigned to remain, which muft be naturally warm and well fheltered, or they will be liable to be deftroyed by the feverity of the following winter. In fuch a fituation they will live for many years; though it may be advifeable to keep a plant or two in the greenhoufe, to keep up the ftock, if a more than common hard winter fhould put a period to thofe that are planted abroad.

A T R I P L E X.

Linnean Clafs and Order, *Polygamia Monoecia* : Hermaphrodite flowers containing five males and one female, and female

male flowers containing one piftil, on the fame plant. There are twelve SPECIES ; two only of which are to our purpofe.

1. ATRIPLEX *Halimus* : BROAD-LEAVED SEA PURSLAIN-TREE : *a non-deciduous fhrub* ; grows naturally upon the fea-coaft of Spain and Portugal ; as alfo in Virginia.

2. ATRIPLEX *Portulacoides* : NARROW-LEAVED SEA-PURSLAIN-TREE ; *a non deciduous fhrub* ; native of our own fea-coaft, and of the North of Europe.

1. The BROAD-LEAVED PURSLAIN-TREE generally grows to about five or fix feet ; and will fend forth its branches fo as to fpread around, and form a large broad head. The young branches are covered with a fmooth white bark ; that of the older is of a light grey colour, which will be peeling length-vays, and falling, efpecially in the fpring. The branches are exceedingly brittle, and their infide is green to the very pith, of which there is very little. The leaves are foft, white, and filvery, and nearly of the fhape of the Greek letter *Delta*. They have their edges entire ; and look well at all times, efpecially in winter, when they caufe as great a variety as poffible among thofe trees that retain their leaves at that time. This fhrub feldom flowers in our gardens ; and when that happens, it is poffeffed of no beauty to recommend it to the florift.

2. The NARROW-LEAVED PURSLAIN-TREE commonly grows to about four feet high. The branches are numerous and grey ; and they naturally fpread abroad in a bufhy manner. The leaves are filvery ; though not fo white as the other fort ; but they are narrower, which occafions its being fo diftinguifhed ; and of an oval figure ; and by them the fhrub receives no fmall orna-ment. The flowers have little beauty.

Thefe fhrubs are PROPAGATED by cuttings ; which will grow, if planted at any time of the year ; though the beft way is to take the cuttings in March, of the ftrongeft former fummer's fhoots, to cut them into lengths about a foot each, and to plant them a third part deep in the mould. Thefe will all readily take root, and be good plants by the autumn following. In fummer, flips and cuttings may be planted ; but then it will be advifeable to plant them pretty clofe together in beds, and afterwards to hoop the beds, and fhade them from the heat at

that

that time. They will foon take root; and after that will require no further trouble : But until that is effected, they fhould be watered and fhaded in the hot weather, and the mats fhould be conftantly taken off in the evening, and alfo in rainy, moift, or cloudy weather; and by this means plenty of plants may be raifed. If it happened to be a dripping day when they were firft planted, much trouble in fhading and watering will be faved, as they may be nearly upon ftriking root before the weather clears up. Thefe fhrubs fhould be always raifed at a diftance from farm-yards, barns, &c. where there are fparrows ; for thefe birds are fo exceedingly fond of the leaves, that when once they find them out, they will never leave nor forfake them until they have entirely ftripped the plants ; and though the fhrub will fhoot out afrefh, yet they will as conftantly repair to their repaft, and will thus continue to prey upon them until they have entirely deftroyed them. I am obliged (continues HANBURY) to give this precaution, becaufe all my plants of thefe forts are thus conftantly eat up by the fparrows in my gardens at Church-Langton, as often as I plant them ; fo that I am obliged to keep them at Gumley, and in my other diftant nurferies, where they remain free from fuch devourers.

These plants require a warm fheltered fituation, being fubject to be cut by the early frofts.

A Z A L E A.

LINNEAN Clafs and Order, *Pentandria Monogynia :* Each flower contains five males and one female. There are fix SPECIES ; two of which are proper for the fhrubbery.

AZALEA *Nudiflora :* The RED AMERICAN UPRIGHT HONEYSUCKLE : or the RED AZALEA ; *a deciduous fhrub* ; native of Virginia.

AZALEA *Vifcofa :* The WHITE AMERICAN UPRIGHT HONEYSUCKLE ; or The WHITE AZALEA ; *a low deciduous fhrub* ; native of Virginia.

1. The RED AZALEA has feveral ftems arifing from the fame root, which will grow to feven or eight feet high. The leaves

are

are of an oval figure, fmooth, entire, and placed alternately on the branches. The flowers are produced in clufters from the fides of he branches, on long naked footftalks : Their colour is red, and they are agreeably fcented ; each compofed of a long naked tube, cut at the top into five fpreading fegments. They will be in blow in July ; but they feldom ripen their feeds in our gardens. There is a variety of this, with yellow flowers.

2. The WHITE AZALEA. From the root of this arife feveral flender brown ftems, to three or four feet high. The leaves are fpear-fhaped, narrow at their bafe, have a rough border, and grow in clufters. The flowers terminate the branches in clufters, coming out between the leaves. They are finely fcented, and each of them has a tube of near an inch long, divided at the top into five fegments, two of which are reflexed. Their colour is white, with a bad yellow on their outfide ; they will be in blow in July, but are never fucceeded by feeds in our gardens.

Thefe forts are PROPAGATED, 1. By layering the young fhoots; and for this purpofe, a flit muft be made on each, as is practifed for carnations : The autumn is the beft feafon for the work. When the layers have ftruck good root, they may be removed into the nurfery, and planted in lines at a fmall diftance from each other ; where after having ftood a year or two at the moft, they will be proper plants to be planted out. 2. Thefe forts alfo propagate themfelves very faft ; for as they throw up many ftems from the fame roots after they have ftood a few years, fome of thefe may eafily be taken off, with fome root at each, and either planted in the nurfery ground, or the places where they are to remain.

B E R B E R I S.

LINNEAN Clafs and Order, *Hexandria Monogynia :* Each flower contains fix males and one female : There are three SPECIES ; two of which are here treated of :

1. BERBERIS *Vulgaris :* The COMMON BERBERY : a well-known *deciduous fhrub,* common in our hedges.

2. BERBERIS *Cretica :* The CRETAN, or BOX-LEAVED BER-BERY : *a low deciduous fhrub ;* native of Crete.

1. The

1. The COMMON BERBERY. This shrub is distinguished by the acidity of its leaves, the sharpness of its spines, the yellowness of its inner bark, and the scarlet colour of its berries, which add a beauty to our hedges in winter, and afford a favourite pickle and garnish for our tables. Hanbury enumerates other uses of the Berbery, particularly in medicine, and recommends the cultivation of it in the warmest manner. There is however an evil attendant of the Berbery-bush which ought to confine it within the pale of our gardens and shrubberies; we mean its poisonous effect upon corn; more particularly upon wheat. This is a circumstance which has been long known to the common farmers in different parts of the kingdom, especially in Norfolk, where the farmers are more observant and much more enlightened than those useful members of society in general are. The idea, nevertheless, has been treated by theoretical writers on Husbandry as chimerical and superstitious; and has been brought forward as one of those vulgar errors of farmers which ought to induce gentlemen and men of genius to rescue so useful a science as that of Agriculture out of the hands of ignorance. Being however always ready to hear the opinion of *professional* men, and having been assured by many sensible farmers of the truth of this matter, we had a few years ago a Berbery-bush planted, in the month of February, in the centre of a large piece of wheat. No obvious effect took place until the corn began to change its colour before harvest, when a long blackening stripe became so conspicuous amongst the growing whiteness of the wheat, that it might have been distinguished at a mile's distance. It resembled the tail of a comet; the bush representing the comet itself; and what rendered the experiment striking, whilst on one side the effect did not reach more than three or four feet, on the opposite side it was obvious to the distance of ten or twelve yards; notwithstanding the top of the shrub planted was not much larger than a man's head. At harvest, the ears which grew in the immediate neighbourhood of the bush, stood erect, the grains shriveled and empty; —— as the distance from the Berbery increased the effect lessened, vanishing imperceptibly: whilst the grain of the rest of the field was of a good quality. We do not mean in this place to comment upon the fact, or to attempt to account for so singular an effect by the help of *reasoning* only; having in our intentions a suite of *experiments*

in

in order to endeavour to come at the caufe. Our motive for
mentioning the fact at prefent is to induce others to make
fimilar experiments, as well as to ftimulate gentlemen to extirpate
from their eftates fo pernicious a plant ; more particularly from
the hedges and borders of arable fields.

There are three *Varieties* of the Common Berbery :

The Berbery with white fruit.

The Berbery with black fruit.

The Berbery without ftones : which laft is the fort principally
cultivated for the berries.

2. The Box-leaved Berbery grows to a yard or four feet
high, and is poffeffed of many fharp fpines at the joints. The
leaves are like thofe of the box-tree, between which the flowers
come out, on flender footftalks. But as this fort never produces
any fruit in England, and being alfo liable to be killed by hard
frofts, it is feldom propagated in our gardens.

The propagation of the Berbery is as follows. 1. When
a quantity of the common Berbery is wanted, the beft way is to
raife it from the feeds, which fhould be fown, foon after they
are ripe, in a bed made in any part of the garden. Thefe will frequently remain till the fecond fpring before the plants come all up;
till which time the beds fhould be weeded as often as the weeds
appear ; for if they are neglected fo as to get ftrong, by pulling
them up many of the feeds will alfo be drawn out of the bed by
their roots. After the plants have grown one year in the feed-
bed, they fhould be planted out in the nurfery, where they may
remain for about two years, when they will be fit to plant out
finally. This is the moft expeditious method of raifing a large
quantity of thefe trees, when wanted. 2. Another method of
propagating the Berbery is by layers ; a method by which all
the forts may be encreafed ; and in the performance of which,
no other art or trouble need be ufed, than laying the branches
down in the ground, without either flit or twift. If this is done
any time in the winter, by the autumn following they will have
taken good root ; the ftrongeft of which layers will be then
fit to plant out ; whilft thofe that are weaker may be planted in the nurfery-ground, to gain ftrength. 3. The cuttings
alfo of thefe trees will grow ; for if they be planted in October,
in a moiftifh good earth, they will moft of them ftrike root ; fo
that the propagation of this tree by any of thefe ways is very
eafy.

eafy. Whoever is defirous of the Box-leaved Berbery muft afford it a warm dry foil, in a well-fheltered place.

B E T U L A.

LINNEAN Clafs and Order, *Monoecia Tetrandria* : Male flowers containing four ftamens, and female flowers containing two piftils, difpofed in feparate cylindrical catkins upon the fame plant. There are five SPECIES :

1. BETULA *Alba* : The COMMON WHITE BIRCH : a well known *deciduous tree* ; native of this country, as alfo of moft of the colder parts of Europe.

2. BETULA *Nigra* : The VIRGINIAN BLACK BIRCH : *a deciduous tree*; native of Virginia, as alfo of Canada.

3. BETULA *Lenta* : The CANADA BIRCH : *a deciduous tree*; native of Canada, and of Virginia.

4. BETULA *Nana* : The DWARF BIRCH : *a deciduous fhrub* ; native of Lapland, Ruffia, Sweden, and Scotland.

5. BETULA *Alnus* : The ALDER : a well known aquatic *deciduous tree* ; common in our fwamps and low grounds; it is alfo common in moft parts of Europe, and in America.

1. The COMMON WHITE BIRCH. This tree is fo common, and its ufes fo well known throughout the kingdom, that any defcription of it feems unneceffary. It is in general of a humble growth; however, in a foil and fituation it affects, it will rife to a great height, and fwell to a confiderable fize. There is a fprucenefs in its general appearance in fummer; and in winter its bark fometimes exhibits, in its variegations of red and white, no inelegant object. Were it not for its being fo commonly feen upon poor foils, and in bleak inhofpitable fituations, as well as for the mean and degrading purpofes to which it is univer-fally put, the Birch would have fome claim to being admitted to a place among the *ornamentals*. Its *ufes* are chiefly for brooms, fuel, and charcoal : if it be fuffered to grow to a proper fize, it will make tolerable gates, hurdles, rails, &c : it is alfo ufed by the patten-makers. HANBURY fays, it is alfo applicable

to larger ufes ; and is highly proper for the fellies of broad-wheel waggons, it being inlocked fo as not to be cleaved. " I have been informed (fays he) by an old experienced wheelwright, that old Birch-trees cannot be cleft, as the grains run croffways, and that he prefers it for feveral ufes in his way to moft wood ; and as I have feen feveral of thefe trees more than two feet fquare, the timber of the Birch may perhaps be of more value than it has hitherto been efteemed." Its ufe in making wine is well known. But although we enumerate the ufes to which the Birch is applicable when it is already in poffeffion, we do not mean to recommend in general terms the planting of Birch ; except in bleak and barren fituations where no other tree will thrive ; and except as a fkreen and guardian to nurfe-up and de-fend from chilling blafts plants of greater value.

The PROPAGATION of the Birch is eafy : it may be raifed either from feeds, or by layering ; and it will flourifh in almoft any foil or fituation.

2. The BLACK VIRGINIA BIRCH will grow to upwards of fixty feet in height. The branches are fpotted, and more fpar-ingly fet in the trees than the common forts. The leaves are broader, grow on long footftalks, and add a dignity to the ap-pearance of the tree ; and as it is naturally of upright and fwift growth, and arrives at fo great a magnitude, HANBURY thinks it ought to have a fhare among our foreft-trees, and to be planted for ftandards in open places, as well as to be joined with other trees of its own growth in plantations more imme-diately defigned for relaxation and pleafure.

There are feveral *varieties* of this fpecies, differing in the colour, fize of the leaves, and fhoots ; all of which have names given them by nurferymen, who propagate the different forts for fale ; fuch as, 1. *The Broad-leaved Virginian Birch*; 2. *The Poplar-leaved Birch*; 3. *The Paper-Birch*; 4. *The Brown Birch*, &c.

3. CANADA BIRCH. This grows to a timber-tree of fixty or more feet in height. The leaves are heart-fhaped, oblong, fmooth, of a thin confiftence, pointed, and very fharply ferrat-ed. They differ in colour ; and the *varieties* of this fpecies go by the names of, 1. *Dufky Canada Birch*; 2. *White-Paper Birch*, 3. *Poplar-leaved Canada Birch*; 4. *Low-growing Canada Birch,*

Birch, &c. The bark of this species is very light, tough, and durable ; and the inhabitants of America use it for canoes.

4. DWARF BIRCH. This is a low branching shrub, about two feet high. The leaves are round, and their edges are serrated. It hardly ever produces either male or female flowers, and is chiefly coveted when a general collection of plants is making.

The method of PROPAGATING all the foreign sorts is, 1. From seeds. We receive the seeds from America, where they are natives ; and if we sow them in beds of fine mould, covering them over about a quarter of an inch deep, they will readily grow. During the time they are in the seminary, they must be constantly weeded, watered in dry weather, and when they are one or two years old, according to their strength, they should be planted in the nursery, in rows, in the usual manner. Weeding must always be observed in summer, and digging between the rows in winter ; and when the plants are about a yard or four feet high, they will be of a good size to be planted out for the shrubbery-quarters. A part, therefore, may be then taken up for such purposes ; whilst the remainder may be left to grow for standards, to answer such other purposes as may be wanted. 2. These trees may also be propagated by layers ; and this is the way to continue the peculiarities in the varieties of the different sorts. A sufficient number of plants should be procured for this purpose, and set on a spot of double-dug ground, three yards distance from each other. The year following, if they have made no young shoots, they should be headed to within half a foot of the ground, to form the stools, which will then shoot vigorously the summer following ; and in the autumn the young shoots should be plashed near the stools, and the tender twigs layered near their ends. They will then strike root, and become good plants by the autumn following ; whilst fresh twigs will have sprung up from the stools, to be ready for the same operation. The layers, therefore, should be taken up, and the operation performed afresh. If the plants designed for stools have made good shoots the first year, they need not be headed down, but plashed near the ground, and all the young twigs layered. Thus may an immediate crop be raised this way ; whilst young shoots will spring out in great plenty below the plashed part, in order for layering the succeeding year. This work, therefore, may be repeated every autumn or winter ; when some of the strongest

layers

layers may be planted out, if they are immediately wanted; whilst the others may be removed into the nursery, to grow to be stronger plants, before they are removed to their destined habitations. 3. Cuttings also, if set in a moist shady border the beginning of October, will frequently grow: But as this is not a sure method, and as these trees are so easily propagated by layers, it hardly deserves to be put in practice.

5. The ALDER. This well-known aquatic will grow to a large timber tree. The Alder, like the Birch, suffers, as an *ornamental*, from an association of ideas; we not only see it very common, but we see it in low, dreary, dirty situations: nevertheless, if the Alder be suffered to form its own head in an open advantageous situation, it is by no means an unsightly tree: in Stow Gardens, in what is called the old part, there are some very fine ones; and in coming round from the house by the road leading to Buckingham, there is one which is truly ornamental. Hacked and disfigured in the manner in which Alders in general are, they have but little effect in doing away the unsightliness of a swamp; but if they were suffered to rise in groups and singlets, open enough to have room to form their full tops, and close enough to hide sufficiently the unseemliness of the surface, even a moor or a morass seen from a distance might be rendered an agreeable object. Many *uses* of Alders have been enumerated by authors: they were, indeed, more numerous than they are at present. Leaden pipes have superseded them as pump-trees and water-pipes, and logwood has rendered their bark of little value. They are however still useful as piles, and make tolerable boards; they are also convenient as poles, and make good charcoal: great quantities are cut up for patten wood, and for the wooden heels and soals of shoes. But upon the whole the consumption is too inconsiderable to make them an object of the planter's notice, except in particular situations. For securing the banks of rivers we know of nothing better than the Alder; its roots are stronger and more interwoven with each other than those of the Salix tribe: also in low swampy situations where the ground cannot be drained but at too great an expence, the Alder may be planted with propriety and advantage: but wherever the soil is or can be made pasturable, the Alder should by no means be permitted to gain a footing. Its suckers and seedlings poison the herbage; and it is a fact well known to the observant husbandman,

man, that the roots of the Alder have a peculiar property of rendering the foil they grow in more moift and rotten than it would be if not occupied by this aquaous plant. Plantations of Alders fhould therefore be confined to fwampy, low, unpafturable places ; except when they are made for the purpofes of ornament ; and in this cafe the native fpecies ought to give place to its more ornamental *varieties*, of which HANBURY makes five ; namely. 1. The Long-leaved Alder. 2. The White Alder. 3. The Black Alder. 4. The Hoary-leaved Alder. 5. The Dwarf Alder.

The PROPAGATION of the Alder, like that of the other aquatic natives, is very eafy : it may be raifed either from fuckers, from cuttings, or by layering ; and no doubt from feed, though this mode of propagation is feldom practifed in this country. EVELYN mentions a peculiar method of raifing this tree from cuttings or truncheons, which he calls the *Jerfey manner :* he fays, " I received it from a moft ingenious gentleman of that country : it is, to take truncheons of two or three feet long at the beginning of the winter, and to bind them in faggots, and place the ends of them in water till towards the fpring, by which feafon they will have contracted a fwelling fpire or knur about that part, which being fet does (like the Gennet-moil Apple-Tree) never fail of growing and ftriking root." MILLAR recommends truncheons of three feet long, two feet of which to be thruft into the ground. HANBURY fays, that truncheons are uncertain, and ftrongly recommends layering ; which for preferving the varieties, as well as for ornamental plantations of Alders in general, is the beft method.

B I G N O N I A.

LINNEAN Clafs and Order, *Didynamia Angiofpermia :* Each flower contains four males and one female ; two cf the males being fhorter than the other two ; and the feeds being inclofed in a pod. There are eighteen SPECIES ; five of which are enured to this climate.

1. BIGNONIA

1. BIGNONIA *Catalpa* : The CATALPA ; *a deciduous tree* ; native of Carolina.

2. BIGNONIA *Sempervirens* : The EVERGREEN BIGNONIA, or the VIRGINIA JASMINE, or the VIRGINIA CLIMBER ; an *evergreen climber* ; native of Virginia.

3. BIGNONIA *Unguis* : The CLAW BIGNONIA, or the QUA-DRIFOLIATE BIGNONIA ; *a deciduous climber* ; native of Barbadoes, and other West-India Islands.

4. BIGNONIA *Capreolata* : The TENDRIL BIGNONIA, or the CAPREOLATE BIGNONIA ; *a deciduous climber* ; native of North America.

5. BIGNONIA *Radicans* : The SCARLET TRUMPET FLOWER ; *a deciduous climber* ; native of Carolina, Virginia, and Canada.

1, The CATALPA will grow to the height of thirty or forty feet ; and as the stem is upright, and the leaves fine and large, it should be planted as a standard in the midst of fine opens, that it may without molestation send forth its lateral branches, and shew itself to every advantage in view. These opens, neverthelefs, should be such as are well sheltered, otherwise the ends of the branches will be destroyed by the severity of the winter's frost, which will cause an unsightly appearance ; and the leaves, being very large, make such a resistance to the summer's high winds, as to occasion whole branches to be split off by that powerful element. The bark of the Catalpa is brown and smooth, and the leaves are cordated. They are about five or six inches in breadth, and as many in length. They stand by threes at the joints, are of a blueish cast, and are late in the spring before they come out. The flowers are tubulous ; their colour is white, having purple spots, and yellowish stripes on the inside. They will be in full blow in August ; but are not succeeded by good seeds in England.

Whoever has the conveniency of a bark-bed may PROPA-GATE this tree in plenty, . By cuttings, which being planted in pots, and plunged into the beds in the spring, will soon strike root, and may afterwards be so hardened to the open air, that they may be set abroad in the shade before the end of summer : in the beginning of October, they should be removed into a green-house, or under some shelter, to be protected from the winter's frost. In the spring, after the bad weather is past,

they

they may be turned out of the pots, and planted in the nurfery-
way, in a well sheltered place ; and if the soil be rich, and rather
inclined to be moist, it will be the better. Here they may stand
for four or five years, the rows being dug in winter, and weeded
in summer, when they will be of a proper size to be planted out
to stand. These cuttings will often grow in a rich, shady, moist
border; so that whoever can have plenty of them, should plant
them pretty thick in such a place, and he may be tolerably sure,
by this way, of raising many plants. 2. From seed, which must be
procured from America, and should be sown in a fine warm
border of light rich mould, or else in pots or boxes ; the seedling plants requiring more than a common care.

2. The EVERGREEN BIGNONIA has almost every perfection
to recommend it as a climber ; for though the plants are small,
yet if they are trained up to a wall, or have bushes or trees on
which to climb, they will mount to a great height, by their
twining stalks, and over-top hedges, and even trees, and will
form at a distance a grand figure from the sway they will bear.
The leaves of Bignonia are single, and of a lanceolate figure
They grow from the joints, are of a fine strong green colour, and
very ornamental : but the flowers constitute the greatest value of
this plant, on account of the fine odour nature has bestowed on
them ; which is to so great a degree as to perfume the circumambient air to a considerable distance. These flowers are of
a yellow colour, and less beautiful than some of the other sorts,
which is sufficiently recompensed by their extraordinary fragrance. They grow in an erect manner, from the wings of the
leaves at each joint, and their figure nearly resembles that of a
trumpet. The pods that succeed these flowers are small.

There is a *variety* of this species, which over-tops whatever
plants are near it, to a great height. The leaves are of a lanceolate figure, and grow from the joints, often four opposite.
They are of a fine green; but their flowers are produced rather
thinly, and stand each on its own footstalk ; and are not possessed of the heightened fragrance of the other.

3. The CLAW BIGNONIA is another noble climber. It
rises by the help of claw-like tendrils, the branches being very
slender and weak ; and by these it will over-top bushes, trees, &c.
twenty or thirty feet high. The branches, however, shew their

G 4

natural

natural tendency to afpire, for they wind about every thing
that is near them ; fo that, together with the affiftance nature
has given them of tendrils, it is no wonder they arrive at fo
great an height. Thefe branches, or rather ftalks, have a fmooth
furface, are often of a reddifh colour, particularly next the fun,
and are very tough. The tendrils grow from the joints ; they are
bowed, and are divided into three parts. The leaves grow in pairs at
the joints, and are four in number at each. Thefe are of an oblong
figure, have their edges entire, and are very ornamental to the
plant ; for they are of an elegant green colour : their under furface is much paler than their upper, and their footftalks, mid-
rib, and veins, alter to a fine purple. The flowers are monope-
talous and bell-fhaped. The tube is very large, and the rim is
divided and fpreads open. They grow from the wings of the
leaves, in Auguft, two ufually at each joint ; and they are fucceeded in the countries where they grow naturally by long pods.

4. The TENDRIL BIGNONIA is another fine climber, which
rifes by the affiftance of tendrils or clafpers. The leaves grow
at the joints oppofite by pairs, though thofe which appear at the
bottom frequently come out fingly. They are of an oblong
figure, and continue on the plant all winter. The flowers are
produced in Auguft, from the wings of the leaves : they are of
the fame nature, and of the fhape nearly of the former ; are
large, of a yellow colour, and fucceeded by fhort pods.

5. The SCARLET TRUMPET FLOWER will arrive to a prodigious height, if it has either buildings or trees to climb up by ;
for it ftrikes root from the joints into whatever is near it, and
thus will get up to the tops of buildings, trees, &c. be they ever
fo high. This fpecies has pinnated leaves, which grow oppofite
by pairs at the joints. Thefe leaves are compofed of about four
pair of folioles, which end with an odd one. They are of a good
green colour, have their edges deeply cut, and drawn out into
a long point. The flowers are produced in Auguft, at the ends
of the branches, in bunches : they are large, and, like the other,
are compofed of one tube ; but they are fhaped more like a
trumpet than any of the forts. They are of a fine red colour,
and make a grand fhow This is the fort chiefly known by the
name of the Scarlet Trumpet Flower.

There is another fort called, the *Smaller Trumpet Flower*. It
differs from the laft only in that the leaves and flowers are
                                                        fmaller,

fmaller, and fome fancy their colour to be a finer red; the colour of the former, in fome fituations, often approaching to that of an orange colour. Thefe two forts are more hardy than any of the others, and confequently more proper to be fet againft old walls, &c. in expofed fituations: they will all, however, bear our climate very well; though it would be advifeable to fet the tender forts in well-fheltered places, as they will otherwife be in danger of fuffering by fevere frofts, efpecially while young, if there be nothing to break them off.

The PROPAGATION of the Climbers. 1. If the fhoots are laid upon the ground, and covered with a little mould, they will immediately ftrike root, and become good plants for fetting out where they are wanted. 2. They will all grow by cuttings. The bottom part of the ftrongeft young fhoots is the beft; and by this method plenty may be foon raifed. 3. They are to be raifed by feeds; but this is a tedious method, efpecially of the pinnated-leaved forts; for it will be many years before the plants raifed from feeds will blow.

B U P L E U R U M.

LINNEAN Clafs and Order, *Pentandria Digynia:* Each flower contains five males and one female. There are fixteen SPECIES; but they are principally herbaceous: There is only one fit for open grounds in this country.

BUPLEURUM *Fruticofum:* The ETHIOPIAN HARTWORT, or the SHRUBBY BUPLEURUM; *an evergreen fhrub;* growing naturally amongft the rocks on the coaft of the South of France, and alfo in fome parts of Italy.

The ETHIOPIAN HARTWORT is of low growth; it feldom rifes more than eight feet high; and will produce plenty of flowers before it gets to the height of one yard. The bark of the oldeft ftems is of a brown, that on the younger fhoots of a reddifh, colour; but this is not conftant, for fometimes it will be greyifh, at others of a purplifh blue. The leaves are of a fine pale green colour, and placed alternately on the branches. They are of an oblong, oval figure, and have their edges entire. They are fmooth, and, being of a delicate pale green, are very ornamental to the fhrub. The flowers are produced from

the

the ends of the branches, in longish umbels. They make no great figure (having but a bad yellow colour), appear in July and August; and are succeeded by seeds, which will often, though not always, ripen with us; and by which, when they do, plenty of plants may be raised.

The method of PROPAGATING this shrub is either from seeds sown in pots of rich light loom in March; or from cuttings, in the following manner: The latter end of July is the time; and if the weather be moist or rainy, so much the better; if not, some beds must be well dug, and made moist by watering. The cuttings should be planted in the evening, and the beds must be hooped, to be covered with mats in the heat of the day. On their being first planted, no sun should come near them; but after they have been set a fortnight, they may have the morning sun until nine o'clock, and afterwards shading; observing always to uncover them in the evening, as also in moist, cloudy, or rainy weather. Many of these cuttings will grow; and in winter it will be proper to protect them from the frost with mats in the like manner: After that they will require no further trouble until they are planted out.

This evergreen is scarcely hardy enough to struggle with our severest weather; whenever therefore it is introduced into plantations, it should always have the advantage of a dry soil and a well-sheltered situation.

B U X U S.

LINNEAN Class and Order, *Monoecia Tetrandria:* Male flowers containing four stamens, and female flowers containing three pistils upon the same plant. LINNEUS makes only one SPECIES of *Buxus:* of this however there are several *varieties*; some of which in their present state have every appearance of distinct species * :

* MILLAR says, "The two sorts of Tree Box have been frequently raised from seeds, and constantly produced plants of the same kind with those the seeds were taken from, and the Dwarf Box will never rise to any considerable height with any culture." (Art. BUXUS.)

1. The

1. The BROAD-LEAVED TREE-BOX : A tall *evergreen shrub* ; native of the southern parts of Europe, and, *it is said*, of this island.

2. The NARROW-LEAVED TREE-BOX.

3. The Gold-striped Box.

4. The Silver-striped Box.

5. The Gold-edged Box.

6. The Curled-leaved Striped Box.

7. The DWARF Box.

1. The BROAD-LEAVED TREE Box. This we will consider as the TRUE BUXUS, and the rest as *varieties*. The Box-Tree will grow to the height of fifteen or twenty feet. The leaves are smooth and shining, and the branches of a yellowish hue. There is a swelling softness and a peculiar delicacy and richness in the general appearance of the Box, which, in winter more especially, affords the eye a delicious repast. As an *ornamental* it stands first among the evergreens ; and its *uses* are very many. Indeed, we know of no shrub or tree whatever, the Oak, the Ash, the Elm, and the Beech excepted, so deserving of the planter's notice as the Box. It will flourish upon barren soils and in bleak situations where scarcely any other plant will thrive. The only extensive *plantations* of Box in this kingdom are those upon Box-Hill ; and the soil there is a poor thin-skinned chalky loam, and the situation high, unsheltered, and bleak in the extreme ; yet the plants thrive with great luxuriance. The Box however is by no means partial to poor land and an open country ; it thrives in every soil and in the closest situations, being remarkably patient of the shade and drip of other trees : we have seen it in a neglected grove, growing under a perfect canopy of foliage with the same healthfulness and luxuriancy as if it had stood in the open air. This naturally points out a situation and use proper for the Box, which does not seem to have been thought of : we mean that of UNDERWOOD TO THE OAK. Thus employed, what an admirable cover to game ; and how friendly to the sportsman ! what a delightful passage in cultivated nature ; and how profitable to the planter ! Box wood is now worth 16s. per cwt.

2. The NARROW-LEAVED TREE-BOX. Of this beautiful plant there are some *sub-varieties*, that differ in the size
of

of their leaves; but it is the smallest-leaved sort that is here meant; and as this sort is not very common, it is valued on that account. It is rather of a lower growth than the former sort, and its branches are more slender and numerous. It forms itself naturally into a regular head, and the whole shrub assumes an air of delicacy. The leaves grow opposite by pairs, as in the other sort; but are produced in great plenty. They are very small and narrow; and their surface is not so shining as the Broad-leaved Box. As the branches and leaves are the only ornament these trees afford, nothing further need be added to the description of this sort.

3. 4. *The two forts with striped leaves* are the Common Tree-Box variegated; though they have a different appearance in their manner of growth, as well as in their striped leaves. They will grow indeed to be as tall; but the branches will be naturally more slender and weaker, and many of them will often hang downwards, which gives the tree a much different appearance from the plain Tree-Box, whose branches are naturally straight and upright. The leaves of these sorts being beautifully striped, makes them coveted by those who are fond of variegated trees.

5. *The Gold-edged Box* is still the Tree-Box, in the same natural upright growth. The branches of this are not so weak as those of the former sorts, but are upright and strong. Their bark is rather yellower than the green sort: in other respects there is no difference, except that the leaves are tipped or edged with yellow; which is thought by many to be very ornamental to the shrub.

6. *The Curled-leaved Striped Box* is so called on account of its leaves being a little waved. This, together with the Narrow-leaved, is the scarcest of all the sorts; and is indeed, like that, a very elegant shrub. It is certainly a variety of the Common Tree-Box; but it seems rather of lower growth. Its leaves are waved; and they are variegated in such a manner as to cause the shrub to have what HANBURY calls a luscious look. It makes a variety from all the other sorts, and is truly beautiful and pleasing.

7. The DWARF Box is a plant so well known as an edging to the borders of flower-gardens, that it needs no description. It may be planted as an evergreen shrub among the lower sorts.

The

The method of PROPAGATING the Box is perfectly eafy : it may be raifed from cuttings, o r from the feed, or by layering. 1. For planting the cuttings, HANBURY fays, the month of Auguft is the beft time, if any rain falls. If none fhould happen, then the work muft be deferred till it does. Indeed the cuttings may be planted with fuccefs any time in the winter, even till- the middle of April ; but it is moft prudent, if the ground is ready, to have this work done as foon as the firft autumnal rains fall. Thefe cuttings ought to be of one and two years wood, fhould be about a foot long, rather more than the half of which muft be planted in the ground. A flip of the laft year's wood, ftripped from an older branch, is an excellent fett, of which there will be little fear of its growing. The cuttings for the firft raifing of thefe trees fhould be at about four inches diftance in the beds ; and, after they are planted, will need no trouble except w ering in dry weather, and keeping clean from weeds, till about the third year after planting; for in all that time they will not be got too big for the *feed-beds.* The feafon for tranfplanting thefe trees from the *feed-beds* to the nurfery is any time from Auguft to April ; though if they are to be tranfplanted early in the autumn, or late in the fpring, moift weather fhould be made choice of for this purpofe. The diftance thefe plants fhould be placed at in the nurfery muft be a foot afunder, and two feet in the rows ; and here they may ftand till they are planted out. 2. The Box-Tree may be alfo propagated from feeds ; and trees raifed this way will often grow to a larger fize. In order to raife this tree from feeds, let them be gathered when they are quite ripe, and juft ready to burft out of their cells, and foon after fow them in a border of light fandy earth, about half an inch deep. In the fpring the plants will appear ; though it fometimes happens that they lie in the beds one whole feafon before they come up, efpecially if they happen to have been kept long before they were fowed after being gathered. If they fhould not appear in the fpring, the beds muft remain undifturbed till the next, only keeping them free from weeds, and now and then giving them a gentle watering in dry weather. After they have made their appearance, they fhould ftand two or three years in the feed-bed, the firft of which will require attendance by watering in dry weather. When they are ftrong enough to

plant

plant out, they may be fet in rows in the nurfery, as was directed for the cuttings. 3. The Box propagates itfelf by *layering*; for whether it be borne down by the weight of its own foliage, or be broken down by a fall of fnow lodging upon its leaves and branches, it no fooner comes into contact with the ground than it fends forth fibres, and the branch layered (whether by nature, by accident, or by art) prefently forms to itfelf a detached roo , which being fevered from the main tree, a feparate plant is produced.

CALLICARPA.

LINNEAN Clafs and Order, *Tetrandria Monogynia* : Each flower contains four males and one female. There are two SPECIES ; one of which is enured to our climate.

CALLICARPA *Americana* : The CALLICARPA ; *a low deciduous fhrub*; native of many parts of America, but particularly of Virginia and Carolina.

The CALLICARPA. The leaves are roundifh, acute, pointed, and are near three inches in length. They are of a hoary caft, being, like the youngeft fhoots, covered with a kind of woolly matter. They ftand oppofite by pairs on moderate footftalks, and their edges are made delicate by beautiful fmall ferratures. The flowers are produced in whirls round the twigs, at the fetting-on of the leaves, and are of a reddifh-purple colour. Each flower feparately is fmall and inconfiderable ; though the whole number of which the whirls are compofed form, together with the leaves and nature of the growth of the tree, a fingular and pleafing afpect. Their appearance is ufually in July, and they are fucceeded by fucculent berries, which are at firft red, and afterwards of a deep purple when ripe.

It is PROPAGATED, 1. By cuttings. When by cuttings, they fhould be planted, in the fpring, in a moift fandy border. As the hot weather comes on, they fhould be conftantly fhaded, and watered if the bed is not naturally very moift ; and by this

means

means many of the cuttings will ſtrike root, and become good plants. 2. By layers, which is a certain method, theſe plants may alſo be encreaſed. If a few plants are obtained for this purpoſe, they ſhould be planted in a warm well-ſheltered ſituation; and if the ſoil be naturally ſandy, it will be the better. The autumn after theſe ſtools have ſhot forth young wood, theſe young ſhoots ſhould be laid in the ground, and by the autumn following they will be fit to take off, either for the nurſery, or where they are to remain. 3. By ſeeds, which ſhould be ſown in a warm border of ſandy earth, a quarter of an inch deep, and ſhould be carefully ſhaded and the ſeedlings ſheltered; theſe plants being tender when young, though afterwards they are ſufficiently hardy.

CALYCANTHUS.

LINNEAN Claſs and Order, *Icoſandria Polygynia :* Each flower contains twenty males and numerous females. There are two SPECIES; namely, CALYCANTHUS *Precox* ; not enured, we believe, to this climate ; and

CALYCANTHUS *Floridus :* The FLOWERING CALYCANTHUS, or CAROLINA ALLSPICE TREE; *a deciduous aromatic ſhrub* ; native of Carolina.

The FLOWERING CALYCANTHUS is a ſhrub which ſeldom grows, at leaſt with us, to more than five feet high. It divides into many branches irregularly near the ground. They are of a brown colour, and being bruiſed emit a moſt agreeable odour. The leaves that garniſh this delightful aromatic are of an oval figure, pointed : They are near four inches long, and are at leaſt two and a half broad, and are placed oppoſite by pairs on the branches. At the end of theſe ſtand the flowers, of a kind of chocolate-purple colour, and which are poſſeſſed of the oppoſite qualities of the bark on the branches. They ſtand ſingle on their ſhort footſtalks, come out in May and June, and are ſucceeded by ripe ſeeds in England.

The

The PROPAGATION of this fhrub is not very difficult; tho' more than a common care muft be taken, after fmall plants are obtained, to preferve them till they are of a fize to be ventured abroad. The laft year's fhoots of this tree, if laid in the ground, the bark efpecially being a little bruifed, will ftrike root within the compafs of twelve months, particularly if the layers are fhaded, and now and then watered in the fummer's drought. In the fpring they fhould be taken off, and planted in pots; and if thefe are afforded a fmall degree of heat in a bed, they will ftrike fo much the fooner and ftronger. After they have been in this bed a month or fix weeks, they fhould be taken out. In the heat of the fummer they fhould be placed in the fhade; and if the pots are plunged into the natural ground, it will be fo much the better. At the approach of the fucceeding winter's bad wea-ther, the pots fhould be removed into the greenhoufe, or fome fhelter, and in the fpring may refume their old ftations: and this fhould be repeated till they are of a proper fize and ftrength to be planted out to ftand. If the pots in which they were firft planted were fmall, they may be fhifted into larger a fpring or two after; and, when they have got to be pretty ftrong plants, they may be turned out, mould and all, into the places where they are to remain. By this care of potting them, and houfing them during the fevere weather in winter, the young crop will be preferved; otherwife, if they were planted immediately abroad, the firft hard froft the enfuing winter would deftroy them all: Tanner's bark about their roots will be the moft proper fecurity; as they are at beft, when full grown, but tender plants, and muft have the warmeft fituation and the drieft foil.

C A R P I N U S.

LINNEAN Clafs and Order *Monoecia Polyandria:* Male flowers containing many ftamens, and female flowers con-taining two piftils difpofed in feparate catkins, upon the fame plant: There are only two SPECIES:

1. CARPINUS *Betulus:* The COMMON HORNBEAM: a *deci-duous tree,* native of Europe and America.

2. CAR-

2. CARPINUS *Oftrya* : The HOP HORNBEAM ; *a low deci-*
duous tree ; native of Italy and of Virginia.

1. The COMMON HORNBEAM. This tree, it is said, will
grow fo high as fixty or feventy feet : we feldom fee it, however,
arrive at fo great a height. Its leaves are of a darkifh green, and
about the fize of thofe of the Beech, but more pointed and
deeply ferrated. Its branches are long, flexible, and crooked ;
yet in their general appearance very much refemble thofe of the
Beech : indeed there is fo great a likenefs between thofe two
trees, efpecially in the fhrubby underwood ftate, that it would
be difficult to diftinguifh them at the firft glance, were it not for
that gloffy varnifh with which the leaves of the Beech are ftrongly
marked. In the days of EVELYN, when topiary work was the
Gardener's idol, the Hornbeam might be confidered as deferving
of thofe *endearing* expreffions which that enthufiaftic writer has
been pleafed to lavifh upon it : neverthelefs, as an *ornamental*
in *modern* gardening it ftands low ; and its *prefent ufes* are few.
As an underwood it affords ftakes and edders, fuel and charcoal.
Its timber ranks with that of Beech and the Sycamore. The
only fuperior excellency of the Hornbeam lies in its fitnefs for
SKREEN-FENCES for fheltering gardens, nurferies, and young
plantations from the feverities of the winter feafon. It may be
trained to almoft any height, and by keeping it trimmed on the
fides it becomes thick of branches, and confequently thick of
leaves ; which being by their nature retained upon the plant
after they wither, a Hornbeam-hedge occafions a degree of
fhelter nearly equal to that given by a brick wall. Indeed, being
lefs reflective than that expenfive fkreen, it affords a more uniform temperature of air to the plants which ftand near it. In this
point of view, too, the Hornbeam is ufeful to be planted promifcuoufly, or in alternate rows, amongft more tender plants in
expofed fituations, in the fame manner as the Birch ; to which
it has more than one preference : namely, it is warmer in
winter,—And, HANBURY fays, the Hornbeam is peculiarly
grateful to hares and rabbits ; confequently it may prevent their
injuring its more valuable neighbours : yet, like EVELYN, he
feems to be of opinion that it is difaffected by deer. If this be
really the cafe, the Hornbeam may upon many occafions be introduced into deer-parks with fingular propriety.

H   The

The Common Hornbeam may be PROPAGATED either by layering, (at almoſt any time of the year) or from ſeeds, in the
following manner : In the autumn the ſeeds will be ripe ; when,
having gathered a ſufficient quantity for the purpoſe, let them
be ſpread upon a mat a few days to dry. After this, they
ſhould be ſown in the ſeminary-ground, in beds four feet wide,
with an alley of about two feet, and from one to two
inches deep. In this bed they muſt remain till the ſecond
ſpring before they make their appearance ; and all the ſummer
they lie concealed, the weeds ſhould conſtantly be plucked up as
ſoon as they peep ; for if they are neglected they will get ſo
ſtrong, and the fibres of their roots will be. ſo far ſtruck down
among the ſeeds, as to endanger the drawing many ſeeds out
with them, on weeding the ground. After the young plants
appear, they ſhould conſtantly be kept clear of weeds during the
next ſummer ; and if they were to be now and then gently re-
freſhed with water in dry weather, it would prove ſerviceable to
them. In the ſpring following they may be taken out of theſe
beds, and planted in the nurſery, in which ſituation they
may remain till they are of a ſufficient ſize to plant out for.
ſtandards.

Of the Common Hornbeam there are three *Varieties:* The
Eaſtern Hornbeam, Flowering Hornbeam, American Hornbeam.

The *Eaſtern Hornbeam* arrives to the leaſt height of all the.
ſorts : about ten feet is the fartheſt of its growth, and it looks.
pretty enough with trees of the ſame growth. The leaves are by
no means ſo large as the common ſort ; and as the branches are
always cloſer in proportion to the ſmallneſs of the leaves, where
a low hedge is wanted of the deciduous kind, this would not be
an improper tree for the purpoſe, either to be kept ſheered, or
ſuffered to grow in its natural ſtate. The bark of this ſort is
more ſpotted than that of the Common.

The *Flowering Hornbeam* is the moſt free ſhooter of any of
the ſorts ; and will arrive to be the higheſt, the Common Hornbeam only excepted. It will grow to be thirty or forty feet
high. The branches of this tree are leſs ſpotted with greyiſh
ſpots than any of the other ſorts. The leaves are very rough,
of a dark-green colour, and are longer than the common ſort.
The property which the Common Hornbeam is poſſeſſed of, of
retaining

retaining its leaves all winter, does not belong to this sort, the leaves of which conftantly fall off in the autumn with other deciduous trees.

American Hornbeam is a more elegant tree than any of the former sorts. The branches are flender, covered with a brownifh fpeckled bark, and are more fparingly fent forth than from any of the others. The leaves are oblong, pointed, and of a palifh green, and are not nearly fo rough as the Common Hornbeam, though the flowers and fruit are produced in the fame manner.

2. Hop Hornbeam is of taller growth than the Eaftern kind. It will arrive to the height of twenty feet, or more. The leaves are nearly the fize of the common fort, and fome people admire this tree on account of the fingular appearance it makes with its feeds, before they begin to fall. There is a *Variety* of this tree, which grows to thirty feet high, fhoots freely, has long rough leaves like thofe of the elm, and longifh yellow-coloured flowers, called the Virginian Flowering Hop Hornbeam.

Thefe different forts of Hornbeam are to be PROPAGATED by layers; for which purpofe a few plants for ftools muft be procured. The ftools of the Eaftern Hornbeam fhould be planted a yard, and the other forts a yard and a half or two yards afunder. After thefe plants have made fome young fhoots, they fhould be layered in the autumn, and by that time twelvemonth they will have ftruck root; at which time, or any time in the winter, or early in the fpring, they fhould be taken off, and planted in the nurfery-way, obferving always to brufh up the ftool, that it may afford fine young fhoots for frefh layering by the autumn following. The diftance the plants fhould be allowed in the nurfery need be no more than one foot, in rows that are two feet afunder; and here they may ftand, with the ufual nurfery care of weeding and digging the rows in winter, until they are to be finally planted out; though the Virginian Hornbeam will frequently fend forth two fhoots, which will feem to ftrive for maftery in the lead. When this is obferved, the weakeft fhould always be taken away, otherwife the tree will grow forked.

H 2   CEANOTHUS.

CEANOTHUS.

LINNEAN Clafs and Order, *Pentandria Monogynia* : Each flower contains five males and one female. There are three SPECIES ; one of which will bear the open air of this climate.

CEANOTHUS *Americana* : The NEW-JERSEY TEA, or the RED TWIG ; *a low deciduous fhrub* ; native of North America.

The NEW-JERSEY TEA. The height to which it will grow in our country is about a yard. The ftem, which is of a pale-brown colour, fends out branches from the bottom. Thefe are thin, flexible, and of a reddifh colour, which may have occafioned this tree to go by the name of *Red Twig*. The leaves which ornament thefe branches ftand on reddifh pedicles, about half an inch in length. They are oval, ferrated, pointed, about two inches and a half long, are proportionably broad, and have three nerves running lengthways. From the footftalk to the point they are of a light-green colour, grow irregularly on the branches, and not oppofite by pairs, as has been afferted. They are late in the fpring before they fhoot. The flowers grow at the ends of the twigs in clufters : They are of a white colour, and when in blow give the fhrub a moft beautiful appearance. Indeed, it feems to be almoft covered with them, as there is ufually a clufter at the end of nearly every twig ; and the leaves which appear among them ferve as ornaments only, like myrtle in a diftant nofegay : nature however has denied them fmell. This tree will be in blow in July ; and the flowers are fucceeded by fmall brownifh fruit, in which the feeds will fometimes ripen in England.

The PROPAGATION of this plant is by layering ; or from feeds fown in pots of compoft confifting of two parts virgin earth well tempered, and one part fand, about a quarter of an inch deep ; being equally careful to defend the young feedlings from an extremity of cold in winter as from the parching drought of the fummer months. The beft time of layering them is in the fummer, juft before they begin to flower : At that time lay the tender twigs of the fpring fhoots in the earth, and nip off the end which would produce the flowers. By the

autumn

autumn twelvemonth fome of them will be rooted. At the ftools, however, the plants fhould remain until the fpring, when they fhould be taken off, and the beft-rooted and the ftrongeft may be planted in the nurfery-way, or in a dry foil and well-fheltered place, where they are to remain ; while the bad-rooted ones and the weakeft fhould be planted in pots ; and if thefe are plunged into a moderate warmth of dung, it will promote their growth, and make them good plants before autumn. In the winter they fhould be guarded againft the frofts; and in the fpring they may be planted out where they are to remain.

CELASTRUS.

LINNEAN Clafs and Order, *Pentandria Monogynia :* Each flower contains five males and one female. There are eleven SPECIES ; two of which are to our purpofe.

1. CELASTRUS *Bullatus :* The STAFF-TREE ; *an uncertain-deciduous fhrub ;* native of Virginia.

2. CELASTRUS *Scandens :* The CLIMBING STAFF-TREE, or BASTARD EVONYMUS; a *climber ;* native of Canada.

1. The STAFF-TREE is a fhrub of about four feet in growth, rifing from the ground with feveral ftalks, which divide into many branches, and are covered with a brownifh bark. The leaves are of a fine green colour, and grow alternately on the branches. They are of an oval figure, and have their edges undivided. The flowers are produced in July, at the ends of the branches, in loofe fpikes. They are of a white colour, and in their native countries are fucceeded by very ornamental fcarlet fruit; but with us this feldom happens.

It is eafily PROPAGATED from feeds fown, about an inch deep, in beds of good frefh mould made fine. They feldom come up until the fecond, and fometimes not before the third fpring. This fpecies is alfo PROPAGATED by layers; and, to be concife, the work muft be performed on the young wood, in the autumn, by a flit at the joint. Thefe layers may be expected to ftrike root

by

by the autumn following; when they may be taken up and planted in the nurſery-ground. This ſhrub muſt have a well-ſheltered ſituation, otherwiſe the leaves are apt to fall off at the approach of froſty weather. And MILLAR ſays, that, growing naturally in moiſt places, it will not thrive well in a dry ſoil.

2. The CLIMBING STAFF-TREE. The ſtalks are woody, twining, and will riſe by the help of neighbouring trees or buſhes to the height of twelve feet. The leaves are oblong, ſerrated, of a pleaſant green colour, pale, and veined underneath, and grow alternately on the branches. The flowers are produced in ſmall bunches, from the ſides of the branches, near the ends. They are of a greeniſh colour, appear in June; and are ſucceeded by roundiſh, red, three-cornered capſules, containing ripe ſeeds, in the autumn.

The plant is exceeding hardy, and makes a beautiful appearance among other trees in the autumn, by their beautiful red berries, which much reſemble thoſe of the Spindle-tree, and will be produced in vaſt profuſion on the tops of other trees, to the height of which theſe plants by their twiſting property aſpire. They ſhould not be planted near weak or tender trees, to climb on; for they embrace the ſtalks ſo cloſely as to bring on death to any but the hardieſt trees and ſhrubs.

It is PROPAGATED, 1. By laying down the young ſhoots in the ſpring. By the autumn they will have ſtruck root, and may then be taken off and ſet in the places where they are deſigned to remain. 2. They are alſo propagated by ſeeds. Theſe ſhould be ſoon ſown after they are ripe, otherwiſe they will be two, and ſometimes three years before they come up. When they make their appearance, nothing more need be done than keeping them clear from weeds all ſummer and the winter following; and in the ſpring the ſtrongeſt plants may be drawn out, and ſet in the nurſery for a year, and then removed to the places where they are deſigned to remain; whilſt the weakeſt, being left in the ſeed-bed one year more, may undergo the ſame diſcipline.

C E L.

C E L T I S.

Linnean Clafs and Order, *Polygamia Monoecia.* Hermaphrodite flowers, containing five ftamens and two piftils, and male flowers containing five ftamens. There are three Species.

1. Celtis *Auftralis:* The Southern Celtis, or the Black-fruited Nettle-Tree, or Lote-Tree; *a deciduous tree;* native of Africa and the South of Europe.

2. Celtis *Occidentalis:* The Western Celtis, or the Purple-fruited or Occidental Nettle-Tree; *a deciduous tree;* native of Virginia.

3. Celtis *Orientalis:* or the Eastern Celtis; or Yellow-fruited or Oriental Nettle-Tree; *a deciduous fhrub;* native of Armenia.

1. The Southern Celtis.

2. The Western Celtis.

Thefe two fpecies grow with large, fair, ftraight ftems; their branches are numerous and diffufe; their bark is of a darkifh grey colour; their leaves are of a pleafant green, three or four inches long, deeply ferrated, end in a narrow point, nearly refemble the leaves of the common ftinging-nettle, and continue on the trees till late in the autumn: So that one may eafily conceive what an agreeable variety thefe trees would make. Add to this, their fhade is admirable. The leaves are late in the fpring before they fhew themfelves; but they make amends for this, by retaining their verdure till near the clofe of autumn, and then do not refemble moft deciduous trees, whofe leaves fhew their approaching fall by the change of their colour; but continue to exhibit themfelves of a pleafant green, even to the laft.

Hanbury fpeaks highly of the Celtis as a timber-tree: he fays, " The wood of the Lote-Tree is extremely durable. In Italy they make their flutes, pipes, and other wind inftruments of it. With us the coach-makers ufe it for the frames of their vehicles." Millar mentions alfo the wood of the *Occidentalis* being ufed by the coach-makers.

The two fpecies of Tree-Celtis are PROPAGATED from feeds, which ripen in England, if they have a favourable autumn; but the foreign feeds are the moft certain of producing a crop. Thefe feeds fhould be fown, foon after they are ripe, either in boxes, or in a fine warm border of rich earth, a quarter of an inch deep; and in the following fpring many of the young plants will appear, tho' a great part often lie till the fecond fpring before they fhew their heads. If the feeds in the beds fhoot early in the fpring, they fhould be hooped, and protected by mats from the frofts, which would nip them in the bud. When all danger from frofts is over, the mats fhould be laid afide till the parching beams of the fun get powerful; when, in the day-time, they may be laid over the hoops again, to fcreen the plants from injury. The mats fhould be conftantly taken off every night, and the young plants fhould never be covered either in rainy or cloudy weather. During the whole fummer, thefe feedlings fhould be frequently watered in dry weather, and the beds kept clean of weeds, &c. In the autumn, they muft be protected from the frofts, which often come early in that feafon, and would not fail to deftroy their tops. The like care fhould be continued all winter, to defend them from the fame enemies. In this feminary they may remain, being kept clean of weeds and watered in dry weather, till the end of June; when they fhould be taken out of their beds, and planted in others at fix inches diftance. And here let no one (continues HANBURY) be ftartled at my recommending the month of June for this work; for I have found by repeated experience, that the plants will be then almoft certain of growing, and will continue their fhoots till the autumn; whereas I have ever perceived, that many of thofe planted in March have frequently perifhed, and that thofe which did grow made hardly any fhoot that year, and fhewed the early figure of a ftunted tree. In June, therefore, let the ground be well dug, and prepared for this work; and let the mould be rich and good: But the operation of removing muft be deferred till rain comes; and if the feafon fhould be dry, this work may be poftponed till the middle of July. After a fhower, therefore, or a night's rain, let the plants be taken out of their beds, and pricked out at fix inches diftance from each other. After this, the beds in

which

which they are planted fhould be hooped, and covered with mats when the fun fhines; but thefe muft always be taken away at night, as well as in rainy or cloudy weather. With this management, they will have fhot to a good height by the autumn, and have acquired fo much hardinefs and ftrength as to need no farther care than to be kept clear of weeds for two or three years; when they may be planted out in places where they are to remain, or fet in the nurfery, to be trained up for large ftandards.

The beft feafon for planting out thefe ftandard trees is the latter end of October, or beginning of November; and in performing that operation, the ufual rules muft be obferved, with care.

The foil for the Lote-tree fhould be light, and in good heart; and the fituation ought to be well defended, the young fhoots being very liable to be deftroyed by the winters frofts.

3. The EASTERN CELTIS. The height to which this fpecies will grow is no more than about twelve feet; and the branches are many, fmooth, and of a greenifh colour. The leaves are fmaller than thofe of the other forts, though they are of a thicker texture, and of a lighter green. The flowers come out from the wings of the leaves, on flender footftalks: They are yellowifh, appear early in the fpring, and are fucceeded by large yellow fruit.

The CULTURE of this fpecies is the fame, and the plants may be raifed in the fame manner as the other two forts, only let this all along have a peculiarly dry foil, and a well-fheltered fituation, otherwife it will not bear the cold of our winters.

CEPHALANTHUS.

LINNEAN Clafs and Order, *Tetrandria Monogynia*: Each flower contains four males and one female. There is only one SPECIES.

CEPHALANTHUS *Occidentalis*: The CEPHALANTHUS, or BUTTON-WOOD; *a deciduous fhrub*; native of North America.

The

The CEPHALANTHUS grows to about five or six feet high. It is not a very bushy plant, as the branches are always placed thinly in proportion to the size of the leaves, which will grow more than three inches long, and one and a half broad, if the trees are planted in a soil they like. The leaves stand opposite by pairs on the twigs, and also sometimes by threes, and are of a light-green colour: Their upper surface is smooth; they have a strong nerve running from the footstalk to the point, and several others from that on each side to the borders: These, as well as the footstalks, in the autumn die to a reddish colour. The flowers, which are aggregate flowers, properly so called, are produced at the ends of the branches, in globular heads, in July. The florets which compose these heads are funnel-shaped, of a yellow colour, and fastened to an axis which is in the middle.

The PROPAGATION of the Cephalanthus is from seeds, which we receive from America. These should be sown as soon as they arrive, and there will be a chance of their coming up the first spring; though they often lie till the spring after before they make their appearance. They may be sown in good garden mould of almost any soil, if somewhat moist the better, and should be covered about a quarter of an inch deep. This shrub is also propagated by layers. If the young shoots are laid in the autumn, they will have struck good root by the autumn following, and may be then taken up, and set in the places where they are designed to remain. Cuttings of this tree, also, planted in the autumn in a rich, light, moist soil will grow: and by that means also plenty of these plants may be soon obtained.

C E R C I S.

LINNEAN Class and Order, *Decandria Monogynia:* Each flower contains ten males and one female. There are only two SPECIES.

1. CERCIS

1. Cercis *Siliquaſtrum* : The Common Judas-Tree, or the
Italian Cercis ; a tall *deciduous flowering ſhrub* ; native of
Italy and other parts of the South of Europe.

2. Cercis *Canadenſis* : The Canadian Judas-Tree, or the
Red Bud, or the Canadian Cercis; *a deciduous flowering
ſhrub* ; native of Canada, Virginia, and other parts of America.

1. The Common Judas-Trees differ in the height of their
growth in different places : In ſome they will arrive to be
fine trees, of near twenty feet high; whilſt in others they will
not riſe to more than ten or twelve feet, ſending forth young
branches irregularly from the very bottom. The ſtem of this
tree is of a dark-greyiſh colour, and the branches, which are few
and irregular, have a purpliſh caſt. The leaves are ſmooth, heart-
ſhaped, and roundiſh, of a pleaſant green on their upper ſurface,
hoary underneath, and grow alternately on long footſtalks. The
flowers are of a fine purple : They come out early in the ſpring,
in cluſters, from the ſide of the branches, growing upon ſhort
footſtalks ; and in ſome ſituations they are ſucceeded by long flat
pods, containing the ſeeds, which, in very favourable ſeaſons,
ripen in England. Some people are fond of eating theſe flowers
in ſallads, on which account alone in ſome parts this tree is propagated. The *varieties* of this ſpecies are, 1. The Fleſh-co-
loured ; 2. The White-flowered ; and, 3. The Broad-podded
Judas-tree.

2. Canada Judas-Tree will grow to the ſize of the firſt ſort in
ſome places. The branches are alſo irregular. The leaves are
cordated, downy, and placed alternately. The flowers uſually
are of a paliſh red colour, and ſhew themſelves likewiſe in the
ſpring, before the leaves are grown to their ſize. Theſe too are
often eaten in ſallads, and afford an excellent pickle. There is
a variety of this with deep red, and another with purple flowers.
The pleaſure which theſe trees will afford in a plantation may
be eaſily conceived, not only as they exhibit their flowers in
cluſters, in different colours, early in the ſpring, before the leaves
are grown to ſuch a ſize as to hide them ; but from the difference
of the upper and lower ſurface of the leaves ; the one being of a
fine green, the other of a hoary caſt ; ſo that on the ſame tree,
even in this reſpect, is ſhewn variety ; an improvement where-
of

of is made by the waving winds, which will prefent them alternately to view.

As thefe fpecies will not take root by layers, they muft be
PROPAGATED by feeds, which may be had from abroad. They
are generally brought us found and good, and may be fown in
the months of February or March. Making any particular compoft for their reception is unneceffary ; common garden mould,
of almoft every fort, will do very well : And this being well
dug, and cleared of all roots, weeds, &c. lines may be drawn for
the beds. The mould being fine, part of it fhould be taken out,
and fifted over the feeds, after they are fown, about half an
inch thick. Part of the feeds will come up in the fpring, and
the others will remain until the fpring following ; fo that whoever is defirous of drawing the feedlings of a year old to plant
out, muft not deftroy the bed, but draw them carefully out, and
after that there will be a fuccceding crop. However, be this as
it will, the feeds being come up, they muft be weeded, and en-
couraged by watering in the dry feafon ; and they will require
no farther care during the firft fummer. In the winter alfo
they may be left to themfelves, for they are very hardy ; tho'
not fo much but that the ends of the branches will be killed by
the froft, nay, fometimes to the very bottom of the young plant,
where it will fhoot out again afrefh in the fpring. Whoever,
therefore, is defirous of fecuring his feedling-plants from this
evil, fhould have his beds hooped, in order to throw mats over
them during the hard frofts. Toward the latter end of
March, or beginning of April, the plants having been in the
feed-bed one or two years, they fhould be taken out, and planted in the nurfery : The diftance of one foot afunder, and two
feet in the rows, fhould be given them. Hoeing the weeds
down in the fummer muft alfo be allowed, as well as digging
between the rows in the winter. Here they may ftand until they
are to be removed finally ; but they muft be gone over in the
winter with the knife, and fuch irregular branches taken off as
are produced near the root ; by which management the tree
may be trained up to a regular ftem. Such, continues HAN-
BURY, is the culture of the fpecies of Cercis ; forts that are
not to be omitted where there are any pretenfions to a collection. Befides, the wood itfelf is of great value ; for it

polifhes

polifhes exceeding well, and is admirably veined with black and green.

***

CHIONANTHUS.

LINNEAN Clafs and Order, *Diandria Monogynia:* Each flower contains two males and one female. There are two Species : CHIONANTHUS *Zeylonica* ; and

CHIONANTHUS *Virginica :* The SNOW-DROP-TREE, or the FRINGE-TREE, or the VIRGINIAN CHIONANTHUS ; *a tall deciduous fhrub*; native of Virginia and other parts of North America.

The SNOW-DROP-TREE. This fhrub will grow to the height of about fifteen feet, and, until late years, was very rarely to be met with in our gardens. The ftem of it is rough, and of a dark-brown colour. The leaves are large, fhaped like a laurel, broad and roundifh, of a fine deep green on their upper furface, but rather hoary. The flowers come out in bunches, in May, from every part of the tree : They are of a pure white ; and, in the places where it grows naturally, this muft be a moft delightful plant ; for at that feafon it exhibits its white flowers in bunches all over it, fo as to refemble a tree covered with fnow. The few trees we have feldom flower ; and even when they do, the flowers are few, and make no great figure. Whoever is defirous of raifing this fhrub muft plant it in a moift part of the garden, which is well defended with other trees ; for there he will have a chance of feeing the flowers (which are fucceeded by black berries, of a moderate fize) in more plenty, and in greater perfection.

The CULTURE of this tree is not very eafy ; for if we attempt to propagate it by layers, thefe are with difficulty made to ftrike root ; and if we obtain good feeds from abroad, great care and management muft be ufed, to make them to be ftrong plants, fit to be fet out to ftand. By layers and feeds, however, this tree may be encreafed ; and, 1. When layers is the method adopted, let the plants defigned for ftools be fet in a very moift place,

place, where the foil is rich and good. After thefe ftools have thrown out young fhoots, they fhould be layered in the autumn. If there be many twigs of the fummer's growth to be layered, different methods may be ufed on the different twigs; for no one particular method can be depended on, and yet they will grow by almoft all. One time the layering has been performed by a fmall flit at the joint; another twig has had a gentle twift, fo as to juft break the bark; a third has been wired. The flit-layers, after three or four years, have only fwelled to a knob, without any fibres; while the twifted parts have fhot out fibres, and become good plants. At other times, the twifted part, after waiting the fame number of years, has ftill remained in the ground as a branch without any root; whilft the flit-twig, in the mean time, has become a good plant. The like uncertainty has been found to attend the other manner of layering. To propagate the Snow-drop tree this way, every method fhould be ufed; and then there will be a greater chance of having fome plants; but, at the beft, you muft not expect them with good roots, until they have lain in the ground about three years; for it is very rarely that they are to be obtained fooner. The layers fhould be taken from the ftools the latter end of March, and planted in pots, Thefe fhould be plunged into a hotbed; and, after they have ftruck root, fhould be ufed to the open air. In May they may be taken out, and plunged in the natural foil, in a moift fhady place. When the froft comes on, they fhould be removed into the greenhoufe, or fet under a hotbed-frame for protection; and in the fpring they may be turned out of the pots, with the mould, into the places where they are to remain, which ought to be naturally moift and well fheltered. 2. From feeds: they muft be fown in large pots, about half an inch deep, in a ftrong fandy loam, plunging the pots into a moift fhady place in fummer, and in winter removing them into the greenhoufe or under hotbed frames.

MILLAR fays, " This fhrub delights in a moift, foft, loamy foil; and is fubject to decay in dry foils and hot feafons."

CISTUS.

C I S T U S.

LINNEAN Clafs and Order, *Polyandria Monogynia* : Each flower contains numerous males and one female. There are no lefs than forty-three SPECIES of this genus of plants, moft of which are herbaceous, or herbaceous-fhrubby ; of the thirteen arborefcent fpecies, twelve are naturalized to this climate.

1. CISTUS *Populifolius* : The POPLAR-LEAVED CISTUS, or ROCK-ROSE ; *an evergreen fhrub*; native of Spain and Portugal.

2. CISTUS· *Laurifolius* : The BAY-LEAVED CISTUS, or ROCK-ROSE ; *an evergreen fhrub* ; native of Spain.

3. CISTUS *Ladaniferus* : The LADANUM CISTUS, or ROCK-ROSE ; *an evergreen fhrub*; native of Spain, Italy, Crete, and the South of France.

4. CISTUS *Incanus* : The HOARY CISTUS, or HOARY-LEAVED ROCK-Rofe ; *a low evergreen fhrub* ; native of Spain and the South of France.

5. CISTUS *Monfpelienfis* : The MONTPELIER CISTUS, or GUM CISTUS of MONTPELIER ; *an evergreen fhrub*; native of the South of France.

6. CISTUS *Albidus* : The WHITE CISTUS, or OBLONG WHITE-LEAVED ROCK-ROSE; *an evergreen fhrub* ; native of Spain, Portugal, and France.

7. CISTUS *Salvifolius* : The SAGE-LEAVED CISTUS, or ROCK-ROSE ; *a low evergreen fhrub* ; native of France, Italy, and Sicily.

8. CISTUS *Crifpus* : The CURLED CISTUS, or WAVED-LEAVED ROCK-ROSE ; *an evergreen fhrub* ; native of Lufitania.

9. CISTUS *Halimifolius* : The HALIMUS-LEAVED CISTUS, or SEA-PURSLAIN-LEAVED ROCK-ROSE ; *a low evergreen fhrub* ; grows common near the fea-fhore in Spain and Portugal.

10. CISTUS *Villofus* : The SHAGGY-LEAVED CISTUS, or SPANISH ROUND-LEAVED ROCK-ROSE ; *a low evergreen fhrub* ; native of Italy and Spain.

11. CISTUS *Creticus* : The CRETAN CISTUS, or ROCK-ROSE; *a low evergreen fhrub* ; native of Crete and Syria.

12. CISTUS

12. Cistus *Libanotis :* The Frankincense Cistus, or Narrow-leaved Spanish Rock-Rose ; *a low evergreen shrub;* native of Spain.

1. The Poplar-leaved Cistus is a shrub of about six feet in height, though it begins its bloom when lower than two feet. The branches have no regular way of growth, and are covered with a brown bark, which will be lighter or darker according to the different foils. The leaves are cordated, smooth, pointed, have footstalks, and a little resemblance to those of the Black Poplar. Old Botanists have distinguished two species of this sort, which they called the Major and the Minor, the one being of larger growth than the other ; but modern improvements shew these to be varieties only. The flowers are white, and produced about Midsummer, in plenty, at the ends and sides of the branches. They are of short continuance ; but there will be a succession kept up for near six weeks, during which time the shrub will have great beauty.

2. Bay-leaved Cistus is an irregular branching shrub, of about the same height with the former. The leaves are oval, pointed, and in the Midsummer months are very clammy. Their upper surface is of a strong green, but their under is white, and they grow on footstalks which join together at their base. The flowers are produced from the ends and sides of the branches, about Midsummer. They are white, and stand on naked footstalks ; and being large, and produced in plenty at that time, make a good figure. This species is rather tender, and requires a warm, dry soil, and a well sheltered situation.

3. The Ladanum Cistus is so called, because the Ladanum of the shops is collected from this shrub. There are many varieties of it differing in the colour of the flowers, or in some respect or other ; and the tree, with its varieties, will grow to be six or more feet high ; though it produces its flowers and exhibits great beauty when very low. It rises with a woody stem ; and tho' it produces its branches in no regular manner, yet it has the appearance of a well-fashioned shrub. The leaves are of a lanceolate figure. Their upper surface is smooth, and of a fine green colour, but their under is whitish and veined. They are scented ; and have footstalks that join together at their base. The flowers are very large and delicate, and are produced all over the shrub in plenty. They exhibit themselves about the usual

time :

time : Many of them are of a pure white, with a deep purple fpot at the bottom of each petal ; whilft others again from thefe afford a variety, being of a purple colour, or having their edges of a reddifh tinge. The beauty of this tree, when in blow, is often over, in very hot weather, by eleven o'clock in the morning ; but that is renewed every day ; and for about fix weeks fucceffively a morning's walk will be rendered delightful by the renewed bounties which they beftow.

4. The HOARY CISTUS is a fhrub of about four feet high, and forms itfelf into a bufhy head. There are four or five varieties of this fort, that have been looked upon by fome authors as diftinct fpecies ; but experience now teaches us better. The leaves of all are hoary ; but they differ often in fhape, fize or figure ; and this has occafioned their being named accordingly, and to be diftinguifhed by the names of Common Hoary-leaved Ciftus ; the Long-leaved Hoary Male Ciftus : the Rounder-leaved Male Ciftus ; the Large Hoary-leaved Male Ciftus, &c. When thefe different forts can be procured, they make the plantations more agreeable. The leaves of thefe forts of Ciftus fit clofe to the branches, are hairy, and rough on both fides. Their figure will be different on the fame plant, and be produced in different manners : thofe on the tops of the branches are fpear-fhaped, and grow fingly ; but the lower ones are oval, and joined together at their bafe. All of them are hoary, though fome of the forts are whiter than others ; and thefe leaves make a good contraft with the ftronger greens during the winter months. Thefe fhrubs produce their flowers earlier than the other forts ; they often fhew fome in May. They are of a purple colour, which, in different forts, will be ftronger or lighter. They fall away in the evening ; but are conftantly renewed, for a month or longer, by a fucceffion every morning.

5. The GUM CISTUS OF MONTPELIER is commonly of about four feet growth, though, like the others, it is very beautiful when no higher than one or two feet. The branches proceed from the bottom of the plant, in plenty ; they are hairy, tough, and flender. Their leaves are lanceolated, exfude a very fragrant matter, are hairy on both fides, have three veins running lengthways, are of a dark-green colour, and fit clofe to the branches.

branches. The flowers are produced in their greateſt plenty about Midſummer, and ſometimes earlier, on long footſtalks, at the ends of the branches. They are white, and the ſucceſſion of the blow will be continued often longer than ſix weeks.

6. WHITE CISTUS will grow to be five or ſix feet high; and the younger branches, which will grow in an upright manner, are tough, and covered with a woolly ſubſtance. The leaves are oblong, very white, downy, trinervous, and ſit cloſe, ſurrounding the ſtalk at the baſe. The flowers are produced from the ends of the branches, at the beginning of June. They are large, of a fine purple colour, and look very beautiful.

7. The SAGE-LEAVED CISTUS is a much lower ſhrub, and the branches are many, ſpreading, and ſlender. The leaves reſemble thoſe of ſome of the ſorts of ſage-plants. They are oval, on both ſides hairy, and have very ſhort footſtalks. The flowers are produced in June, from the wings of the leaves. They are white, and ſtand on naked footſtalks; and though they are ſmaller than ſome of the other ſorts, yet being produced all over the ſhrub, they make a fine ſhow.

8. The CURLED CISTUS is of about four or five feet growth. The branches are very many, and ſpreading. The leaves are ſpear-ſhaped, waved, hairy, naturally bend backwards, and grow oppoſite by pairs on the braches. The flowers are produced from the wings of the leaves in June. Their colour is white. The ſucceſſion will be kept up for a month, or though

9. The SEA-PURSLAIN-LEAVED CISTUS is a ſhrub of about four feet growth, and ſends forth many branches in an upright pretty manner. The younger branches are downy, and the leaves have ſome little reſemblance to the Sea-Purſlain; though there are varieties of this ſpecies with broader and narrower leaves; ſome that approach to an oval, and others that are ſharp-pointed. They grow oppoſite by pairs, and make a good variety by their white and hoary look. The flowers are produced in June and July on very long, naked footſtalks, which ſupport others alſo with ſhorter footſtalks. They are of a fine yellow colour, and make a good figure when in blow. This is the moſt tender of all the ſorts, and is generally treated as a greenhouſe plant; but if the ſoil be naturally dry and warm, and the ſituation well-ſheltered, it will do very well abroad in

our

our tolerably open winters. It may be adviſeable, however, to ſecure a plant or two in the greenhouſe, that, in caſe a very ſevere winter ſhould happen to kill thoſe abroad, a freſh ſtock may be raiſed from the thus-preſerved plants.

10. SPANISH ROUND-LEAVED CISTUS. This is a branching ſhrub, of about a yard or four feet high. The leaves are oval, round, hairy, and placed on footſtalks on the branches. The flowers come out in plenty from the tops and ſides of the branches, in July. Their colour is purple; and though they are very fugacious, yet there will be a ſucceſſion of them for a long time.

11. CRETAN CISTUS. This is a branching ſhrub, of about the ſame height with the former. The leaves are ſpatulated, oval, enervous, rough, and grow on footſtalks on the branches. The flowers are red; and they make their appearance about the ſame time with the former.

12. SPANISH NARROW-LEAVED CISTUS. This riſes with a ſhrubby, naked, purple-coloured ſtalk, to about four feet high. The leaves are narrow, light, reflexed on their ſides, and grow oppoſite to each other without any footſtalks. The flowers grow in ſmall umbels, and come out from the ends and ſides of the branches, on long ſlender footſtalks. Their colour is white; and their appearance is about the ſame time with the former.

All the ſorts of Ciſtus are PROPAGATED by ſeeds and cuttings. 1. Seeds is the beſt way, as by them the moſt handſome plants are produced, though they will not always afford ſo great a plenty of flowers as the plants raiſed from cuttings. When they are to be raiſed by ſeeds, a moderate hotbed ſhould be in readineſs for their reception by the beginning of March; and they ſhould be ſown in drills a quarter of an inch deep. A dry day ſhould be made choice of for the purpoſe, and pegs ſhould be ſtuck to ſhew the extremity of the drills. The drills may be made two inches aſunder; and the bed being neated up, no other covering will be neceſſary than an old mat, to guard the plants, when coming up, from the ſpring froſts which may happen; for if the ſeeds are good, you may expect many plants to appear in leſs than a month; at which time they ſhould be covered in the night, but be always kept uncovered in open and fine weather. As the dry weather comes on, they muſt be watered

moderately

moderately every other morning, and the weeds conftantly cleared off; and as the fummer heat encreafes, the mats ufed to guard
them from the froft in the night, muft change their office : They
muft never come near them in the night, but only protect them
from the fcorching heat in the middle of the day. By the latter
end of Auguft many of the plants will be four or five inches high;
when they may be thinned, and thofe drawn out either pricked
in the nurfery-ground, in beds at fmall diftances, in well-fheltered places, or planted in pots, to be fecured in the winter, and
turned out at leifure. Of all the forts, the Bay-leaved and the
Sea Purflain-leaved fpécies, with all their varieties, require this
treatment. The reft are all very hardy. Thofe that are pricked out in ·rows in the nurfery will immediately ftrike root : and,
as well as thofe left in the old hotbed, if they are in well-fheltered places will do without any protection. If the place is not
well defended, either by trees or hedges, it will be proper to
prick fome furze-bufhes all around, to break the keen edge of
the fevere frofts. Thofe left in the old bed fhould be
planted out in the fpring in the nurfery-ground ; and in a
fpring or two after this, they fhould all be planted out where
they are to remain ; for none of thefe plants fucceed fo well if
removed when grown old and woody. 2. Thefe plants are
eafily raifed by cuttings ; and plants raifed this way are often
the beft flowerers, though their manner of growth is not always
fo upright and beautiful. Auguft is the month for this work ;
and if a dripping day happens in that month, it muft be made
choice of ; if not, a bed of fine mould muft be prepared, and the
cuttings fhould be planted a few inches afunder ; and after
that fhould be watered to fettle the mould to them. The beds
fhould be hooped ; and the next day, as the heat of the fun
comes on, they fhould be covered with mats : This covering
fhould be repeated, obferving always to uncover them in the
evenings, and alfo in moift and cloudy weather. Thefe cuttings will take root in a very little time ; and their after management may be the fame as the feedlings.

CLEMATIS.

C L E M A T I S.

NNEAN Clafs and Order, *Polyandria Polygynia*: Each flower contains many males and many females. There are thirteen Species, ten climbing, and three erect: Eight of the former have been introduced into this country.

1. Clematis *Viticella*: The Virgin's Bower; *a deciduous climber*; native of Italy and Spain.

.2. Clematis *Viorna*: The Virginia Climber, or the Purple Climber; *a deciduous climber*; native of Virginia and Carolina.

3. Clematis *Crifpa*: The Carolina Climber, or the Curled Purple Climber; *a deciduous climber*; native of the Eaft,

4. Clematis *Orientalis*: The Oriental Climber; *a deciduous climber*; native of the Eaft.

5. Clematis *Vitalba*: The Traveller's Joy, or Old Man's Beard, or Bind-with; *a deciduous climber*; growing naturally in the hedges of England, and moft of the northern parts of Europe; alfo in Virginia and in Jamaica.

6. Clematis *Cirrhofa*: The Evergreen Clematis, or Evergreen Spanish Climber; *an evergreen climber*; native of Spain and Portugal.

7. Clematis *Flammula*: The Creeping Climber; *a deciduous climber*; native of the South of Europe.

8. Clematis *Virginiana*: The Sweet-Scented Clematis, or the Sweet-Scented American Climber; *a deciduous climber*; native of North America.

1. Virgin's Bower. Of this fpecies of Clematis there are the following *Varieties*:
Double Purple Virgin's Bower.
Single Purple Virgin's Bower.
Single Blue Virgin's Bower.
Single Red Virgin's Bower.
Double Purple Virgin's Bower. This fort ftands firft on the lift, not only becaufe it is an admirable climber, but alfo is poffeffed of a large double flower. It will grow to the
I 3   height

height of twenty or thirty feet, if fupported ; and is very proper to cover arbours, as well as walls, hedges, &c. The branches are of a dark-brown or dufky colour, angular and channelled. The younger branches are of a fine green colour, and nearly fquare : They are very numerous, and grow from the joints of the older ; and thus they multiply in that manner from the bottom to the top of the plant. The leaves alfo grow from the joints : They are both compound and decompound * The folioles, of which each is compofed, are of an oval figure, and their edges are entire ; and in fummer, when the plant is in full leaf, if fet alone to form an arbour, after it is faid to be grown ftrong, the branches and large leaves will be produced in fuch plenty, as not only effectually to procure fhade, but even to keep off a moderate fhower ; fo excellently is this plant adapted to this purpofe ; and more particularly fo, as it will grow, when it has properly taken to the ground, fifteen or fixteen feet in one year. The flowers are double, and of a purple colour They blow in July and Auguft, and are fucceeded by no feeds, the multiplicity of the petals entirely deftroying the organs of generation.

The *Single Purple Virgin's Bower* is rather a ftronger fhooter than the Double, and will climb to rather a ftill greater-height. The Double is only a fub-variety of this, which ought not to be neglected ; for this exhibits a fair flower, compofed of four large petals, in the center of which are feated the numerous ftamina.

The *Single Blue Virgin's Bower* produces its fhoots, leaves, and flowers, in the fame manner as the other ; and makes a variety only in that the flowers are of a blue colour.

The *Single Red* is of much lower growth, and feems of a more delicate and tender nature ; not but it is hardy enough to endure any weather ; but its fhoots are weak, and fhort in proportion. They are angular, and channelled in the manner of the other ; but they are of a reddifh colour. The leaves are fmaller than the other forts, and the flowers alfo are fmaller, though they make a fine variety, by their colour being red. Thefe all flower at the fame time ; but are fucceeded by no ornamental feeds

* Doubly compound.

2. VIRr

2. Virginia Climber. The branches are slender and numerous; and the leaves, as in the Virgin's Bower, are both compound and decompound. The folioles grow by threes, and these are often multiplied to form a decompound leaf of nine in number. They are nearly cordated, of a good green, and some of them are trifid. The flowers are produced in July and August, from the wings of the leaves. They are a kind of blue colour; and the petals (which are four in number), of which each is composed, are of a thick coriaceous substance. This sort will sometimes ripen its seeds in England.

3. Carolina Climber. This is by some called the Curled-flowering Climber; and indeed by that name it is chiefly distinguished in our gardens. It is one of the lower kind of climbers; seldom arising, by the assistance of its claspers, to more than six feet. The stalks are very weak and slender. The leaves afford great variety, being sometimes trifoliate and sometimes single. The folioles also differ much; for some of them are found whole and entire, whilst others again are divided into three lobes. These leaves are of a dark-green colour, and are produced opposite, from the joints of the stalks. The flowers are produced in July and August, on short footstalks, below which a pair or more of oblong pointed leaves often grow. These flowers are composed of four thick, coriaceous, purple, curled petals. This species will for the most part produce ripe seeds in our gardens.

4. Oriental Climber is no great rambler; for, notwithstanding its slender stalks are well furnished with claspers, it is seldom found to climb higher than about ten feet. The leaves of this sort are compound. The folioles are cut angularly, and the lobes are shaped like a wedge. They are of a good green colour, and are very ornamental to the plant. The flowers are produced from the wings of the leaves early; for it will often be in blow in April. They are of a kind of yellowish-green colour, and the petals naturally turn backwards. These flowers differing in colour from the above sorts, and coming earlier in the spring, make it more desirable, as it testifies how many months in the summer are ornamented with the blow of some one or other species of Clematis. The seeds of this sort also will often ripen with us.

I 4

5. Tra-

5. TRAVELLER's JOY is a noble climber, and well known in many parts of England; the hedges where it abounds being frequently covered with it: But its greateſt ſingularity is in winter; at which time it more peculiarly invites the traveller's attention. The branches of this ſpecies are very thick and tough, ſufficient to make withs for faggots; and for this purpoſe it is always uſed in the woods where it can be got. Theſe are ſo numerous, and produce ſide-branches in ſuch plenty, which divide alſo into others, that they will over-top hedges, or almoſt any thing they can lay hold of to climb by. Beſides the claſpers with which it is furniſhed, the very leaves have a tendency to twine round plants. Theſe leaves are pinnated; and a variety is occaſioned by them; for the folioles of ſome ſorts are indented at their edges, whilſt others are found with their edges entire. They are of a blueiſh-green, and moderately large. The flowers are produced in June, July, and Auguſt, all over the plant, in cluſters. They are ſucceeded by flat ſeeds, each of which, when ripe, is poſſeſſed of a white hairy plume, and growing in cluſters will exhibit themſelves in winter all over the tops of buſhes, hedges, &c. which at that time will look beautiful and ſingular. This is the *Viorna* of old Botaniſts; and is called Traveller's Joy from its thus ornamenting hedges, buſhes, &c. to the entertainment of the traveller.

6. EVERGREEN CLEMATIS. This is but a low climber, ſeldom growing higher than ſix or eight feet. The branches are very numerous, weak, and ſlender; but it riſes by claſpers, which naturally lay hold on any thing near them. The footſtalks of the leaves, alſo, will twine round twigs, &c. ſo that they become claſpers, and enſure the hold of the plant. Nay, if there be no hedge or plant near, by which they may hold and riſe, they will twine among themſelves; and as the branches are produced in great plenty, they will be ſo mixed one amongſt another, as to form a low thicket, which makes this plant well adapted to produce variety in evergreen ſhrubbery quarters, where, if planted ſingly, at a diſtance from other trees, it will naturally form itſelf into a thick buſh. Theſe leaves are ſometimes cut into three lobes, ſometimes into two, and many of them are undivided. The lobes when moſt perfect are nearly lanceolate, have their edges indented, and are of as fine a ſhining green as can be conceived

ceived. The flowers are produced in the midft of winter, from the fides of the branches : They are of a greenifh colour, though inclined to a white ; but the petals being pretty large, and blooming at that unufual feafon, makes this plant highly valuable.

7. CREEPING CLEMATIS, or *Flammula*, will mount by the affiftance of other plants to a good height, fometimes near twenty feet. The ftalks are flender and numerous ; and the leaves are in this refpect fingular ; for the lower ones are pinnated, and their edges are jagged ; but the upper ones grow fingle. They are of a lanceolate figure, and their edges are entire. The flowers of this fpecies are exhibited in June, July, and Auguft. They are white, and by fome admired.

8. SWEET-SCENTED CLEMATIS. This fort will rife, by the affiftance of neighbouring bufhes and trees, to a great height. The branches are many, fpread themfelves all around, and lay hold of every thing that is near them. The leaves are ternate. The folioles are heart-fhaped, angular, and nearly cut into three lobes. The flowers are white, and, being poffeffed of a moft greeable fragrance, render this climber highly proper for arbours, and to be ftationed near feats and places of refort.

Thefe are all the hardy climbing fpecies of this genus yet known. The *varieties* of the firft kind are notable, and afford as much diverfity in a garden as if they were diftinct fpecies. The other forts alfo admit of varieties; but the difference is very inconfiderable, and makes little variety, as they nearly agree with fome or other of the above forts.

The PROPAGATION of all thefe forts is by layers ; and this is beft done in fummer on the young fhoots as they grow. As foon, therefore, as they have fhot about a yard or four feet in length, let the ground be well dug about each ftool, and made fine, and a gentle hollow made about a foot from the ftool. In this hollow let the young fhoots be preffed, and covered with mould, leaving their ends out to continue growing. In a very little time they will be a yard or more in length ; when a fecond hollow may be made, at a diftance from the other, and the fhoots preffed down and covered with mould as before, the ends being ftill left out to grow. On fome of the long fhooting forts this may be repeated again, and even again ; and thefe fhoots,

thus

thus layered, will ftrike root. Many of the forts will have good root by the autumn; and others muft be waited for until the autumn following. This fummer-method of layering is highly neceffary; becaufe fome of the forts, particularly the Virgin's Bower, if layered in winter in the common way, will be often two whole years, nay fometimes three, before they will ftrike root. Any time from autumn to fpring the layers may be taken up; and from one ftool fome fcores are often obtained. Thofe with good roots may be fet out to remain; and every bit that has a fibre fhould be cut off below that fibre, and fhould be headed to one eye or joint above the part that had been out of the ground; and thus all the layers being collected together fhould be planted in the nurfery at fmall diftances, and in a year or two they alfo will be good plants for ufe.

The TRAVELLER's JOY may be layered at any time, for the roots will eafily ftrike; nay, they will grow by cuttings.

The EVERGREEN SPANISH CLIMBER requires no art or trouble to encreafe it; for it will encreafe itfelf if the ground is left undifturbed a year or two, and will throw out plenty of fuckers, which will have roots, and be good plants.

C L E T H R A.

LINNEAN Clafs and Order, *Decandria Monogynia:* Each flower containing ten males and one female. There is only one known SPECIES:

CLETHRA *Alnifolia:* The ALDER-LEAVED CLETHRA, or the AMERICAN ALDER; *a deciduous aquatic fhrub;* growing naturally in Virginia, Carolina and Pennfylvania, like our Alder, by the fides of rivers and watery places.

The CLETHRA is fhrub, with us, about four or five feet high, tho' in its native foil it is fometimes found fo high as eight or ten feet. The branches it fends forth are not numerous, and thefe are garnifhed with leaves, which are fpear-fhaped and ferrated.

ferrated. They are about three inches long, an inch and a half broad, and have ſhort footſtalks. The *Clethra* uſually flowers in July. The flowers are produced at the ends of the branches, in long ſpikes : They are white, and poſſeſſed of a ſtrong ſcent. This plant, at preſent, is not very common in our gardens.

The CULTURE of this ſhrub is by layers, ſeeds, and ſuckers. 1. The plants deſigned to be encreaſed by layers ſhould be ſet in the moiſteſt part of the garden, and managed like thoſe of the CHIONANTHUS. 2. By ſeeds,—which alſo ſhould be ſown and managed the ſame as CHIONANTHUS. 3. Theſe ſhrubs will very often ſend out ſuckers, by which they may likewiſe be propagated. Theſe may be taken off in the autumn, if they have good roots, and planted out in the nurſery-way : if they have not, they ſhould be let alone till March ; then taken up, and planted in pots of good loamy ſoil, and afterwards plunged into a moderate warmth of dung; which will promote their growth. The autumn following they will be fit to be planted out to ſtand.

C N E O R U M.

LINNEAN Claſs and Order, *Triandria Monogynia* : Each flower contains three males and one female. There is only one SPECIES :

CNEORUM *Tricoccon* : The CNEORUM, or WIDOW-WAIL : *a low evergreen ſhrub* ; native of dry gravelly places in Spain, Italy, and France.

The CNEORUM, or WIDOW-WAIL, is a ſhrub of about a yard in growth, and is an excellent one for the front of evergreen quarters, where the loweſt ſhrubs are to be placed. The wood of this tree is very hard, and the older branches are covered with a brown bark. The ſtem naturally divides into many branches ; and the bark on the youngeſt is ſmooth, and of a pale-green colour. The leaves are ſmooth, of a fine dark-green colour, and conſtitute the greateſt beauty of this ſhrub. They are of an oblong figure, and very long in proportion to the breadth :

They

They will be two inches or more long, and about half an inch in breadth. Their under furface is of rather a paler green than their upper, and their bafe joins to the young branches without any footftalk. The flowers are yellow, and make no great fhow. A healthy plant may be expected to be in blow moft part of the fummer. They grow from the wings of the leaves, towards the ends of the branches ; and are fucceeded by the feeds, which grow together by threes ; which will be of a dark-brown or black when they are ripe.

CNEORUM may be PROPAGATED by feeds or by cuttings. 1, By feeds. Thefe fhould be gathered in October, and be thofe which have grown from the firft flowers of the fhrub that fummer, and which will be then black, or nearly fo, if ripe. They fhould be fown, in a bed of common garden-mould made fine, about half an inch deep. One may expect to fee the plants come up in the fpring ; though it often happens that the greateft part of them remain until the fecond fpring before they appear. 2. Thefe plants may be encreafed by cuttings ; but they never make fuch beautiful fhrubs ; neither is the method worth practifing if feeds can be obtained. The cuttings may be planted in fpring ; then it will be neceffary to fet them in pots, and give them the affiftance of an hotbed ; and this will fet them a-growing. The beginning of Auguft is a very good time for planting thefe cuttings or flips. They fhould be planted in beds of good fine mould ; and thefe fhould be hooped, and matted from nine o'clock in the morning until near funfet. Then they fhould be uncovered, and remain fo in all cloudy and rainy weather. Moft of thefe cuttings will grow ; and there they may remain without removing until they are fet out for good.

When thefe fhrubs are to be planted out, the moft dry and gravelly fpots muft be chofen for them ; and in thefe places they will bid defiance to our fevereft weather ; though in fuch a foil they will not grow fo high as in a moift fat foil, by a foot or more, which is confiderable in a fhrub of fuch a natural low growth ; but it is neceffary for them to be planted in a dry or gravelly foil, becaufe there they will be fecure from injury by frofts.

COLUTEA.

C O L U T E A.

Linnean Clafs and Order, *Diadelphia Decandria*: Each flower contains ten males and one female, the males ftanding in two divifions. There are three Species; one of which is herbaceous, and another a fhrub, too delicate for the open air of this climate; the third has long been an ornament to the Englifh garden.

Colutea *Arborefcens*: The Bladder-Sena; a well known *deciduous fhrub*; native of the South of Europe, paiticularly about Mount Vefuvius.

The Bladder-Sena fports in the following *varieties*; all of which are beautiful in their kind, and afford delight both by their flowers and leaves; viz. 1. The Common Bladder-Sena. 2. The Oriental Bladder-Sena. 3. Pocock's Bladder-Sena. 4. The Red-podded Bladder-Sena.

The *Common Bladder-Sena* is the talleft grower of all the forts. It will arrive to the height of about ten or twelve feet. The branches are of a whitifh colour, which diftinguifh it in the winter, and the leaves in the fummer have a pleafing effect. They are pinnated; the folioles are oval, and indented at the top; they confift of fometimes four, fometimes five pair, placed oppofite, and are terminated by an odd one. The flowers are of the butterfly kind: They are produced in June, July, and Auguft, in clufters; are numerous, of a yellow colour, and the footftalk that fupports them is long and flender. The flowers are fucceeded by large inflated pods, like bladders, which catch the attention of thofe who have never before feen them. This tree has variety enough of itfelf to make it efteemed; but it fhould always be planted among other trees of the fame growth, to break the force of the ftrong winds; not but that it is hardy enough to refift our fevereft winters, but the branches will eafily fplit, which will make it unfightly, unlefs they are fheltered in fome degree by other trees. This fort will ripen its feeds in the autumn.

The *Oriental Colutea* will grow to the height of about five or fix feet. The branches of this tree alfo are greyifh, and the leaves pinnated, as well as terminated by an odd one, and the lobes are obverfely cordated and fmall. The flowers are reddifh,

fpotted

spotted with yellow, and grow from the fides of the branches on footftalks, each of which is formed fometimes with two, fometimes with three flowers. This tree is extremely hardy; and as it does not grow to the fize of the common fort, nor in fo luxuriant a manner, the branches will not be fo liable to be fplit off by the winds; and therefore the precaution neceffary for that, in this fort may be the lefs obferved.

Pocock's Bladder-Sena is another variety, of lower growth than the common fort. The leaves are pinnated, and the folioles ftand oppofite by pairs in both the kinds. They are indented in the fame manner at the top; neither can I perceive any other difference between this and the Common Bladder-Sena, only that the one is larger than the other, and the flowers come out earlier in the year.

The *Red-podded Bladder-Sena* is alfo a variety, which will happen in common to all the forts, more or lefs, when raifed from feeds.

Thefe trees are all very eafily PROPAGATED. 1. By feeds. Any time in the fpring will do for the work, though the month of March is the beft feafon; and no other compoft will be required than garden-mould of almoft any fort, dug and raked fine. If the feeds are fown about half an inch deep, they will come up like corn in a month or two after. Keep the beds weeded until the fpring following; and then plant them out in the nurfery-way, obferving always to fhorten the tap-root which they often have. In a year or two they will be good and proper plants for the fhrubbery. 2. Thefe trees may alfo be propagated by layers; and that is the method generally practifed with Pocock's fort, to continue it in its low growth.

C O R N U S.

LINNEAN Clafs and Order, *Tetrandria Monogynia* : Each flower contains four males and one female. There are eight SPECIES; three of which are adapted to ornamental gardening.

1. CORNUS *Mafcula* : The CORNELIAN CHERRY; *a tall deciduous fhrub*; growing naturally in the hedges of Auftria.

2. CORNUS

2. CORNUS *Sanguinea* : The COMMON DOGWOOD, or BLOODY-TWIG ; *a deciduous shrub* ; common in our hedges, and is natural to most parts of Europe, Afia, and America.

3. CORNUS *Florida :* The VIRGINIAN DOGWOOD, or FLOWER-ING CORNUS ; *a deciduous shrub* ; native of Virginia.

1. The CORNELIAN CHERRY will rise to twenty feet high. Its principal merit as an *ornamental* lies in its flowering early in the fpring, and in exhibiting its beautiful fcarlet berries in autumn. Its *ufes* are held out as numerous. Its fruit was for-merly in good efteem ; and its wood is faid to be ufeful for wheel-work, pins, hedges, &c. It is arranged by EVELYN and HANBURY among Foreft-trees.

2. The COMMON DOGWOOD is well-known all over England, as it grows naturally in moft parts of the kingdom ; a few of thefe trees are neverthelefs admiffible into the fhrubbery, if they are not already too common in its neighbourhood ; for the young twigs are red, efpecially in winter, which look well at that feafon, as do alfo its flowers in the fummer, and its leaves in the autumn. The rednefs of thefe young fhoots has occafion-ed this fort to go by the name *Bloody-Twig.* The leaves are about two inches long, and an inch and a half broad : Thefe have large nerves, which terminate in a point, and they often die in the autumn to a reddifh colour. The flowers are white, produced in umbels at the ends of the branches, and are fuc-ceeded by black berries, like thofe of the Buckthorn, but have in each only one ftone. The wood, it is faid, makes the beft kind of charcoal in the world for gunpowder. It is brittle, ex-ceedingly white, and when growing is covered with a dark-brown bark, the twigs being red.

3. VIRGINIAN DOGWOOD will grow rather higher than our Common Dogwood. The twigs are of a beautiful red. The leaves are obverfely cordated. The flowers are produced in large bunches fomewhat like thofe of the Elder : Their colour is white ; they come out in May and June, and the berries ripen in autumn.

From thefe fpecies, the following beautiful *varieties* figure in our nurferies ; viz. Female Virginian Dogwood ; American Blue-berried Dogwood ; White-berried Dogwood of Pennfylvania ; and Swamp Dogwood.

Female

Female Virginian Dogwood, during the winter months, ex-hibits its branches of so beautiful a red colour, as to distinguish itself to all at that season. It grows to eight or ten feet high; the leaves are somewhat spear-shaped, acute, nervous, and in the autumn die to a fine red. The flowers come out in umbels, at the ends of the branches: They appear in May and June, and the berries ripen in the autumn.

The *American Blue-berried Dogwood* arrives at the height of about eight or nine feet. The twigs of this tree also are of a delightful red. The leaves are large, oval, and hoary on their under-side. The flowers are white, come out in umbels from the extremity of the branches, and are succeeded by large, oval, blue berries, which make a fine appearance in the autumn.

White-berried Dogwood arrives at the same size with the others. The young shoots, like those of the former, are of a beautiful red colour during the winter. Like them, also, it produces its white flowers in large umbels in May; but they are succeeded by white berries in the autumn.

Swamp Dogwood grows naturally in moist places, almost all over America; and it will grow with us in almost any soil or situation. The leaves of this are of a much whiter colour than any of the other sorts; though the flowers and fruit are pro duced in the same manner.

One method of PROPAGATION is common to all these sorts of *Cornus*; though this may be effected three ways; by seeds, layers, and cuttings. 1. The seeds of the common sort should be sown in the autumn, soon after they are ripe; and these will come up in the spring. The seeds of the American sorts we generally receive in the spring: These should be sown directly; but they will not come up till the spring following; nor would those of our common sort, if they were kept until the spring before they were sown. No particular art is required for these seeds. They will grow in common garden-mould of almost any sort, though the richer it is the better. This must be made fine, cleared of all roots, weeds, &c. and the seeds should be sown about half an inch deep. The spring after the plants come up, they should be planted in the nursery, at a small distance from each other, where they may stand for two or three years, and then be planted out to stand. 2. These trees may be easily propagated by layers; for after having obtained

some

fome plants for the purpofe, if the fhoots that were made the preceding fummer be only laid in the ground in the autumn, they will have good roots by the autumn following. Thefe may be taken off, and planted in the nurfery for a year or two, as the feedlings; and the ftools being cleared of all ftraggling branches, and refrefhed with a knife, they will make ftrong fhoots for a fecond operation by the autumn next enfuing. 3. By cuttings likewife thefe forts may be propagated. This work fhould be done in October; and the cuttings for the purpofe fhould be the ftrongeft part of the laft year's fhoot, that had fhot vigoroufly from a healthy foil. If thefe are cut into lengths of about a foot long, and planted in a moiftifh foil, three parts deep, they will grow, and make good fhoots the fummer following; and thefe will require no removing before they are planted out finally.

C O R I A R I A.

LINNEAN Clafs and Order, *Dioecia Decandria* : Male flowers containing ten ftamina, and female flowers containing five piftils upon diftinct plants: There are two SPECIES; one of which will bear the open air of this climate.

CORIARIA *Myrtifolia*: The MYRTLE-LEAVED SUMACH, or TANNER's SUMACH; *a deciduous fhrub*; grows naturally about Montpelier in France, where it is faid to be ufed by the tanners in tanning of leather.

The MYRTLE-LEAVED SUMACH is a fhrub of lowifh growth, feldom arriving to more than four or five feet high. The bark is of a greyifh colour, and fpotted. The wood is very brittle, and very full of light pith. The young fhoots are produced in great plenty from the bottom to the top: They are fquare, and come out three or four together, from one fide of the ftem, whilft the other fide is often furnifhed with an equal number. The leaves refemble fome of the forts of Myrtle, which gave occafion for its being called the Myrtle-leaved Sumach: They are oblong, pointed, of a bright-green, and ftand oppofite by pairs on the twigs. The flowers grow in fpikes, at the ends and

K   fides

fides of the branches, and have little beauty to recommend them. The tree is planted, however, as a flowering fhrub, amongft others of its own growth ; but the place in which it is fet fhould be well fheltered ; for notwithftanding this is a very hardy fhrub, yet the ends of the branches are often killed in the winter, which makes the plant unfightly in the fpring.

The PROPAGATION of the *Coriaria* is very eafy. No other art need be ufed, than, after having obtained a few plants, to plant them in a lightifh foil of any fort. Here they will propagate themfelves in great plenty ; for they will (what gardeners call) *fpawn* ; *i. e.* their creeping roots will fend forth many young plants, at more than three yards diftance from the real plant. The ftrongeft of thefe may be taken up, and planted where they are to remain, whilft the weaker may be fet in the nurfery-way, to gain ftrength, before they are fet out for good. In this eafy manner may plenty of thefe fhrubs be obtained ; and every winter after they are taken up, if the mould about the mother-plant be raked fmooth, and weeded in fummer, fhe will afford you a frefh crop by the autumn following, which may be taken off and planted as before.

C O R O N I L L A.

LINNEAN Clafs and Order, *Diadelphia Decandria* : Each flower contains ten males and one female, the males being divided at the bafe into two fets. There are eleven SPECIES ; two of them herbaceous, the reft of a ligneous nature ; but only one of them has been introduced into our fhrubberies.

CORONILLA *Emerus:* The SCORPION SENA, or JOINTED-PODDED COLUTEA; *a deciduous fhrub;* native of the South of Europe.

The SCORPION SENA fends out numerous irregular branches from the root and on all fides ; the oldeft and moft woody of which are of a greyifh colour, whilft the youngeft are fmooth, and of a dark brown. The leaves are pinnated, and conftitute a great beauty in this fhrub, being of a pleafant green, and are compofed of three pair of folicles, which are terminated by an

odd

odd one; thefe ftand oppofite on the midrib, and each has an indenture at the top. Thefe leaves, by a proper fermentation, will afford a dye nearly like that of indigo. However, beautiful as the leaves are, it is the flowers which conftitute the beauty of thefe fhrubs; and, indeed, of all the fhrubby tribe, there is none more ftriking or pleafing than this when in full blow. This ufually happens in May; when it will be covered all over with bloom, the fhrub itfelf appearing as one large flower divided into many loofe fpikes; for the flowers come out all along the fides of the branches by the leaves, on long footftalks, each fupporting two or three flowers, which are butterfly-fhaped, of a yellowifh colour, and large in proportion to the fize of the fhrub. They are fucceeded by longifh pods, in which the feeds are contained. This fhrub often flowers again in the autumn.

There is a *variety* of lower growth, called *Dwarf Scorpion Sena.*

This beautiful fhrub is very readily PROPAGATED, either by feeds, layers, or cuttings; any of which may be eafily made to grow. 1. By feeds. Thefe fhould be fown, in the fpring, in beds of common garden-mould made fine, and cleared of the roots of all weeds, &c. They fhould be covered about half an inch deep; and, if a very dry fpring does not enfue, they will be up in about a month or fix weeks. If this fhould happen, the beds muft be now and then watered, and fhaded from the heat of the fun, which fometimes is very intenfe and parching, even at the beginning of May. They may ftand in the feed-bed two years before they are taken up; all which time they will want no other care than weeding; and if they have watering the firft fummer, fhould it prove a dry one, they will grow the fafter. After this, they may be taken out of the feed-bed, planted in the nurfery-way, and in about two or three years will be good plants to join in the fhrubbery. 2. By layers. This bufinefs may be performed any time in the winter; but as the fhrub fends forth numerous branches, many of them fhould be taken off, and only fuch a number left, as that they may be laid into the ground without crowding one another. The branches fhould be of the laft year's fhoot; and the operation fhould be performed by a gentle twift, fo as juft to break the bark; for, fays HANBURY, without this I have found them in the autumn juft as they were when layered; and with this, they have always

K 2   ftruck

struck root, fo as to be fit to take off the winter following. Thefe
layers fhould be planted out in the nurfery; and after having
ftood about two years, they alfo will be grown to be good
plants. 3. By cuttings. The cuttings fhould be the ftrongeft of
the laft year's fhoots. They fhould be planted clofe, in October,
in a fhady border of good fine mould. If the fpring and fummer prove dry, watering muft be afforded them every other day;
and by this means many plants may be raifed. If the cuttings
are planted clofe, and moft of them grow, they fhould be
thinned, by taking up fo many as may leave the others at a
foot or more afunder; and thefe plants alfo, thus taken up,
fhould be fet out in the nurfery-ground a foot afunder, in rows
at a foot and a half diftance; where they may ftand until they
are finally taken up.

C O R Y L U S.

LINNEAN Clafs and Order, *Monoccia Polyandria:* Male
and female flowers upon the fame plant: The males, containing
ten ftamina each, are collected in cylindrical catkins; the
females, containing two piftils each, iffue from the point of the
leaf-bud: There are two SPECIES.

1. CORYLUS *Avellana:* The HAZEL; a well-known *tall
deciduous fhrub*; very common in this country, and in moft
parts of Europe.

2. CORYLUS *Colurna:* The BYZANTINE NUT, or DWARF
NUT TREE; *a low deciduous fhrub*; growing naturally near
Conftantinople.

The HAZEL will grow to twenty feet high and upwards. A
particular defcription of it here would be fuperfluous. LINNEUS
confiders the various kinds of FILBERTS as *Varieties* of the
common Hazel, improved by culture. MILLAR was of a different
opinion: he fays, " I have feveral times propagated both from
the nuts, but never have found them vary from the other,
though they have altered in the fize and colour of their fruit
from the forts which were fown;" he therefore divides them into
two diftinct fpecies: But HANBURY on the other hand fays,

that

that they " are varieties only of the fame fpecies ; for I have planted the nuts of all the forts, and forts ot all kinds have been produced from them." (Page 111.) As an *Ornamental*, the Hazel is of an inferior clafs ; neverthelefs, in reclufe quarters, the Filbert may be introduced with propriety : the idea of utility affociated with that real ornament which is undoubtedly given by the various tints of the leaves of the different kinds and colours of Filberts, may probably afford more real fatisfaction, efpecially to the owner, than the tranfient glare of a ufelefs exotic. Be this as it may, the Hazel in point of *ufe* ftands high ; as an underwood it has no fuperior : indeed, the Oak and Afh excepted, the hufbandman knows not fo ufeful a wood as the Hazel. For ftakes, edders and withs it is in ufe every where. In Surry, Kent, and other fouthern counties where numerous flocks of fheep are kept, the Hazel alone fupplies the farmer with folding-hurdles ; and in Yorkfhire and other parts of the North of England, from whence great quantities of butter are fent to the London market, the hoops or firkin-rods are gathered almoft wholly from this ufeful fhrub.

2. The BYZANTINE NUT. This is diftinguifhed from the other fpecies chiefly by the ftipulæ, which are very narrow and acute whereas thofe of the common nut are oval and obtufe. It differs alfo in the fize of its growth, the true Bizantine Nut-tree feldom growing higher than four or five feet ; and hence the name *Dwarf Nut-tree* has been ufed for this plant. In other refpects, it is like our common nut-tree ; it flowers at the fame time, the fruit is produced in clufters, and it ripens accordingly.

The method of PROPAGATING the Hazel kind is from feeds, by layering, or from the fuckers, which it fpontaneoufly fends up in great plenty. The *Nuts* fhould be fown about two inches deep, in February ; until which time they fhould be kept in a cool, moift place to prevent the kernels from becoming dry and fhrivelled, yet fufficiently airy to prevent their growing mouldy. The *Varieties* are beft preferved by layering ; for which purpofe a few plants fhould be procured of the moft valuable kinds, and planted for ftools. They will grow on almoft any foil ; and the young twigs being laid in the ground in the autumn, will have ftruck root by the autumn following, Thefe fhould be taken off, and planted in the nurfery, a foot

afunder, and two feet diſtant in the rows ; and if there be any
young ſhoots made the intermediate ſummer, they alſo may be
laid down, or the plant headed within half a foot of the
ground, to ſend forth young ſhoots for a ſecond operation the
autumn following. By this means the ſorts may be propagated,
and kept diſtinct ; for the ſeeds ſown of any of them will not in
general come to good ; though it is obſervable, that from the
beſt nuts there will be the beſt chance of having good nuts
again ; and " I have (ſays HANBURY) ſometimes known ſome
few trees, raiſed from ſeeds, which have produced nuts better
than thoſe they were raiſed from. This may, perhaps, induce
a gardener deſirous of obtaining a great variety to try this
method, when he may extirpate the worſt ſorts, and, if any
ſhould be worthy of it, may propagate the others in the manner
directed."

C R A T Æ G U S.

LINNEAN Claſs and Order, *Icoſandria Digynia* : Each
flower contains about twenty males and two females : There are
ten SPECIES ; eight of which add conſiderable beauty to the
modern garden.

   1. CRATÆGUS *Oxicanthus* : The HAWTHORN, or WHITE
THORN ; a well-known *deciduous tree or ſhrub* ; common with
us, and growing naturally all over Europe.

   2. CRATÆGUS *Azarolus* : The AZAROLE ; *a tall deciduous
ſhrub* ; native of Italy and the South of France.

   3. CRATÆGUS *Aria* : The WHITE LEAF ; or the WHITE
BEAM, or the ARIA, or the ARIA THEOPHRASTI ; *a deciduous
tree or ſhrub* ; grows naturally upon the hills of Kent and Surry,
particularly near Box-hill ; and in moſt of the cold parts of
Europe.

   4. CRATÆGUS *Torminalis* : The WILD SERVICE, or the
MAPLE-LEAVED SERVICE TREE ; *a deciduous tree* ; native of
England, Germany, Switzerland, and Burgundy.

   5. CRATÆGUS *Coccinea* : The VIRGINIA AZAROLE ; *a
tall deciduous ſhrub* ; native of Virginia and Canada.

6. CRA-

6. CRATÆGUS *Crus Galli* : The COCKSPUR HAWTHORN ; *a tall deciduous shrub* ; native of Virginia.

7. CRATÆGUS *Tomentosa* : The GOOSEBERRY-LEAVED VIR-GINIA HAWTHORN ; *a deciduous shrub* ; native of Virginia.

8. CRATÆGUS *Viridis* : The GREEN-LEAVED VIRGINIA HAWTHORN ; *a deciduous shrub* ; native of Virginia.

1. The HAWTHORN, in the state in which we are used to ob-serve it, is nothing better than a tall, uncouth, irregular shrub ; but trained up as a standard, it swells to a large timber size, with a tall stem and a full spreading head ; though we believe it seldom rises to a great height ; perhaps not often so high as thirty feet. We have measured the stem of a youthful thriving Hawthorn eight feet high, and five feet and a half in circum-ference, with a head proportionable. Mr. Marsham * mentions one near Bethel-Church, in the neighbourhood of Norwich, which, at four feet high, girted, in the year 1755, nine feet one inch and a quarter, one of its arms extending more than seven yards. The Standard Hawthorn, whether we view its flowers in the spring, its foliage in the summer, or its fruit in the autumn and winter, is one of the most *ornamental* plants, stand-ing singly, that can be scattered over a park or lawn. Its *uses* will be explained when we come to treat of HEDGES.

In order to PROPAGATE a quantity of *Quick*, one method is generally practised ; namely, first burying the haws, and taking them up to sow the October following ; though, says HANBURY, there is another way more preferable ; namely, to prepare the beds, and sow the haws soon after they are gathered. Whoever pursues the former method, having gathered what quantity of haws will answer his purpose, should in some bye-corner of the kitchen-garden or nursery dig an hole or pit ca-pacious enough to receive them ; some of the earth which came out of the hole, after the haws are put in it, should be laid upon them ; and, being thus carefully covered down, they may remain there till October. Then, having ground well dug, and cleared of the roots of all troublesome weeds, and the mould being fit for working, the beds should be made for the haws. Four feet is a very good width for these beds, as they may be

* Of Norfolk, in a Letter published in the First Volume of the Papers of the Bath Agriculture Society.

eafily reached over to be weeded; and if the alleys between be each one foot and a half wide, they will be of a good fize. The beds being marked out with a line, fufficient mould muft be raked out to cover the haws an inch and a half deep. This being done, and the bottom of the beds being made level and even, the haws fhould be fown, and afterwards gently tapped down with the back of the fpade; and then the fine mould, which had been raked out of the beds, muft be thrown over them, covering them an inch and a half deep. In the fpring the plants will come up, and in the fummer following fhould be kept clear of weeds; though it does fometimes happen, that few of them will appear till the fecond fpring after fowing. Sometimes the young plants are planted out from the feed-beds at one, two, or three years old; but the beft plants are obtained by tranfplanting them into frefh mould the firft or fecond year, letting them remain in the nurfery two or three years longer. The practice of the London Nurferymen is this: The ftrongeft of the feed-bed plants having been drawn at two or three years old for fale, they clear the beds entirely by drawing the remaining weak underling plants, and tranfplanting them into frefh beds in this manner (which they call *bedding* them): The ground having been trenched, and the tips of the plants as well as the lower fibres of their roots having been taken off with a fharp knife, they ftrain a line along one fide of the bed; and, by chopping with a fpade by the fide of the line, leave a cleft or drill, of a depth proportioned to the length of the plants to be laid in; and, drawing the loofe mould fomewhat towards them, leave the fide of the drill next to the line with a fmooth polifhed face. Againft this face the plants are fet up, leaning towards the line, about three inches afunder, leaving their heads about an inch above the mould, and placing their roots at fuch a depth as to bury their ftems from two to three inches deeper than they ftood in the feed-bed. The loofe mould being returned and preffed gently to the roots with the foot, the line is removed, and another row planted in the fame manner, about a foot from the firft.

The Common Hawthorn fports in the following *Varieties*:

The Large Scarlet Hawthorn.

The Yellow Hawthorn.

The White Hawthorn.

The

The Maple-leaved Hawthorn,
The Double-bloſſomed Hawthorn.
The Glaſtonbury Thorn.

The Large Scarlet Hawthorn is no more than a beautiful variety of the Common Haw. It is exceedingly large, oblong,
perfectly ſmooth, and of a bright ſcarlet ; and, from the additional ſplendor it acquires by the berries, it is propagated to
cauſe variety in plantations for obſervation and pleaſure.

Yellow Haw is a moſt exquiſite plant, The buds, at their
firſt coming out in the ſpring, are of a fine yellow, and the
fruit is of the colour of gold. The tree is a great bearer, and
retains its fruit all winter, cauſing a delightful effect in plantations of any kind. It was originally brought from Virginia,
is greatly admired, and no collection of hardy trees ſhould be
without it.

White Haw is but a paltry tree, compared with the former.
It hardly ever grows to the height of the Common Hawthorn,
is an indifferent bearer, and the fruit is ſmall, and a very bad
white.

Maple-leaved Hawthorn will grow to be near twenty feet
high, and has very few thorns. The leaves are larger than the
Common Hawthorn, reſemble thoſe of the Maple, and are of a
whitiſh-green colour. The flowers are produced in large
bunches, in June, and are ſucceeded by remarkable fruit, of a
ſhining red, which looks beautiful in the winter.

Double-bloſſomed Hawthorn produces a full flower, and is one
of the ſweeteſt ornaments in the ſpring. Nature ſeems to have
peculiarly deſigned this ſort for the pleaſure-garden ; for though
it be the Common Hawthorn only, with the flowers doubled, yet
it may be kept down to what ſize the owner pleaſes , ſo that it
is not only ſuitable for wilderneſs-quarters, ſhrubberies, and the
like, but is alſo uſeful for ſmall gardens, where a tree or two
only are admitted. Theſe beautiful double flowers come out in
large bunches in May, and the tree is ſo good a bearer, that it
will often appear covered with them. Their colour, at their firſt
appearance, is a delicate white : They afterwards die to a faint
red colour, and are frequently ſucceeded by ſmall imperfect fruit.

Glaſtonbury Thorn differs in no reſpect from the Common
Hawthorn, only that it ſometimes flowers in the winter. It is
ſaid to have originally been the ſtaff of Joſeph of Arimathea,
that

that noble counfellor who buried Chrift. He, according to the
tradition of the abbey of Glaftonbury, attended by eleven com-
panions, came over into Britain, and founded, in honour of the
Bleffed Virgin, the firft Chriftian Church in this ifle. As a
proof of his miffion, he is faid to have ftuck his ftaff into the
ground, which immediately fhot forth and bloomed. This tree
is faid to have bloffomed on Chriftmas-Day ever fince, and is
univerfally diftinguifhed by the name of the Glaftonbury Thorn.
HANBURY fays, I have many plants that were originally propagated from this thorn; and they often flower in the winter,
but there is no exact time of their flowering; for in fine feafons
they will fometimes be in blow before Chriftmas, fometimes
they afford their bloffoms in February, and fometimes it fo
happens that they will be out on Chriftmas-Day.

2. L'AZAROLE. The Azarole Thorn will grow to be
fifteen or fixteen feet high. The leaves are large, nearly trifid,
ferrated and obtufe. The flowers are large, come out in May,
and, in the different varieties, are fucceeded by fruit of different
fize, fhape, and relifh.

The principal *varieties* of this fpecies are, The *Azarole with
ftrong thorns*; the *Azarole with no thorns*; the *Jagged-leaved
Azarole*; the *Oriental Medlar*.

3. The WHITE LEAF. The *Aria Theophrafti*, called the
White-leaf-tree, will grow to be more than twenty feet high.
This tree is engaging at all times of the year, and catches the
attention, even in the winter; for then we fee it ftand, though
naked of leaves, with a fine ftrait ftem, with fmooth branches,
fpotted with white, at the end of which are the buds, fwelled
for the next year's fhoot, giving the tree a bold and fine appearance. In the fpring the leaves come out of courfe, and
look delightfully, having their upper furface green, and the
lower white. Their figure is oval; they are unequally ferrated,
about three inches long, and half as wide. Several ftrong
nerves run from the mid-rib to the border, and they are placed
alternately on the branches, which appear as if powdered with
the fineft meal. The flowers are produced at the end of the
branches, in May; they are white, grow in large bunches,
having meally footftalks, and are fucceeded by red berries, which
will be ripe in autumn.

4. The

4. The WILD SERVICE. The Maple-leaved Service is a large-growing tree. It will arrive to near fifty feet, and is worth propagating for the fake of the timber, which is very white and hard. This tree grows naturally in feveral woods in England; and it is the fruit of this fpecies that is tied in bunches, and expofed for fale in the autumn: It is gathered in the woods, and by fome perfons is much liked. The leaves in fome degree refemble thofe of the Maple-tree in fhape; their upper furface is a fine green, their under hoary; and they grow alternately on the branches. The flowers come out in May, exhibiting themfelves in large clufters at the ends of the branches: They are white, and are fucceeded by the aforefaid eatable fruit, which, when ripe, is of a brown colour, and about the fize of a large haw.

5. VIRGINIAN AZAROLE. This fpecies will grow to be near twenty feet high. The ftem is robuft, and covered with a light-coloured bark. The branches are produced without order, are of a dark brown colour, and poffeffed of a few long fharp thorns. The leaves are fpear-fhaped, oval, fmooth, and ferrated; of a thickifh confiftence, and often remain on the tree the greateft part of the winter. Each feparate flower is large; but as few of them grow together, the umbels they form are rather fmall. They come out in May, and are fucceeded by large dark-red-coloured fruit, which ripens late in the autumn.

The *varieties* of this fpecies are, The *Pear-leaved Thorn*; the *Plum-leaved Thorn with very long ftrong fpines and large fruit*; the *Plum-leaved Thorn with fhort fpines and fmall fruit*.

6. COCKSPUR HAWTHORN. The Virginian Cockfpur Thorn will grow to about twenty feet high. It rifes with an upright ftem, irregularly fending forth branches, which are fmooth, and of a brownifh colour, fpotted thinly with fmall white fpots. It is armed with thorns, that refemble the fpurs of cocks, which gained it the appellation of Cockfpur Thorn. In winter, the leaf-buds appear large, turgid, and have a bold and pleafant look among others of different appearances. In fummer, this tree is very delightful. The leaves are oval, angular, ferrated, fmooth, and bend backwards. They are about four inches long, and three and a half broad; have five or fix pair of ftrong

nerves

nerves running from the midrib to the border; and die to a brownifh-red colour in the autumn. The flowers are produced in very large umbels, making noble fhow, in May; and are fucceeded by large fruit, of a bright red colour, which have a good effect in the winter.

The principal *varieties* of this fpecies are, The *Cockfpur Haw-thorn with many thorns* ; the *Cockfpur Hawthorn with no thorn* ; the *Cockfpur with eatable fruit.* The latter was fent me, fays HANBURY, from America with that name, and I have raifed fome trees of the feed ; but they have not yet produced any fruit, fo that I cannot pretend to fay how far it may be defirable ; though I have been informed it is relifhed in America by fome of the inhabitants there.

7. GOOSEBERRY-LEAVED VIRGINIA HAWTHORN. This fpecies grows to about feven or eight feet high. The branches are flender, and clofely fet with fharp thorns. The leaves are cu-neiform, oval, ferrated, and hairy underneath. The flowers are fmall, and of a white colour : They are produced from the fides of the branches, about the end of May; and are fucceeded by yellow fruit, which ripens late in autumn.

There is a *variety* of this, called the *Carolina Hawthorn,* which has longer and whiter leaves, larger flowers and fruit, and no thorns.

8. GREEN-LEAVED VIRGINIA HAWTHORN. The ftem and branches of this fpecies are altogether deftitute of thorns. The leaves are lanceolate, oval, nearly trilobate, ferrated, fmooth, and green on both fides. The flowers are white, moderately large, come out the end of May, and are fucceeded by a roundifh fruit, which will be ripe late in the autumn.

The refpective fpecies are all PROPAGATED by fowing of the feeds ; and the varieties are continued by budding them upon ftocks of the White Thorn. This latter method is generally practifed for all the forts ; though, when good feeds can be procured, the largeft and moft beautiful plants are raifed that way. 1. In order to raife them from feeds, let thefe be fown foon after they are ripe, in beds of frefh, light, rich earth. Let alleys be left between the beds, for the conveniency of weeding, and let the feeds be covered over with fine mould, about an inch deep,

The

The fummer following, the beds muft be kept clean of weeds, and probably fome few plants will appear : But this is not common in any of the forts ; for they generally lie till the fecond fpring after fowing before they come up. At the time they make their appearance they muft be watered, if the weather proves dry ; and this fhould be occafionally repeated all fummer. They fhould alfo be conftantly kept clean from weeds ; and in the autumn the ftrongeft may be drawn out, and fet in the nurfery-ground, a foot afunder, in rows that are two feet diftant from each other ; while the weakeft may remain until another year. During the time they are in the nurfery, the ground between the rows fhould be dug every winter, and the weeds conftantly hoed down in the fummer ; and this is all the trouble they will require until they are planted out for good, which may be in two, three, or more years, at the pleafure of the owner, or according to the purpofes for which they are wanted. 2. Thefe trees are eafily propagated by budding alfo ; they will all readily take on one another ; but the ufual ftocks are thofe of the Common Hawthorn. In order to have thefe the beft for the purpofe, the haws fhould be got from the largeft trees, fuch as have the feweft thorns and largeft leaves. After they are come up, and have ftood one year in the feed-bed, the ftrongeft fhould be planted out in the nurfery, a foot afunder, and two feet diftant in the rows ; and the fecond fummer after, many of them will be fit for working. The end of July is the beft time for this bufinefs ; and cloudy weather, night and morning, are always preferable to the heat of the day. Having worked all the different forts into thefe ftocks, they may be let alone until the latter end of September, when the bafs matting fhould be taken off. In the winter the ground between the rows fhould be dug, and in the fpring the ftock fhould be headed about half a foot above the bud. The young fhoots the ftocks will always attempt to put out, fhould be as conftantly rubbed off ; for thefe would in proportion ftarve the bud, and ftop its progrefs. With this care, feveral of the forts have been known to fhoot fix feet by the autumn ; and as they will be liable to be blown out of their fockets by the high winds which often happen in the fummer, they fhould be flightly tied to the top of the ftock that is left on for the purpofe, and this will help to preferve them.

C. U.

CUPRESSUS.

LINNEAN Clafs and Order, *Monoccia Monadelphhia* : Male flowers containing four ftamens connected at the bafe, and female flowers containing many piftils ; the males being difpofed in oval catkins ; and the females, collected in roundifh cones upon the fame plant. There are five SPECIES (one of them lately difcovered in Japan) :

1. CUPRESSUS *Sempervirens :* The COMMON CYPRESS ; *a deciduous tree* ; native of Italy, Spain, Portugal and Crete.

2. CUPRESSUS *Thyoides :* The AMERICAN CYPRESS ; or the ARBOR-VITÆ-LIKE CYPRESS, or the SMALL BLUE-BERRIED CYPRESS ; *an evergreen tree or fhrub* ; native of Maryland and Canada.

3. CUPRESSUS *Juniperoides :* The AFRICAN CYPRESS, or the JUNIPER-LIKE CYPRESS, or the CAPE CYPRESS ; *a deciduous tree or fhrub* ; native of the Cape of Good Hope.

4. CUPRESSUS *Difticha .* The DECIDUOUS CYPRESS; *a deciduous tree* ; native of North America.

1. The COMMON CYPRESS. There are two ftriking *varieties* of this plant (MILLAR makes them two diftinct SPECIES) ; namely,

The Upright or Female Cyprefs, and
The Spreading or Male Cyprefs.

There is alfo a third *variety* (which the fame profeffional writer confiders likewife as a diftinct fpecies) ; namely,

The Small-fruited Cyprefs.

The *Upright Cyprefs* is a moft elegant plant, and, notwithftanding it has of late years been fomewhat unfafhionable, it certainly merits a place amongft *ornamental* evergreens. Its conical, or rather fomewhat obelifcal, form makes an agreeable variety with fuller-headed plants. It afpires to a confiderable height, though we believe it feldom fwells to a large girt. However, EVELYN and HANBURY fpeak of this kind of Cyprefs

as

as a timber-tree; but both of them seem to give preference to

The *Spreading Cyprefs.* This grows with a fuller and lefs regular head than the upright fort. MILLAR tells us, that ir the Levant this is the common timber; and recommends the planting of it in England very ftrongly; efpecially upon hot, fandy, or gravelly foils.

The *Small-fruited Cyprefs* is ftill more fpreading than the other, and produces its boughs in an irregular manner. If it is not crowded by other trees, and is lef to nature, it will be feathered from the top to the bottom. It vill grow to about the height of the Common Cyprefs, and is a fort that looks well if planted fingly on grafs-plats, &c. as well as when affifting to form clumps, or larger quarters of evergreens.

2. AMERICAN CYPRESS. This is the loweft grower of all the forts with us; though in America, where it grows naturally, it arrives to timber, which ferves for many excellent purpofes. The talleft of thefe trees feldom rife much higher than fifteen feet; and as this tree is encreafed by cuttings, thofe plants raifed this way feldom rife higher than about nine or ten feet. The branches ftand two ways, and are pretty numerous; and the tree naturally forms itfelf into a regular head. The leaves of this fort are imbricated, like the *Arbor Vitæ*, though fmall, and are of a browner kind of green than the Common Cyprefs. The fruit is very fmall, and of a blue colour, and will be produced in great plenty all over the plant. They are of the fize of the juniper-berry, and much refemble it; though they are cones, and like the other fpecies of this genus, but much fmaller. When thefe plants are raifed from feeds, they will afpire to a greater height, efpecially if planted in a moift foil; but thofe raifed by cuttings generally have the appearance of fhrubs. They are all, however, very beautiful, and greatly embellifh thofe parts of the evergreen plantations where they are ftationed.

3. AFRICAN CYPRESS. The branches of this fpecies are numerous, flender, and fpread themfelves all around. The leaves are narrow, awl-fhaped, about an inch long, of a light-green colour, and grow oppofite to each other on the branches. The flowers come out from the fides of the branches, like the

Common

Common Cyprefs, and they are fucceeded by black fruit; but the feeds never ripen in England.

The method of PROPAGATING the Common Cyprefs is from feeds fown in a warm border, or well-fheltered beds, of light fandy loam, near half an inch deep, in the month of March; and by the beginning of May the plants will be up. After they have come up, if the fummer fhould not prove very dry, they will require little watering; and even in the greateft drought twice a week will be fufficient for them, provided it be done in the evenings. This is the only care they will require the firft fummer, except being kept clean from weeds. In the winter, if the place where they are fown be tolerably well fheltered, they will ftand it very well, though it fhould prove fevere; but where the fituation is not well fheltered by plantations to break the violence of the frofty black winds, they muft be fcreened, otherwife many will be loft. It is the black frofts, attended by high winds, which will deftroy thefe plants; fo that where there is not fhelter enough to break their edge, the beds fhould be hooped over, and covered with mats during that fevere weather. The enfuing fummer the plants may remain undifturbed, when they will require no watering, and no farther care except weeding. The fpring following, being then two years old, they fhould be fet out in the nurfery, exactly at two feet fquare. In taking them out of the feed-bed, fome earth fhould be taken with the roots. The latter end of March is the moft proper time for this work; and if the weather fhould prove dry and cold, as it often happens, the March winds blowing, the work muft be deferred till rainy or cloudy weather; for without thefe precautions, you will find this a difficult plant to remove. After they are planted out in the nurfery, they may be now and then watered in dry weather, kept clean from weeds, and thus may ftand till they are of a fufficient fize to be planted out. They will grow in almoft any *foil*; but above all affect fandy gravelly ground.

With regard to the African and the American Cyprefs, the *feeds* fhould be fown in pots or boxes. We receive them from abroad: They are very fmall, and feldom come up before the fecond fpring; fo that there will be lefs danger of their being loft if they are fown in pots or boxes, which may be fet in the fhade in fummer, and removed into well-fheltered places during the winter.

winter. In the fpring the plants will come up ; and after that the Blue-berried Cyprefs may have the fame treatment as the young feedlings of the Common fort. With refpect to the Cape Cyprefs, the plants muft be fet in pots, to be houfed in winter, until they are grown to be a yard high. When they are turned out into the open air, they fhould have a dry, warm foil, and a well-fheltered place, and even thefe will not enfure their fafety ; fo that whoever is defirous of having thefe trees in his plantations, fhould have fome wooden fconces made, to cover them in frofty weather ; and if this is obferved until they are grown of a tolerable fize, there is no doubt but they will live, in a warm well-fheltered place, through our common winters.

4. The DECIDUOUS CYPRESS will grow to be near fixty feet high, if ftationed in a place fuitable to its nature. It is very hardy in refpect to cold ; and a fhare of the moifteft part of the plantation muft be allotted it. In Virginia and feveral parts of America, where this tree is a native, it is a real aquatic ; being found growing to a very large fize in places wholly cover- ed with water ; and with us, if planted in watery places, by the edges of rivers, ponds, fprings, &c. it will be more luxuriant, and will proportionally rife to a greater height and bulk than if planted in a dry foil. This tree in the fummer has a little the refemblance of an evergreen, and the leaves have a pleafing effect, appearing in fome refpect like fome forts of the Acacias ; and thefe are the chief inducements for its admiffion into the pleafure-ground.

This fpecies may be PROPAGATED from *feeds* in the fame manner as the Common Cyprefs ; alfo from *cuttings* planted in October in a moift fandy foil. Many of them will grow ; though a general crop can feldom be obtained ; and they fhould be kept clean of weeds the fummer following, as well as the fum- mer after that. In the autumn, or any part of the winter, they fhould be planted out in the nurfery ; and, if they are to ftand there a confiderable time, they fhould be allowed a good diftance ; for they will grow, with proper care, when removed at a large fize. If any part of the nurfery-ground is moifter than the other, they muft have a fhare of it. The ground fhould be conftantly dug between the rows every winter, the weeds hoed down in fummer, and, when planted out, thefe

L   trees

trees fhould have moift places, in confequence of what has
been before obferved.

CYNANCHUM.

LINNEAN Clafs and Order, *Pentandria Digynia* : Each
flower contains five males and two females : There are fourteen
SPECIES ; moft of them climbing : Three are fufficiently hardy
for this climate.

1. *Cynanchum Acutum:* The ACUTE-LEAVED CYNANCHUM,
or the ACUTE-LEAVED MONTPELIER SCAMMONY ; *an herbaceous climber* ; native of the South of Europe.

2. *Cynanchum Monfpeliacum* : The ROUND-LEAVED CYNAN-
CHUM, or ROUND-LEAVED MONTPELIER SCAMMONY ; *an herbaceous climber* , native of Spain and the South of France.

3. *Cynanchum Suberofum* : The CAROLINA CYNANCHUM,
or the CAROLINA PERIPLOCA ; *a ligneous climber* ; native of
Carolina and other parts of America.

1. The ACUTE-LEAVED CYNANCHUM. The root is ftrong,
creeping, and fpreads itfelf to a confiderable diftance. The
ftalks are herbaceous, twift about every thing that is near them,
will grow to be fix feet long, but always die to the ground in
the autumn , and frefh ones are put forth from the roots in the
fpring. The leaves are oblong, heart-fhaped, acute-pointed,
fmooth, and grow oppofite by pairs on long footftalks. The
flowers come out from the wings of the leaves in fmall bunches ;
they are of a dirty white colour, appear in June and July, but
are not fucceeded by good feeds in our gardens. This plant, on
being wounded, emits a milky juice.

2. ROUND-LEAVED CYNANCHUM. The root of this fpecies
is large, thin, juicy, and fpreads itfelf to a confiderable diftance.
The ftalks are herbaceous, and twine to fix or feven feet high
about whatever is near them. The leaves are broad, reniform,
roundifh, and grow oppofite, on long footftalks. The flowers
come out from the wings of the leaves, in fmall bunches ; they
are of a bad white colour, appear in June and July, and are
                                                            rarely

rarely fucceeded by good feeds in our gardens. The ftalks die to the ground in the autumn, and frefh ones arife again in the fpring. On wounding any part of this plant, a milky juice immediately flows.

3. CAROLINA CYNANCHUM. The ftalks of this fpecies are flender, ligneous, fhrubby, and will twift about any thing to the height of about feven feet. They are hairy, and their lower part is covered with a thick, fungous, cloven, cork-like bark. The leaves are oval, heart-fhaped, pointed, and grow oppofite at the joints on long hairy footftalks. The flowers come out from the wings of the leaves in fmall bunches.They are greenifh on their firft appearance, but die away to a bad purple. They exhibit themfelves in July and Auguft; but are not fuc-ceeded by good feeds in our gardens.

This fort is PROPAGATED by laying down the young fhoots as they advance in the fummer, and covering them over with fome fine mould. Thefe will foon put out roots, by the autumn will be good plants, and may then be removed to the places where they are defigned to remain. This fpecies is rather tender; and the foil in which it is planted fhould be naturally dry, warm, light, and fandy, and the fituation well defended. Being thus ftationed, it will live abroad, and continue for many years; but if the foil is moift, rich, and ill-defended, the chance will be very great but it will be deftroyed the firft winter.

The two firft forts are exceedingly hardy, will grow in any foil or fituation, and will over-run any fmall plants that are near them. Their fituation, therefore, fhould be among fuch trees as have ftrength enough to admit their embraces; and their propagation is by cutting the roots in the autumn. Every cut will grow; and, when planted, will call for no trouble except keeping them clear from weeds, when they firft fhoot up in the fpring.

C Y T I S U S.

LINNEAN Clafs and Order, *Diadelphia Decandria :* Each flower contains ten males and one female; the males rifing in

L 2

two

two divifions : There are fourteen SPECIES; five of which afford confiderable ornament to the Englifh garden.

1. *Cytifus Seffilifolius :* The SESSILE-LEAVED CYTISUS (or TREFOIL TREE, or BASE TREE TREFOIL), or CYTISUS SE-CUNDUS CLUSII, or the SMOOTH ROUND-LEAVED CYTISUS ; *a deciduous fhrub* ; native of France, Italy, and Spain.

2. *Cytifus Nigricans:* The BLACK CYTISUS ; or the BLACKISH SMOOTH CYTISUS ; *a deciduous fhrub* ; native of Auftria, Bohe-mia, Italy and Spain.

3. *Cytifus Auftriacus :* The TARTARIAN CYTISUS; or the AUSTRIAN CYTISUS ; *a low deciduous fhrub* ; native of Auftria, Siberia and Italy.

4. *Cytifus Laburnum :* The LABURNUM ; *a deciduous tree* ; native of Switzerland, Savoy, and moft parts of Europe.

5. *Cytifus Hirfutus :* The EVERGREEN CYTISUS ; or the EVERGREEN CYTISUS OF NAPLES ; or the ITALIAN CYTISUS WITH HAIRY LEAVES ; *an evergreen fhrub* ; native of Italy, Spain, Auftria and Siberia.

1. The SESSILE-LEAVED CYTISUS will grow to the height of about five or fix feet. The branches are numerous, erect, very brittle, and covered over with a fmooth brown bark. The leaves are fmall, and of a fine green : They are nearly of an oval figure, and grow by threes on the twigs ; on fome branches they fit quite clofe, on others they grow on very fhort footftalks. The flowers grow at the ends of the branches, in fhort fpikes : They are of a fine yellow, come out the begin-ning of June, and when in full blow the fhrub will appear al-moft covered with them. The feeds ufually ripen in Auguft.

2. BLACK CYTISUS will arrive to about the height of the former, and naturally divides into many branches. The bark is brown, and the young fhoots are of a greenifh-red. The leaves refemble Trefoil : They are fmooth, and grow three together on brownifh footftalks ; the folioles are of an oblong oval figure, and their upper furface is of a dark-green, but they are paler underneath. The flowers are produced in long, erect, clofe fpikes, at the ends of the branches : They are of a beautiful yellow colour, come out in July, and when in full blow make a fine appearance. The feeds ripen in the autumn.

TAR-

3. TARTARIAN CYTISUS. The ſtalks are ſhrubby, branching, green, and grow to three or four feet high. The leaves are oval, oblong, ſmooth, and of a whitiſh-green colour. The flowers come out in cloſe heads from the ends of the branches, in May : They are of a light-yellow colour, and have a cluſter of leaves under them; they are ſometimes ſucceeded by ſhort woolly pods, containing the ſeeds.

There is a *variety* of this ſpecies, with naked ſtalks, ſmaller leaves and flowers, rather earlier in the ſpring, uſually called the *Siberian Cytiſus.*

4. The LABURNUM is a large-growing tree : It will aſpire to the height of near forty feet, and is one of the moſt noble trees our gardens afford. It will form itſelf into a fine head; its branches are ſmooth, of a pale-green colour, and poſſeſſed of a few greyiſh ſpots. The leaves ſtand by threes on long ſlender footſtalks : Each of theſe is oblong and entire; their upper ſurface is ſmooth, and of a ſhining green, but their under ſurface is more inclined to be downy. The time of this tree's flowering is May ; and the effect can hardly be conceived which it will have, when it appears covered with its long pendulent bunches of flowers, of a delightful yellow. Each flower that helps to compoſe one ſet is tolerably large of itſelf, and the common ſtalk to which they adhere by their own ſeparate footſtalks is often a foot or more in length ; ſo that the appearance muſt be moſt noble, when it exhibits theſe long ſeries of flowers hanging down from almoſt every part of the whole head : HANBURY continues, " But this is not all ; the timber when felled is exceeding valuable. It will arrive in bulk in proportion to its height ; and the timber is both heavy and hard, and of a fine colour, inclined to yellow. The very branches of this tree are ſo ponderous as to ſink in water. It poliſhes extremely well, and is ſo much like to green ebony, that it is called by the French, *Ebony of the Alps,* where the tree grows naturally. And as the timber is ſo valuable for many ſorts of rich furniture, this ſhould arouſe the timber-planter's attention ; for it will grow to be a timber-tree of more than a yard in girt, in almoſt any poor and ſorry ſoil, where other trees will hardly grow, let the ſituation be what it will : And how enchantingly ornamental muſt large quarters or clumps of theſe trees appear,

either

either by the borders of other woods, or in parks, and at the same time the expectation of the timber-crop retained !"

There are some other forts of LABURNUMS, of equal or more beauty than the preceding : One is called the *Scotch Laburnum*, another the *Italian*. The leaves of thefe are larger, and the bunches of flowers longer ; and the individual flowers of which the bunches are compofed proportionally larger. There is alfo another fort, with fmaller leaves, and bunches longer than the common, which difference it always preferves from feeds ; and thefe being planted among the common fort, will afford the greater variety.

One method of PROPAGATION is common to all thefe forts : It is to be performed both by feeds and cuttings. 1. When by feeds, common garden-mould, when dug, and cleared of the roots of all weeds, will do for their reception. They fhould be fown in the fpring, in beds neated up, about half an inch deep, and in about fix weeks the young plants will appear. Nothing more will be neceffary than keeping them clean from weeds during the fummer, unlefs the weather proves very dry ; if it does, a little watering fometimes will be proper. The fpring following, the *Laburnums* fhould be planted out in the nurfery ; but the other forts fhould ftand in the feed-bed two years, to gai ftrength, before they are taken up. Thefe fhould be planted a foot afunder, and two feet diftant in the rows ; but the *Laburnums* ought to have a rather greater diftance, efpecially if they are defigned to be trained up for ftandards. 2. Another method of encreafing thefe forts is by cuttings. October is the beft month for the work ; and the cuttings may be planted either a foot afunder, and two feet diftant in the rows, fo that they need not be removed till they are taken up for good ; or they may be fet very thick, and thofe which live taken up the winter following, and planted out in the nurfery-way, at diftances wide in proportion to the time they are to ftand.

5. The EVERGREEN CYTISUS. This fhrub is naturally of an upright growth, and its common height is about fix or feven feet. It may be trained up to a fingle ftem, for two, three, or four feet high, and will naturally fend out many branches, which will form themfelves into a fine head. The bark on the

ftem

ftem is of a grey colour; the branches alfo are grey with a
green caft at a diftance; and many of them will have the appearance of being channelled, the bottom of the grooves being
of a dufky green, but their upper edges white. The younger
fhoots are green and ftreaked, and their furface is hairy. The
leaves alfo have this property, and ftand three upon a fhort footftalk. They are nearly of an oval figure, and have a ftrong
mid-rib running the whole length. They are of a fine green
colour, and clothe the fhrub with great beauty. The flowers
are of a clear yellow colour, and are fhaped like thofe of the
other forts : They appear in June, and are produced from the
fides of the branches, all over the fhrub, in fhort bunches ; fo
that its golden head at that time is both beautiful and ftriking.
Neither is June the only time of its flowering ; for it will often
flower-again in October ; and, if the winter continues open
and mild, it will fometimes fhew its bloffoms in November and
December. The flowers that appeared in June, which is its
regular time of blow, will be fucceeded by fmall hairy pods,
in which the feeds are contained, and which ripen with us very
well in the autumn.

This fort fhould be PROPAGATED by feeds, which fhould be
fown in the fpring, and managed as directed for the deciduous
forts ; only it may not be amifs to obferve, that it will be neceffary to plant the feedlings in the nurfery when they have ftood
one year in the feed-bed. They fhould be fet about a foot
afunder, in rows at two feet diftance ; and here they may ftand
for about two years, when they fhould be planted out.

D A P H N E.

LINNEAN Clafs and Order, *Octandria Monogynia :* Each
flower contains eight males and one female : There are fifteen
SPECIES ; eight of which are proper for our collection.

1. *Daphne Mezereum :* The MEZEREON, or SPURGE-OLIVE ;
a low deciduous fhrub ; native of Germany ; and has been difcovered in this country in fome woods near Andover, in Hamp-
fhire.

2. *Daphne*

2. *Daphne Gnidium:* The FLAX-LEAVED DAPHNE, or FLAX-LEAVED THYMELÆA ; *a low deciduous shrub* ; native of Italy, Spain, and about Montpelier.

3. *Daphne Cneorum:* The SPEAR-LEAVED DAPHNE,, or the CNEORUM ; or the CLUSTER-FLOWERING SPEAR-LEAVED DAPHNE ; *a very low deciduous shrub* ; native of Switzerland, Hungary, the Alps and the Pyrenean Mountains.

4. *Daphne Tartonraira:* The OVAL-LEAVED DAPHNE, or the TARTON-RAIRE ; or CLUSTER-FLOWERING OVAL-LEAVED DAPHNE ; *a very low deciduous shrub* ; native of France and Italy.

5. *Daphne Alpina:* The ALPINE DAPHNE ; or the ALPINE CHAMELÆA ; a *low deciduous shrub* ; native of the Alps, Geneva, Italy and Austria.

6. *Daphne Thymelæa:* The MILKWORT-LEAVED DAPHNE, or the THYMELÆA ; a *low deciduous shrub* ; native of Spain and the South of France.

7. *Daphne Villosa:* The HAIRY-LEAVED DAPHNE, or the SMALL HAIRY PORTUGAL DAPHNE ; a *very low deciduous shrub* ; native of Spain and Portugal.

8. *Daphne Laureola:* The SPURGE LAUREL, or the EVER-GREEN DAPHNE ; a *low evergreen shrub* ; common in some parts of this kingdom, also in Switzerland and France.

1. The MEZEREON. Of this elegant plant there are four *Varieties:* 1. The *White.* 2. The *Pale-red.* 3. The *Crimson.* and, 4. The *Purple-flowering.*—HANBURY is very lavish of his praise of these shrubs ; he says, " they have each every per-fection to recommend them as flowering-shrubs. In the first place, they are of low growth, seldom arising to more than three or four feet in height, and therefore are proper even for the smallest gardens. In the next place, they will be in bloom when few trees, especially of the shrubby tribe, present their honours. It will be in February, nay, sometimes in January ; then will the twigs be garnished with flowers, all around, from one end to the other. Each twig has the appearance of a spike of flowers of the most consummate lustre ; and as the leaves are not yet out, whether you behold this tree near or at a distance, it has a most enchanting appearance. But this is not all ; the sense of smelling is peculiarly regaled by the flowers ; their spicy sweetness is diffused around, and the air is perfumed with their odours to a considerable distance. Many flowers, deemed sweet, are not liked by all ; but the agreable inoffensive sweetness

of the *Mezereon* has ever delighted the senfe of smelling, whilst the lustre of its blow has feasted the eye. Neither is this the only pleasure the tree bestows; for besides the beauty of the leaves, which come out after the flowers are fallen, and which are of a pleasant green colour and an oblong figure, it will be full of red berries in June, which will continue growing till the autumn. Of these berries the birds are very fond; so that whoever is delighted with those songsters, should have a quantity of them planted all over the outsides of his wilderness quarters."

PROPAGATION. This sort ripens its feeds with us, and may at any time be easily obtained, if they are secured from birds. Previous therefore to sowing, the healthiest and most thriving trees of the White, the Pale, and the Deep-red sorts should be marked out; and as soon as the berries begin to alter from green, they must be covered with nets, to secure them from the birds, which would otherwise devour them all. The berries will be ripe in July; and due observance must be had to pick them up as they fall from the trees, and to keep the sorts separate. As soon as they are all fallen, or you have enough for your purpose, they may then be sown. The best soil for these plants is a good fat black earth, such as is found in kitchen-gardens that have been well manured and managed for many years. In such soil as this they will not only come up better, but will grow to a greater height than in any other. No particular regard need be paid to the situation; for as this tree is a native of the northern parts of Europe, it will grow in a north border, and flourish there as well as in a south; nay, if there be any difference, the north border is more eligible than the south. The ground being made fine, and cleared of roots of all sorts, the feeds should be sown, hardly half an inch deep. The mould being riddled over them that depth, let the beds be neated up, and they will want no other attention until the spring. These feeds will sometimes remain in the ground two years; but for the most part they come up the spring after sowing; and the feedlings will require no other care during the summer than weeding, and gentle watering in dry weather. After they have been in the feed-bed one year, the strongest may be drawn out, and planted in the nursery, to make room for the others; though if they do not come up very close, it would be as well to let them remain

in

in the feed-bed until the fecond autumn: when they fhould be taken up with care, and planted in beds at a foot afunder each way. This will be diftance enough for thefe low-growing fhrubs. October is the beft month for planting them out finally; for although they will grow if removed any time between then and fpring, yet that will certainly be a more proper feafon than when they are in full blow. Such is the culture of this fhrub. The other fpecies of this genus require a different management.

2. FLAX-LEAVED DAPHNE feldom grows higher than three feet. The branches are very flender, and ornamented with narrow, fpear-fhaped, pointed leaves, much like thofe of the Common Flax. The flowers are produced in panicles at the ends of the branches: They are fmall, come out in June, but are rarely fucceeded by feeds in England.

3. SPEAR-LEAVED DAPHNE, or CNEORUM. This rifes with a fhrubby, branching ftalk, to about a foot or a foot and a half high. The leaves are narrow, fpear-fhaped, and grow irregularly on the branches. The flowers are produced in clufters at the ends of the little twigs: They make their appearance in March, are of a purple colour, and poffeffed of a fragrance little inferior to that of the *Mezereon*; but they are feldom fucceeded by feeds in England.

4. OVAL-LEAVED DAPHNE, or TARTON-RAIRE. This rifes with a woody ftalk to the height of about two feet. The branches are numerous, irregular, tough, and covered with a light-brown-coloured bark. The leaves are oval, very fmall, foft to the touch, and fhining. The flowers are produced in clufters from the fides of the ftalks: They are white, come out in June, and are fucceeded by roundifh berries, which feldom ripen in England. This fort fhould have a dry foil and a warm fituation.

5. The ALPINE DAPHNE, or CHAMELÆA, will grow to the height of about a yard. The leaves are fpear-fhaped, obtufe, and hoary underneath. The flowers come out in clufters from the fides of the branches, and are very fragrant: They appear in March, and are fucceeded by red berries, that ripen in September.

6. MILKWORT-LEAVED DAPHNE, or THYMELÆA, will grow to the height of a yard. The ftalks of this fpecies are upright, branched, and covered with a light-brown bark. The leaves are fpear-fhaped, fmooth, and in fome refpect refemble thofe of Milk-
wort.

wort. The flowers are produced in clusters from the sides of the stalks : They are of a greenish colour, have no footstalks, appear in March, and are succeeded by small yellowish berries, which will be ripe in August. This sort requires a dry soil and a warm situation.

7. HAIRY-LEAVED DAPHNE. The stalks are ligneous, about two feet high, and send forth branches alternately from the sides. The leaves are spear-shaped, plane, hairy on both sides, and grow on very short footstalks. The flowers have very narrow tubes, are small, and make no great show : They come out in June, and are not succeeded by ripe seeds in England. This shrub, in some situations, retains its leaves all winter in such beauty as to cause it to be ranked among the low-growing evergreens ; but as in others it is sometimes shattered with the first black winds, it is left to the Gardener whether to place this shrub among the Deciduous Trees or Evergreens.

All these sorts are with some difficulty PROPAGATED and retained. They will by no means bear removing, even when seedlings ; and if ever this is attempted, not one in an hundred must be expected to grow. They are raised by seeds, which we receive from the places where they grow naturally ; and he who is desirous of having these plants, must manage them in the following manner : Let a compost be prepared of these equal divisions ; one-fourth part of lime rubbish ; one-fourth part of drift or sea sand ; another of splinters of rocks, some broad and others smaller ; and the other part of maiden earth, from a rich pasture. Let these be mixed all together, and filled into largish pots. In each of these pots put a seed or two, about half an inch deep, in the finest of the mould. We receive the seeds in the spring ; so that there is little hopes of their coming up until the spring following : Let, therefore, the pots be set in the shade all the summer, and in the autumn removed into a warm situation, where they may enjoy every influence of the sun's rays all winter. In March let them be plunged into a moderate hot-bed, and the plants will soon after appear. This bed will cause them to be strong plants by the autumn ; and when all danger of frost is over, they may be uncovered wholly, and permitted to enjoy the open air. In the autumn, they should be removed into the greenhouse, or set under an hot-bed frame all winter ; and in spring they should be placed

where

where they are to continue, moulding them up the height of the pot; the pots being fufficiently broken to make way for their roots, as they fhoot, and then left to Nature.—The fituation of the four tenderer forts muft be well fheltered, and if it be naturally rocky, fandy, and dry, it will be the better; for in the places where they grow naturally, they ftrike into the crevices of rocks, and flourifh where there is hardly any appearance of foil.

This is one method of obtaining thefe fhrubs. Another way is, by fowing the feeds in the places where they are to remain. The fituation and nature of the foil fhould be as near that above defcribed as poffible; and the mould fhould be made fine in fome places, and a feed or two fown in each. After this, pegs fhould be ftuck down on each fide of them, to direct to the places where they are fown. The exacteft care muft be obferved, all fummer, to pull up the weeds as often as they appear; for if they are permitted to get ftrong, and have great roots, they will pull up the feeds with them. In the fpring following, if the feeds are good, the plants will appear. During the fummer, they fhould be watered in dry weather; and, for the firft winter or two, fhould have fome furze-bufhes pricked all round them, at a proper diftance, which will break the keen edge of the frofty winds, and preferve the young plants until they are ftrong enough to defend themfelves.

The CNEORUM and the ALPINE CHAMELÆA are very hardy, and will grow in the coldeft fituation; but the other forts fhould have a warm foil and a well-fheltered fite, or they will be fubject to be deftroyed in bad weather.

8. The SPURGE-LAUREL, or EVERGREEN DAPHNE, is a low fhrub, feldom growing more than a yard or four feet high; it fends out many branches from the bottom, and thefe are covered with a fmooth light-brown bark, that is very thick. The bark on the younger branches is fmooth and green; and thefe are very clofely garnifhed with leaves of a delightful ftrong lucid green colour. Thefe leaves fit clofe to the branches, and are produced in fuch plenty, that they have the appearance, at a fmall diftance, of clufters at the ends of the branches. They are fpear-fhaped, fhining, fmooth, and thick; their edges are entire. HANBURY extols this plant with a degree of enthufiafm; continuing, " and this is another excellent property of this tree, that it is thus poffeffed of fuch delightful leaves for its ornament. Thefe

leaves,

leaves, when growing under the drip of trees, spread open, and exhibit their green pure and untarnished, in its natural colour : when
planted singly in expofed places, they naturally turn back with a
kind of twift, and the natural green of the leaf is often alloyed
with a brownish tinge. This fhrub is alfo valuable on account
of its flowers ; not becaufe they make any great fhow, but from
their fragrance, and the time they appear; for it will be in blow
the beginning of January, and will continue fo until the middle
or latter end of April before the flowers fall off ; during which
time they never fail to diffufe abroad their agreeable odours,
which are refrefhing ard inoffenfive. In the evenings efpecially,
they are more than commonly liberal ; infomuch that a few plants
will often perfume the whole end of a garden ; and when this
happens early, before many flowers appear, the unfkilful in
flowers, perceiving an uncommon fragrancy, are at once ftruck
with furprize, and immediately begin enquiring from whence it
can proceed. Neither are its odours confined to a garden only ;
but, when planted near windows, they will enter parlours, and
afcend even into bed-chambers, to the great comfort of the poffeffor, and furprize of every frefh vifitor." Thefe flowers make
but little fhow ; for they are fmall, and of a greenifh-yellow.
They are produced amongft the leaves from the fides of the
ftalks, in fmall clufters, and will often be fo hid by them, as to
be unnoticed by any but the curious. They are fucceeded by
oval berries, which are firft green, and afterwards black when ripe.
Thefe berries will be in fuch plenty as to be very ornamental ;
but will foon be eaten up by the birds ; which is another good
property of this tree, as it invites the different forts of whiftling
birds to flock where it is planted in great plenty.

This fhrub is PROPAGATED by feeds, in the fame manner as
the Common *Mezereon*. The feeds muft be preferved from the
birds by nets, until they are ripe. Soon after, they muft be fown
as is directed for the *Mezereon*. They will often be two years
before they come up ; during which time, and afterwards, they
may have the fame management as has been laid down for the
Common *Mezereon*, until they be finally fet out.

This fhrub will grow in almoft any foil or fituation, but
flourifhes moft under the fhade and drip of taller plants, giving
a peculiar chearfulnefs to the bottoms of groves and clumps in
winter.

D I O S.

D I O S P Y R O S.

LINNEAN Clafs and Order, *Polygamia Dioecia:* Some of the plants of this genus bear hermaphrodite and female flowers upon the fame individual, whilft others bear male flowers only ; each of which contains eight ftamina. There are five SPECIES ; three of which are of late difcovery : The other two are,

1. *Diofpyros Lotus:* The INDIAN DATE-PLUM ; *a low deci-duous tree* ; native of Africa and the South of Europe.

2. *Diofpyros Virginiana:* The PISHAMIN-PLUM ; *a tall de-ciduous fhrub* ; native of Virginia, Carolina, and many parts of North-America.

1. The INDIAN DATE-PLUM will arrive at the height of more than twenty feet, and is an excellent tree for fhade. It afpires with an upright ftem, and the young branches are covered with a fmooth whitifh bark. The youngeft twigs ftand alter-nately on thofe of the preceding year, and the buds for the next year's fhoot begin to fwell foon after the fall of the leaf. The leaves are of two colours ; their upper furface is of a delightful green, and their lower of a whitifh caft. They are of an oblong figure, end in a point, and are in length about four inches and a half, and near two inches broad. They are placed alternately on the branches, and feveral ftrong veins run alternately from the mid-rib to the borders, which are entire. Thefe leaves will be of a deep green, even when they fall off in the autumn. The flowers have little beauty to recommend them : they are pitcher-fhaped, and grow fingly on fhort foot-ftalks, on the fides of the branches : they are of a reddifh colour, and are fucceeded by largifh black berries, which are eatable, like the medlar, when in a ftate of decay.

2. The PISHAMIN-PLUM will not afpire to the height of the former fpecies, though it will fometimes grow to near twenty feet. The branches of this tree are whitifh, fmooth, and produced in an irregular manner. The leaves are very large and beautiful ; about five or fix inches long, and three broad. Their upper fur-face is fmooth, and both fides are of a beautiful green. They are of an oblong figure, end in a point, grow irregularly on the branches, and have feveral veins running from the mid-ribs to

the

the borders, which are entire. They fall off in the autumn, at
the coming-on of the firft frofts, when their colour will be that
of a purplifh red. The flowers, like thofe of the other fort, make
no great appearance ; but are fucceeded by a fruit, which is eatable, when, like medlars, it is in a ftate of decay.

Both thefe forts are PROPAGATED from the feeds, which we
receive from abroad, in the fpring. The compoft proper for their
reception is maiden earth, from a rich pafture, dug up fward and
all a year before, and three or four times turned in order to rot
the fward. This being made fine, a fourth part of drift or fea
fand fhould be added ; and being all well mixed, the feeds fhould
be fown in pots or boxes, three quarters of an inch deep. The
pots fhould afterwards be placed in a fhady place during the fummer ; for the feeds rarely come up until the fecond fpring ; and
in the autumn they fhould be removed into a well-fheltered
place, where they may enjoy the benefit of the fun all winter.
In the fpring the plants will come up ; and if they are affifted by
plunging the pots into a moderate hotbed, it will make them
fhoot ftronger ; though this is not abfolutely neceffary. All the
fummer they fhould ftand in a fhady place, where they may
have free air ; and, if the weather proves dry, they fhould be
watered every other evening. At the approach of winter, they
fhould be removed into the greenhoufe, or placed under an hot-
bed-frame, or fome fhelter ; and, when all danger of froft is over,
they muft be put in the fame fhady fituation as in the former
fummer. In the winter alfo they fhould be hooped as before ;
and in fpring may be planted in the nurfery-ground. Thefe
plants, when they get tolerably ftrong, are very hardy ; though
even then the ends of the branches are fubject to be killed; fo
that when they are feedlings, or very young, they will be in
danger of being deftroyed by the frofts, which makes the above-
directed care and protection neceffary till they have gained
ftrength.

E L Æ A G N U S.

LINNEAN Clafs and Order *Tetrandria Monogynia* : Each flower contains four males and one female : There are four SPECIES ; two of which have been introduced into this country ; one of them requiring a ftove heat ; the other fufficiently hardy to bear the open air ; namely,

Elæagnus Anguftifolia : The NARROW-LEAVED ELÆAGNUS, or the OLEASTER, or the WILD OLIVE ; a *tall deciduous fhrub* ; native of Bohemia, Spain, Syria, and Cappadocia.

The NARROW-LEAVED ELÆAGNUS, or the OLEASTER, will grow to be near twenty feet high. Whilft the leaves of moft trees are poffeffed of a verdure, and occafion variety by the dif_ference of greens they exhibit, the leaves of the plant under con- fideration are white, efpecially the under-fide, and ftand upon white twigs. The branches are of a brown colour ; but the pre- ceding year's fhoots are white and downy, the filvery leaves being placed irregularly upon them : Thefe are of a fpear-fhaped figure, about two, and fometimes three inches long, and three quarters of an inch broad, and are as foft as fattin to the touch. Neither is fummer the only time the leaves afford us pleafure : They continue on the tree great part of the winter ; fo that the effect they caufe, when other trees are defpoiled of their honours, may be eafily conceived. The flowers appear in July, but make no figure : They are fmall, and come out at the footftalks of the leaves ; their colour is white, and they are poffeffed of a ftrong fcent. The fruit that fucceeds them much refembles a fmall olive.

This fhrub has a *variety*, with yellow flowers.

The culture of both the forts is very eafy. They are PROPAGATED by cuttings, which muft be of the laft fummer's fhoot. But in order to have them proper for the purpofe, a fuffi- cient number of trees muft be fixed on, from which the family is to be encreafed. They muft be headed near the ground in the winter ; which will caufe them to make ftrong fhoots the fucceed- ing fummer, and thefe fhoots afford the cuttings. They fhould be taken off in the autumn, and cut into lengths of about a foot each, three parts of which fhould be fet in the ground. They

may

may be planted very clofe, and in the autumn following removed
into the nurfery, where they fhould be fet a foot afunder, and
two feet diftant in the rows ; or, if there be ground enough, they
may be planted thinner, and fo will want no removing until they
be finally fet out. The beft foil for thefe cuttings is a rich
garden mould, inclined to be moift, and lying in a fhady place ;
in fuch a foil and fituation almoft every cutting will grow. The
tree itfelf is exceedingly hardy, and will afterwards fhoot vigoroufly, in almoft any foil or ftation.

E P H E D R A.

LINNEAN Clafs and Order, *Dioecia Monadelphia:* Male
flower, containing feven ftamens connected at the bafe, and female flowers containing two piftils, fituated upon diftinct plants.
There are two SPECIES ; one of them of a hardy nature ;—

EPHEDRA DISTACHYA : The EPHEDRA, or SHRUBBY HORSE
TAIL ; *a fub-evergreen fhrub* ; native of rocky mountains, near
the fea-coaft of Italy, France, and Spain.

The EPHEDRA will grow to three, four, five, or fix feet high,
according to the nature of the foil in which it is 'aced ; for if
it be a fat moift foil, it will arrive to double the height it will
attain in that of a contrary nature, and will be more tree-like ;
it will alfo have much larger leaves, and be more beautiful.
The bark on the old ftem is rough, and of a dark, dirty colour.
Thefe ftems or branches are few ; but they have joints at fhort
intervals. Many of them are protuberant, and fend forth younger
fhoots and leaves in prodigious plenty, fo as to caufe the fhrub to
have a clofe bufhy look. The older branches will have bark
that is fmooth, and of a brown, reddifh, or yellowifh colour; whilft
that on the younger fhoots will be of a fine green. The larger
branches are jointed and hollow, though they have fometimes in
them a kind of reddifh pith ; thofe fend forth fmaller, which are
called the leaves. Thefe leaves are jointed, grow oppofite by pairs,
are alternately produced at every joint in oppofite directions, and
will thus branch out in a fingular and horfe-tail manner, in
a fuitable foil, to a great length. The leaves and fhoots of this
fhrub being bruifed in the winter, emit a very fœtid difagreeable

M　　　　　　　　　fcent ;

scent ; but in the spring when the juices begin ɔ flow, they are
possessed of a different quality, emitting a fine odour, by many
supposed or fancied to be like that of the pine-apple ; and on account of this scent alone, in the spring, this tree is by many much
coveted and admired. The flower-buds will appear in May opposite at the sides of the joints ; they grow by pairs, and by the
middle of June will be in full blow, each standing on very short
green foot-stalks. Male and female flowers will be found on different plants; they are small, and of a yellow colour, and afford
pleasure only to the nice observer of the wonderful structure of
the minute parts of the vegetable world. This shrub should
always have a moist, fat soil ; and in those places it will appear
more luxuriant and beautiful. It is very hardy, and, although
it has been used to be preserved in pots in greenhouses, will bear
the cold of our severest winters. In the winter the leaves, or rather the young shoots or joints, are of a dark, dusky green ; but as
the spring approaches, that goes off, and a fine, lively, chearful
green possesses the whole plant. The old leaves fall off the latter
end of April, or beginning of May ; at which time the tree will
send forth young ones, and will continue to do so until late in the
autumn.

This shrub is very easily PROPAGATED ; it will, indeed, propagate
itself in great plenty, especially if planted in a light, moist soil :
So that where a quantity is wanted, some plants are to be procured for breeders ; and these being planted in good light earth,
will soon spread their roots, and produce plenty of *suckers*, which
may be taken off, and planted in the nursery-ground, to gain
strength, for a year or two ; or they may be immediately, especially the strongest plants, finally set out. As these shrubs naturally spawn, and produce suckers in great plenty, after they are
planted out in the shubbery-quarters, the spawn should be
every year taken off, and the ground dug about the roots ; otherwise they will not only appear rambling and irregular, but they
will diminish the beauty of the mother-plants, which will by no
means appear to be luxuriant and healthy.

This plant merits a place amongst evergreens, rather for the
sake of variety, or as a foil to more elegant species, than for any
intrinsic beauty or elegance of its own. MILLAR says, it rarely
flowers in gardens.

E U O N Y M U S.

Linnean Clafs and Order, *Pentandria Monogynia* : Each flower contains five males and one female : There are four Species; three of which are cultivated in this country; one of them, however, requires a ftove-heat : The other two are,

1. *Euonymus Europæus :* The Common or European Euonymus, or the Spindle Tree ; *a deciduous fhrub or tree* ; native of fome parts of England, and of Europe in general.

2. *Euonymus Americanus :*—The Evergreen or American Euonymus ; or the Evergreen Spindle ;—*an evergreen fhrub* ; native of Virginia, Carolina, and other parts of North-America.

1. The Common Euonymus. There are of this fpecies five *varieties :*

The Deep Red-berried Narrow-leaved Spindle Tree.
—-- Pale Red-berried Spindle Tree.
—-- White-berried Narrow-leaved Spindle Tree.
—-- Broad-leaved Spindle Tree.
—-- Variegated Spindle Tree *.

The *Narrow-leaved Spindle Tree* will grow to be fixteen or eighteen feet high, will afpire with an upright ftem to a confiderable height, naturally forming itfelf into a regular head. The bark of the ftem is of a dark-brown ; but that of the firft and fecond years fhoots is fmooth, and of a fine green, the White-berried fort efpecially, which differs from the Red-berried in this refpect, as the fhoots of that are browner. The leaves are fpear-fhaped, of a fine deep green colour, about three inches long, and an inch and a half broad, very flightly ferrated, and placed nearly oppofite on the branches. The flowers have little beauty to recommend them : They are fmall, and of a greenifh colour, produced in fmall bunches from the fides of the branches, the latter end of May, the bunches hanging on long footftalks ; and are fucceeded by fruit, which conftitutes the greateft beauty of thefe plants.

* Millar makes the two laft diftinct *Species*; but Hanbury fays, " I have raifed thoufands of them for fale (there being hardly any fhrub more called for), and ever found the feeds of the Broad-leaved Spindle Tree to come up the Common Narrow-leaved fort."

The feeds are of a delightful fcarlet; four are contained in each veffel; and thefe opening, expofe them to view all over the head of the plant, fome juft peeping out of their cells, others quite out, and fticking to the edge; and thefe veffels being in bunches on long pendulent footftalks, have a look which is fingularly beautiful. The feed-veffels of the firft-mentioned fort are of the fame deep fcarlet with the feeds; thofe of the fecond, of a paler red; thofe of the third are white, which, together with the twigs of the latter being of a lighter green, conftitute the only difference between thefe forts; for the feeds themfelves of all the forts are of a deep fcarlet.

The *Broad-leaved Spindle Tree* is a *variety* of the Common Spindle Tree, though it will grow to a greater height than either of the other forts. It will arrive at near five-and-twenty feet high; and the branches are fewer, and the leaves broader. The young fhoots are fmooth, and of a purplifh colour; and the buds at the ends of them, by the end of October, will begin to be fwelled, and be near an inch long, preparing for the next year's fhoot. The leaves are much larger than thofe of the other forts, being, on a thriving plant, near five inches long and two broad. Their figure is like the other, though rather inclined to an oblong oval: Some are moft flightly ferrated, of a light-green, ftand oppofite by pairs, and fall off much fooner in the autumn, before which their colour will be red. The flowers make an inconfiderable figure, though they are rather larger than the other forts: The feeds that fucceed them with their veffels alfo are proportionably larger; and many of the common footftalks to each bunch will be four inches, which caufes a more noble look in the autumn; though the others are equally pleafing, as the flowers are produced on the Narrow-leaved forts in greater plenty: Add to this, the berries of the Broad will fall off long before the others.

The wood of the Common Spindle Tree is fpoken of by MILLAR and HANBURY as being very valuable. The mufical-inftrument-makers, fay they, ufe it for keys of organs, and other purpofes. Tooth-picks, fkewers, and *fpindles* of the beft kind are alfo made from this wood; hence *Spindle Tree.*

There is but one good method of PROPAGATING the Common Spindle Tree, and that is by feeds; though it may eafily be done by layers or cuttings for if the young fhoots be laid in the ground

in

In the autumn, they will have ftruck root by the autumn following; and if cuttings are planted in the autumn in a moift rich
earth, that is fhaded, many of them will grow; but neither of
thefe methods will produce fuch fine upright plants, or that will
grow to fuch a height as thofe raifed from feeds, though they will
be every whit as prolific of flowers and fruit. Whoever has not
the convenience of procuring the feeds, let him improve thefe
hints, if he has got a plant or two, which will be fufficient for his
purpofe: Whoever can get the feeds, had better never attempt
thofe arts. The feeds fhould be fown in the autumn, foon after
they are ripe. They will thrive in a'moft any foil or fituation,
if it be made fine, and clear of the roots of all weeds, &c. though
if it be a fine garden-mould, it will be the better. They fhould
be fown three-fourths of an inch deep. It feldom happens that
more than a few odd plants come up the firft fpring; the beds
muft, therefore, remain untouched until the fpring twelvemonth
after fowing; only conftant weeding muft be obferved. At that
time the plants will come up very thick, and all the fummer they
muft be weeded. In this feed-bed they may ftand two years, and
be then planted out in the nurfery, where they may remain, with
no other care than weeding and digging between the rows in
winter, until they are finally planted out.

The Broad-leaved fort will take very well b budding it on the
Common. The ftocks for this purpofe fhould be planted out
when they are one year's feedlings, and by the fummer twelvemonth after they will be fit for working; fo that whoever has
young plants of the Common fort, and only one of the other, may
encreafe his number this way.

2. The EVERGREEN EUONYMUS. Befides the genuine fpecies,
there is a *variety* of the *Americanus*, having its leaves beautifully ftriped with yellow. Thefe forts grow to the height of
about feven feet. The branches are flender, covered with a
fmooth green bark, and grow oppofite by pairs at the joints. The
leaves alfo grow oppofite, are fpear-fhaped, and have a ftrong
mid-rib running their whole length. The upper furface is of a
fine ftrong green colour, but their under is paler. They are
fmooth, are lightly indented, acutely pointed, and juftly entitle this
fhrub to be called a fine evergreen. The flowers are produced in
July, from the fides and ends of the branches, in fmall bunches.

They

They make no great fhow; but they will be fucceeded by rough, warted, red, five-cornered capfules, containing the feeds.

This fpecies is to be PROPAGATED in the fame manner as the other forts.   1. The beft way is from feeds, which we receive from Virginia.   Thefe will be two, and fometimes three years before they appear ; fo that a perfon fhould not be too hafty in difturbing the beds ; and after this precaution, what has been already faid rerating to the management of raifing the common forts of Spindle Trees from feeds, muft conftantly be cbferved in this fpecies.   2. By layers alfo, and cuttings, it may be encreafed ; but when the latter way is to be practifed, it will be proper to plant each cutting feparately in a fmall pot, and plunge them into a bark-bed, otherwife it is very feldom that they will grow.   After they have taken root, the pots may be fet in the natural mould up to the rims for about two years ; then the plants fhould be turned out into the places where they are to remain, and they will be fure of growing

---

F A G U S.

LINNEAN Clafs and Order, *Monoecia Polyandria* : Male flowers and female flowers upon the fame plant ; the males containing about twelve ftamina, and the females three piftils each : There are three SPECIES :

1. *Fagus Sylvatica* : The BEECH ; a well-known *tall deciduous tree* ; common in England and moft parts of Europe, alfo in Canada.

2. *Fagus Caftanea* : The CHESNUT, or the SPANISH or SWEET CHESNUT ; *a tall deciduous tree* ; natural to the mountainous parts of the South of Europe.

3. *Fagus Pumila* : The DWARF CHESNUT, or the CHINQUEPIN ; *a deciduous fhrub* ; native of North-America.

1. The BEECH.   In ftatelinefs, and grandeur of outline, the Beech vies with the Oak.   Its foliage is peculiarly foft and pleafing to the eye ; its branches are numerous and fpreading ; and its ftem waxes to a great fize.   The bark of the Beech is remarkably

ably

ably fmooth, and of a filvery caft ; this, added to the fplendor and
fmoothnefs of its foliage, gives a ftriking neatnefs and delicacy to
its general appearance. The Beech, therefore, ftanding fingly, and
fuffered to form its own natural head, is highly *ornamental* ; and
its leaves varying their hue as the autumn approaches, renders it
in this point of view ftill more defirable. In point of actual *Ufe*
the Beech follows next to the Oak and the Afh : it is almoft as
neceffary to the cabinet-makers and turners (efpecially about the
Metropolis), as the Oak is to the fhip-builder, or the Afh to the
plough and cart-wright. EVELYN neverthelefs condemns it in
pointed and general terms; becaufe " where it lies dry, or wet
and dry, it is exceedingly obnoxious to the worm :" He adds,
however, " but being put ten days in water, it will exceedingly
refift the worm." The natural foil and fituation of the Beech is
upon dry, chalky, or limeftone heights : It grows to a great
fize upon the hills of Surry and Kent ; as alfo upon the declivities of the Cotfwold and Stroudwater hills of Gloucefterfhire, and
flourifhes exceedingly upon the bleak banks of the Wye, in Here-
ford and Monmouth fhires ; where it is much ufed in making
charcoal. In fituations like thofe, and where it is not already
prevalent, the Beech, whether as a timber-tree or as an underwood, is an object worthy the planter's attention.

The method of PROPAGATING the Beech is from feeds :
EVELYN is brief upon this head. For woods, he fays, the Beech
muft be governed as the Oak :—In nurferies, as the Afh ; fowing
the mafts " in autumn, or later, even after January, or rather
nearer the fpring, to preferve them from vermin, which are very
great devourers of them. But they are likewife to be planted of
young feedlings to be drawn out of the places where the fruitful
trees abound." MILLAR fays, the feafon for fowing the mafts
" is any time from October to February, only obferving to fecure
the feeds from vermin when early fowed, which, if carefully done,
the fooner they are fown the better, after they are fully ripe."
HANBURY orders a fufficient quantity of mafts to be gathered
about the middle of September, when they begin to fall : Thefe
are to be " fpread upon a mat in an airy place fix days to dry ;
and after that you may either proceed to fow them immediately,
or you may put them up in bags in order to fow them nearer the
fpring ; which method I would rather advife, as they will keep
very well, and there will be lefs danger of having them deftroyed

by

by mice or other vermin, by which kinds of animals they are
greatly relished." They must be sown in beds properly prepared (as directed under the article PROPAGATION FROM SEED, in
the Introductory Part of this Work) about an inch deep. In
the first spring many of the young plants will appear, whilst
others will not come up till the spring following. Having stood
two years in the seminary, they should be removed to the
nursery, where they may remain till wanted. More will be
said of the Beech under WOODLANDS.

2. The CHESNUT. This is a tree of the first magnitude;
growing to a great height, and swelling to an immense size.
Mr. Brydone, in a tour through Sicily and Malta, measured the
ruins of a celebrated Chesnut, called *Castagno de cento Cavalli*,
standing at the foot of Mount Etna, and made it " two hundred
and four feet round!" The largest we know of in this country
stands at Tortworth, near Berkley, in Gloucestershire. Sir
Robert Atkins, in his History of Gloucestershire, says, " By tradition, this tree was growing in King John's reign;" and Mr.
Marsham calculates it to be " not less than eleven hundred years
old." Sir Robert makes it nineteen yards, and Mr. Marsham
forty six feet six inches in circumference. With great defer-
ence however to the authority and veracity of these gentlemen,
we have every reason to believe that what is called the Tortworth Chesnut is not one, but two trees: supposing them to be
only one, its dimensions are by no means equal to what are
given above. We have the highest opinion of Mr. Marsham's
ingenuousness and accuracy; and fortunately, in this case, he
has furnished us with a proof of his candour, in saying, " As
I took the measure in a heavy rain, and did not measure the
string till after I returned to the inn, I cannot so well answer
for this as the other measures." We will venture to add, that
had the day been fine, and Mr. Marsham had viewed the *field-*
side as well as the *garden* side of this venerable ruin; had he
climbed upon the wall, and seen the gable of the old building,
adjoining, clasped in between the two stems; and had further
ascended to the top of the old stump, which is not more than
twelve feet high, and, looking down its hollowness, seen its
cavity tending not to the centre of the *congeries*, but to the
centre of the *old Tree*, we are convinced he would not have suffered so inaccurate an account to have been published with his
                                                    signature,

signature, as that which appears in page 81 of the First Volume of Papers of the Bath Agriculture Society. The leaves of the Chefnut are long, fomewhat large, ftrongly marked by the nerves, and of a dark and fomewhat glofly appearance, in fummer; but, in autumn, change to a yellow hue. In open uncrouded fituations, the Chefnut throws out large fpreading arms, forming a magnificent ftrongly-featured outline; whilft in a clofe-planted grove the ftem will fhoot up clean, and ftraight as an arrow, to a great height.

As an *Ornamental*, the Chefnut, though unequal to the Oak, the Beech, and the Efculus, has a degree of greatnefs belonging to it which recommends it ftrongly to the gardener's attention. Its *Ufes* have been highly extolled; and it may deferve a confiderable fhare of the praife which has been given it. As a fubftitute for the Oak, it is preferable to the Elm: For door-jambs, window-frames, and fome other purpofes of the houfe carpenter, it is nearly equal to Oak itfelf; but it is very apt to be *fhaly*, and there is a deceitful brittlenefs in it which renders it unfafe to be ufed as beams, or in any other fituation where an uncertain load is required to be borne. It is univerfally allowed to be excellent for liquor cafks; as not being liable to fhrink, nor to change the colour of the liquor it contains: it is alfo ftrongly recommended as an underwood for hop-poles, ftakes, &c. Its fruit too is valuable, not only for fwine and deer, but as a human food: Bread is faid to have been made of it. Upon the whole, the Chefnut, whether in the light of ornament or ufe, is undoubtedly an object of the planter's notice.

The PROPAGATION of the Chefnut is chiefly from feeds: EVELYN fays, " Let the nuts be firft fpread to fweat, then cover them in fand, a month being paft, plunge them in water, and reject the fwimmers; being dried for thirty days more, fand them again, and to the water-ordeal as before. Being thus treated until the beginning of fpring, or in November, fet them as you would do Beans; and, as fome practife it, drenched for a night or more in new milk; but with half this preparation they need only to be put into the holes with the point upmoft, as you plant tulips."—" If you defign to fet them in winter or autumn, I counfel you to inter them in their hufks, which being every way armed, are a good protection againft the moufe, and a providential integument."—" Being come up, they thrive beft
unremoved,

unremoved, making a great ftand for at leaft two years upon
every tranfplanting; yet if needs you muft alter their ftation,
let it be done about November;—thus far EVELYN.—MILLAR
cautions us againft purchafing foreign nuts that have been kiln-
dried, which, he fays, is generally done to prevent their fprouting in their paffage; therefore, he adds, " if they cannot be procured frefh from the tree, it will be much better to ufe thofe
of the growth of England, which are full as good to fow for
timber or beauty as any of the foreign nuts, though their fruit
it much fmaller." He alfo recommends preferving them in
fand, and proving them in water. In fetting thefe feeds or nuts,
he fays, " The beft way is to make a drill with a hoe (as is commonly practifed for kidney-beans) about four inches deep, in
which you fhould place the nuts, at about four inches diftance,
with their eye uppermoft; then draw the earth over them with
a rake, and make a fecond drill at about a foot diftance from
the former, proceeding as before, allowing three or four rows
in each bed."—" In April" (he does not mention the time of
fowing) " thefe nuts will appear above-ground; you muft
therefore obferve to keep them clear from weeds, efpecially
while young : in thefe beds they may remain for two years,
when you fhould remove them into a nurfery at a wider diftance.
The beft time for tranfplanting thefe trees is either in October,
or the latter end of February, but October is the beft feafon : the
diftance thefe fhould have in the nurfery is three feet row from
row, and one foot in the rows. If thefe trees have a downright tap-
root, it fhould be cut off, efpecially if they are intended to be removed again; this will occafion their putting out lateral fhoots,
and render them lefs fubject to mifcarry when they are removed
for good. The time generally allowed them in the nurfery is
three or four years, according to their growth, but the younger
they are tranfplanted the better they will fucceed. Young trees
of this fort are very apt to have crooked ftems; but when they
are tranfplanted out and have room to grow, as they increafe
in bulk they will grow more upright, and their ftems will become ftraight, as 1 have frequently obferved where there have
been great plantations."—HANBURY follows MILLAR almoft
literally; except that he mentions February as the time of
fowing; and recommends that the young plants, a year after
they have been planted in the nurfery, be cut down to within

an

an inch of the ground; which, he fays, " will caufe them to fhoot vigoroufly with one ftrong and ftraight ftem." There is one material objection againft fowing Chefnuts in drills, which are well known to ferve as guides or conductors to the field-moufe, who will run from one end to the other of a drill without letting a fingle nut efcape her : we rather recommend fetting them with a dibble, either promifcuoufly or a quincunx, at about fix inches diftance.

EVELYN fays, that coppices of Chefnuts may be thickened by layering the tender young fhoots ; but adds, that " fuch as fpring from the nuts and marrons are beft of all." There is a ftriped-leaved variegation which is continued by budding; and the French are faid to graft Chefnuts for their fruit ; but MIL-LAR fays, fuch grafted trees are unfit for timber.

The Chefnut will thrive upon almoft any foil which lies out of the water's way ; but difaffects wet moory land. See more of this tree under WOODLANDS.

3. The DWARF CHESNUT grows to about eight or ten feet high. The ftem is of a brown colour, and divides into feveral branches near the top. The leaves are of an oval, fpear-fhaped figure, acutely ferrated, with a hoary caft on their under fide. The flowers come out in the fpring, in flender knotted catkins : They are of a greenifh-yellow colour, and are very feldom fucceeded by ripe feeds in England. This tree is hardy, and thrives beft in a moift foil and fhady fituation.

The method of PROPAGATING the Dwarf Chefnut is from feeds, which we receive from America. Thefe fhould be planted in drills, as foon as they arrive, in a moiftifh bed of rich garden-mould. If the feeds are good, they will come up pretty foon in the fpring. After they appear, they will require no trouble, except keeping them clean from weeds, and watering them in dry wea-ther. They may ftand in the feed-bed two years, and be after-wards planted in the nurfery-ground, at a foot afunder and two feet diftance in the rows ; and here when they are got ftrong plants, they will be fit for any purpofe.

F R A X I N U S.

LINNEAN Clafs and Order, *Polygamia Dioecia:* Hermaphto-
dite flowers and female flowers upon diftinct plants : the former
containing two males and one female each ; the latter one piftil-
ium only : There are three SPECIES.

1. *Fraxinus Excelfior :* The COMMON ASH ; a well-known *tall
deciduous tree;* common throughout England and moft parts of
Europe.

2. *Fraxinus Ornus:* The FLOWERING ASH ; *a low deciduous
tree ;* native of Italy and other fouthern parts of Europe.

3. *Fraxinus Americana :* The AMERICAN ASH , *a low deciduous tree;* native of Carolina and Virginia.

1. The COMMON ASH is one of the loftieft of our foreft-trees.
In a clofe grove and in a foil it affects it lengthens out into a
beautifully clean ftem, and rifes to an aftonifhing height : But
ftanding fingly, it throws out large arms, forms a full fpreading
head, and fwells out into a ftem proportionable : Mr. Marfham
mentions a very flourifhing one, growing in Benel church-yard,
three miles north of Dunbarton, in Scotland, which, in 1768, mea-
futed, at five feet high, fixteen feet nine inches in circumference.
The l aves of the Afh, too well known to require defcription, are
amongft the laft which foliate in the fpring, and amongft the firft
which fall in autumn. This alone depreciates its value very
much as an *Ornamental,* efpecially near gardens and gravel-walks :
and planted fingly or in hedges, it becomes an utter nuifance
in the neighbourhood it ftands in : every hufbandman knows the
injury it does to corn ; and there are few dairy-women who are
not well acquainted with the evil effects of its leaves, in autumn,
upon the produce of her dairy ; befides, being large and numerous, they foul and injure the after-grafs by rotting amongft
it. Clofe groves are the only proper fituation for the Afh ; its
ufes require a length and cleannefs of grain ; and it would be well
for the occupiers of land, and, indeed, for the community at large,
if a fevere penalty was laid upon planting it in any other fituation.
To enumerate the *Ufes* of the Afh would require a feparate volume :
in this point of view it undoubtedly ftands next to the Oak : The
Farmer would find it difficult to carry on his bufinefs without it :

and,

and, indeed, the cooper and the coach-maker would be equally at
a lofs with the wheel-wright, fhould a fcarcity of Afh take place ;
and we know of no fpecies of timber fo likely to be worn out in this
country as the Afh. The juft complaints of the Hufbandman are
expelling it very properly from our hedges ; and we are concerned
to fee, amongft the numerous plantations which have of late years
been made, fo few of this neceffary tree : it is therefore more than
probable that no tree will pay better for planting ; not, however,
in fingle trees and hedge-rows, but in clofe plantations, in the
manner which will be pointed out when we come to fpeak generally of WOODLANDS.

The method of PROPAGATING the Afh is from feeds; which
are peculiarly prone to vegetation, and frequently catch under
or near the tree they are produced upon, from whence tolerable
plants may fometimes be collected ; but in general they are
either cropt by cattle, or are drawn up flender and ill-rooted, and
feldom make fo good plants as thofe raifed by the gardener's
affiftance in a prepared feed-bed. EVELYN directs us to
gather the keys from a young thriving tree in October or November, and having laid them to dry, fow them " any time
betwixt then and Chriftmas ; but not altogether fo deep as
your former mafts." (meaning thofe of Beech, Hornbeam, &c.)
" Thus they do in Spain, from whence it were good to procure
fome of the keys from their beft trees." He recommends the
young plants' ftanding two years in the feminary, and cautions us,
in removing them into the nurfery, " not to cut their head at all,
which being young is pithy, nor by any means the fibrous part
of the roots ; only that down-right or tap-root, which gives our
hufbandmen fo much trouble in drawing, is to be totally abated ;
but this work ought to be in the increafe of October or November, and not in the fpring. We are, as I told you, willing to
fpare his head rather than the fide branches (which whilft young
may be cut clofe) becaufe being yet young, it is but of a fpungy
fubftance ; but being once fixed, you may cut him as clofe to
the earth as you pleafe ; it will caufe him to fhoot prodigioufly,
fo as in a few years to be fit for pike-ftaves."— " Young Afhes are
fometimes in winter froft-burnt, black as coals ; and then to ufe
the knife is feafonable, though they do commonly recover of
themfelves flowly." He adds, " You may accelerate their fpringing by laying the keys in fand, and fome moift earth, *ftratum*
fuper

super stratum ;" but does not say that this preparation will cause
them to vegetate the first spring. MILLAR says, " the seeds
should be sown as soon as they are ripe, and then the plants
will come up the following spring ; but if the seeds be kept out
of the ground till spring, the plants will not come up 'till the
year after."—" If they make good progress in the seed-bed,
(he says) they will be fit to transplant by the following
autumn," ——" as soon as their leaves begin to fall." Great
care (he says) is necessary in taking them up : they should not
be drawn, but taken up with a spade ; clearing the whole bed at
once, placing the larger together in rows, and the smaller by
themselves. " The rows should be three feet asunder, and the
plants a foot and a half distance in the rows : in this nursery
they may remain two years, by which time they will be strong
enough to plant where they are to remain ; for the younger
they are planted the larger they will grow."——HANBURY is
very deficient upon the subject of raising Ashes in the nursery
way : he does not even tell us the depth at which the keys are
to be sown ; nor, except in general terms, when they are to be
sown ; namely, " soon after they are gathered."——We beg
leave, however, to differ from these three great authorities. In-
stead of sowing the keys in autumn, presently after they are
gathered, we venture to recommend their being sown in the
spring, in the first favourable opportunity in February or
March ; for being sown in autumn some few may, and in general will, vegetate the first spring, whilst much the greatest
part will lie in the ground until the spring following : the few
that come up will be an incumbrance upon the beds, and will
render the expence of clearing them the first summer unnecessa-
rily great ; whereas, on the contrary, if the sowing be deferred
'till spring, the hoe and rake will have free range over the beds,
and the expence of cleaning them the first summer will be comparatively trifling. If the keys be well cured by spreading them
thin in an airy place, and keeping them turned for a few days
after gathering, they may be kept in a heap (moving them now
and then) until spring, with safety. The depth proper for sowing
Ashen keys is from one inch to an inch and a half or two inches,
according to the stiffness or the lightness of the soil of the seed-
bed. If they be sown too deep in a close-textured soil, they will
be apt to be smothered ; and if too shallow in a porous one,
													the

the drought has too much power over them, and they àre liable to be difturbed by the hoe and rake in clearing them the firft fummer after fowing. Upon the approach of the fecond fpring, the furface of the beds fhould be made as light and pulverous as pollible, in order to give to the embryo plants a free admiffion of air, and to facilitate their rifing : if part of the ftale mould be raked off, and a little frefh earth be fifted over in its ftead, it will add confiderable vigour to the young plants ; which may be removed into the nurfery whenever the croudednefs of the beds, the ftrength of the plants, or the conveniency of the planter, may render it requifite. For raifing groves of Afh fee WOODLANDS.

There are three *Varieties* of the Common Afh : The *Silver-ftriped*; the *Gold-ftriped*; and the *Yellow-coloured Afh*. Thefe *varieties* may be continued by *budding*.

EVELYN tells us, that " Afh may be propagated from a bough *flipt* off with fome of the old wood, a little before the bud fwells, but with difficulty by *layers*.

The Afh will thrive in almoft any foil ; but delights moft in a moift fituation, fo that it ftand above the level of ftagnant water ; in marfhes, half-drained bogs, and by the fides of rivers it flourifhes extraordinarily, outgrowing even many of the aquatics themfelves.

2. The FLOWERING ASH. Of this fpecies there are two kinds or *varieties*: The *Virginia Flowering Afh* ;—and the *Dwarf Afh of Theophraftus.*

The *Virginia Flowering Afh* when in blow is inferior in beauty to few of our flowering trees. It will grow to near thirty feet in height. The branches of this fort, in the winter, have nearly the fame appearance with the Common ; only they are, efpecially the youngeft, more inclined to a black caft : The buds alfo, which will begin to fwell in the autumn, are of that hue. The branches will not burn, when green, fo well as thofe of the Common Afh. The leaves are of a fine green, fmooth, ferrated, and confift of about three or four pair of folioles, placed a good way afunder along the mid-rib ; and they are ufually terminated by an odd one. The mid-rib is long, but not ftraight ; fwelling where the leaves, which fall off early in the autumn, come out. The flowers are white, produced in May, in large bunches, at the ends of the branches. HANBURY fays, " I have had this tree,
the

the fecond year from the bud, produce, on the leading fhoot, a tuft of flowers; and although this is not common, yet, when it gets to be about ten feet high, almoft every twig will be terminated with them. The flowers exhibit themfelves not in a gaudy drefs, but in a loofe eafy manner, all over the tree, which, together with the green leaves peeping from amongft this white bloom, makes the appearance extremely pleafing. I have never yet known the flowers to be fucceeded by feeds."

Dwarf Afh of Theophraftus is, as the name imports, a low tree for the Afh tribe; about fourteen or fifteen feet is the height it generally afpires to. The branches are fmooth, and of a darkifh green. The leaves are pinnated, of a dark green, and ferrated on the edges, but proportionably fmaller than thofe of the Common Afh. The flowers of this fort make no fhow, though they are poffeffed of the petals neceffary to complete a flower, which are denied the Common Afh.

3. AMERICAN ASH. The *varieties* of this fpecies are, Manna Afh, White Afh, Red Afh, Black Afh, and New-difcovered Afh.

Manna Afh will grow to about twenty feet high. It will fometimes fhoot eight feet the firft year from the bud, though it feldom fhoots more than two feet in a fummer afterwards. The bark of the young fhoots is fmooth, of a brownifh-green, and has a few greyifh fpots. The leaves are compofed of four or five pair of folioles, placed on a ftraight mid-rib; they are of a fine pleafant green, and more acutely and deeply ferrated than any of the other forts. The flowers make no fhow: They are partly the colour of thofe of the Common Afh, and are produced, like them, early in the fpring, before the leaves appear.

White Afh is fo called from the whitifh colour of the young branches in winter. They are fpotted all over with many white fpots, which makes their colour that of a lightifh-grey. This fort will arrive to about thirty feet high; and the branches are ftrong, and produced in an irregular manner. The folioles which compofe the leaves are of a light green, and obtufely fawed on the edges: They feldom confift of more than three pair, with the ufual odd one, which has a long point; and thefe are placed far afunder, on the mid-rib. Thefe leaves fall off early in the autumn, when they are of a light colour. This, together with the grey branches, make the tree have a whitifh look. The flowers are

produced

produced in the spring, and make no show. This sort is commonly called the New-England Ash.

Red Ash. The Red Ash is a stronger shooting tree than any of the former, the Common Ash excepted. The branches, which are fewer, are smooth, and the young shoots are of a reddish colour in the autumn. The leaves of this sort make the most noble figure of any of the others ; for although they are seldom composed of more than three pair of folioles, besides the odd one, yet these are exceedingly large, especially the odd one, which will be sometimes six inches long, and three and an half broad. The pair next it, also, will be fine and large ; though they diminish in size as they get nearer the base of the footstalk. These folioles are distinctly sawed on their edges, are of a fine light-green during the summer, and in the autumn die to a red colour ; from which circumstance, together with that of their red twigs, this sort takes the denomination of the Red Ash. It has its seeds very broad, and is commonly called the Carolina Ash.

Black Ash we receive from abroad by that name ; though it is difficult to see the propriety of its being so called. The colour of the shoots is nearly like that of the White Ash ; but they shoot stronger, and promise to form a larger tree. The leaves are large, and ribbed underneath ; of a very dark green, and die to a still darker in the autumn. The folioles are not so large as those of the Red sort, but they quit the tree about the same time. The keys are very broad, and, when we receive them, of a blackish colour.

" *New-discovered Ash* I received from Pennsylvania; where it was discovered growing in the woods near Philadelphia. The keys are very small and flat, and come up in a fortnight after being sown. The young shoots of this sort are covered with the same kind of bark as the White Ash, and the leaves nearly resemble those of the Black Ash, tho' they are not quite so large." HANBURY.

All the sorts of foreign Ashes are easily PROPAGATED. 1. By seeds, if they can be procured from abroad. We often have them in February ; and if they are sown directly, they will sometimes come up the beginning of May, though they generally lie, or at least the greatest part of them, until the spring following. · The beds may be made in any part of the garden ; and almost any sort of garden-mould, made fine, will do for the purpose. After

N   the

the feeds are fown, they will want no other care than weeding, until the plants are a year or two old in the feed-bed, when they may be taken up, and planted in the nurfery, at the ufual diſtance of a foot afunder, and two feet in the rows, which will be fufficient for them till they are finally taken up. 2. Budding is another good method of propagating thefe trees ; fo that thofe who have not the convenience of a correfpondence in the countries where they grow naturally, fhould procure a plant or two of a fort, and raife young Afhes of the Common fort for ſtocks. Thefe ſtocks fhould be planted out in the nurfery, a foot afunder, and two feet diftant in the rows. When they are one year old, and grown to be about the thicknefs of a bean-ſtraw, they will be of a proper fize for working. A little after Midfummer is the time for the operation ; and care muſt be obferved not to bind the eye too tight. They need not be unloofed before the latter end of September. In March, the head of the ſtock fhould be taken off, a little above the eye ; and by the end of the fummer following, if the land be good, they will have made furprifing ſtrong fhoots, many of them fix feet or more.

G E N I S T A.

LINNEAN Clafs and Order, *Diadelphia Decandria:* Each flower contains ten males and one female ; the males ſtanding in two divifions: There are fourteen SPECIES ; feven of which come under our notice :

1. GENISTA *Tridentata :* The PORTUGAL BROOM; *a deciduous fhrub*; native of Portugal and Spain.

2. GENISTA *Tinctoria:* The DYER's BROOM, or WOOD-WAXEN; *a low deciduous fhrub* ; native of England and Germany.

3. GENISTA *Pilofa :* The BRANCHING BROOM; *a deciduous fhrub* ; native of Hungary, Germany, and France.

4. GENISTA *Anglica :* The DWARF ENGLISH BROOM, or PETTY WHIN ; *a deciduous fhrub* ; natural to moiſt, heathy grounds in feveral parts of England.

5. GENISTA *Germanica* : The PRICKLY GERMAN BROOM ; *a low deciduous shrub* : native of Germany.

6. GENISTA *Hispanica* : The PRICKLY SPANISH BROOM ; *deciduous shrub* ; native of Spain and France.

7. GENISTA *Candicans* : The ITALIAN BROOM, or The CYTI-SUS OF MONTPELIER ; *a low deciduous shrub* ; native of Italy and about Montpelier in France.

 * For another Clafs of BROOMS, fee SPARTIUM.

1. The PORTUGAL BROOM is one of the larger growers : It will arrive to be five or fix feet high : the branches are very flender, tough, and for the most part three-cornered and jointed. The leaves end in three points, and are small ; though fome of them will be produced by threes, in fuch a manner as to be entirely trifoliate leaves ; whilst others again are often found fingle. By the beginning of May, this fhrub will be in blow. The flowers, which are yellow and of the butterfly kind, are each very large : They grow from the fides of the branches, and wings of the leaves, fingly, on fhort footftalks, and are produced in fo free and eafy a manner, that they may not improperly be faid to have a genteel appearance. They are fucceeded by pods, in which are contained kidney-fhaped feeds, that will be ripe in autumn.

There are two *varieties* of this fpecies of Broom, one with larger, the other with narrower leaves, both of which are fought after by thofe who are fond of having great varieties. Thefe forts are the leaft kinds, and require a fheltered fituation.

2. The DYER's BROOM. Of this fpecies there are two varieties, one of which has a narrower leaf, and grows more upright ; the other is more fpreading in its branches. Their natural growth is about two or three feet high, and their branches are taper and channelled. The leaves are of a lance-like figure, and placed alternately on the branches. Thefe branches will produce fpikes of yellow flowers in June, in fuch a manner, that though each individual flower is but fmall for thofe of the butterfly kind, the whole fhrub will appear covered with them to the pleafure of all beholders. Thefe flowers are fucceeded by pods, which will have ripe feeds in the autumn.

3. BRANCHING BROOM, as the name indicates, is a plant whofe branches fpread abroad, and decline towards the earth's

furface. The main ftalk is befet all over with tubercles, and the leaves that ornament the flender branches are obtufe and fpear-fhaped. The flowers, which are yellow, are produced at the ends of the branches, in fpikes, in June ; and they are exhibited in fuch profufion as to make a delightful fhow. They are fucceeded by pods that ripen their feeds in autumn.

4. DWARF ENGLISH BROOM has many beauties to recommend it to the gardener, though it grows common in many of our barren heaths. In thefe places, it goes by the cant name of *Petty-Whin.* All the forts of our choiceft cultivated plants grow wild in fome parts of the globe, but lofe nothing of their value becaufe they appear thus fpontaneoufly : Why then fhould this, becaufe it is common in fome parts of England, be denied admittance into gardens, efpecially thofe that are at a remote diftance from fuch places, as it has many natural beauties to recommend it ? It is a low plant, feldom growing to be more than two feet high ; on which account no garden is fo fmall but it may be there planted, if the commonnefs of it be no objection to the owner. This fhrub has fome fingle, long fpines, though the flower-branches are entirely free from them. The leaves, like the fhrub, are proportionally fmall, of a lanceolated figure, and grow alternately on the branches. The flowers, which are of a fine yellow, are produced the beginning of May, in clufters, at the ends of the branches ; and are fucceeded by thick fhort pods, in which the feeds are contained.

5. GERMAN PRICKLY BROOM will grow to be about a yard high. The fhrub is armed with many compound fpines ; the branches are flender and numerous, though thofe that produce the flowers are entirely free from fpines. The leaves of this fort, alfo, are fmall, and of a lanceolate figure, and grow alternately on the branches. The flowers are produced in plenty at the ends of the branches, in June : They are of the colour and figure of the others, and are fucceeded by pods, in which the feeds are contained.

6. PRICKLY SPANISH BROOM will grow to be five or fix feet high. This fhrub is poffeffed of many compound fpines ; though the branches that produce the flowers are entirely free from them. The leaves are exceeding narrow, many of them being no wider than a thread, but very hairy. The flowers are
yellow,

yellow, produced in May, in clusters, at the ends of the branches; and are succeeded by hairy compressed pods, in which the seeds are contained.

7. ITALIAN BROOM rises, with an erect, shrubby, branching, striated stalk, to the height of about a yard. The leaves are trifoliate, oval, and hairy underneath. The flowers come out on leafy foot-stalks, from the sides of the branches: They are of a bright yellow colour, appear in June, and are succeeded by hairy pods, containing ripe seeds, in September.

The best way of PROPAGATING all these sorts is by seeds; and if these are sown soon after they are ripe, they will come up earlier in the spring, and make better plants by the autumn. They should only stand one year in the seed-bed before they are transplanted. They should be taken up in the spring, and planted out finally, in ground properly prepared for such small plants; for the less they are removed, so much the faster will they thrive; as they naturally grow with long strong stringy roots, that do not love to be disturbed; on which account, if places in the plantations were to be marked out, the mould made fine, a few seeds of the different sorts sown, and sticks set as guides to prevent their being hoed or dug up; plants that have been thus raised, without removing, will shoot stronger, and flower better, than any that have been brought from the seed-bed or nursery. After they are come up, if there be too many in a place, the weakest may be drawn out, and only two or three of the strongest left, which will cause them to flower better and stronger.

GLEDITSIA.

LINNEAN Class and Order, *Polygamia Dioecia:* Hermaphrodite flowers and male flowers upon one plant, and female flowers upon a different plant. There are two SPECIES: one of them a Stove-plant; the other

GLEDITSIA *Triacanthos:* The TRIPLE-THORNED ACACIA; *a deciduous tree;* native of Virginia and Pennsylvania.

The GLEDITSIA. Its growth is naturally upright, and its trunk is guarded by thorns of three or four inches in length, in a remarkable manner. Thefe thorns have alfo others coming out of their fides at nearly right angles : Their colour is red. The branches are fmooth, and of a white colour. Thefe are likewife armed with red thorns, that are proportionally fmaller : They are of feveral directions, and at the ends of the branches often ftand fingle. The young fhoots of the preceding fummer are perfectly fmooth, of a reddifh green, and retain their leaves often until the middle of November. Although there is a peculiar oddity in the nature and pofition of the fpines, yet the leaves conftitute the greateft beauty of thefe trees : They are doubly pinnated, and of a delightful fhining green. The pinnated leaves that form the duplication, do not always ftand oppofite by pairs on the middle rib ; the pinnæ of which they are compofed are fmall and numerous ; no lefs than ten or eleven pair belong to each of them ; and as no lefs than four or five pair of fmall leaves are arranged along the middle rib, the whole compound leaf confifts often of more than two hundred pinnæ of this fine green colour : They fit clofe, and fpread open in fine weather; though during bad weather they will droop, and their upper furfaces nearly join, as if in a fleeping ftate. The flowers are produced from the fides of the young branches, in July : They are a greenifh katkin, and make little fhow; though many are fucceeded by pods, that have a wonderful effect; for thefe are exceedingly large, more than a foot, fometimes a foot and a half in length, and two inches in breadth, and of a nut-brown colour when ripe ; fo that the effect they occafion, when hanging on the fides of the branches, may eafily be guefled.

There is a *variety* of this fpecies, with fewer thorns, fmaller leaves, and oval pods. It has nearly the refemblance of the other; though the thorns being not fo frequent, and the pods being fmaller, each containing only one feed, this fort lofes that fingular effect which the other produces by them.

The PROPAGATION of thefe trees is not very difficult. We receive the feeds from America in the fpring, which keep well in the pods, and are for the moft part good. They generally arrive in Febuary ; and, as foon as poffible after, they fhould be fown in a well-fheltered warm border of light fandy earth. If no border is to be found that is naturally fo, it may be improved by applying drift fand, and making it fine. The feeds fhould be fown about half an inch deep ; and they will for the moft part

come

come up the firſt ſpring. If the ſummer ſhould prove dry, they muſt be conſtantly watered; and if ſhade could be afforded them in the heat of the day, they would make ſtronger plants by the autumn. A careful attention to this article is peculiarly requiſite; for as the ends of the branches are often killed, if the young plant has not made ſome progreſs, it will be liable to be wholly deſtroyed by the winter's froſt, without protection: And this renders the ſowing the ſeeds in a warm border, under an hedge, in a well-ſheltered place, neceſſary; for there theſe ſhrubs will endure our winters, even when ſeedlings, and ſo will require no farther trouble; nay, though the tops ſhould be nipped, they will ſhoot out again lower, and will ſoon overcome it. It will be proper to let them remain two years in the ſeed-bed, before they are planted out in the nurſery. The ſpring is the beſt time for the work. Their diſtances ſhould be one foot by two; the rows ſhould be dug between every winter; and, being weeded in ſummer, here they may remain, with no other particular care, until they are ſet out to remain. Theſe trees are late in the ſpring before they exhibit their leaves, but keep ſhooting long in the autumn.

G L Y C I N E.

LINNEAN Claſs and Order, *Diadelphia Decandria:* Each flower contains ten males riſing in two diviſions, and one female. There are fourteen SPECIES; five of which have been enured to the open air of this country.

1. GLYCINE *Fruteſcens:* The CAROLINA KIDNEY BEAN; *a climber*; native of Carolina and Virginia.

2. GLYCINE *Apios:* The ASH-LEAVED MILK-VETCH; *a climber*; native of Virginia.

3. GLYCINE *Tomentoſa:* The CLIMBING REST-HARROW; *a climber*; native of Virginia.

4. GLYCINE *Comoſa:* The MARYLAND KIDNEY BEAN; *a climber*; natural to moiſt ſhady places in Virginia.

5. GLYCINE *Monoica:* The VIRGINIA GLYCINE; *a climber*; native of moiſt ſhady places in Virginia.

1. CARO-

1. Carolina Kidney Bean does not rife by the affiftance of clafpers, but by the twining branches, which naturally twift round any adjacent tree ; nay, if trees are ten feet or more diftance from the root of the plant, its branches, being too weak to fupport themfelves, will trail along the ground until they reach thefe trees, and then they will twine their branches with theirs, and arrive to a great height : Indeed, where trees are near at hand, and they begin by the firft fpring-fhoot to twift about them, they will twine up to them to the height of near twenty feet. This climber is poffeffed of noble large pinnated leaves, very much like thofe of liquorice. The folioles are about three pair in number, arranged on their common mid-rib, and they always end with an odd one. Their colour is for the moft part of a lightifh hoary caft, with a blueifh tinge. The flowers are very large and ornamental : Their colour is that of a blueifh purple, and their general characters indicate their ftructure. They are produced from the wings of the leaves, in July and Auguft ; and are fucceeded by long pods, like Kidney Beans.

This fine climber is eafily PROPAGATED, 1. By feeds, if there is a conveniency of procuring them from abroad ; for they never ripen with us. In the fpring, as foon as we receive them, they fhould be fown, in fine beds of light fandy earth, half an inch deep. They will readily come up, and all fummer muft have frequent waterings ; and if the beds be fhaded in hot weather, it will be the better. In winter the beds fhould be hooped, and covered with mats in frofty weather : And in fpring the ftrongeft may be drawn out, which will thin the bed, and make way for the others, which fhould ftand until the next fpring. Plants thus drawn fhould be fet in the nurfery, at fmall diftances, and in a year or two after, they will be good plants for any place where they are wanted. 2. This plant is alfo eafily encreafed by layers ; for if the young fhoots of the preceding fummer be laid in the ground in the autumn, by the autumn following they will have ftruck root ; when the beft-rooted and ftrongeft layers may be planted out to ftand where they are wanted, whilft the weaker, or thofe with hardly any root, may be fet in the nurfery, like the feedlings, to gain ftrength.

2. The

2. The Ash-leaved Milk-Vetch will twine from six to twelve feet high, according to the nature of the soil ; for in a rich fat mould it will grow near double the length it will in a foil of an oppofite nature. The ftalks die to the ground every autumn : and in the fpring new ones are iffued forth from the roots, which are compofed of many knobs, that encreafe in number, the longer the plant is fuffered to remain. The leaves fomewhat refemble thofe of the Afh-tree, being pinnated almoft in the fame manner. The folioles, which confift of three pair befides the odd one, are of an oval lanceolate fhape ; and being arranged oppofite along the mid-rib, and terminated with a fingle one, form a fine leaf. The flowers are produced from the fides of its twining ftalks, in Auguft. They grow in fmall fpikes, are of a reddifh colour, and being of the butterfly or pea-blof-fomed kind, make a pretty good fhow. Thefe flowers are fometimes fucceeded by pods, which never perfect their feeds with us.

3. Climbing Rest-Harrow is but a low plant for a climber, feldom arifing higher than five feet. The ftalk dies to the ground every autumn ; and the lofs is repaired by a natural fucceffion prefented from the root every fpring. The leaves are trifoliate and very downy. Every one knows the beauty that arifes from leaves of an hoary nature, amongft the variety of greens of different tinges. The flowers are of the pea-bloom kind, and are produced in fhort bunches, in June and July, from the fides of the ftalks. They are of a yellow colour, and, though they are rather fmall in proportion, are very beautiful. They are fucceeded by pods, in which two feeds only are contained, and which will be ripe with us in September.

4. The Kidney Bean Plant of Maryland has a flender, annual, twining ftalk, which will arife to be three or four feet high. The leaves are trifoliate, and fit clofe to the ftalks. They are hairy, and the folioles are of an oval lanceolate fhape ; and being of a good green, make the whole ornamental enough. But the greateft ornament this plant receives is from the flowers, which are alfo of the pea-bloom kind, and are of a clear blue. They are produced in June, from the fides of the ftalks, in fine
                                                              recurved

recurved bunches; and thefe are fucceeded by pods, which will have ripe feeds in Auguft or September.

5. The VIRGINIAN GLYCINE will arife with its flender branches to a degree higher than the other. The ftalks are hairy, and the leaves with which they are ornamented are trifoliate and naked. The flowers are produced from the fides of the ftalks, in June and July. They grow in pendulent bunches, and are alfo of the butterfly kind. They are very beautiful, and each exhibits a variety of colours; for the wings and the keel are white, whilft the ftandard is of a pale violet colour. Thefe flowers are fucceeded by compreffed half-rounded pods, hanging by lengthened peduncles; and the feeds will often be ripe in September.

All thefe forts are PROPAGATED by the feeds; and this may be in the places where they are to remain, or in warm well-fheltered beds, or in pots, to be houfed for the firft winter, if it fhould prove fevere. They will very readily come up; and if they are fown in the open ground, the beds fhould be hooped at the approach of winter, to be covered with mats, in cafe it fhould prove bad. It will be proper to plunge thofe fown in pots, immediately after, up to the rims in the natural mould; this will keep them cool and moift: At the approach of hard frofts, they may be removed into the greenhoufe; and in fpring may be turned out into the places where they are defigned to remain. Thofe in the beds, alfo, fhould be tranfplanted to fuch places: Their after-management will be only to part the roots about every three or four years; and by this method alfo they may be all encreafed. The fpring is the beft time for parting the roots; and by this way they may be multiplied faft enough. As to the firft fort, this method is chiefly practifed for its propagation, as it does not ripen its feeds here, unlefs there is a conveniency of procuring them from abroad. The roots of this fort are compofed of feveral knobs; and thefe being taken up and divided readily grow, and become good plants.

Thefe perennials are all proper to be planted amongft fhrubs in warm and well-fheltered places; for they are rather of a tender nature, and are often deftroyed by fevere frofts. As the ftalks are all annual, as foon as they decay at the approach of winter, they fhould be cut up clofe to the ground, and cleared off fuch plants as are near them, by which they have afpired,

otherwife

otherwife they will have a dead paltry look, and render the place inelegant; for, even in the dead of winter, neatnefs and elegance muft be obferved, which will not only fhew a more promifing expectation of a refurrection, but the clearing away old ftalks, &c. will be better for the plants themfelves, as they would in fome degree hinder and choke the young fhoots as they advance in the fpring.

GUILANDINA.

LINNEAN Clafs and Order, *Decandria Monogynia:* Each flower contains ten males and one female: There are five SPECIES: one of which will ftand our winter.

GUILANDINA *Dioica:* The CANADA NICKAR-TREE; *a deciduous tree or fhrub;* native of Canada.

The CANADA NICKAR TREE. The ftem is erect, firm, often twenty feet high, and fends forth feveral branches, which are covered with a fmooth, bluifh, afh-coloured bark. The leaves are bipinnated; and the folioles are large, fmooth, entire, and ranged alternately on the mid-rib. The flowers appear in July or Auguft; but are very rarely fucceeded by feeds in England.

This fpecies is PROPAGATED, 1. By feeds, which muft be procured from the places where the tree naturally grows. The feeds are very hard, and often lie two years before they make their appearance; fo that if they are fown in common ground, the beds muft all the time be kept clean from weeds. In the autumn it will be proper to ftir the furface of the mould, but not fo deep as to difturb the feeds. In the fpring the plants will come up: All fummer they muft be kept clean from weeds, watered in dry weather, and in the autumn the ftrongeft may be planted out in the nurfery, at the ufual diftance, while the weakeft may remain another year in the feed-bed to gain ftrength. The feeds alfo may be fown in pots, and plunged into a hot-bed: This will bring the plants up the firft fpring.

After

After they make their appearance, they muſt be hardened by degrees to the open air. 2. This tree may likewiſe be propagated by layers. Theſe muſt be the ſmaller ſhoots of the laſt year's wood. The operation muſt be performed by making a ſlit, as is practiſed for carnations; and the beſt time for the buſineſs is the autumn. 3. By cutting the root, alſo, this tree may be encreaſed. In order to this, bare away the earth from the top of the root; then with the knife cut off ſome parts of it, leaving them ſtill in the ground, and only directing their ends upwards : Then cover the whole down lightly with mould. The parts that have been ſeparated will ſhoot out from the ends, and come up as ſuckers all round the tree. If dry weather ſhould happen, you will do well to water them all the ſummer; and in the autumn they may be removed to the place where they are deſigned to remain ; which ought always to be in a light dry ſoil, in a well-ſheltered place.

H A M A M E L I S.

LINNEAN Claſs and Order, *Tetrandria Digynia:* Each flower contains four males and two female : There is only one SPECIES :

HAMAMELIS *Virginica :* The DWARF HAZEL ; *a low deciduous ſhrub ;* native of Virginia.

The DWARF HAZEL is a ſhrub of about four feet in growth, and will conſtitute a variety among other trees, though there is no great beauty in it, except what is afforded by the leaves. Theſe are placed on the branches, which are numerous and ſlender, in an alternate manner, and much reſemble thoſe of our Common Hazel, that are known to all. The flowers make no ſhow ; but perhaps the time of their appearing, which happens in winter, in November or December, when they will be produced in cluſters from the joints of the young ſhoots, may make the plant deſirable to ſome perſons. Nothing farther need be ſaid to the gardener concerning this ſhrub, which ature ſeems

to have defigned for the ftriéter eye of the botanift ; fo that we
fhall proceed to its culture.

It is PROPAGATED, 1. By feeds, which muft be procured from
America, for they do not ripen here. An eafterly border, well
defended from the north and wefterly winds, is beft for their reception ; for thefe plants, when feedlings, are rather tender;
when older, they are hardy enough. They will grow in almoft
any kind of good garden-mould, made fine ; and they fhould be
covered about half an inch deep. They will not come up before
the fecond, and fometimes the third fpring. 2. This tree may
alfo be PROPAGATED by layers ; fo that whoever has not the
conveniency of procuring the feeds from abroad, having obtained
a plant or two, may encreafe them this way. The operation
fhould be performed on the twigs of the preceding fummer's
fhoot : Thefe fhould be flit at the joint, and a bit of chip, or
fomething, put in to keep the flit open. If thefe ftools ftand in a
moiftifh place, which thefe fhrubs naturally love, and are layered
in the autumn, they will have fhot root by the autumn following ;
and may be then either planted out in the nurfery, or where they
are to remain.

H E D E R A.

LINNEAN Clafs and Order, *Pentandria Monogynia :* Each
flower contains five males and one female : There are only two
SPECIES :

1. HEDERA *Felix :* The COMMON IVY ; a well-known *evergreen climber* ; native of England and moft parts of Europe.

HEDERA *Quinquefolia :* The DECIDUOUS IVY, or the VIR-
GINIAN CREEPER ; *a deciduous climber* ; native of Virginia and
Canada.

1. The COMMON EVERGREEN IVY. Befides the genuine
fpecies there are three *varieties :* namely, the Yellow-berried
Ivy ; the Gold-ftriped Ivy ; and the Silver-ftriped Ivy.

The

The *Common Ivy* is well known all over England, and how
naturally it either trails on the ground, or rifes with walls or
trees, ftriking its roots all along the fides of the branches for its
fupport. It chiefly delights in old houfes or walls; and when
it has taken poffeffion of any outfide of the outer buildings, will
foon cover the whole. It will make furprifing progrefs when it
reaches old thatch; and will foon, if unmolefted, climb above
the chimney itfelf. Neither are old houfes or walls what it
chiefly likes to grow on; for it will ftrike its roots even into the
bark of trees. But above all, it chiefly affects old rotten trees or
dodderels; for thefe it will almoft cover, and rear its head with
a woody ftem above the trunk, and will produce flowers and
fruit in great plenty. There, as well as on the fides of old walls
and buildings, it becomes a habitation for owls and other birds.
The ufefulnefs of Ivy, then, in Gardening, is to over-run caves,
grottos, old ruins, &c. to which purpofe this plant is excellently
adapted; and were it not for its commonnefs, it would be
reckoned inferior to few evergreens; for the older grey ftalks
look well, whilft the younger branches, which are covered with
a fmooth bark of a fine green, are very beautiful. The leaves,
alfo, are of a fine ftrong green, are large and bold, and make a
variety among themfelves; for fome are compofed of lobes,
whilft others are large, and of an oval figure. The flowers are
nothing extraordinary, unlefs it be for the figure in which they
grow. This is ftrictly the *Corymbus*; and all flowers growing in
fuch bunches are called by Botanifts Corymbofe Flowers. The
fruit that fucceeds them, however, is very beautiful; for being
black, and growing in this round regular order, and alfo continuing on all winter, it makes the tree fingular, and, were it not
for its commonnefs, defirable. It is obfervable, if Ivy has no
fupport, but is left to creep along the ground only, it feldom
flowers; but having taken poffeffion of rails, hedges, trees, or
buildings, from thefe it fends out woody branches, which produce the flowers and fruit.

The *Yellow-berried Ivy* differs from the Common Ivy in that
its berries are yellow. It grows common in the iflands of the
Archipelago; and is at prefent rare with us. This is the *Hedera
Poetica* of old authors.

The

The *Gold-ftriped Ivy* is the Common Ivy with yellow-blotched leaves ; though it is obfervable, that this fort has very little inclination to trail along the ground, or up trees or buildings, as it naturally rifes with woody branches, and forms itfelf into a bufhy head : So that this fort may be planted amongft variegated trees, or evergreens, as a fhrub. Let it be fet where it will, it is very beautiful ; for the leaves will be a mixture of yellow and green ; and fometimes they will have the appearance of being all yellow, thereby caufing a very fingular and ftriking look at a diftance.

The *Silver-ftriped Ivy* is a variety of our Common fort, though the branches are naturally more flender. The leaves alfo are fmaller ; and of all the forts this creeps the clofeft to walls or buildings, or is of ftrength fufficient to form its ligneous branches, when got to the top, to any head. " This plant (continues HAN-BURY, with whom it feems to be a very great favourite) is of all others to be planted againft walls for ornament ; for its leaves are ery finely ftriped with ftreaks of filver, and the fets being firft planted at fmall diftances, will foon cover them all over, fo as to have a delightful look. A more beautiful ornament to a wall cannot be conceived, than what belongs to a wall of Charles Morris, Efq. of Loddington. It confifts of thefe plants, which having firft taken properly to the ground, and afterwards to the mortar-joints, have fo over-fpread the furface as to be a fight, of the kind, fuperior to any I ever beheld ; and I am perfuaded there are few people of tafte, who had feen any thing of this nature, but would be induced to have the like, even againft their choiceft walls. And here let it always be remembered, that whereas our Common Green Ivy is to hide and keep from view all old and unfightly walls, fo the Silver-ftriped Ivy is to orna-ment all walls, even thofe of the fineft furface."

2. The DECIDUOUS AMERICAN IVY is a real fpecies of the *Hedera*. It fheds its leaves in the autumn ; and will fpread itfelf over pales, walls, buildings, &c. in a very little time. It puts forth roots at the joints, which faften into mortar of all forts ; fo that no plant is more proper than this to hide the unfightly furface of an old barn end, or any other building which cannot be concealed from the view by trees being planted at fome diftance ;

diſtance ; as in one year it will ſhoot often near twenty feet, and, let the building be ever ſo high, will ſoon be at the top of it. The bark on the ſhoots is ſmooth, and of a brown colour; and the buds in the ſpring, as they are beginning to open, will be of a fine red. The leaves are large and well-looking. Each is compoſed of five ſmaller, which are ſerrated at their edges. Their common foot-ſtalk is proportionably ſtrong, and they die to a fine red in the autumn.

All the ſorts are to be PROPAGATED by cuttings; for theſe being ſet any time in the winter, in almoſt any ſoil, will ſtrike root by the autumn following; and if they are permitted to remain another year, they will then be ſtrong plants, fit to be ſet out for good. The Common Ivy is alſo to be raiſed from ſeeds.

H I B I S C U S.

LINNEAN Claſs and Order *Monadelphia Polyandria :* Each flower contains numerous males and one female ; the males being joined together at the baſe. There are thirty-ſeven SPECIES ; one of which adds great beauty to our grounds and ſhrubberies, in autumn.

HIBISCUS *Syriacus :* The ALTHEA FRUTEX, o the SYRIAN MALLOW, or the SYRIAN HIBISCUS ; *a deciduous ſhrub ;* native of Syria.

The ALTHEA FRUTEX, or the SYRIAN HIBISCUS. Of this ſpecies there are ſeveral *varieties :*

The *White Althea Frutex.*

Red-flowering Althea Frutex.

Yellow-flowering Althea Frutex.

Pale Purple-flowering Althea Frutex.

Deep Purple Althea Frutex.

All

All thefe, though fuppofed to be only forts of one fpecies of *Hibifcus*, afford wonderful varieties to the gardener. They will grow to the height of about fix feet. Their branches are not very numerous ; they are fmooth, and of a whitifh colour. The leaves are of a pleafant green, and grow on fhort footftalks, irregularly on the branches. They are of an oval, fpear-fhaped figure, ferrated at the edges, and many of them are divided at the top into three diftinct lobes. The flowers have longer footftalks than the leaves, and come out from the fides of the young fhoots with them ; infomuch that the young fhoots are often garnifhed with them their whole length. The Common Mallow produces not a bad flower, did not its commonnefs render it unnoticed. The flowers of thefe fpecies fomewhat refemble it in fhape, but by far exceed it both in fize and fplendor of colour ; and each has a greater variety ; infomuch that though they are termed Red, White, Purple, &c. from the colour of the upper part of the petals, yet the lower part of all of them is very dark, and feems to fhoot out in rays in directions towards the extremity of each petal. Auguft is the month we may expect to be entertained with this bloom ; though in ftarved cold foils, the flowers rarely ever appear before September.

This beautiful fhrub may be PROPAGATED by two methods. 1. By feeds, which we receive from abroad. Thefe fhould be fown in a bed of light fandy earth ; and if it is not naturally fo, drift fand muft be added ; and if fome old lime-rubbifh, beat to powder, be alfo mixed with it, it will be the better. Having worked them all together, and made the bed fmooth and fine, the feeds fhould be covered about a quarter of an inch deep. The fituation of this bed muft be in a warm well-fheltered place, that the young plants may not fuffer by frofts the firft winter. Any time in March will do for the work ; and in about fix weeks the young plants will come up. In the heat of fummer, it will be proper to fhade them ; and if conftant waterings are afforded them in dry weather, they will acquire greater ftrength and vigour by the autumn. At the beginning of November, befides the natural fhelter of thefe beds, it will be proper to prick furze-bufhes at a little diftance all around, to break the keen edge of the black frofts, which otherwife would deftroy many of them the firft winter : After that, they will be hardy enough for our fevereft weather. They fhould ftand in thefe feed-beds two years,

O   and

and all the while be weeded and watered in dry weather. The
fpring is the beft time for planting them out in the nurfery,
where no more diftance need be allowed them than one foot.
2. Thefe plants may be propagated by layers ; for which purpofe the ftools fhould be headed near the ground, to throw out
fome good ftrong fhoots the following fummer. Thefe fhould
be laid in the ground, the bark being broken, or cut at one or
two of the joints, and they will have ftruck root by the autumn
following, when they may be taken up and planted in the
nurfery, like the feedlings ; and a fecond operation performed on
the ftools. 3. Thefe plants may be raifed alfo by cuttings ; for
by planting them in a fhady border, many of them will grow ;
though this is not a certain method.

H I P P O P H A E.

Linnean Clafs and Order, *Dioecia Tetrandria :* Male flowers
containing four ftamina, and female flowers containing one piftil,
upon diftinct plants : There are only two Species :

1. Hippophae *Rhamnoides :* The European Sea-Buckthorn ; *a tall deciduous fhrub* ; native of the fea-fhores of this
country and moft parts of Europe.

2. Hippophae *Canadenfis :* The American Sea-Buckthorn ; *a tall deciduous fhrub* ; native of Canada.

1. The European Sea-Buckthorn will grow to the height
of about twelve feet, and fends forth numerous branches in an
irregular manner. Their colour is that of a dark brown ; and
on them a few ftrong and long fharp fpines are found, nearly like
thofe of the Common Buckthorn. This tree is chiefly admired
for its fingular appearance in winter ; for the young fhoots of
the preceding fummer are then found thickly fet on all fides with
large, turgid, uneven, fcaly buds, of a darker brown, or rather a
chocolate colour, than the branches themfelves : Thefe give the
tree fuch a particular look, that it catches the attention, and occafions it to be enquired after, as much as any fhrub in the plantation.

HIP

tation; About the end of February thefe turgid buds will be much larger ; and a little before their opening, upon ftriking the tree with a ftick, a yellow duft, like brimftone, will fall from them. Though fome think the beauty of this fhrub to be diminifhed after the leaves are opened, yet thefe have their good effect ; for they are of two colours : Their upper furface is of a dark green, their under hoary ; they are long and narrow, entire, have no footftalks, nearly like thofe of the rofemary, though rather longer and broader ; and they are placed alternately all around, without any footftalks, on the branches. They continue on the tree green and hoary late, fometimes until the beginning of December, and at length die away to a light brown. The flowers are of no confequence to any but Nature's ftrict obfervers. They are produced in July, by the fides of the young fhoots ; the male flowers appear in little clufters, but the females come out fingly. They are fucceeded by berries, which, in the autumn, when ripe, are either of a red or yellow colour, for there are both thofe forts.

2. CANADA SEA-BUCKTHORN will grow to about the fame height as the other fpecies ; nearly the fame dark brown bark covers their branches ; and, except the figure of their leaves, which are oval, this plant differs in few refpects from the European-Sea-Buckthorn.

Both thefe forts may be PROPAGATED, 1. By cuttings of the young fhoots, planted in a fhady border, in October ; though the moft certain method is by layers. If the trees to be encreafed are of fome years growth, the ground fhould be dug and made fine, as well as cleared of the roots of bad weeds, &c. all round. The main branches may be plafhed, and the young twigs that form the head laid in the ground ; taking off their ends with a knife, that they may only juft peep. If this work be performed in the autumn, they will be good rooted plants by the autumn following, when they may be taken off, and either planted in the nurfery; or where they are to remain. 2. Both thefe forts are fubject to fpawn, and throw out many fuckers, fometimes at a good diftance from the plants ; fo that by this method they propagate themfelves.

O 2   H Y-

H Y D R A N G E A.

LINNEAN Clafs and Order, *Decandria Digynia* : Each flower contains ten males and two females : There is only one SPECIES :

HYDRANGEA *Arborefcens* : The HYDRANGEA; *a deciduous fhrub*; native of Virginia.

The HYDRANGEA feldom grows to more than a yard or four feet high, and affords as much pleafure to thofe who delight in fine flowers as it does to the botanift. It forms itfelf into no re-gular head; but the branches of which it is compofed fhoot chiefly from the root. Thefe, when young, are four-cornered and green; when old, of a fine brown colour : They are very large for their height, as well as very full of pith. The leaves are a great ornament to thefe plants; being alfo very large, and having their upper furface of a fine green, and their under rather downy. Their figure is nearly fhaped like a heart, but ends in an acute point; and their fize will prove according to the nature of the foil they grow in. On a dry foil, they will often be no more than two inches long, and fcarcely an inch and a half broad; but, in a moift rich foil, they will frequently grow to near four inches long, and two and three quarters broad in the wideft part. They are ferrated at their edges, and are placed on long footftalks, oppofite to each other, on the branches. But the flowers conftitute the greateft beauty of thefe plants; for they are produced in very large bunches, in Auguft : Their colour is white, and the end of every branch will be ornamented with them. They have an agreeable odour, and make fuch a fhow all together as to diftinguifh themfelves even at a confiderable diftance. With us, however, they are feldom fucceeded by any feeds.

The PROPAGATION of this plant is more eafy than to keep it within bounds; for the roots creep to a confiderable diftance, and fend up ftalks which produce flowers; fo that thefe being taken off, will be proper plants for any place. It likes a moift foil.

HYPE-

HYPERICUM.

Linnean Clafs and Order, *Polyadelphia Polyandria :* Each flower contains many males and about three females ; the males being joined at the bafe in many fets : There are forty Species; two of which are proper for ornamental fhrubberies.

1. Hypericum *Hircinum :* The Shrubby St. John's Wort, or the Stinking or Goat-scented St. John's Wort; *a deciduous fhrub ;* grows naturally by the fides of rivers in Sicily Calabria and Crete.

2. Hypericum *Canarienfe :* The Canary St. John's Wort ; *a deciduous fhrub ;* native of the Canaries.

1. The Shrubby St. John's Wort. Of this there are feveral *varieties.* The Common is a beautiful fhrub, near four feet in height. The branches are fmooth, of a light brown, and come out oppofite by pairs from the fide of the ftrongeft ftalks ; and thefe alfo fend forth others, which alternately point out dif-ferent directions. The leaves are of an oblong, oval figure, grow oppofite by pairs, and fit very clofe to the ftalks. Thefe being bruifed, emit a very ftrong difagreeable fcent. The flowers are yellow, and make a good fhow in June and July; for they will be produced in fuch clufters, at the ends of the young fhoots, that the fhrub will appear covered with them. They are fuc-ceeded by oval black-coloured capfules, containing ripe feeds, in the autumn.

There is a *variety* of this fpecies, which will grow to be eight feet high : The ftalks are ftrong, the leaves broad, and the flowers large ; and being produced in great plenty, caufes it to be a valuable fhrub for the plantation. There is another variety with variegated leaves, which is admired by thofe who are fond of fuch kinds of plants. There is alfo a variety difpoffeffed of the difagreeable fmell, which caufes it to be preferred by many on that account.

2. The Canary St. John's Wort is a fhrub of about fix or feven feet high : The branches divide by pairs, and the leaves, which are of an oblong figure, grow oppofite by pairs, without any footftalks. The flowers come out in clufters from the ends of the branches : They are of a bright yellow, have

O 3 numerous

numerous ſtamina, which ·are ſhorter than the petals, and three
ſtyles. They appear in July and Auguſt, and are ſucceeded by
oval roundiſh capſules, containing the ſeeds.

No art need be uſed in PROPAGATING theſe ſhrubs; for, 1.
Having obtained a plant or two of each, they will afford encreaſe
enough by ſuckers. Having ſtood about three years, the whole
of each plant ſhould be taken up, and the ſuckers and ſlips
with roots that this may be divided into, may reaſonably be
ſuppoſed to be twenty in number. The ſtrongeſt of theſe may
be planted where they are to remain, while the weaker may be
ſet out in the nurſery to gain ſtrength. 2. Theſe ſhrubs may
alſo be propagated by ſeeds, which ripen well with us, and will
come up with common care; nay, they will often ſhed, their
ſeeds, which will come up without ſowing, eſpecially the laſt
ſort.

J A S M I N U M.

LINNEAN Claſs and Order, *Diandria Monogynia:* Each
flower contains two males and one female: There are ſix
SPECIES; three of which are hardy enough for our purpoſe.

1. JASMINUM *Officinale:* The COMMON WHITE JASMINE;
a deciduous ſhrub or climber; native of India.

2. JASMINUM *Fruticans:* The COMMON YELLOW JAS-
MINE; *a deciducus ſhrub or climber;* native of the South of
Europe, and of the Eaſt.

3. JASMINUM *Humile:* The ITALIAN JASMINE; *a deciduous
ſhrub or climber;* native of Italy.

1. The COMMON WHITE JASMINES have uſually been
planted againſt walls, &c. for the branches being ſlender, weak,
and pithy, by ſuch aſſiſtance they have arrived to a good height;
though this ſhrub is not the moſt eligible for that purpoſe, as
its branches, which are numerous, are covered with a brown
dirty-looking bark, and afford ſhelter for ſnails, ſpiders, and
other inſects, which in winter, when the leaves are fallen, will
give

give them an unfightly look; and if they are clipped and kept up to the wall, as the flowers are produced from the ends and wings of the fhoots, thefe muft of courfe be fheared off; fo that little bloom will be found, except what is at the top of the tree. It is not meant, however, to diffuade thofe people who are fond of it from planting it againft walls : It naturally requires fupport, though attended with thofe defects. It may neverthelefs be planted among fhrubs in the fhrubbery, to appear to great advantage. It fhould keep company with the lower kinds of fhrubs ; and whenever the branches grow too high to fuftain themfelves without nodding, and difcover their rufty ftems, thefe fhould be taken off from the bottom. There will always be a fucceffion of young wood; and thefe young fhoots, which are covered with a fmooth bark, of a delightful green colour, alfo exhibit the leaves and bloom. The leaves are pinnated, and very beautiful : They grow oppofite by pairs ; and the folioles are ufually three pair in number, befides the odd one with which each leaf is terminated. They are all of a dark ftrong green colour, are pointed, and the end one is generally the largeft, and has its point drawn out to a greater length. The flowers are produced from the ends and joints of the branches, during moft of the fummer months : They are white, and very fragrant; but are fucceeded by no fruit in England.

There is a *variety* of this fort with yellow, and another with white ftriped leaves.

2. The YELLOW JASMINE is often planted againft walls, pales, &c. as the branches are weak and flender; and it will grow to be ten or twelve feet high, if thus fupported. It may, however, be planted in fhrubbery-quarters, in the fame manner as the other. The young fhoots are of a fine ftrong green colour, angular, and a little hairy. The leaves are trifoliate, though fometimes they grow fingly : They are placed alternately on the branches, are of a thick confiftence, fmooth, and of a fine deep green colour. Thefe leaves, in well-fheltered places, remain until the fpring before they fall off; fo that this plant may not improperly be ranked among evergreens, efpecially as the young fhoots are always of a ftrong green. The flowers are yellow, and do not poffefs the fragrance of the preceding fpecies : They are produced in June, and the blow is foon over; but they are fucceeded by berries, which, when ripe, are black.

Thefe

Thefe have occafioned this fort to be called by fome perfons the Berry-bearing Jafmine.

3. The ITALIAN JASMINE is, of all the forts, beft adapted to a fhrubbery, becaufe it lofes part of its beauty if nailed to a wall. It is naturally of lower growth, and the branches are ftronger, fewer in number, able to fupport themfelves in an upright pofition, and are angular. The bark is fmooth, and of a fine deep green colour. The leaves grow alternately : They are chiefly trifoliate, though fome pinnated ones are found upon this fhrub. The folioles are fmooth, and of a fine ftrong green : They are much broader than the preceding forts, and often continue till fpring before they drop off; fo that this fhrub, on account of the beautiful green colour of the young fhoots, might have a place among evergreens. The flowers are yellow, and much larger than thofe of the other forts : They are produced in July, and are fometimes fucceeded by berries; but thefe feldom if ever come to perfection. This fpecies is very hardy, and has grown in the moft expofed places, refifting the fevereft frofts for many years.

Little need be faid concerning the PROPAGATION of thefe plants; for they will all grow by layers or cuttings; fo that if either way be purfued in the winter, you will have plenty of plants by the autumn following. The cuttings, however, muft have a moift good foil, and fhould be fhaded and watered as the hot weather comes on, the beginning of fummer. The Common Yellow Jafmine may be propagated by the feeds; but it naturally fends forth fuch plenty of fuckers as to render it needlefs to take any other method for its encreafe; for thefe being taken off, will be good plants; nay, if it is planted in borders, they muft be annually taken for ufe, or thrown away, or they will overfpread every thing that grows near them. The Yellow and White ftriped-leaved Jafmines are propagated by grafting, budding, or inarching, into ftocks of the Common White : They are rather tender, efpecially the White, therefore muft have a warm fituation. The Yellow-ftriped is the moft common and leaft beautiful, and may be encreafed by layers and cuttings, like the plain fort.

I L E X.

I L E X.

Linnean Clafs and Order, *Tetrandria Tetragynia* : Each flower contains four males and four females : There are three Species ; one of Europe, one of Afia, and one of America : the laft has been introduced into our gardens and fhrubberies, making with our own Holly two fpecies.

1. Ilex *Aquifolium* : The Common Holly, or the European Holly ; a well known *evergreen tree or fhrub* ; native of moft parts of Europe, particularly of England.

2. Ilex *Caffine* : The American Holly, or the Dohoon Holly ; *an evergreen tree or fhrub* ; native of Carolina.

1. The European or Common Holly will grow to thirty or forty feet high, with a proportionable ftem. In its natural foil and fituation, namely, a high, chalky, marly, or limeftone loam, the ftem frequently fhoots up naked and filvery, fix or eight feet high, fupporting a clofe, fnug, elliptical head : This may be called its tree ftate. But the Holly, almoft as frequently, puts on a very different appearance ; feathering from the ground and rifing with an irregular, loofe, elegant outline ; forming one of the moft *Ornamental* evergreens which nature has furnifhed us with. What renders it in this point of view peculiarly valuable,—it is not only highly ornamental in finglets or groups ftanding in the open air, but will flourifh with great beauty under the fhade and drip of the more lofty deciduous tribes. Befides, the blufhing fruit of the Holly renders it moft ornamental at a time when the face of nature is in a manner divefted of every other ornament : In this light it is fuperior to the Box ; and, indeed, taken all in all, the Holly is undoubtedly entitled to take rank amongft the firft clafs of Ornamentals. In refpect of *Utility*, the Holly gives place to the Box ; except for the purpofe of hedges, and for this purpofe it ftands unrivalled ;—but of this under the Article Hedges. Its wood, however, is in good efteem among the inlayers and turners ; it is the *whiteft* of all woods ; its colour approaching towards that of Ivory.

The

The PROPAGATION of the Holly is principally from feeds.
EVELYN tells us that young feedlings, collected from the
woods, and planted in a nurfery, in a few years will make tolerable plants. MILLAR fays, the feeds " never come up the firft
year; but lie in the ground as the Haws do: therefore the
berries fhould be buried in the ground one year, and then taken
up and fown at Michaelmas, upon a bed expofed only to the
morning fun; the following fpring the plants will appear."—
" In this feed-bed, he fays, the plants may remain two years;
and then fhould be tranfplanted in the autumn, into beds
at about fix inches afunder, where they may ftand two years
longer, during which time they muft be conftantly kept clean
from weeds; and if the plants have thriven well, they will be
ftrong enough to tranfplant where they are defigned to remain;
for when they are tranfplanted at that age, there will be lefs
danger of their failing, and they will grow to a larger fize than
thofe which are removed when they are much larger."—He
alfo tells us, " the beft time for removing Hollies is in autumn,
efpecially in dry land; but where the foil is cold and moift,
they may be tranfplanted with great fafety in the fpring; if
the plants are not too old, or have not ftood long unremoved;
for if they have, it is great odds of their growing when removed." HANBURY differs with MILLAR in regard to fowing : he thinks the beft way is to fow them as foon as they are
ripe, and then, he fays, " they will undoubtedly come up the
fpring twelvemonth following."—" However, he adds, if the
feeds have been buried, let them be taken up in October; and
having fome fine light foil for the feminary, let them be fown
half an inch deep, and carefully covered from the mice." He
recommends, when the feedlings are two years old, that " in
the fpring they fhould be taken out of the beds and planted in
the nurfery in rows, a foot afunder, and two feet diftance between the rows. Here they may ftand until they are of a fufficient fize to be finally planted out." He follows MILLAR
as to the time of tranfplanting; recommending autumn if the
land be naturally dry : but if of a moift nature, he fays, " the
planter need not be very anxious about the time of the winter
in which he makes his plantations of Hollies." It is fomewhat
extraordinary that men, practical as MILLAR and HANBURY
undoubtedly

undoubtedly were, fhould not have been acquainted with the proper time of removing fo prevalent and fo ufeful a plant as the Holly : and it is ftill more remarkable, that the profeffional nurferymen of the prefent day fhould, in general, be involved in the fame darknefs. Spring is the very worft time for performing this bufinefs ; winter and autumn may be fomewhat more eligible ; but SUMMER is of all others the moft proper feafon for tranfplanting the Holly. At this time of the year, if the plants be young and well rooted, it matters not much how dry the foil is ; for they will, notwithftanding, fucceed with great certainty. MILLAR neverthelefs is right in faying that large Hollies which have not lately been moved are difficult to tranfplant ; more efpecially fuch as have ftood in a thicket, or under the fhade of other trees. If, however, even thefe be taken up with with good roots, together with a large quantity of native mould adhering to them : their heads leffened by pruning them in the conoidic manner, and be planted during the fummer months in a well-tempered pafte, agreeably to the directions given under the Article TRANSPLANTING, fuccefs, though it cannot be infured, may with great probability be expected. Thus far the Common Holly ;—which will be refumed again under the Article HEDGES.

But befides the genuine fpecies, there are of the European Holly almoft endlefs *varieties :* Millar mentions fome eight or ten ; and Hanbury enumerates upwards of forty ! Five of them are fufficiently diftinct to merit feparate defcriptions ; the reft, diftinguifhed chiefly by the variegation or mottled appearance of their refpective leaves, would only form a long lift of uncouth names, wholly uninterefting upon paper, how elegant foever they themfelves may appear in a group of Evergreens.

The Smooth-leaved Holly.

The Green-leaved Yellow-berried Holly.

The Box-leaved Holly.

The Hedge-Hog Holly.

The Saw-leaved Holly.

The *Smooth-leaved Holly* refembles in general appearance the Common fort. Of the two it feems to be the ftrongeft fhooter, and bids fair for the largeft growing tree. The leaves are nearly oval, and moft of them are entirely free from prickles, only they

end

end in acute points. This fort is commonly called the Carolina Smooth-leaved Holly : But it is a native of England, and is found growing amongft the others in many parts.

The *Green-leaved Yellow-berried Holly* differs in no refpect from the Common Holly, only the berries are yellow ; and as this tree produces berries in plenty, which are thought by moft people to be uncommon and curious, this fort, on their account, is deferving of a place, either in fmall or large gardens, in fhrubbery-quarters, or plantations of any kind.

The *Box-leaved Holly* has but little claim to be fo called ; for though fome of the leaves be fmall, pretty free from prickles, and nearly oval, yet there will be fo many nearly as prickly as the Common Holly as to merit no claim to that appellation. The leaves, however, are fmall ; and by them, on that account, the chief va riety is occafioned.

The *Hedge-Hog Holly* has the borders of the leaves armed with ftrong thorns, and the furface befet with acute prickles, a little refembling thofe of an hedge-hog, which gave occafion to this fort being fo called by the Gardeners. This, together with the Striped forts of it, is juftly ranked among our Hollies of the firft rate.

The *Saw-leaved Holly* is a kind very different from any of the other forts. The leaves are of the ordinary length, but very narrow, and of a thick fubftance. Their edges are formed into the likenefs of a faw; though hey are not very fharp and prickly. This is a very fcarce and valuable Holly, and is by all admired.

Thefe fix forts of themfelves form a Collection truly valuable to our evergreen ornamental plantations : if the variegated forts are alfo to have a place, which they may properly enough, we introduce then a frefh Collection, which for variety and beauty far exceeds not only the variegated forts belonging to any one genus, but perhaps all the variegated forts of trees and fhrubs put together.

But let us proceed to the CULTURE of thefe forts. We have already fhewn how the Common Englifh Holly may be raifed from the berry. That method is to be practifed, and plenty of that fort may be raifed. Thefe are to be ftocks, on

which

which the others are to be budded or grafted : for though they
will take by layers, yet plants raifed that way are of little or no
value; and if the berries of the variegated forts be fown, the
plants will come up plain, and be our Common Englifh Holly
(tho' from Hedge-Hog berries plants of the Hedge-Hog Holly are
frequently raifed). By grafting or budding, then, thefe forts muft
be propagated ; and for this purpofe young ftocks muft be raifed
of the Common Holly, as has been already directed. After thefe
have ftood two years in the feed-bed, they fhould be taken up,
have their roots fhortened, and be planted out in the nurfery, a
foot afunder, in rows at two feet diftance. The fummer 1-
lowing they will probably make few fhoots; but the fummer
after that, they will fhoot ftrongly ; and when the operation is to
be performed by grafting, thefe will be proper ftocks for the
purpofe by the fpring following. The firft week in March is a
good time for the work. Whip-grafting is the method to be
practifed ; and it muft be performed on the young wood, namely,
on that of the preceding fummer's fhoot. The cions being cut
true and even, and well jointed to the ftock, many of them will
grow ; and this is a very good method of encreafing thefe trees.
They may alfo be multiplied at pleafure by inoculation. This
operation is beft performed about ten days after Midfummer, in
cloudy weather; and for want of this, evening fhould be the
time : and if much work is to be done, morning too may be
added ; nay, it may be practifed all day in the hotteft feafons,
with tolerable fuccefs ; but this is never fo eligible, unlefs when
the multiplicity of work obliges us to lofe no time. The young
wood of the preceding fummer's fhoot is proper for the purpofe;
and the operation is to be performed in the ufual way. In the
autumn the bands fhould be loofed, and in the fpring the ftocks
dreffed up, and headed two or three inches above the bud; the
buds will be as early in fhooting out as any of the fhoots of the
growing trees, and will foon become good plants for any place.

1. The DOHOON HOLLY is an American plant, particularly of
Carolina, where it grows to be nearly as large a tree as our Holly
does with us. It naturally rifes with an upright ftem, which is
covered with a brown bark, and this affords plenty of younger
branches, whofe bark is green and very fmooth. The leaves are
pretty large, and of an oval lanceolated figure ; they are of a
thickifh

thickifh compofition, of a fine green, and grow alternately on the branches. Their edges are ferrated, though altogether different from the Common Sawed Holly, their ferratures towards the upper end of the leaf being fmall and fharp. The leaf, on the whole, is of a fine compofition, and grows on fhort footftalks on the branches. The flowers are fmall and white, and a little refemble thofe of the Common Holly. They are produced from the fides of the branches, in fhort thick clufters; and are in their native climate fucceeded by red berries, equalling thofe of our Common fort in beauty.

The Dohoon Holly may be PROPAGATED by feeds, which we receive from the countries where it grows naturally; for the berries will not ripen, and indeed are very feldom produced, in England. The beft way is to fow them in pots filled with light fandy earth, as foon as they arrive, and then plunge them up to the rims in the natural mould, where they may remain until the fpring following; for they rarely ever come up the firft fummer. The fpring after that the plants will appear; and if they have then the affiftance of a hotbed, it will greatly help them forward. They muft be ufed to the open air foon. The pots muft be taken up and plunged in a fhady place, and in October they fhould be removed into the greenhoufe for the winter. In the fpring the plants in the pots may be thinned by drawing out the ftrongeft; and thofe thus drawn fhould be planted each in a feparate pot, and muft be fet forward with a hotbed as before. The others, alfo, may be taken out at two or three years growth, planted in pots, and affifted in the fame manner. Every October they fhould be removed into the greenhoufe, fet out in the fpring, and treated as greenhoufe plants, until they are at leaft five or fix years old; for before then they will be hardly woody enough to venture the planting them out to ftand. The latter end of March; when the danger of bad weather is chiefly over, is the beft time for the purpofe; and if they have a dry foil and a warm fituation, they will bear the cold of our common winters; though if a very fevere winter fhould happen before they are got very ftrong and woody, it is more than probable that all of them will be deftroyed.

I T E A.

I T E A.

Linnean Clafs and Order, *Pentandria Monogynia :* Each flower contains five males and one female : There is only one Species :

Itea *Virginica :* The Itea'; *a deciduous fhrub ;* native of Virginia.

The Itea is a plant of about five or fix feet in height. The branches are numerous, and are produced irregularly all round. The leaves with which they are ornamented are of a fine green colour, gentle ferratures poffefs their edges, their figure is that of a fpear, and they grow alternately on the twigs. But the flowers conftitute the greateft beauty of thefe fhrubs ; for they are produced in July, at the ends of the young fhoots, in large erect fpikes : Their colour is white ; and as moft of the branches will be terminated by them, the tree itfelf appears at a diftance like one large bunch of white flowers : So delightful is the variety which Nature furnifhes for our contemplation and pleafure.

The propagation of this beautiful fhrub is not very eafy ; though it may be propagated by feeds and layers. 1. We receive the feeds from abroad. They fhould be fown in pots or boxes of fine loamy earth, mixed with drift or fea fand ; and thefe fhould be plunged up to the brim in the moifteft part of the garden, where they may remain till the fpring after ; for the feeds feldom come up the firft year. In March, therefore, the pots fhould be taken up, and plunged into an hotbed, which will promote the growth of the feeds, and make them become ftronger by the autumn. After the heat of the bed is over, they may be put in the fame moift places again. The plants ought to be conftantly weeded and watered ; and in the autumn fhould be removed into the greenhoufe, or placed under an hot-bed-frame, to be protected in fevere weather. This care fhould be continued through the next winter alfo. In the fpring, a damp day being made choice of, and a moift part of the nurfery being well prepared, they fhould be taken out of the pots or boxes, and planted at about a foot afunder, which will be

diftance

diftance enough for their ftanding two or three years, when they will be of a fufficient fize to be finally planted out. 2. Thefe trees are alfo propagated by layers ; for which purpofe, fome of them fhould be planted for ftools in a moift rich foil. The young fhoots of the preceding fummer fhould be laid in the ground in the autumn ; and in order to make them ftrike root, a little wire fhould be twifted pretty clofe round the bud, where the root is defired to be : This wire impeding the motion of the fap the fucceeding fummer, will occafion them to fwell in thofe parts, and ftrike root. There are other methods by which the operation may be performed ; but this has been found the moft expeditious and fureft.

J U G L A N S.

LINNEAN Clafs and Order, *Monoecia Polyandria :* Male flowers containing many ftamina, and female flowers containing two piftils, upon the fame plant ; the males being collected in oblong catkins, the females fitting in clufters clofe to the branches : There are five SPECIES, four of which are fufficiently hardy for this climate.

1. JUGLANS *Regia :* The COMMON WALNUT ; *a deciduous tree;* whofe native country is uncertain.

2. JUGLANS *Nigra :* The BLACK VIRGINIA WALNUT ; *a deciduous tree ;* native of Virginia, Carolina, and Maryland.

3. JUGLANS *Alba :* h e HICKERY-NUT, or WHITE VIRGINIA WALNUT ; *a low deciduous tree ;* native of Virginia. .

4. JUGLANS *Cinerea :* The PENNSYLVANIA WALNUT ; *a low deciduous tree ;* native of Pennfylvania and other parts of North-America.

1. The WALNUT-TREE. This as a fruit-tree is univerfally known. We fometimes fee it rife to a confiderable height, and grow to great fize ; in general, however, it does not rife higher than fortyfeet ; fpreading out into a globular inelegant head ; this added to the late feafon at which it puts forth its leaves, and the ftiff uncouth appearance it takes after their fall, renders it of low value as an *Ornamental* ; unlefs indeed its

general

general appearance be heightened, by the imagination, with the idea of *Utility*. In this point of view, whether we confider its fruit, or the timber it produces, the Walnut ranks high, and is no doubt an object of notice to the planter. We are far, however, from being fuch enthufiafts to the Walnut as EVELYN was: indeed its ufes as a timber are greatly leffened fince his day: Mahogany has fuperfeded it in the more elegant kinds of furniture; and the Beech, being raifed at lefs expence, and, from the cleannefs of its texture, being worked with lefs trouble, has been found more eligible for the commoner forts; chairs more efpecially. Neverthelefs, the Walnut is ftill a ufeful wood: it takes a fine polifh, and is in good efteem among the cabinet-makers, turners, and gunfmiths. Were the importation of Mahogany to be obftructed, the Walnut it is probable would become a very valuable wood.

The method of PROPAGATING the Walnut is from feeds: EVELYN recommends the Black Virginia fort for timber; and fays the nuts fhould be kept in the hufks, or fhucks, until March; when they fhould be planted in the hufks; for, he fays, " the extreme bitternefs thereof is moft exitial and deadly to the worm; or it were good to ftrew fome furzes, broken or chopt fmall, under the ground among them, to preferve them from mice and rats when their fhells begin to wax tender." He recommends their being planted in the place where they are to abide; the Walnut being very impatient of tranfplanting. If, however, it be neceffary to remove the plants, he cautions us not to touch the head with the knife; nor even the tap-root, except when very young. Speaking of experienced hufbandmen, he fays, " what they hint of putting a tile-fhard under the nut, when firft fet, to divaricate, and fpread the roots (which are otherwife apt to penetrate very deep) I like well enough." And from the fame fource he was informed, " that if they be tranfplanted as big as one's middle, it may be done fafer than when young:" he adds, however, " I do only report it." MILLAR alfo recommends the Black Virginia fort for timber, which he fays " is much more inclinable to grow upright than the Common fort; and the wood being generally of a more beautiful grain, renders it preferable to that, and better worth cultivating." " I have feen fome of this wood," continues he, " which hath been beautifully veined with black and white,

P which

which when polifhed has appeared at a diftance like veined
marble." The nuts, he fays, fhould be kept in the hufks till
February, the proper time of fowing. If the trees be intended
for timber, the nuts fhould be planted where they are to remain;
but if for fruit, in a feed-bed; becaufe tranfplantation checks
their upward growth, and renders them more fruitful. For
timber, " they fhould be planted in lines at a diftance you in-
tend them to remain; but in the rows they may be placed
pretty clofe, for fear the nuts fhould mifcarry; and the young
trees where they are too thick may be removed, after they have
grown two or three years, leaving the remainder at the diftance
they are to ftand." He alfo cautions againft too free a ufe of
the pruning-knife, either to the roots or the branches; but
when there is a neceffity, he fays, of cutting any of their
branches, " it fhould be done early in September."—He adds,
" The beft feafon for tranfplanting thefe trees is as foon as the
leaves begin to decay, at which time, if they are carefully
taken up, and their branches preferved entire, there will be little
danger of their fucceeding, although they are eight or ten
years old; though thefe trees will not grow fo large or continue fo long as thofe which are removed young." HANBURY
likewife follows EVELYN in recommending the Black Virginia
Walnut in preference to the Common fort for timber. His further
directions concerning the choice of the feed and the method of
propagation are as follow : " If the fruit of thefe trees are
greatly coveted, the utmoft care fhould be taken to gather
the nuts from thofe trees which produce the beft forts;
and although the varieties of Walnuts are only feminal
variations, yet there is the greater chance of having a fucceffion of good nuts, if they are gathered from trees that
produce good fruit. This maxim holds good in animals : The
fineft breed would degenerate, if attention was not paid to the
forts for breeding; and the like care muft be extended throughout the whole fyftem of planting, whether for fruit or timber.
If for timber, we fhould be folicitous to gather the feeds from
the healthieft, the moft luxuriant and thriving young trees : If
for fruit, from thofe which produce the richeft and beft kinds.
Having marked the trees that produce the fineft nuts, either for
thinnefs of fhell or goodnefs of tafte, when they have begun to
fall they will be ripe enough for gathering. But as collecting
them

them by the hand would be tedious, they may be beat down by
long poles prepared for that purpofe. Having procured the
quantity wanted, let them be preferved, with their husks on,
in fand till the beginning of February, which is the time for
planting them. , This is to be done in the following manner :
Let drills be made acrofs the feminary, at one foot afunder,
and about two inches and a half deep, and let the nuts be put
in thefe at the diftance of about one foot. In the fpring the
young plants will come up ; and here they fhould continue for
two years, being conftantly kept clear of weeds ; when they
will be of a proper fize to plant out in the nurfery. The
ground fhould be prepared, as has been always directed, by
double digging; and the trees being taken out of the feminary,
and having their tap-roots fhortened, fhould be planted therein,
in rows two feet and a half afunder, and the plants at a
foot and a half diftance. Here they may remain, with the
fame culture as has been all along directed for the management
of timber trees, till they are of a proper fize for planting out
for good. If they are defigned for ftandards to be planted in
fields, &c. before they are taken out of the nurfery they fhould
be above the reach of cattle, which may otherwife wantonly
break their leading fhoots, though they do not care to eat them
on account of their extraordinary bitternefs. They ought likewife to be removed with the greateft caution, and the knife
fhould be very fparingly applied to the roots. They muft alfo
be planted as foon as poffible after taking up; and this work
fhould be always done foon after the fall of the leaf."

EVELYN tells us, that the Walnut-Tree may be propagated
" by a branch flipped off with fome of the old wood, and fet in
February ;" and in another place, " it is certain they will receive their own cyons being grafted, and that it does improve
their fruit."

It is agreed on all hands, that the Walnut requires a dry,
found, good foil, and will make but little progrefs, as a timber
tree, in a cold barren fituation.

2. The BLACK VIRGINIA WALNUT. This is the fort recommended for timber, and will grow to a large tree. The
young fhoots are fmooth, and of a greenifh-brown. The leaves
are produced irregularly : They are large and finely pinnated,
being compofed of about eight, ten, twelve, and fometimes

fourteen pair of fpear-fhaped, fharp-pointed folioles, which are terminated by an odd one, fawed at the edges, and the bottom pair are always the leaft. The flowers give pleafure only to the curious botanift. They blow early in the fpring; and the females are fucceeded by nuts of different fizes and fhapes. The nuts of the Common fort have a very thick fhell, inclofing a fweet kernel. They are furrowed, and of a rounder figure than thofe of the Common Walnut.

There are many *varieties* of this fpecies, and nuts of different fizes, like thofe of the Common Walnut, will always be the effect of feed : Some will be fmall and round ; others oblong, large, and deeply furrowed. You muft expect alfo to find a variety in the leaves ; fome will have no fcent, others will be finely perfumed. Hence the names, *Common Virginian Walnut, Aromatic Walunt, Deeply-furrowed-fruited Walnut,* &c. have been ufed to exprefs the different varieties of this fpecies.

3. The WHITE VIRGINIAN WALNUT, called the Hickery-Nut, is a tree of lower ftature, feldom rifing more than thirty or thirty-five feet high ; though the fort called the Shag-Bark is the ftrongeft fhooter. The young fhoots of all are fmooth. The leaves are alfo pinnated, though fome of them are fmall, the number of folioles being from two or three pair to fix or feven, befides the odd one with which they are terminated. The folioles are of a pleafant green colour, narroweft at their bafe, and ferrated at their edges. The flowers are no ornament ; and the nuts are fmall, hard, and of a white colour.

The *varieties* of this fpecies go by the various names of *Common Hickery-Nut, Small-fruited Hickery-Nut, Shag-Bark Hickery-Nut,* &c.

4. PENNSYLVANIA WALNUT. This fpecies grows to about the height of the former. The leaves are very long, being compofed of about eleven pair of folioles, befides the odd one with which they are terminated. The flowers are yellowifh, come out at the ufual time with the others, and are fucceeded for the moft part by fmall, roundifh, hard-fhelled fruit ; though the nuts will be of different fizes in the different varieties.

The method of PROPAGATING thefe trees is from the nuts, which we receive from America, where they grow naturally. Thefe muft be fown as foon as they arrive, in the manner directed,

rected for raifing the Common Walnut. Their after-management muft alfo be the fame.

J U N I P E R U S.

LINNEAN Clafs and Order, *Dioecia Monadelphia* : —— Male
flowers containing three ftamina, and female flowers containing
three piftils, fituated on diftinct plants. There are ten
SPECIES ; nine of them as follow:

1. JUNIPERUS *Communis* : The COMMON JUNIPER ; *an
evergreen fhrub*; native of England, and of many of the northern
parts of Europe.

2. JUNIPERUS *Oxycedrus* : The SPANISH JUNIPER; *an
evergreen fhrub or tree* ; native of Spain and South of France.

3. JUNIPERUS *Virginiana :* The VIRGINIA CEDAR, or
the RED CEDAR ; *an evergreen tree* ; native of Virginia and
Carolina.

4. JUNIPERUS *Bermudiana:* The BERMUDIAN CEDAR; *an
evergreen tree* ; native of Bermudas and America.

5. JUNIPERUS *Barbadenfis* : The JAMAICA CEDAR ; *a tall
evergreen tree* ; native of Jamaica and other Weft-India Iflands.

6. JUNIPERUS *Thurifera* : The SPANISH CEDAR ; *an evergreen tree or fhrub* ; native of Spain.

7. JUNIPERUS *Lycia :* The LYCIAN CEDAR; *an evergreen tree or fhrub* ; native of Spain, Italy and France.

8. JUNIPERUS *'Phœnicea :* The PHENICIAN CEDAR; *an
evergreen fhrub or tree* ; native of Portugal, South of France and
the Eaft.

9. JUNIPERUS *Sabina:* The SAVIN ; *a very low evergreen fhrub* ; native of Italy, Siberia, Mounts Olympus and
Ararat, and of Lufitania.

1. The COMMON JUNIPER. This fpecies is divided into two
varieties :

> The Englifh Juniper.
> The Swedifh Juniper.

The

The *Englifh Juniper* grows common upon the challiy hills
about Banftead and Leatherhead in Surry, where it appears in
a low weak ftate, as if cropped and kept down by the browfing of
fheep, feldom rifing higher than two or three feet ; but when
planted in a good foil, it will grow to the height of fifteen or
fixteen feet, and will produce numerous branches from the
bottom to the top, forming a large well-looking, bufhy plant.
Thefe branches are tough, and covered with a fmooth bark of a
reddifh colour, with a gentle tinge of purple. The leaves are
narrow, and fharp-pointed : They grow by threes on the
branches ; their upper furface has a greyifh ftreak down the
middle, but their under is of a fine green colour, and they
garnifh the fhrub in great plenty. This tree flowers in April
and May. The flowers are fmall, of a yellowifh colour, and
make no figure. They are fucceeded by the berries, which are of
a kind of a blueifh purple when ripe, which will not be before
the autumn twelvemonth following.

The *Swedifh Juniper* has a natural tendency to grow to a
greater height, and confequently has more the appearance of a
tree than the former fort ; fixteen or eighteen feet, however, is
the higheft it commonly grows to ; and the plants raifed from
its feeds have, for the moft part, a tendency to grow higher,
and become more woody and ramofe. The leaves, flowers, and
fruit, grow in the fame manner, and are of the fame nature,
which fhews it to be a variety only. Old Botanifts mention it
as a diftinct fpecies : CASPAR BAUHINE afferts this, and calls
one the Shrubby Juniper, and the other Tree Juniper ; and he
alfo mentions another fort, which he calls the Leffer Mountain
Juniper, with a broader leaf and a larger fruit. This is ftill a
variety of the Common Juniper. The leaves, flowers, and fruit,
however, are much the fame ; though there may be fome difference in the fize of their growth. From what has been faid,
the Gardener will know, when he meets with them by thofe
different names, where to plant them in fuitable fituations. It
is obfervable of both thefe forts, that in the beginning or middle
of May, when they will be in full blow, the farina of the male
flowers is difcharged in fuch plenty, that upon ftriking the
fhrub with a ftick, it will rife up, in a ftill air, like a column
                                                          of

of white fmoke, and like that will be wafted with the gentleft wind, until it is loft or out of fight.

2. Spanish Juniper will grow to be rather an higher tree than the Swedifh, in fome foils. It will be feathered from the bottom to the top, if left untouched from the firft planting, or if not crowded with other trees. The leaves are awl-fhaped, and finely fpread open. They are very fhort, fharp-pointed, and give the tree a fine look. The flowers are fucceeded by large reddifh berries, which are very beautiful when ripe.

3. Virginia Cedar. This tree is held out by Evelyn and Hanbury as being proper to be planted as a foreft or timber-tree. It grows to near forty feet high, the branches forming a beautiful cone, and, if left unpruned, the tree will be feathered to the very bafe; confequently it is highly *ornamental*. And Hanbury fays, it " is valuable for many excellent and rare *ufes:* It will, he fays, continue found and uncorrupt for many ages, being poffeffed of a bitter refin, which prevents the worms from attacking it. The wood may be converted into utenfils of moft forts, as well as applied to great part of the ufes to which the cyprefs is adapted. It is remarkable, however, for being of a very brittle nature, and is therefore not proper to be introduced into buildings where any great weight is to be lodged. Never-thelefs, in Virginia and Carolina, where they abound, thefe trees are ufed in ftructures of all kinds, with this precaution; and the inhabitants prefer the timber to moft other wood for wainfcoting their rooms, and building of veffels."

4. Bermudian Cedar. In the ifland from which this tree takes its name, it grows to a timber fize; but in this country it is a very tender plant, and requires not only a dry warm foil, but open mild winters to make it continue thro' them; fo that when a perfon is defirous of having an extenfive collection, then and then only is this fort to be fought after; for, when planted abroad, even in the warmeft quarters, the plants require fheds to preferve them from the winter's froft. It is the wood of the Bermudian Cedar of which pencils are made; alfo drawers of cabinets; and formerly wainfcotting was made of this wood. In the ifland of Bermudas (which is in a manner covered with this tree), as well as upon the Continent of America, fhips are built of it: its fcent is peculiarly aromatic; but though agree-

able

able to moſt people, it is offenſive to ſome ; and this may be one reaſon why leſs of this wood has of late years been imported into this country than was formerly.

5. JAMAICA CEDAR. This, in Jamaica, is a very large timber-tree ; but in our climate it is ſtill more delicate than the Cedar of Bermudas ; both of which ought to bè conſidered as greenhouſe rather than as ſhrubbery plants.

6. SPANISH CEDAR grows plentifully in the country by whoſe name it is diſtinguiſhed : it is a handſome, regular-growing tree, riſing in a conical form, if the branches are untouched, to the height of thirty or more feet. The leaves are imbricated, and lie over each other four ways ; they are acute, and of a fine green colour : From theſe properties only, an idea of a fine tree may be had. The flowers are inſignificant to a common obſerver ; but they are ſucceeded by berries which make a good ſhow when ripe ; for they are very large, and of a fine black colour, and adorn the young branches in great plenty.

7. LYCIAN CEDAR, alſo common in Spain, will riſe to the height of about twenty-five feet ; the branches have naturally an upright poſition, and their bark is of a reddiſh hue. The leaves are every where imbricated, and each is obtuſe and of an oval figure. They reſemble thoſe of the Cypreſs, and are very beautiful. The flowers are ſucceeded by large oval berries, of a brown colour, and will be produced in plenty from the ſides of the younger branches all over the tree.

8. PHENICIAN CEDAR ſeldom grows higher than twenty feet, and is a beautiful upright ſort, forming a kind of pyramid, if untouched, from the bottom. It has both ternate and imbricated leaves ; the under ones grow by threes, and ſpread open ; and the upper ones are obtuſe, and lie over each other like the Cypreſs. The flowers are produced from the ends of the branches ; and the fruit that ſucceeds them is rather ſmall, and of a yellow colour. It is commonly called the Phenician Cedar, though it is found growing naturally in moſt of the ſouthern parts of Europe.

9. SAVIN. Of this ſpecies there are three ſorts :
    Spreading Savin,
    Upright Savin, and
    Striped Savin.

Spreading

Spreading Savin is a low-fpreading fhrub; the branches have a natural tendency to grow horizontally, or nearly fo; fo that it muft be ranked amongft the loweft growing fhrubs; infomuch that unlefs it is planted againft a wall, or.fupported in an upright-pofition, we feldom have it higher than two feet. When it is to be planted and left to Nature, room muft be firft allowed for its fpreading; for it will occupy a circle of more than two or three yards diameter, and will choke any other lefs powerful fhrub that is placed too near it. The bark on the older fhoots is of a light-brown colour; but the younger, which are covered with leaves running into each other, are of as fine a green as any fhrub whatever. Thefe leaves àre erect, and acute-pointed. They are placed oppofite, and grow a little like thofe of the French Ta-marifk. This fhrub feldom produces flowers or berries; but when any berries do appear, they are fmall and of a blueifh colour. It deferves a place amongft low-growing evergreens, on account of the fine ftrong green of its leaves both in winter and fummer; but it is valuable for nothing elfe; for it produces neither flowers nor fruit ornamental, and is poffeffed of a very ftrong fmell; infomuch that, being ftirred by whatever runs amongft it, the whole air is filled with a foetid fcent, which is emitted from its branches and leaves, and which to moft people is difagreeable. It is in great requeft with horfe-doctors and cow-leeches, by which they much benefit thofe creatures in many diforders. The juice of it, mixed with milk and honey, is faid to be good to expel worms from children; as well as, without that mixture, to deftroy thofe in horfes, for which purpofe it is ftrongly recommended.

Upright Savin is a delightful tree; it will grow to be twelve or fourteen feet high. The branches are numerous and flender, and give the tree a genteel air. The leaves are nearly of the fame nature with the other, though they are of a darker green. The flowers, though produced in plenty, make no fhow; but they are fucceeded by berries in fuch plenty as to caufe a good effect. The upright tendency of growth of this tree, together with the very dark green of the leaves, which caufes a good contraft with others that are lighter, together with its not being poffeffed of that ftrong difagreeable fcent of the other fort, makes it valuable for evergreen plantations.

Variegated

Variegated Savin is a variety of the former; it has not that tendency to spread like the Common, neither does it grow quite so upright as the Berry-bearing Savin. It is a fine plant, and at present rather scarce. The ends of several of the young shoots are of a fine cream-colour; nay, all the smaller branches appear often of that colour, and at a distance will have the appearance of flowers growing on the tree. In short, to those who are fond of variegated plants, this shrub has both beauty and scarcity to recommend itself.

The method of PROPAGATING this genus of plants varies in some degree with the respective species.

> The Common Juniper,
>
> The Spanish Juniper,
>
> The Virginia Cedar,
>
> The Spanish Cedar,
>
> The Lycian Cedar, and
>
> The Phenician Cedar,

are raised from seeds procured from the respective places of their growth, and sown the latter end of February or the beginning of March, in beds of light sandy earth, about half an inch deep. In about twelve months after sowing the plants will appear. Having stood two years in the seed-bed, they may be removed to the nursery, planting them from one to two feet distance in the rows, with two-feet intervals; and here they may remain until wanted for use. They may be transplanted either in autumn or in spring, care being had to perform the removal in moist weather, preserving as much mould as possible amongst the fibres. HANBURY recommends that the seeds of the COMMON JUNIPER be sown as soon as possible after they are ripe; for if this precaution, he says, is observed, they will come up the spring following; whereas if they are neglected till the spring, they will not appear till the spring after that; and sometimes a great part of them will remain till the second and even third season before they come up. The COMMON JUNIPER may also be increased by layers.

> The Bermudian Cedar, and
>
> The Jamaica Cedar,

require

require that the feeds be fown in pots ; that the young feedlings be planted out into feparate pots ; and that thefe be always houfed in the winter.

The COMMON SAVIN is to be increafed by flips, which if planted almoft at any time, or any how, will grow. The Upright Savin alfo is to be encreafed by flips planted in moift weather, in Auguft, and kept fhaded and watered in dry weather afterwards. This is the beft way of treating cuttings of the Upright Savin, though they will often grow if planted at any time, either, in winter or fummer. The Striped Savin alfo is to be encreafed this way ; though care muft be always ufed to take off thofe branches that are moft beautifully variegated, and fuch alfo as are entirely of a cream-colour ; for this will be the moft probable method of continuing it in its variegated beauties. This plant is alfo to be raifed by berries ; and if thefe have the fame treatment as the other forts, it will be very proper; and by thefe the moft upright and beft plants are raifed.

K A L M I A.

LINNEAN Clafs and Order, *Decandria Monogynia* : Each flower contains ten males and one female: There are two SPECIES :

1. KALMIA *Latifolia* : The BROAD-LEAVED KALMIA ; *an evergreen fhrub* ; native of Maryland, Virginia, and Pennfylvania.

2. KALMIA *Anguftifolia* : The NARROW-LEAVED KALMIA ; *an evergreen fhrub*; native of Pennfylvania and Carolina.

1. The BROAD-LEAVED KALMIA feldom rifes to more than four or five feet high ; and the branches, which by no means are regularly produced, are hard, and of a greyifh colour. The leaves are of an oval, fpear-fhaped figure, and of a fine fhining green colour. Their confiftence is rather thick in proportion to their footftalks, which are but flender, and grow irregularly on the branches. The flowers are produced at the ends of

the

the branches, in roundifh bunches. They are firft of a fine deep red, but die away to a paler colour. Each is compofed of a fingle petal, which is tubular at the bottom, fpreading open at the top, and has ten permanent corniculæ furrounding them on their outfide. They generally flower with us in July ; and are fucceeded by roundifh capfules, full of feeds, which feldom ripen in England. In fome places this is a fine evergreen ; and in others, again, it often lofes its leaves, and that fometimes before the winter is far advanced.

2. The Narrow-leaved Kalmia is rather of lower growth than the other, and the branches are more weak and tough. The leaves are very beautiful, being of a fine fhining green ; they are of a lanceolate figure, and in all refpects are fmaller than thofe of the former fort, and ftand upon very fhort footftalks. They are produced in no certain regular manner, being fometimes by pairs, at other times in bunches, growing oppofite at the joints. The flowers are produced from the fides of the branches in roundifh bunches ; they are of a fine red colour, and each is compofed of one petal, that has the property of fpreading open like the former. They flower in July, and are very beautiful ; but are not fucceeded by ripe feeds with us.

Both thefe forts are to be propagated three ways ; by feeds, layers, and fuckers. 1. By feeds. Thefe we receive from abroad ; and for their reception we fhould prepare a compoft, confifting of half frefh foil from a rich pafture, taken from thence a year before, and half drift or fea fand : thefe being well mixed, will be proper for the reception of the feeds, which fhould be fown in pots or boxes, half an inch deep. As foon as they are fown, they fhould be removed into a fhady place, to remain until the fpring following, and all this time nothing but weeding will be wanted ; for they feldom if ever come up the firft fummer. About the beginning of March it will be proper to plunge thefe pots into an hotbed, and this will fetch the plants up, and make them grow ftrong. They muft be hardened by degrees to the air, and then fet in a fhady place. Watering muft be now and then given them, if the feafon proves dry ; and at the approach of winter they may be removed into the greenhoufe, or fet under an hotbed-frame, but fhould always have the free air in open weather. In thefe pots or boxes they fhould remain until

they

they are two-years-old feedlings; when they should be shaken out, and planted in a separate pot. They should then be forced, by plunging the pots into an hotbed. Afterwards, they may be removed into the shade; and if they are kept growing in the pots, and removed under shelter in hard weather for a year or two, they may be afterwards planted out finally. 2. These shrubs are propagated by layering. It should be done in the autumn; and the young wood of the preceding summer's shoot is proper for the purpose. If the soil is free and light, they will strike root pretty readily; though we must sometimes wait two years before we find any: But by this way the strongest plants are obtained in the least time. 3. They are also encreased by suckers; for if the soil be light and fine, and is what agrees with them, after standing a few years, they naturally send out suckers in plenty. These should be taken off in the spring: and those with bad roots should be set in pots, and plunged into an hotbed, to make them grow.

L A V A T E R A.

LINNEAN Class and Order, *Monadelphia Polyandria*: Each flower contains many males and many females; the males being joined together at the base: There are nine SPECIES; four of which are cultivated in our open grounds:

1. LAVATERA *Arborea*: The COMMON LAVATERA, or MALLOW-TREE; *a deciduous shrub*; native of Italy.

2. LAVATERA *Triloba*: The THREE-LOBED LAVATERA, or MALLOW-TREE; *a deciduous shrub*; native of Spain.

3. LAVATERA *Olbia*: The FIVE-LOBED LAVATERA, or MALLOW-TREE; *a deciduous shrub*; native of the South of France.

4. LAVATERA *Micans*: or the GLITTERING LAVATERA; or the SHINING-LEAVED MALLOW-TREE; *a deciduous shrub*; native of Spain and Portugal.

1. The COMMON LAVATERA is a well-known plant: It usually grows to eight or ten feet high, and in a rich soil will

grow

grow to twelve, or more. The ftem is thick and ftrong, and divides near the top into feveral branches, which are clofely ornamented with large downy leaves; they are foft to the touch, plaited, and their edges are cut into many angles. The flowers are produced in clufters, from the wings of the leaves, in June, and
there will be a fucceffion of them until late in the autumn. Each
flower has its feparate footftalk: Their colour is purple; their
fhape like that of the Common Mallow; and they would make a
great fhow, were they not much obfcured by the largenefs of the
leaves. The whole tree has a noble look; and its continuing for
about three months in flower makes it very valuable. But tho
its fhort-lived continuance is much to be regretted, yet Nature
feems to have made fome amends for this, by furnifhing it with
good feeds in very great plenty; for by thefe thoufands of plants
may be foon raifed; nay, they will fometimes fhed themfelves,
and come up without any art. But when they are to be regularly
fown, let it be done in April, in the places where they are defigned to remain, and they will flower the fummer after. Tho'
this plant is called a biennial, in fome warm dry fituations the
ftalks become hard and woody, and the plants will continue to
produce flowers and feeds for many years.

There are feveral *varieties* of this fpecies; the leaves of fome
being round and indented, others acutely cut, others waved:
Thefe among old gardeners go by the names of the *Round-leaved,*
Waved-leaved Mallow-Tree, &c.

2. The THREE-LOBED LAVATERA. This fpecies is very ornamental in the front, or among the low fhrubs in the wildernefs-quarters, or when ftationed in large borders in pleafure-
grounds, as it is naturally of low growth, feldom rifing to above
four or five feet high. It has rather a large fpreading root, in
proportion to the fize of the fhrub. The branches are numerous,
and of a palifh green colour; and the leaves are of different
figures, though chiefly trilobate, or compofed of three lobes, that
are indented on their edges. They vary much in their fize, fome
being larger, fome fmaller, and fome more divided than others.
Their colour, when the plant is in perfect health, is a very pleafant green; but they will often fhew themfelves a little variegated; at which time you may be affured the fhrub is in a fickly
ftate. This often does not continue long, and the plant will
assume

affume its former verdure, and as frequently and very fpeedily relapfe into its weak ftate ; which fhews that, though hardy with refpect to cold, it is rather of a fickly nature in this country. The flowers are produced fingly, on fhort footftalks : They grow from the joints, at the bofoms of the leaves ; three or four of them will appear at each joint ; and being large, they make a fine fhow in Auguft, the time of flowering.

There are *varieties* of this fpecies, differing in the fhape of the leaves and fize of the flowers, which ftill have names among old botanifts.

3. The FIVE-LOBED LAVATERA is a diftinct fpecies from the preceding, though it differs little from it, except in the nature of the leaves, each of which is compofed of five lobes, that are haftated, or pointed like a fpear, and in the flowers of this fhrub being fmaller. They will be in full blow in Auguft, and there will often be a fucceffion of them till the early frofts advance. The leaves of this fpecies vary : Some are fhaped like Briony ; others are nearly round ; and the lobes of others are very acute. Hence the names Briony-leaved, Round-leaved, Acute-leaved Mallow-Tree, &c. have been ufed to exprefs them.

4. GLITTERING LAVATERA grows to about the height of the former. The leaves are large, feptangular, plaited, downy, white, and gliften towards the fun. The flowers are produced in bunches, from the ends of the branches : They are fhaped like thofe of the Common Mallow, come out in July, and continue in fucceffion until the end of autumn.

The *varieties* of this fpecies go by the names of *Waved-leaved*, the *Common Spanifh*, the *Sulphur-leaved Mallow-Tree*, &c. Thefe three forts are eafily PROPAGATED by cuttings, which fhould be planted, early in the fpring, in a fhady border of ligh rich earth. Many of them will grow ; and the plants may ftand two or three years before they are removed to the places where they are defigned to remain.

The beft method of PROPAGATING all the forts is by feeds ; and by this way frefh varieties may be obtained. The feeds fhould be procured from Spain, where the plants naturally grow ; for none, except the firft fort, ripen well here. Having got a fufficient quantity, fow them in a border of light, fine, rich earth, about the middle of March. They will eafily come up,

and

and nothing but weeding and watering in dry weather will be required until the fpring after : when they fhould be planted in nurfery-lines, there to remain until they are fet out to ftand. The leaves of all the forts continue until the frofts come on ; fo that if an open winter happens, they will continue in verdure the greateft part of the feafon.

The Lavatera tribe affect a warm fandy fituation and foil, in which they will fometimes continue to exhibit their beauties for many years; but in general they are fhort-lived, continuing only two or three years : this renders them peculiarly eligible to be fcattered plentifully in a newly made fhrubbery ; they will add warmth to young plants, and will die away themfelves before the fpaces they occupy will be required by the furrounding fhrubs.

L A U R U S.

LINNEAN Clafs and Order, *Enneandria Monogynia :* Each flower contains nine males and one female : There are eleven SPECIES ; four of which are adapted to our purpofe :

1. LAURUS *Nobilis :* The EVERGREEN BAY, or the COM-MON BAY ; *an evergreen fhrub or tree* ; native of Italy, Greece, and Afia.

2. LAURUS *Æftivalis :* The DECIDUOUS BAY; *a tall deciduous fhrub*; native of fwampy places, and the fides of brooks and rivers in Virginia.

3. LAURUS *Benzoin :* The BENZOIN TREE, or BENJAMIN TREE ; *a deciduous tree or fhrub* ; native of Virginia and Pennfylvania.

4. LAURUS *Saffafras :* The SASSAFRAS TREE ; *a deciduous tree or fhrub* ; native of Virginia, Carolina and Florida.

1. The EVERGREEN BAY. EVELYN fays, he has feen Bay Trees near thirty feet high and almoft two feet in diameter ; and enumerates the Bay amongft *ufeful* trees. HANBURY catches at this idea, and tells us in general terms, that " it will grow to

thirty

thirty feet in height, with a trunk of two feet in diameter," and, accordingly, he arranges it amongst his Forest Trees : he acknowledges however at the same time that the wood is of little value. The Bay is neverthelefs a fine aromatic and a beautiful evergreen : It is faid to be the true *Laurus* or Laurel of the antients, with which they adorned the brows of their fuccefsful generals. The leaves ftand clofe, are about three inches long and two broad ; are hard, rigid, and of a deep green colour. The Bay, too, like the Holly, Box, and Laurel, will bear the fhade and drip of taller trees, and it is upon the whole a very defirable, as being a very *ornamental*, evergreen. There are feveral *varieties :* as the Broad-leaved Bay, the Narrow-leaved Bay, and the Wave-leaved Bay.

This tree is PROPAGATED by layers, or by the berries. In order to raife a quantity of thefe trees by layers, fome ftools fhould be planted for the purpofe ; and after thefe are fhot about a yard high, the branches muft be brought down to the ground in the winter, all the preceding fummer's fhoots laid on it, and pegged down (being firft flit in the joint), and the leaves taken off, which would otherwife be under ground. In one year's time thefe layers will have taken root ; and in the fpring they fhould be taken up, and planted in the nurfery a foot afunder, in rows two feet diftance. After they are planted out, if the weather fhould prove dry, they muft be conftantly watered ; for without fuch care, it is difficult to make this tree grow. After they have taken well to the ground, they will require no farther trouble than keeping them clean from weeds, and digging between the rows each winter, till they are finally planted out. 2. In order to raife this tree from the berries, they ought to hang on the trees till about January before they are gathere . A well-fheltered fpot of ground for the feminary muft be made choice of ; and having the mould fmooth and fine, they fhould be fown foon after they are gathered in beds, or drills, rather more than half an inch deep. Towards the clofe of the fpring the plants will come up, and during fummer muft be duly attended, by watering and weeding. In the winter following, their fheltered fituation muft not be trufted to, to defend them from the froft : Furze-bufhes, or fome fuch things, ought to be ftuck in rows, between the beds or drills, to guard them from the black frofts. Indeed, without this precaution, if the winter fhould prove very

Q   frofty,

frofty, few of the young feedlings will be alive in fpring. During the following fummer, weeding and watering muſt be obſerved, and the winter after that they ſhould be defended with covering as before ; for they will be ſtil! in danger of being deſtroyed by fevere froſts. In the enſuing fpring, the ſtrongeſt may be taken out of the feed-beds, and planted in the nurſery way ; though if they have not by that time made good ſhoots, it will be adviſeable to let them remain in their beds till the third fpring ; for a ſmall plant of this kind is with more difficulty made to grow than one which is larger. When they are planted in the nurſery, the diſtance which ſhould be allowed them is the fame as the layers, a foot afunder and two feet diſtance in the rows ; and this will not be found too cloſe ; for notwithſtanding the greateſt care is exerted in planting them in the nurſery, even making choice of rainy and cloudy-weather, which muſt always be obferved in fetting them out, many of them will be loſt by being tranſplanted. After they are thus planted out in the nurſery, whether layers or feedlings, they muſt be ſtill watered in dry weather, kept free from weeds, and the rows dug between every winter. You will even find, that thoſe plants which ſuffer leaſt by being tranſplanted will have met with a check, which they will not recover in two or three years ; and till they have ac uired new ſtrength they ſhould not be taken from the nurſery ; but when they appear to be good ſtiff plants, having the year before made a vigorous ſhoot, they will be then proper plants for planting out where they are to remain. Holes ſhould be got ready for their reception ; and as foon as the firſt autumnal rains fall, the work ſhould be fet about, eſpecially if the land be gravelly or dry ; but if it be moiſt, the fpring will do as well. Being now planted at one yard diſtance, they will make a poor progreſs for two or three years more ; but after this, when they have overcome all theſe difficulties, they will grow very faſt, and arrive to be good trees in a few years.

Although this tree flouriſhes beſt in old gardens, where the foil has been made rich and deep, and loves the ſhade, HANBURY tells us, " it thrives neverthelefs exceedingly well in our hotteſt gravels and fands ; and, after it has furmounted the hardſhips of tranſplanting, will grow in ſuch ſituations extremely faſt, and arrive to a larger bulk. "

2. The

2. The Deciduous Bay, in a moiſt rich ſoil, in which it principally delights, will grow to be about ſixteen feet high ; but in ſome ſoils, that are poſſeſſed of the oppoſite qualities, it will hardly arrive at half that height. The branches are not very numerous, but they are ſmooth, and of a purpliſh colour, look well in winter, and in ſummer exhibit their leaves of an oval ſpear-ſhaped figure : They are about two or three inches in length, are proportionally broad, and placed oppoſite to each other on the branches. Their upper ſurface is ſmooth, and of a pleaſant green colour, whilſt their under is rough and veined. The flowers are ſmall and white, make no figure, come out from the ſides of the branches in May, and are ſucceeded by large red berries, which never ripen in England : So that, notwith-ſtanding the leaves in ſummer are very pretty, and the colour of the bark makes a variety in winter, it is principally the ſcarcity of this plant which makes it valuable.

3. The Benzoin Tree will grow to a much larger ſize than the other, and its branches are rather numerous : They are ſmooth, and of a fine light-green colour. The leaves are oval, acute, near four inches long, and two broad ; their upper ſurface is ſmooth, and of a fine light green colour, but their under ſur-face is venoſe, and of a whitiſh caſt : When bruiſed, they emit a fine fragrance. The flowers make no figure : They are ſmall and yellowiſh, come out from the ſides of the branches in little cluſters, and are ſucceeded by large blackiſh berries, which never ripen in England.

4. The Sassafras Tree. The wood of the Saſſafras is well known in the ſhops, where it is ſold to be made into tea, being eſteemed an excellent antiſcorbutic and purger of the blood. A decoction of the leaves and bark is alſo ſaid to poſſeſs the ſame virtues, and is drank by many perſons for thoſe purpoſes. This tree will grow to nearly the height of the others, though the branches are not ſo numerous. Its bark is ſmooth, and of a red colour, which beautifully diſtinguiſhes it in winter · whilſt the fine ſhining green of its leaves conſtitutes its greateſt beauty in ſummer. In theſe, indeed, there is a variety, and a very extraordinary one. Some are large, and of an oval figure ; others are ſmaller, and of the ſame ſhape ; whilſt others, again, are ſo divided into three lobes, as to reſemble the leaves of ſome ſorts of the Fig-tree. Their edges are entire ; their

under furface is of a whitifh caft ; their footftalks are pretty long, placed alternately on the branches, and die to a red co- lour in the autumn. The flowers are fmall and yellowifh : They are produced in clufters on longifh pedicles, and are fuc- ceeded by blackifh berries, which never ripen in England.

The PROPAGATION of thefe three forts of trees may be per- formed two or three ways, 1. By the feeds. Thefe we re- ceive, from the places where the trees grow naturally, in tho fpring. They fhould be preferved in fand ; and, as foon as they arrive, fhould be fown in largifh pots, an inch deep. The foil for their reception fhould be taken from a rich pafture at leaft a year before, with the fward. It fhould alfo be laid on an heap, and frequently turned, until the fward is grown rotten, and the whole appears well mixed and fine. If the pafture from whence it was taken near the furface is a fandy loam, this is the beft compoft for thefe feeds ; if not, a fmall addition of drift or fea fand fhould be added, and well mixed with the other mould. After filling the pots with this foil, the feeds fhould be fown an inch deep; and then they fhould be plunged into common mould up to the rim. If the foil be naturally moift, it will keep them cooler, and be better ; and if the place be well- fheltered and fhaded, it will be better ftill. Nothing more than weeding, which muft be conftantly obferved during the fummer, will be neceffary ; and in this ftation they may remain until the March following ; about the middle of which month, having prepared a good hot bed, the pots fhould be taken up and plun- ged therein. Soon after the feeds will come up ; and when the young plants have fufficiently received the benefit of this bed, they fhould be enured by degrees to the open air. Weeding and watering muft be obferved during the fummer ; and, at the ap- proach of the cold weather in the autumn, they fhould be re- moved under an hot-bed frame, or fome cover, to be protefted from the frofts during the winter. In the fpring, when this danger is over, they fhould refume their firft ftation ; namely, the pots fhould be plunged up to the rim, as when the feeds were firft fown ; and if this place be well fheltered, they may remain there all winter; if not, and fevere frofts threaten, they fhould be taken up and placed under cover as before. After they have been thus managed three years from the feeds, they

fhould

should be taken out of the pots with care, and planted in the nursery-ground, at small distances; where they may remain until they are strong enough to be finally set out. By sowing the seeds in pots, and assisting them by an hotbed, a year at least is saved; for they hardly ever come up, when sown in a natural border, under two years from the seeds; nay, they have been known to remain three, and even some plants to come up the fourth year after sowing; which at once shews the preference of the former practice, and should caution all who have not such convenience, not to be too hasty in disturbing the beds when the seeds are sown in the natural ground; as, especially if they are not well preserved in mould or sand, these may be some years before they appear. Indeed, it is the long time we are in obtaining these plants, either by seeds, layers, &c. that makes them at present so very scarce amongst us. 2. These plants may also be encreased by layers; but very slowly; for they will be two, and sometimes three, or even four years, before they have struck out good roots; though the Benjamin Tree is propagated the fastest by this method. The young twigs should be laid in the ground in the autumn; and it will be found that twisting the wire round the bud, so as in some degree to stop the progress of the sap, and taking away with a knife a little of the bark, is a more effectual method of obtaining good roots soon than by the slit or twisting, especially when practised on the Sassafras Tree. 3. Plants of these sorts are likewise sometimes obtained by suckers, which they will at all times throw out, and which may be often taken off with pretty good roots; but when they are weak, and with bad roots, they should be planted in pots, and assisted by a moderate heat in a bed: With such management they will be good plants by the autumn, and in the spring may be planted out any where. 4. Cuttings of these trees, when planted in a good bark-bed, and duly watered, will also oftentimes grow. When this method is practised, and plants obtained, they must be enured by degrees to the open air, till they are hardy enough to be finally planted out.

LIGUS-

L I G U S T R U M.

LINNEAN Clafs and Order, *Diandria Monogyuia :* Each flower contains two males and one female : There is only one SPECIES:

LIGUSTRUM *Vulgare :* The PRIVET; a well-known *deci-duous or evergreen fhrub ;* common in the woods and hedges of many parts of England, and almoft all Europe.

The PRIVET is divifible into two *varieties :*

The Deciduous or Common Privet, and

The Evergreen Privet,

The *Deciduous Privet* will grow to the height of about ten or twelve feet. The branches are very numerous, flender, and tough; covered with a fmooth grey bark ; and, when broken, emit a ftrong fcent. The young twigs are generally produced oppofite, and alternately of contrary directions on the older branches. The leaves alfo are placed oppofite by pairs in the fame manner. They are of an oblong figure, fmall, fmooth, of a dark green colour, have a naufeous difagreeable tafte, and continue on the trees very late. The flowers are produced in clofe fpikes. at the ends of the branches, in May, June, and often in July : They are white, very beautiful, and fucceed-ed by black berries, which in the autumn will conftitute the greateft beauty of this plant ; for they will be all over the tree, at the ends of the branches, in thick clufters. They are of a jet black ; and will thus continue to ornament it in this fingular manner during the greateft part of the winter.

The PROPAGATION of the Privet is eafy ; for it may be en-creafed, 1. By the feeds ; and by this way the ftrongeft plants may be obtained. The feeds, foon after they are ripe, fhould be fown in any bed of common garden-mould made fine. They ought to be covered about an inch deep ; and all the fucceed-ing fummer fhould be kept clean of weeds ; for the plants never, at leaft not many of them, come up until the fpring after. After they are come up, they will require no other care than weeding ; and in the fpring following may be planted in the nurfery-ground, where they will require very little care befides keeping the weeds down, until they are taken up to plant. 2. Thefe plants

may

may be encreafed by layers; for the young fhoots being laid in the ground in the autumn, will by that time twelvemonth have taken good root; the largeft of which may be planted out to ftand, and the fmalleft fet in the nurfery, to gain ftrength. 3. Cuttings, alfo, planted in October, will ftrike root freely; and if the foil is inclined to be moift, and is fhaded, it will be the better for them, efpecially if the fucceeding fummer fhould prove a dry one. If thefe cuttings are thinly planted, they will require no other removing till they are finally fet out. If a large quantity is defired, they may be placed clofe, within about two or three inches of each other, and then taken up and planted in the nurfery the autumn following, to remain there until they are wanted for the above purpofe.

The Privet, of all others, will thrive beft in the fmoke of great cities; fo that whoever has a little garden in fuch places, and is defirous of having a few plants that look green and healthy, may be gratified in the Privet, becaufe it will flourifh and look well there. It will alfo grow very well under the fhade and drip of trees.

The *Evergreen Privet*. This variety has a tendency to grow to be a taller and a ftronger tree than the common deciduous fort. The leaves are rather larger, more pointed, of a thicker confiftence, of a dark-green colour, and they continue on the fame plant fo long as to entitle it to the appellation of Evergreen; though it may often be obferved to be almoft deftitute of leaves early in winter, efpecially thofe that were on the ends of the higheft branches, which are often taken off by the firft cutting winterly winds. In order to have this tree keep up the credit of an evergreen, it fhould have a well-fheltered fituation; for although it be hardy enough to bear with impunity the fevereft cuts of the northern blafts, on the tops of hills, craggy rocks, &c. yet without fome fhelter the leaves are feldom preferved all winter, and with protection it is generally allowed to be an handfome evergreen. As it is a variety of the deciduous fort, the fame flowers and fruit may be expected.

It is to be raifed in the fame manner, by layers or cuttings; and the feeds of this fort fometimes produce plants of the like fort, that retain their leaves.

LIQUIDAMBER.

LINNEAN Clafs and Order, *Monoecia Polyandria :* Male flowers containing many ftamina, and female flowers containing two piftils, fituated upon the fame plant ; the males being collected into long conical loofe katkins, and the females forming a globe fituated at the bafe of the male fpike : There are only two SPECIES :

1. LIQUIDAMBER *Styraciflua :* The VIRGINIA LIQUIDAMBER, or the MAPLE-LEAVED LIQUIDAMBER ; *a deciduous tree ;* native of the rich moift parts of Virginia and Mexico.

2. LIQUIDAMBER *Peregrinum :* The CANADA LIQUIDAMBER, or the SPLEENWORT-LEAVED GALE ; *a deciduous tree ;* native of Canada and Pennfylvania,

1. The VIRGINIA LIQUIDAMBER will fhoot in a regular manner to thirty or forty feet high, having its young twigs covered with a fmooth, light-brown bark, while thofe of the older are of a darker colour. The leaves grow irregularly on the young branches, on long footftalks : They refemble thofe of the Common Maple in figure ; the lobes are all ferrated ; and from the bafe of the leaf a ftrong mid-rib runs to the extremity of each lobe that belongs to it. They are of a lucid green, and emit their odoriferous particles in fuch plenty as to perfume the circumambient air ; nay, the whole tree exfudes fuch a fragrant tranfparent refin, as to have given occafion to its being taken for the Sweet Storax. Thefe trees, therefore, are very proper to be planted fingly in large opens, that they may amply difplay their fine pyramidal growth, or to be fet in places near feats, pavilions, &c. The flowers are of a kind of faffron colour : They are produced at the ends of the branches the beginning of April, and fometimes fooner ; and are fucceeded by large round brown fruit, which looks fingular, but is thought by many to be no ornament to the tree.

2. CANADA LIQUIDAMBER. The young branches of this fpecies are flender, tough, and hardy. The leaves are oblong. of a deep green colour, hairy underneath, and have indentures

on

on their edges alternately, very deep. The flowers come out
from the fides of the branches, like the former; and they are
fucceeded by fmall roundifh fruit, which feldom ripens in
England.

The PROPAGATION of both thefe fpecies is the fame, and may
be performed by feeds or layers; but the firft method is the beft.
1. We receive the feeds from America in the fpring. Againft
their arrival a fine bed, in a warm well-fheltered place, fhould be
prepared. If the foil is not naturally good, and inclined to be
fandy, it fhould be wholly taken out near a foot deep, and the
vacancy filled up with earth taken up a year before, from a frefh
pafture, with the fward and all well rotted and mixed by being
often turned, and afterwards mixed with a fixth part of drift or
fea land. A dry day being made choice of, early in March, let
the feeds be fown, and the fineft of this compoft riddled over
them a quarter of an inch deep. When the hot weather in the
fpring comes on, the beds fhould be fhaded, and waterings given
often, but in very fmall quantities, only affording them a gentle,
nay, a very fmall fprinkling at a time. MILLAR fays, the feeds
of thefe plants never come up under two years. But, continues
HANBURY, with this eafy management, I hardly ever knew it
longer than the end of May before the young plants made their
appearance. The plants being come up, fhading fhould ftill
be afforded them in the parching fummer, and a watering every
other night; and this will promote their growth, and caufe them
to become ftronger plants by the autumn. In the autumn, the
beds fhould be hooped to be covered with mats in the fevere
frofts. Thefe mats, however, fhould always be taken off in
open weather; and this is all the management they will require
during the firft winter. The fucceeding fummer they will require no other trouble than weeding; though, if it fhould prove
a very dry one, they will find benefit from a little water now and
then. By the autumn they will be grown ftrong enough to
refift the cold of the following winter, without demanding the
trouble of matting, if the fituation is well fheltered; if not, it
will be proper to have the hoops prepared, and the matts ready,
againft the black northern frofts, which would endanger at leaft
their lofing their tops. After this, nothing except weeding will
be wanted; and in the fpring following, that is, three years from

their

their firſt appearance, they ſhould be taken up (for they ſhould not be removed before, unlefs ſome of the ſtrongeſt plants be drawn out of the bed) and planted in the nurſery, a foot aſunder, and too feet diſtant in the rows. Hoeing the weeds in the rows in the ſummer, and digging them in the winter, is all the trouble they will afterwards occaſion until they are finally planted out. 2. Theſe plants are eaſily encreaſed by layers. The operation muſt be performed in the autumn, on the young ſummer's ſhoots; and the beſt way is by ſlitting them at a joint, as is practiſed for carnations. In a ſtrong dry ſoil, they will be often two years or more before they ſtrike root; though, in a fine light ſoil, they will be found to take freely enough. By this method good plants may be obtained, though it is not ſo eligible as the other, if we have the conveniency of procuring the ſeeds.

LIRIODENDRON.

LINNEAN Claſs and Order, *Polyandria Polygynia :* Each flower contains many males and many females : There are two SPECIES ; one of them bearing a tulip-like, the other a lily-like flower ; the former is not uncommon in our open grounds :

LIRIODENDRON *Tulipifera :* The TULIP-TREE, or the VIR-GINIA TULIP-TREE ; *a deciduous tree* ; native of moſt parts of America.

The TULIP-TREE. In thoſe parts of America where it grows common, it will arrive to a prodigious bulk, and affords excellent timber for many uſes ; particularly, the trunk is frequently hollowed, and made into a canoe ſufficient to carry many people; and for this purpoſe no tree is thought more proper by the inhabitants of thoſe parts. With us, it may be ſtationed among trees of forty feet growth, The trunk is covered with a grey bark. The branches, which are not very numerous, of the two-years-old wood, are ſmooth and brown ; whilſt the bark of the ſummer's ſhoots is ſmoother and ſhining, and of a bluciſh colour. They

are

are very pithy. Their young wood is green, and when broken emits a ſtrong ſcent. The leaves grow irregularly on the branches, on long footſtalks. They are of a particular ſtructure, being compoſed of three lobes, the middlemoſt of which is ſhortened in ſuch a manner, that it appears as if it had been cut off and hollowed at the middle : The two others are rounded off. They are about four or five inches long, and as many broad. They are of two colours ; their upper ſurface is ſmooth, and of a ſtronger green than the lower. They fall off pretty early in autumn ; and the buds for the next year's ſhoots ſoon after begin to ſwell and become dilated, infomuch that, by the end of December, thoſe at the ends of the branches will become near an inch long, and half an inch broad. The outward lamina of theſe leaf-buds are of an oval figure, have ſeveral longitudinal veins, and are of a blueiſh colour. The flowers are produced with us in July, at the ends of the branches : They ſomewhat reſemble the tulip, which occaſions its being called the Tulip-tree. The number of petals of which each is compoſed, like thoſe of the tulip, is ſix ; and theſe are ſpotted with green, red, white, and yellow, thereby making a beautiful mixture. The flowers are ſucceeded hv large cones, which never ripen in England.

The PROPAGATION of the Tulip-tree is very eaſy, if the ſeeds are good ; for by theſe, which we receive from abroad, they are to be propagated. No particular compoſt need be ſought for ; neither is the trouble of pots, boxes, hotbeds, &c. required : They will grow exceeding well in beds of common garden-mould, and the plants will be hardier and better than thoſe raiſed with more tenderneſs and care. Therefore, as ſoon as you receive the ſeeds, which is generally in February, and a few dry days have happened, that the mould will work freely, ſow the ſeeds, covering them three quarters of an inch deep ; and in doing of this, obſerve to lay them lengthways, otherwiſe, by being very long, one part, perhaps that of the embryo plant, may be out of the ground ſoon, and the ſeed be loſt. This being done, let the beds be hooped ; and as ſoon as the hot weather and drying winds come on in the ſpring, let them be covered from ten o'clock in the morning till ſun-ſet. If little rain happens, they muſt be duly watered every other day ; and by the end of May the plants will come up. Shade and watering in the hotteſt ſummer

muſt

muſt be afforded them, and they will afterwards give very little
trouble. The next winter they will want no other care than, at
the approach of it, ſticking ſome furze-buſhes round the bed, to
break the keen edge of the black froſts; for it is found that the
ſeedlings of this ſort are very hardy, and ſeldom ſuffer by any
weather. After they have been two years in the ſeed-bed, they
ſhould be taken up and planted in the nurſery, a foot aſunder,
and two feet diſtant in the rows. After this, the uſual nurſery
care of hoeing the weeds, and digging between the rows in the
winter, will ſuffice till they are taken up for planting out.

L O N I C E R A.

Linnean Claſs and Order, *Pentandria Monogynia :* Each
flower contains five males and one female : There are fourteen
Species ; eleven of which will bear the open air of this
country.

1. Lonicera *Caprifolium :* The Italian Honeysuckle ;
a deciduous or evergreen climber ; native of Italy and the South of
Europe.

2. Lonicera *Periclymenum :* The English Honeysuckle,
or Woodbine ; *a deciduous or evergreen climber* ; native of
England, Germany, and the midland parts of Europe.

3. Lonicera *Sempervirens :* The Trumpet Honeysuckle ;
a deciduous or evergreen climber ; native of America.

4. Lonicera *Diervilla :* The Diervilla, or Acadian
Honeysuckle ; *a deciduous ſhrub* ; native of Acadia and
Nova-Scotia.

5. Lonicera *Symphoricarpus :* St. Peter's Wort, or the
Virginia Honeysuckle ; *a deciduous ſhrub* ; native of Virginia and Carolina.

6. Lonicera *Cærulea :* The Blue-berried Honeysuckle,
or the Upright Blue-berried Honeysuckle ; *a deciduous
ſhrub* ; native of Switzerland.

7. Lonicera *Alpigena*: The Red-berried Honeysuckle; or the Upright Red-berried Honeysuckle; or the Alpine Honeysuckle; *a deciduous shrub*; native of Savoy, and the Helvetian and Pyrenean Mountains.

8. Lonicera *Nigra*: The Black-berried Honeysuckle; or the Black-berried Upright Honeysuckle; *a deciduous shrub*; native of Switzerland and the Alps.

9. Lonicera *Xylosteum*: The Fly Honeysuckle; *a deciduous shrub*: native of most of the coldest parts of Europe.

10. Lonicera *Pyrenaica*: The Pyrenean Honeysuckle, or Dwarf Cherry; *a deciduous shrub*; native of the Pyrenean mountains.

1. Lonicera *Tartarica*: The Tartarian Honeysuckle, or Dwarf Cherry, or the Dwarf Cherry with Heart-shaped Leaves; *a deciduous shrub*; native of Tartary.

1. The Italian Honeysuckle. The *varieties* of this species are, Early White Italian Honeyfuckle, Early Red Italian Honeyfuckle, Yellow Italian Honeyfuckle, Late Red-flowered Italian Honeyfuckle, Evergreen Italian Honeyfuckle.

The *Early White Italian Honeyfuckle* is that which first makes its appearance in May. The leaves of this fort are oval, and placed opposite by pairs, close to the branches, at the extremity of which the leaves quite furround it. The flowers grow in bunches round the ends of the branches, and have a very fine fcent. Their blow will be foon over; and they are fucceeded by red pulpy berries, which will be ripe in the autumn.

The *Early Red* differs from the preceding in that the leaves are narrower, the fibres of the flowers are more flender, and it blows a little later in the fpring.

The *Yellow Italian Honeyfuckle* does not blow quite fo early as the other, and the flowers are yellow: In other refpects it is very much like the former.

Late Red-flowered Italian Honeyfuckle is one of the beft we have. The ftem is tolerably firm; the branches are few, and the leaves large; the flowers are alfo large, of a deep-red colour, though lefs fcented than the earlier forts,

Evergreen Italian Honeyfuckle. This is a ftronger fhooter than any of the forts. The joints are more diftant from each other. The leaves are large, of a thick confiftence, unite, and furround

the

the ftalk with their bafe, and continue all winter. The flowers are large, of a good red colour, with fome paler ftripes, and often continue to blow to the end of autumn.

2. ENGLISH HONEYSUCKLE. The *varieties* of this fpecies are, The Common Woodbine of our Hedges, The Oak-leaved Honeyfu kle. Red Dutch Honeyfuckle, Midfummer Honey-fuckle, Late German Honeyfuckle, Long-blowing Honeyfuckle, Evergreen Honeyfuckle.

The *Common Woodbine* is known all over England, in our woods and hedges. There are ftill varieties of this fort, in its' wild ftate; fome having prodigious weak trailing branches ; others again with tolerably woody ftems. Some of the flowers are whitifh, others are of a greenifh caft; whilft others are poffeffed of a reddifh tinge. As the flowers of none of thefe are nearly fo beautiful as thofe of the cultivated forts, only a plant or two of them fhould be introduced ; which will caufe fome variety, and ferve as a foil to fet the others off. There is a *fub-variety* of this fort, with ftriped leaves.

Oak-leaved Honeyfuckle is an accidental variety of our Common Woodbine. It differs in no refpect from it, only that fome of the leaves are fhaped like thofe of the Oak-tree, on which account it is valuable, and makes a pretty variety in collections.

There is alfo a *fub-variety* of this fort, with leaves beautifully variegated, called *Striped Oak-leaved Honeyfuckle.*

Red Dutch Honeyfuckle is a very good fort. It flowers in June, and will often continue in blow a month or two. The branches have a fmooth purplifh bark, and may be known from the others even in winter, when they will appear with their fwelled buds alfo of that colour. The leaves are of an oblong oval figure, and ftand oppofite by pairs on the branches, on fhort footftalks. The flowers are produced in bunches at the ends of the branches : Their outfide is red, but within they are of a yellowifh colour, and poffeffed of a delightful odour.

The *Midfummer Honeyfuckle* is very much like the former, only the ftalks are more flender, of a lighter-brown colour, and the tubes of the flowers are fmaller, neither are they fo red. It will be in blow about Midfummer ; and the plant, whether fet againft a wall, pales, a hedge, or in the ground, will be all over covered with bloom, making an enchanting appearance to the eye,

eye, and perfuming the air all around to a confiderable diftance.

Late German Honeyfuckle is very much like the Red Dutch,
only it blows later. It will flower in July and Auguft; and has
all the properties of the other forts, as to fragrance and beauty.

The *Long-blowing Honeyfuckle* is ftill another variety of the
Dutch. It will often exhibit flowers in June, July, and Auguft,
though the profufion will not be fo. great as that of the other
forts.

Evergreen Honeyfuckle is another variety which retains its
leaves all winter. It often flowers late in the autumn; and
fometimes, in mild feafons, retains its bloom until Chriftmas,
which makes it ftill more valuable.

3. TRUMPET HONEYSUCKLES. Of thefe are the following
varieties: Virginian Trumpet Honeyfuckle, Carolina Trumpet
Honeyfuckle, Evergreen Trumpet Honeyfuckle.

Virginian Trumpet Honeyfuckle is the moft beautiful of all the
forts, though Nature has denied it fmell. The branches are
flender, fmooth, and of a reddifh colour. The leaves fit clofe to
the branches by pairs. They are of an oblong oval figure, and
their lower furface is not of fo fhining a green as the upper.
Thofe at the extremity of the branches near the flowers furround
the ftalk, through which it comes. The flowers grow in
bunches, at the ends of the fhoot, and are of a bright fcarlet colour. They will often be in blow from June to October; but
the flowers have no fcent.

Carolina Trumpet Honeyfuckle differs in no refpect from the former, only that the branches are more flender, and the leaves
and flowers alfo are proportionably fmaller, thereby making a
pretty variety. This fort was introduced into our gardens from
Carolina, as was the preceding from Virginia.

Evergreen Trumpet Honeyfuckle. The leaves are of a thicker
fubftance, and continue on the plants all winter; but the flowers
are of a deep fcarlet, like the other, and are poffeffed of little or
no fragrance.

The PROPAGATION of thefe forts is very eafy. 1. The young
branches being laid in the ground any time in the winter, with
no other art, will become good plants by the autumn following,
and may be then taken off for ufe. 2. But our common method

of

of propagating thefe forts is by cuttings. The beſt month for this work is Oſtober. By this way prodigious quantities of plants may be raiſed, and hardly any of them will fail growing. So eaſily may thefe delightful plants be multiplied, when a plant of each fort is once obtained.

The Evergreen Italian Honeyſuckle (the beſt of the evergreen forts) does not readily take by *cuttings*; fo that in order to make fure of this plant, the young branches muſt be *layered*, any time in the autumn or winter, and by the autumn following they will have plenty of roots, and be good plants fit for removing to any place.

The Evergreen Honeyſuckles, though climbing plants, ſhould occaſionally be ſtationed in the evergreen quarters, as ſhould all the other forts among the deciduous trees and ſhrubs; being fo managed that their appearance may agree with thofe of upright growth. This is done by nipping off the young ſhoots (which would foon get rambling and out of reach), that the plants may be kept within bounds, and made to join in the collection with great beauty. Neither may they only be kept low, to almoſt what height is required; but they may, by fixing a ſtake for their ſupport, be trained up to a ſtem, which will every year grow more and more woody and firm; fo that in this cafe the eye muſt frequently overlook the tree, to take off the young ſhoots as they grow out, and not permit the head to grow too large and ſpreading for the ſtem, which it foon would do without this care; and with it, the head may be fo kept in order as to bear good proportion to the ſtem, thereby cauſing the tree to have the appearance of an upright ſhrub.

4. DIERVILLA is a ſhrub of about the height of three or four feet. The branches are few, and larger in proportion than the height of the ſhrub; they are very full of pith, and when broken emit a ſtrong ſcent. The leaves are placed oppoſite by pairs, on ſhort footſtalks: They are near three inches long, and about half as broad; and of an oblong heart-ſhaped figure, finely ferrated, and end in acute points: Their upper ſurface is ſmooth, and of a fine green colour; their under is lighter, and has five or fix pair of ſtrong nerves running irregularly from the mid-rib to the borders. The flowers are produced in loofe bunches, both at the ends and at the fides of the branches:

Each

Each is formed of one leaf; the tube is long, and the top is divided into five parts, which turn backward. They are of a yellow colour, and will be in blow in May, and sometimes most of the summer months. These flowers are succeeded, in the countries where they grow naturally, by black oval berries, each containing four cells. *Diervilla* forms an agreeable variety amongst other shrubs of its own growth, though the flowers make no great figure. It is very hardy with respect to cold; and may be planted in any part of the nursery where it is wanted.

No art is required to PROPAGATE this plant; it spawns, and thus propagates itself in great plenty. These suckers should be taken up in autumn, and planted out in the nursery : After remaining there a year or two, they may be finally taken up. This tree may be also encreased by cuttings. They should be planted in October, very close, if a quantity are wanted. By the autumn following, they will have good roots. They may be taken up and planted in the nursery, like the spawn, for a year or two, and then set out to stand. Plants raised this way will not be quite so subject to throw out suckers as the others.

5. St. PETER's WORT. St. Peter's Wort will arise to the height of about four or five feet. The main stems are ragged, and of a dirty dark brown. The branches are numerous and short, though oftentimes it sends out some trailing slender branches, which will grow to a great length. The leaves of this shrub constitute its greatest beauty : They are very numerous, small, about half an inch long, and of an oval figure. Their footstalks are exceeding short, and they stand opposite by pairs on the slender branches : These die in the autumn to a dark brown. The time of this plant's flowering is August. The flowers grow round the stalks : They are small, of an herbaceous colour, and make no figure.

The PROPAGATION is very easy. 1. If a spade-full of mould be thrown over each of the trailing branches, any time in the winter, they will by the autumn following have struck root; and these may be planted out in the nursery, to stand until they are of a proper size to be planted out for good. 2. This shrub may be also propagated by cuttings; and in order to obtain good cuttings for the purpose, the year before the plants should be headed near the ground, which will make them shoot vigorously

R the

the fummer following. Thefe young fhoots muft be the cuttings to be planted. October is the beft month for the work; and if they are planted in a moiftifh foil, and have a fhady fituation, they will have taken good root by the autumn. If they are planted very thick, as cuttings commonly are, they fhould be all taken up and planted in the nurfery a foot afunder, and two feet diftant in the rows; but if the living cuttings are no nearer than about a foot, they may remain without removing until they are planted out.

6. The BLUE-BERRIED HONEYSUCKLE is a fhrub of about four feet in growth. The branches are round, fmooth, and of a reddifh-purplifh colour. The leaves are oblong, fpear-fhaped, of a fine green, and ftand oppofite by pairs on the branches. The flowers, which are white, are produced in May from the fides of the branches, and are fucceeded by blue berries, that will be ripe in Auguft.

7. The RED-BERRIED HONEYSUCKLE will grow to the height of about five feet. The branches are very upright; the young fhoots are angular, and covered with a brown bark. The leaves are tolerably large, fpear-fhaped, a little refembling thofe of the mock Orange, and grow oppofite to each other. The flowers are produced from the fides of the branches, on long footftalks : They are of a red colour, come out in April, and are each fucceeded by a pair of red berries, which will be ripe the end of July or early in Auguft.

8. BLACK-BERRIED HONEYSUCKLE differs from the Blueberried only in that the feeds of this are black, and grow two together; whereas thofe of the Blue-berried are fingle and diftinct. Except this, there is hardly any difference to be perceived.

9. FLY HONEYSUCKLE will grow to the height of about feven or eight feet. The bark on the branches is of a whitifh colour, which caufes a variety, and makes it diftinguifhed in the winter feafon. The leaves, which are placed oppofite by pairs, are downy, and of an oblong oval figure. The flowers are white and erect : They are produced from the fides of the branches in June, and are fucceeded by two red berries, which will be ripe in September.

10. The

10. The Pyrenean Honeysuckle, or Dwarf Cherry, is but a low shrub: It seldom arrives to more than a yard in height. The branches are produced irregularly. The leaves are smooth, oblong, and placed opposite by pairs. The flowers are white, produced from the sides of the branches, on slender footstalks, in April; and are succeeded by roundish berries, which will be ripe in September.

11. Tartarian Honeysuckle, or Dwarf Cherry with Heart-shaped Leaves, is a shrub of about three or four feet high. Its branches are erect, like the upright sorts; and it differs in few respects from them, except that the leaves are heart-shaped. It exhibits its flowers in April; and these are succeeded by twin red berries, which will be ripe in August.

These are the *Upright* sorts of the *Lonicera*; to which one method of propagation is common; and that may be performed two ways. 1. By seeds. Common garden-mould, dug fine, and cleared of the roots of all weeds, will serve for their reception. In this the seeds should be sown soon after they are ripe, about half an inch deep. After the beds are neated up, they will require no other care until the spring; when the weeds should be picked off as fast as they appear. Some of the plants by this time will have come up; but the far greater part will remain until the second spring before they shew themselves; so that the beds must be entirely untouched until at least two years after sowing. They will require no care all this time, except being kept clear of weeds; though if watering be afforded them in dry weather, it will be the better. After they are all up, and have stood a year or two in the seed-bed, they may be taken up and planted in the nursery, at small distances; and in two or three years they will be of a proper size to plant out to stand. 2. All these sorts may be also propagated by cuttings. These should be planted in October, in any sort of garden-mould that is tolerably good. If a quantity is wanted, they may be placed very close; and a small spot of ground will hold thousands. If the place be shaded, it will be a great advantage, as most cuttings are in danger of suffering by the violence of the sun's rays before they have struck, or whilst they are striking root. The winter following, they may be all taken up and planted out in the nursery, a foot asunder, and two feet distant in the rows, where they may stand until they are finally taken up for planting.

L Y C I U M.

LINNEAN Clafs and Order, *Pentandria Monogynia :* Each flower contains five males and one female : There are eight SPECIES ; one only of which is hardy enough to ftand a fevere winter in our climate.

LYCIUM *Barbarum :* The BOXTHORN ; *a deciduous creeper ;* native of Afia, Africa, and Europe.

The BOXTHORN. This fpecies affords two *varieties :*
The Broad-leaved Boxthorn,
The Narrow-leaved Boxthorn.

The Broad-leaved Boxthorn is a rambling plant, and will, if let alone, in a few years overfpread every thing that is near it. The branches are very many, and fpread about in all directions. They will lie upon the ground, if unfupported, and will fhoot, in a good foil, fixteen feet in length in one fummer. Thofe branches that lie upon the ground will ftrike root ; fo that from every part frefh fhoots will be fet forth the next fpring ; and thus in a few years they will occupy a large compafs of ground ; fo that whenever this plant is defired, they fhould be conftantly kept within bounds. Indeed, from its exceeding rambling nature, not above a plant or two for variety or obfervation fhould be admitted in hardly any place. The branches of this plant are covered with a grey or whitifh bark. The leaves are of a light, whitifh-green, and of a thick confiftence. They grow on the branches, on all fides, by threes. This plant, of all the forts, is poffeffed of the longeft fpines (fome of which are a foot or more in length). Thefe fpines are garnifhed with leaves ; and on thefe they for the moft part ftand fingly in an alternate manner. On the branches where they grow by threes, the middle one is always the largeft. They are all of an oval, fpear-fhaped figure, are very fmooth, a little gloffy, and often continue till the middle of winter before they fall off. Befides the long leafy thorns beforementioned, it produces many fhort fharp fpines, of a white colour, near the ends of the fhoots. The flowers are produced in Auguft, and there will be often a fucceffion of blow until the frofts come on. They grow fingly at the joints, on fhort footftalks.

ftalks. They are of a purplifh colour, fmall, and are fucceeded
by no fruit with us, as I could obferve.

The *Long Narrow-leaved Boxthorn* is alfo a very great rambler,
The branches are many, and are produced irregularly on all
fides. It is poffeffed of fpines, but thefe are very fhort, and the
bark with which they are all covered is pretty white. The leaves
are of a lanceolate figure, and are narrow and long. Their colour is that of a whitifh-green, and they grow alternately on the
branches. The flowers are fmall, and appear in July; and are
fucceeded by red berries, which ripen in September, and at that
time are very beautiful.

The PROPAGATION of thefe forts is by cuttings; for they will
grow, if planted at any time, in any manner, and in almoft any
foil or fituation, except a white clay. In a black rich earth, they
will be the moft healthful and moft vigorous fhooters; and
though the cuttings will grow at all times, yet the winter
months are to be preferred for the purpofe.

M A G N O L I A.

LINNEAN Clafs and Order, *Polyandria Polygynia:* Each
flower contains many males and many females: There are four
SPECIES:

1. MAGNOLIA *Glauca:* The SEA-GREEN MAGNOLIA, or the
BAY-LEAVED TULIP-TREE, or the SMALL MAGNOLIA; *a tall
fub-evergreen fhrub;* native of Virginia and Pennfylvania.

2. MAGNOLIA *Acuminata:* The LONG-LEAVED MAGNOLIA;
a fub-evergreen fhrub or tree; native of Pennfylvania.

3. MAGNOLIA *Tripetala:* The UMBRELLA-TREE; *a fub-evergreen fhrub or tree;* native of Carolina and Virginia.

4. MAGNOLIA *Grandiflora:* The EVERGREEN MAGNOLIA,
or LAUREL-LEAVED TULIP-TREE; *an evergreen tree;* native
of Florida and Carolina.

1. The SEA-GREEN or SMALL MAGNOLIA grows with us to
about the height of ten or twelve feet. The wood is white; and

the branches, which are not very numerous, are covered with a smooth whitish bark. The leaves are tolerably large, and of two colours ; their upper surface being smooth, and of a fine green, whilst their under is hoary. They are of an oval figure, have their edges entire, and often continue the greatest part of the winter before they fall off the trees. The flowers are produced at the ends of the branches, in May : Their colour is white ; and the petals of which they are composed are concave and large ; so that, together with the numerous stamina in the center, they present 'a beautiful appearance. They are also remarkable for their sweet scent ; and are succeeded by conical fruit, which never ripens in England ; but in the places where they grow naturally, a singular beauty and oddity is added to these trees by the fruit; for the seeds are large, and lodged in cells all around the cone. When quite ripe, these are discharged from their cells ; and hang each by a long narrow thread, causing thereby an uncommon and pleasing effect.

2. LONG-LEAVED MAGNOLIA will grow to be near twenty feet high. The wood of this sort is yellow, and the branches are covered with a smooth light bark. The leaves are very large, being near ten inches long ; their figure is oval, spear-shaped, and all end in points. The flowers, which are produced in May, are white, and composed of twelve obtuse petals, which, together with the number of stamina, make a good show. These also are succeeded by conical fruit, which never ripens in England.

3. The wood of the UMBRELLA-TREE, which grows to about twenty feet in height, is more spongy than any of the other species of *Magnolia*. It is called the *Umbrella-Tree*, from its manner of producing the leaves ; for these are exceeding large, and so produced as to form the appearance of an umbrella. The flowers of this sort also are white, and the number of petals of which each is composed is about ten : They are succeeded by fruit of a conical figure, with many cells all round for the seeds, which never ripen in England.

All these sorts may be PROPAGATED by seeds, layers, and cuttings. By the first of these methods the best plants are raised, though it is a very tedious way, and must be followed with great

patience

patience and trouble. We receive the feeds from thofe parts of America where they grow naturally. Thefe are always preferved in fand, but, neverthelefs, will not always prove good. As foon as poffible after they arrive, which is generally in February, they fhould be fown in pots, about half an inch deep. The beft compoft for them is a frefh loamy earth, mixed with a fourth part of drift fand ; and the feeds fhould be thinly fown in each pot. After this is done, the pots fhould be plunged up to the rims in the natural mould, under a warm hedge, where they may reap the benefit of the fun during the month of March and part of April ; but when the rays of the fun begin to be ftrong and powerful, drying the mould in the pots very faft, they fhould be taken up and plunged again up to the rims in a fhady border. By the end of May, if the feeds were good, the plants will come up ; and all the fummer they muft be conftantly attended with weeding and watering. At the approach of winter, they fhould be removed into the greenhoufe, or placed under fome cover ; but in mild weather fhould always have the benefit of the open air and gentle fhowers. In March, the pots with their feedlings fhould be plunged into an hot-bed to fet them forwards. Tanners bark is what the hotbed fhould be compofed of ; and as much air as the nature of the bed will allow, fhould always be afforded them. Water alfo muft be given pretty often, though in fmall quantities, and the glaffes muft be fhaded in the heat of the day. After this, about June, they fhould be inured to the open air ; watering muft ftill be afforded them ; and this is what they require during the fecond fummer. It has been a practice to plunge the pots into an hotbed foon after the feeds are fown ; but this is a very bad method, for the young plants being thereby forced, grow thin and flender, and are feldom made to live longer than the firft year. The fecond fummer's management alfo has ufually been, to plant the feedlings in March, in little pots, and then plunge them into a hotbed ; but this is alfo a very bad way ; for thefe feedlings, whether raifed on hotbeds or the common ground, will be fmall, and not of confiftence fufficient to draw the juices, though the powers of vegetation are affifted by an hotbed : Thus, hardly any of them furvive this early tranfplanting. This having been the general practice, thefe plants have been always thought very difficult to preferve the fecond year ; whereas all

R 4   thofe

thofe difficulties vanifh, by obferving the above-directed method; for by letting the feeds have only the natural foil, they will the firft fummer be formed into young plants, which, though fmall, will neverthelefs be plants, and healthy. Thus being in the fpring in their natural ftate, with their pores open to receive the nutritious juices, and not having fuffered by being tranfplanted, the hotbed will fo help them, that they will be pretty plants by the autumn. At the approach of winter, they muft be removed again under cover, and the former affiftance of an hotbed fhould be afforded them ; and this fhould be repeated until the plants are grown to be a foot or more in length. The fpring following, the mould fhould be turned out of the pots and fhaken from the roots, and each plant put into a feparate pot. For thefe, an hot-bed of tanners bark fhould be ready, which will promote their growth, and make them healthy and fine. During the time they are in the bed, they fhould be fhaded ; and about Midfummer the pots may be taken out and placed in a fhady border. The winter following, it will be proper to houfe them in fevere frofty weather ; but always obferve to place them abroad in mild fea-fons. In March they may be turned out of the pots, the mould hanging to the roots, and planted with that in the places where they are to remain. 2. Thefe plants may be alfo propagated by layers. The young fhoots in the autumn are moft proper for the purpofe ; and it is found that a gentle twift, fo as juft to break the bark about the joint, is a better method than any other in practice. Thefe will fometimes ftrike root in one year, and fometimes you muft wait more than two before you find them with any. After they have ftruck root, and are taken up, the beft time for which is March, it is moft eligible to plant each feparately in a pot, and plunge them into an hotbed, as di-rected for the feedlings ; and by the fpring following they will be ftrong good plants for any place. 3. Thefe plants may like-wife be encreafed by cuttings ; by which they may be procured in plenty, if a perfon has the conveniency of a good ftove ; and without one this method fhould not be attempted. Thefe cut-tings fhould be planted in pots ; and after they are fet in the ftove, muft be duly watered and fhaded : By obferving thefe di-rections, many of them will grow. After this, they fhould be brought by degrees to the open air; the winter following they

fhould

fhould be placed under an hotbed-frame, or fome fhelter ; and in the fpring planted out to remain.

Thefe plants often retain their leaves, efpecially when young all winter, or the greateft part of it, in fome fituations; and in fuch they pafs for evergreens.

4. The EVERGREEN LAUREL-LEAVFD MAGNOLIA. In the countries where it grows naturally, it arrives to the height and bulk of a timber-tree. Thofe countries are adorned with woods that are chiefly compofed of this plant ; and indeed, a wood of fo noble a tree, luxuriantly fhooting, flowering, and feeding, healthy and ftrong, in foil and fituation wholly adapted to its nature, muft be a fight of which we can hardly form an adequate idea, or have a juft conception of its beauty or grandeur ; for the tree naturally afpires with an upright ftem, and forms itfelf into a regular head. Many other trees do the fame ; but its moft excellent properties confift of the fuperlative beauties of the leaves, flowers, and feeds. The leaves much refemble thofe noble leaves of the Laurel, from which it is fo called, only they are larger, and of a thicker confiftence : Many of them will be ten inches or more in length, and four broad, and all are firm and ftrong. Their upper furface is of a fhining green, but their under is lighter, and often of a brownifh colour. This tinge, which is not always found in all trees, is by fome thought a great beauty, and by others an imperfection ; fo various is the tafte of different people. Thefe leaves are produced without any order on the tree, and fit clofe to the branches, having no feparate footftalks. The idea we can form of a tree, of feventy or eighty feet high, plentifully ornamented with fuch large and noble leaves, muft be very great, and will induce us on their account only to endeavour to naturalize fo noble a plant to our country. But let us confider their flowers. Thefe we find large, though fingle, and of a pure white. They are produced at the ends of the branches, in July, and each is compofed of about nine or ten large fpreading petals. They have the ufual properties of thofe that are broad and rounded at their extremity, of being narrow at the bafe, and their edges are a little undulated or waved. In the center of thefe petals are fituated the numerous ftamina, which the Botanift will be more curious in obferving than the Gardener. But what affects all equally alike that have the fenfe

of

of fmelling is, their remarkable fragrance, which indeed is of fo
great a degree, as to perfume the air to fome diftance ; and if one
tree, when in blow, is fufficient to effect this, what conception
fhould we form of the odours diffufed in the countries where
there are whole woods of this tree in full vigour and blow ! The
fruit is nearly of the fhape and fize of a large egg ; but what
make it moft fingular and beautiful are the pendulous feeds, of a
fine fcarlet, which being difcharged from their cells, hang by
long threads, and have an effect both ftriking and uncommon.

Rules have been given above for PROPAGATING deciduous
Magnolias: the fame rules obferved, whether for feeds, layers,
or cuttings, will raife plenty of this fort ; neither need any thing
be added, except hinting to the Gardener, that this is more tender than the other forts, and that from thence he fhould learn
not to be over-hafty in committing thefe plants to the winter's
cold, and planting them finally out. Snow is peculiarly injurious to them while young ; fo that, at the approach of fuch
weather, they muft be particularly covered ; and if fnow fhould
happen to fall unawares, it fhould be carefully cleared off the
leaves and ftems. When thefe plants are fet abroad to remain, if
the place is not exceedingly well fheltered, it will be proper to
have a fhed at hand, which the Gardener may put together, to
fcreen them from the fevere northern frofts, and the black eafterly
winds, from which this fhrub is moft likely to fuffer damage ;
and thefe frofty winds are the moft deftructive to it when they
come early in the winter, while the fhoots are rather tender ; for
then they are often deftroyed, and the tree rendered unfightly for
fome time, though it will fhoot out again. When this fhrub is
to be encreafed by layers, it will be neceffary, after the operation
is performed, to make an hedge of reeds, or fomething, at a little
diftance round it, to keep off the ftrong winds, and prevent them
from blowing the layers out of the ground ; for without fome
guard this will be in danger of being done ; fince the leaves
being very large and ftrong, the wind muft have great power over
them.

M E.

M E D I C A G O.

Linnæan Claſs and Order, *Diadelphia Decandria* : Each flower contains ten males and one female ; the males being conneɕted at the baſe in two diviſions : There are twenty-four Species ; one of which, only, is admiſſible into our collection ; the reſt being herbaceous plants.

Medicago *Arborea :* The Tree Lucerne, or Tree Medick, or Moon-Trefoil ; *an evergreen ſhrub* ; native of Italy, Crete, and the iſlands of the Archipelago.

The Tree Lucerne will grow to be ſix or ſeven feet high, and divides without any order into many branches, which are covered with a grey bark. There is a delicacy in the young ſhoots beyond what is found in moſt trees ; for they are white and ſilvery, and at the ſame time covered with the fineſt down. Theſe young ſhoots are plentifully ornamented with leaves, many of which come out from a bud. They are trifoliate, and grow on long ſlender footſtalks. One of the folioles is cuneiform, or ſhaped like a wedge ; the others grow out more into a lanceolate figure, have alſo a whitiſh look, and are downy, though not to ſo great a degree as the young twigs on which they grow. They have a large mid-rib, which contracts the borders in the evening, and this alters their poſition of ſides on the alteration of weather. The flowers are produced from the ſides of the branches, in cluſters, on long footſtalks. Each of theſe cluſters will be compoſed of ten or twelve flowers, which are of a beautiful yellow. They are of the butterfly kind ; and are ſucceeded by moon-ſhaped pods, that ripen their ſeeds very well. One or other of theſe trees is to be found in blow almoſt at all times. The beginning of the blow is generally ſaid to be in April or May ; and indeed then we may expect to ſee the flowers largeſt and in the greateſt perfection ; but the flowers of theſe trees may be ſeen in July, Auguſt, and September ; and in greenhouſes have been known to blow all winter ; which makes the tree more valuable to thoſe who are deſirous of ſeeing flowers in unuſual months.

This ſhrub is by many ſuppoſed to be the true *Cytiſus* of Virgil. It grows plentifully in Italy, in the iſlands of the Archipelago,

chipelago, and many other parts, where it is efteemed excellent fodder for cattle. "For this purpofe, continues HANBURY, the raifing of it has been recommended in England, but there feems no probability of fuch a fcheme being brought to bear here; neither is it any way neceffary to give ourfelves the trouble to try experiments of this kind, as, fhould it even fucceed to our utmoft wifhes, we have many forts of fodder that will exceed it in quantity and quality, without any proportion to the extraordinary expence which muft attend the raifing any quantity of thefe fhrubs, to cut for that ufe. The flowers, leaves, and top-fhoots have, however, a fine peafe-like tafte, which is what, I make no doubt, moft cattle would be fond of, and of which the inhabitants of fome countries where it grows naturally reap the advantage; for the goats that feed on it yield a greater quantity as well as a more excellent kind of milk, from which good cheefe is at length obtained, where thefe creatures have plenty of thefe fhrubs to brouze upon.

"In our wildernefs-quarters we muft give this tree a very dry foil and a well-fheltered fituation; for with us it is rather a tender fhrub, and has been frequently treated as a greenhoufe-plant; and this is another argument againft any attempt to raife thefe fhrubs for fodder in England: They are too tender to bear our fevere winters without fhelter; and fhould we proceed in raifing fixty or feventy acres, a thorough frofty winter would deftroy the greateft part of them; or, if the winter fhould not be fo fevere as totally to kill them, yet their end fhoots would be fo nipped and damaged, that it would be late in the fummer before they would fhoot out and recover this injury, and confequently fmall crops muft be expected."

This plant is eafily PROPAGATED by feeds or cuttings. 1 The feeds fhould be fown in the fpring, a quarter or half an inch deep, in beds of fine light garden-mould. After they are come up, the ufual care of weeding muft be afforded them; and if they are fhaded and now and then watered in hot weather, it will be fo much the better. The beds muft be hooped againft winter, and plenty of mats muft be ready to cover the plants when the froft comes on; and if this fhould be very fevere, their covering fhould be encreafed, or there will be danger of lofing them all. In the fpring the ftrongeft may be drawn out, and planted in pots, to be houfed for a winter or two,

until

until they are got ftrong ; but where a quantity is wanted, and there is no fuch conveniency, it may be proper to let them remain in the feed-bed another winter, for the conveniency of being covered in bad weather ; and then in the fpring they may be planted out in the nurfery, in lines two feet afunder, and at one foot diftance. This nurfery fhould be in a well-fheltered warm place, and they will be ready for tranfplanting whenever wanted. 2. Thefe plants may be raifed by cuttings. If a few only are wanted for ornamenting a fhrubbery, the beft way will be to plant thefe in pots, and fet them up to the rims in a fhady place, that they may have the conveniency of being houfed in winter. When a quantity is wanted, they muft take the chance of wind and weather, and the moft we can then do is to plant them in fine light foil in a well-fheltered place. The latter end of March is the beft time for the purpofe ; they will ftrike root freely, efpecially if they are fhaded and watered in dry weather ; and from this place they need not be removed until they be finally fet out.

M E L I A.

LINNEAN Clafs and Order, *Decandria Monogynia :* Each flower contains ten males and one female : There are two SPECIES ; one of which will bear the open air ; the other is a hot-houfe plant.

MELIA *Azedarach :* The BEAD-TREE, *a deciduous tree ;* native of Syria.

The BEAD-TREE is a large plant : in its native country it will grow to the fize of one of our pear-trees ; and there is no doubt, if our foil and fituation fuited it, that it would arrive to near that magnitude with us. The trunk is covered with a grey bark ; and the young branches, which are not very numerous, are quite fmooth and green. The leaves are a very great ornament to this tree : They are compound, and very large, the whole leaf being a foot and a half, and fometimes near two feet long. Each is compofed of a great number of

foliis,

folioles, which are all terminated by an odd one. Thefe little leaves have their upper furface of a ftrong fhining green ; their under is paler ; and their edges are indented. The flowers are produced in July, from the fides of the branches, in long cluf-ters : They are, feparately, fmall, of a blueifh colour, very fra-grant, and each ftands on a long footftalk. The flowers are fucceeded by a yellow fruit, tolerably large, in which fome nuts are enclofed, ufed in the Catholic countries to compofe fome forts of rofaries ; on which account this tree is called the Bead-Tree.

" It is generally preferved in winter as a greenhoufe plant ; and indeed a few plants of this fine fhrub ought always to be in-troduced in fuch places defigned for trees as are proper for them. The reafon of its being treated as a greenhoufe plant is, be-caufe it is rather of a tender nature ; and as the plants are not yet very plentiful in England, to this may be added, the defire of preferving thofe few a perfon has obtained. But notwith-ftanding the Bead-Tree's being looked upon as a greenhoufe-plant, fome gardeners have ventured to fet them abroad againft warm walls, where they have ftood the winter, and flourifhes ex-ceeding well ; others have planted them out in well fheltered places only, where they have flourifhed and ftood the brunt of many winters. What inclines me to introduce the *Melia* amongft our hardy trees is, that I have planted it in an open cold expanfe, in a naturally damp and moift foil, where it has flourifhed for more than feven years, and difplayed its beautiful foliage every fummer, to the great pleafure of all beholders. This treat-ment and practice, however, muft be ufed with caution ; and whoever ventures to plant them abroad muft have a dry foil, as well as a warm and well-fheltered fituation, and then nothing but our hardeft frofts will deprive the owner of thefe treafures. But, were they more tender, and if a perfon has no green-houfe, it will be worth while to venture the planting a few abroad, though there fhould be little chance of his keeping them longer than two or three winters, as they are fcarce plants with us, and the leaves, the only beauties the tree can afford in that time, are compounded in fuch a manner as to afford admi-ration and pleafure." HANBURY.

PROPAGATION. Care and trouble muft be ufed before we can raife thefe plants to be of fufficient ftrength and hardinefs to
defend

defend themfelves, when planted finally out· They are all to be raifed from feeds ; and thefe are to be procured from the places where they commonly grow, which is in moft of the Catholic countries. Thefe feeds muft be fown in pots, filled with light fandy earth, half an inch deep, the end of March. This done, the pots fhould be plunged into a bark-bed, which will caufe them to come up. When the plants appear, they muft have plenty of air and water; and the open air muft be afforded them pretty foon in the fummer, that they may be hardened before winter. After they are taken out of the beds, they fhould be fet in a fhady place, and every other day watered till the autumn : and at the approach of winter, they fhould be removed into the greenhoufe, with the hardieft of thofe plants. In April following, the plants fhould be taken out of the pots, and each planted in a feparate fmall pot ; and after this is done, they fhould have the benefit of the bark-bed as before, to fet them a-growing. Care muft be taken to give them fufficient air, and not to draw them too much ; and after they are well entered upon a growing ftate, they muft be hardened to the open air as foon as poffible, and the pots taken out, and plunged up to the rims in a fhady border, which will prevent the mould in the pots drying too much. They will require little watering, if this method be ufed, during the fummer ; and at the approach of winter, they muft be removed into the greenhoufe as before, or placed under a hot-bed-frame, or fome fhelter. The next fpring they muft be fet out with other green-houfe-plants, and managed accordingly, and removed into the houfe again with them. Every other year, they fhould be fhifted out of their pots, with the earth to their roots, and planted in larger ; and by thus treating them as greenhoufe-plants, and letting them have larger pots as they encreafe in fize, till they are fix or eight years old, they will arrive to be good ftrong trees. Then in April, having made choice of the drieft, warmeft, and beft-fheltered fituation, there they may be planted, taking them out of the pots with all their mould ; which if done with care, they will never droop on being removed.

M E N I-

MENISPERMUM.

LINNEAN Clafs and Order, *Dioecia Dodecandria :* Male
flowers containing twelve ftamina, and female flowers containing
two piftils, are fituated upon diftinct plants : There are eight
SPECIES ; three of them as follows :

1. MENISPERMUM *Canadenfe :* The CANADA MOONSEED ;
a ligneous climber ; native of Canada and Virginia.

2. MENISPERMUM *Virginicum :* The VIRGINIA MOONSEED ;
a ligneous climber ; native of the fea-fhore of Virginia and Carolina.

3. MENISPERMUM *Carolinum :* The CAROLINA MOONSEED ;
an herbaceous climber; native of Carolina.

1. The CANADA MOONSEED will twine round trees to the
height of fifteen or fixteen feet ; and if there be no trees near for
it to afpire by, its almoft numberlefs branches will twift and run
one among another, fo as to form a thick clofe-fet bufh. Thefe
twining ftalks are covered with a fmooth green bark, though in
fome places they are often reddifh, and in winter often of a
brown colour. The leaves are very large, and ftand fingly upon
long green footftalks, which alfo have a twining property, and
affift the plant to climb. Thefe leaves have their upper furface
fmooth, and of a ftrong green colour, but are hoary underneath.
They are what are called peltated leaves : The footftalk is not
near the middle of the leaves, but within about a quarter of an
inch of the bafe, and from thence it branches into feveral veins
unto the extremity. Thefe peltated leaves are of a roundifh
figure in the whole, though they are angular, and being large,
and of a good green, make it a valuable climber. The flowers
are produced in July, from the fides of the ftalks. They grow
in bunches, and are of a greenifh colour. They are fucceeded
by feeds, which often ripen well here.

2. The VIRGINIA MOONSEED differs very little from the
other, except in the fhape of the leaves ; for it has the fame
kind of twining ftalks, produced in great plenty, and the flowers
and fructification are the fame ; fo that nothing more need be
obferved of this, only that the leaves are often heart-fhaped, and
many of them have lobes like thofe of the common Ivy.

3. The

3. The Carolina Moonseed is an herbaceous climber, and will, by the affiftance of trees, rife to be ten or twelve feet high. The twining ftalks are garnifhed with heart-fhaped leaves, which do not divide into lobes like the others. Thefe leaves, which are of a good ftrong green colour, have their under furface hairy, and are much fmaller than either of the other forts; the fpecies itfelf being of all the leaft valuable, as it is fcarcely ever known to produce flowers here.

All thefe forts PROPAGATE themfelves very faft. 1. If they are planted in a light foil, their roots will fo fpread and multiply the fhoots, that in a few years after planting, each of them being wholly taken up, they may be parted, often into fome fcores of plants, which will be fit to fet out, the weakeft in the nurfery to gain ftrength, and the ftrongeft where they are to remain. Any time from October to March will do for taking off the fuckers or parting the roots. 2. The young fhoots, alfo, being covered with mould, will grow, and be good plants in one year. 3. They may be likewife raifed by feeds; for if thefe are fown in the fpring, in a bed of light earth, half an inch deep, they will come up, and require no other trouble than weeding until they are finally planted out, which may be two years after their appearance, and which may be done very well from the feed-bed, without previous planting in the nurfery.

M E S P I L U S.

Linnean Clafs and Order, *Icofandria Pentagynia*: Each flower contains about twenty males and five females. There are nine Species; feven of which are here treated of:

1. Mespilus *Germanica*: The German Medlar, or Dutch Medlar; *a deciduous tree*; native of the fouth of Europe.

2. Mespilus *Arbutifolia*: The Arbutus-leaved Medlar, or the Virginia Wild Service Tree; *a deciduous fhrub*; native of Virginia.

3. Mespilus *Amelanchier* : The Amelanchier ; *a deciduous shrub* ; native of Austria, France and Italy.

4. Mespilus *Canadensis* : The Canada Medlar ; *a deciduous shrub* ; native of Canada and Virginia.

5. Mespilus *Cotoneaster* : The Dwarf Quince ; *a deciduous shrub* ; native of the Pyrenees, Ararat, and many of the cold parts of Europe.

6. Mespilus *Chamæ Mespilus* : The Bastard Quince ; *a deciduous shrub* ; native of the Austrian and Pyrenean mountains.

7. Mespilus *Pyracantha* : The Pyracantha, or Evergreen Thorn ; *an evergreen shrub or climber* ; native of Italy and the south of France.

1. The German Medlar in some situations grows to be a moderately large tree. It grows irregularly, and the branches are frequently crooked. The leaves are spear-shaped, large, entire, downy underneath, and grow on very short channeled footstalks. The flowers, which grow singly from the sides of the branches, are very large, and of a white colour. They come out the end of May, and are succeeded by that well-known fruit called the *Medlar*.

The *varieties* of this species are, *The Pear-fruited Medlar*, and *The Nottingham Medlar*. These are plants of more upright growth than the Dutch Medlar. Their leaves are narrower, and their flowers and fruit smaller.

2. Arbutus-leaved Medlar. This is frequently called Virginia Wild Service-tree with an Arbutus Leaf. It is a shrub about six feet high, frequently sending forth many suckers from the root, and branches from the sides of the plant. The leaves are spear-shaped, downy underneath, and indented. They grow alternately on very short footstalks. Their upper surface is a fine green colour; though white below; and they die to a purple colour in the autumn. The flowers are produced in bunches from the ends and sides of the branches : They are small, white, come out in May, and are succeeded by a dark-brown fruit, like the common Haw, which will sometimes be ripe in the autumn.

3. Amelanchier. The stalks of this species are slender, branching a little, and grow to about four feet high. The young branches are of a reddish-purple colour, and the whole plant

plant is altogether deftitute of thorns. The leaves are oval and ferrated, about three quarters of an inch loug, half an inch broad, green on their upper furface, and woolly underneath. The flowers are produced in bunches from the ends of the branches : Their colour, is white ; and they are fucceeded by fmall black fruit, of a fweetifh tafte, which will be often ripe in the autumn. This is a beautiful fhrub, and in different parts goes by the various names of the *Dwarf Black-fruited Medlar*, the *New-England Quince*, *Vitis Idæa*, &c. The young fhoots which fupport the flowers are woolly underneath ; but this by degrees wears off, and they foon become of a purple colour, which remains all winter.

4. CANADA MEDLAR. This fhrub, which rifes to about five feet high, is free from thorns, and divides into a few branches, which are fmooth, and of a purplifh colour. The leaves are oval, oblong, fmooth, flightly ferrated, and grow on long flender footftalks. The flowers are white, and terminate the branches in fmall bunches : They come out in May ; and are fucceeded by a purplifh fruit, hardly fo large as the common Haw.

5. DWARF QUINCE grows to about four or five feet high. The branches are few, fmooth, and of a reddifh-purple colour. The leaves are oval, entire, and grow on very fhort footftalks. The flowers are produced, two or three together, from the fides of the branches, without any footftalks. They are fmall, of a purplifh colour, come out in May, and are fucceeded by round fruit, of a bright red colour when ripe, in the autumn.

6. BASTARD QUINCE. This fpecies grows to about four or five feet high. The branches are few, fmooth, flender, and covered with a purplifh bark. The leaves are oval, fmooth, ferrated, of a yellowifh green, and grow on pretty long footftalks. The flowers are produced in fmall heads, from the wings of the ftalks ; and between them are long narrow bracteæ, which fall off before the flowers decay. Both flowers and bracteæ are of a purplifh colour. The fruit is fmall, and of a red colour when ripe.

All thefe forts are to be PROPAGATED from the feeds, from layers, and by budding them upon Hawthorn ftocks. 1. The feeds fhould be fown in the autumn, foon after they are ripe, in a bed of good earth, in a moift part of the garden. They

ufually lie two years before they make their appearance; during
which time the bed muft be kept clean from weeds. When the
plants come up, they muft be frequently watered, if dry weather fhould happen; and this fhould occafionally be repeated all
the fummer. Weeds muft be eradicated as they arife; and in
the autumn, winter, or fpring, the ftrongeft plants may be drawn
out, and fet in the nurfery ground, a foot afunder, in rows two
feet diftant from each other; whilft the others may remain in
the feed-beds a year longer, to gain ftrength. In the nurfery
the Medlars fhould be trained for ftandards, if defigned for
fruit; or they may be headed to any height if for other purpofes, while the lower kinds will require no other management
than keeping them clean from weeds, and digging the ground
between the rows in winter. 2. Thefe plants may be alfo
raifed by layers, efpecially the five laft forts. The young
branches fhould be laid early in the autumn; and by the
autumn following many of them will have ftruck root, when
they fhould be taken up, and planted in the nurfery-ground, like
the feedlings, to remain there for a year or two, before they are
finally fet out. 3. But the moft expeditious, and by far the beft
way of raifing thefe forts is, by budding them upon ftocks of
the White-thorn. The Haws to raife the ftocks fhould be gathered from fuch trees as are largeft, fhoot freeft, and have the
largeft leaves and feweft thorns. When the ftocks are one year
old, they fhould be fet in the nurfery at the beforementioned
diftance. By the end of July, many of them will be ready for
working; when they fhould be budded in the ufual way, and
they will eafily take. Seldom any other method than this is
practifed for raifing Medlars; and the other forts, when growing
on fo firm a bafis as the White-thorn, will be larger, have a
better look, and be more fertile in flowers and fruit.

7. The PYRACANTHA, OR EVERGREEN THORN, has been
chiefly ufed to ornament or hide the ends of houfes, barns,
ftables, or other buildings that break in upon the view; and for
this purpofe no plant is better adapted, as by its evergreen leaves,
clofely fet it will not only keep from fight whatever cannot re-
gale that fenfe, but will be to the higheft degree entertaining
by the profufion of berries it will produce, and which will be
in full glow all winter. But though the hiding as well as ornamenting of walls, &c. has been the chief ufe for this tree, it
is

is with very good reafon planted as an evergreen in fhrubbery-quarters, where, notwithftanding its branches againft walls, &c. are very flexible, it will become ftronger and more woody, and will diffufe its leafy branches in an agreeable manner. The branches will be terminated with its fine fruit, which will glow in the quarters all winter, if they are not eaten by the birds ; fo that the tree before us is proper for any place. A further account of this fhrub is almoft needlefs, as it is well known ; there being few towns which have not an houfe or two whofe front is ornamented with them, being trained up to a great height ; but when planted fingly in quarters, though their ftems naturally become ftronger, they feldom grow higher than twelve or fourteen feet; and they will fpread abroad their flender branches, and will often have a bufhy, though not un-pleafing form. Thefe branches are covered with a fmooth bark, which is of a dark-greenifh brown colour, and often fpotted with greyifh fpots; and they are often poffeffed of thorns, which, though not numerous, are fharp and ftrong. The leaves are fpear-fhaped, oval, and their edges are crenated. Their upper furface is fmooth, and of a fine fhining green; their under is paler ; and they are produced in much plenty all over the fhrub. The flowers are produced in bunches, like thofe of the common Hawthorn ; though they are fmall, and not of fo pure a white. They are often later before they are produced ; and are fucceeded by thofe large delightful bunches, or berries, which are of a fiery red, and which are as ornamental in the winter as any that are produced on trees of the berry-bearing tribe.

This plant is eafily PROPAGATED by the berries, or from layers. 1. The berries fhould be fown in any common garden-mould made fine, an inch deep ; and thefe will remain two years before they appear: though if the berries are old ones (for they will often remain on the tree two years) they will frequently come up the fucceeding fpring. After the plants have ftood one or two years in the feed-bed, in the fpring they fhould be planted out in the nurfery, at fmall diftances ; and in about two years more they will be good plants, fit for any place. 2. They are eafily propagated by layers ; and this bufinefs fhould be performed in the autumn, on the young fhoots. A

S 3

gentle

gentle twift may be given them; though, if they are only laid
down, and covered with earth, they will ftrike root by the next
autumn; nay, continues HANBURY, " I have known that, by
fome mould being accidentally thrown on a branch which was
near the ground, roots have fhot from almoft every joint." Thefe
layers fhould be taken off any time in the winter; the ftrongeft
will be fit for immediate ufe, while the weaker may be fet in
the nurfery, like the feedlings, and in a very little time they
will grow to be good plants.

M O R U S.

LINNEAN Clafs and Order, *Monoecia Tetrandria* : Male
flowers containing four ftamina, and female flowers containing
two piftils, upon the fame plant; the male flowers being collected in a katkin. There are feven SPECIES; four of which
are proper for our collection :

1. MORUS *Alba*: The WHITE MULBERRY, or the SILK-
WORM MULBERRY; *a deciduous tree*; native of China, and
cultivated almoft univerfally for the feeding of filkworms.

2. MORUS *Nigra*: The BLACK MULBERRY, or the COMMON
GARDEN MULBERRY; *a deciduous tree*; native of Perfia,
and the maritime parts of Italy.

3. MORUS *Papyrifera*: The PAPER MULBERRY; *a low
deciduous tree*; native of Japan.

4. MORUS *Rubra*: The VIRGINIA MULBERRY; *a low de-
duous tree*; native of Virginia.

1. The WHITE or SILKWORM MULBERRY will grow to a
large fize : Its leaves are of a clear light green; and open confiderably earlier in the fpring than thofe of the other fpecies of
Mulberry : Its fruit is alfo paler-coloured than that of the other
forts, which makes this take the name of the White Mulberry
" This tree (fays HANBURY) poffeffes the peculiar property of
breeding no vermin either growing or cut down; neither does it
harbour any fort of caterpillar, the Silk-worm only excepted,
whofe

whofe food is its leaves. The Mulberry-tree was very earneft-
ly recommended by King James to be planted in great quantities to feed thefe worms, in order to have filk of our own working: and, indeed, if we confider what vaft fums the produce of
filk brings into other ftates, we might find an undertaking of
this nature worthy of a princely care and affiftance." The
Mulberry delights moft in a light drv foil; but there is very
little land in this kingdom, generally fpeaking, which might
not be planted with thefe trees, and probably to great national
advantage. Be this as it may, it is fufficiently *ornamental* to be
admitted into a large collection: And, befides the *ufes* of its
leaves to the Silk-worm, EVELYN and HANBURY recommend it
very ftrongly as a foreft or timber-tree, and enumerate fome of
the ufes of its wood; none of them, however, fufficiently ftriking
to induce us to recommend it to the planter's notice merely as
a timber-tree.

2. The BLACK or GARDEN MULBERRY is principally
cultivated for the fruit; and in ornamental plantations a few of
them will be fufficient, to make the collection general, as well
as to be ready at all feafons for the notice and obfervation of
the Botanift.

There is a *variety* of it, with jagged leaves, which makes it
efteemed on that account; but the fruit is fmaller than that of
of the common fort.

3. The PAPER MULBERRY is fo called, becaufe the inhabitants where the trees grow naturally make paper of the bark.
It will grow to the height of about thirty feet; and exhibits
its fine large leaves of different fhapes, many of them being
divided into feveral lobes, whilft others again are entire. They
are of a fine ftrong green colour, though the under furface is
paler than the upper. The flowers, as has been obferved, are
male and female; and the females are fuccceeded by fmall black
fruit. It is the bark of the young fhoots of which the paper is
made; and for this ufe it is cultivated much in China, as well as
Japan, where large plantations are raifed. The plants are headed to within about a foot of the ground; and every year the
crop of the fummer's fhoots is taken.

4. The VIRGINIA MULBERRY TREE will grow to be
thirty or more feet high. It fends forth many large branches;

and

and the bark of the young fhoots is of a blackifh colour. The leaves are larger than the Common Mulberry, and rougher; though in other refpects they fomewhat refemble them. It produces plenty of katkins, in fhape like thofe of the Birch-tree; and the female flowers are fucceeded by a dark reddifh fruit. This is a very fcarce plant at prefent; and is coveted by none but thofe who are defirous of making their collection general.

These feveral fpecies of Mulberry may be PROPAGATED from feeds, by layers, and from cuttings. 1. Where the feeds can be procured, it is the moft expeditious way of raifing great quantities; and whoever has a correfpondence in the South of France, or in Italy, may through that channel obtain them. Having the feeds ready, let a fine warm border of rich mellow earth be prepared, and let this border be hooped, in order to fupport mats to defend the young plants, when they appear, from frofts. If no fuch border can be eafily had, it will be proper to make a gentle hot-bed, and cover it with fat mould: This alfo muft be hooped, as the border. Then fow the feeds in little drills, about a quarter of an inch deep. The middle of March is the beft time for this work; and when the young plants appear, which will be in about fix weeks, they muft be conftantly covered with the mats in the night, if any appearance of frofts prefents itfelf, as there often is at that feafon. During the fummer they fhould be kept clear from weeds, and covered from the extreme heat of the fun while the hot months continue. Whenever any cloudy or rainy weather approaches, the mats fhould be always taken off, that the plants may enjoy the benefit of it. By thus carefully nurfing the beds, keeping them clear from weeds, watering the plants in dry feafons, covering them from the parching fun, and uncovering them again in the night, cloudy or rainy weather, the plants by autumn will be got pretty ftrong; tho' not fo ftrong as to be left to themfelves. The following winter they will require fome care. When the frofts approach they muft be carefully covered with the mats, as in the fpring; for without this protection, many of them would be deftroyed, and the greateft part killed, at leaft down to the ground. In this bed they may ftand two years, when they will be ftrong enough to plant out in the nurfery. The ground for this purpofe being double dug, the young plants fhould be fet in rows, at two feet and a half diftance,

and

and one foot and a half afunder in the rows. Here they may remain till they are of a fufficient fize to be finally planted out. 2. Another method of propagating this tree is y layers. Who ever has not the conveniency of obtaining the feeds, muſt procure a number of plants to be planted for ſtools. The ground on which thefe ſtools are to ſtand ſhould be double dug, and the trees may be planted for this purpofe two yards afunder. The fize of the ground, and the quantity of trees for the ſtools, muſt be proportioned according to the number of plants wanted; though the reader ſhould obſerve, that a few ſtools will foon produce many layers, as they throw out plenty of young branches, when the head is taken off. Having a fufficient quantity of ſtools that have ſhot forth young wood for layering, in the beginning of winter perform this bufinefs, as follows : Let the earth be excavated around each ſtool, and let the preceding fummer-ſhoot be ſlit at a joint, and laid therein ; a peg would be proper, to keep them from being torn up, and the fine mould ſhould fill the interſtices ; the ground muſt be levelled, and the young twigs cut down to one eye above the furface, that it may juſt appear above the ground. Such is the method of layering this tree ; and whoever performs the operation in this manner, will find in the autumn following, that the plants will have all taken good root, and made a confiderable ſhoot in the ſtem. Thefe plants will be now ready for the nurfery-ground, in which they ſhould be planted and managed in the fame way as the feedlings. The ſtools, the fecond year after, will have exhibited a freſh crop of young wood for layering : And thus may this operation be performed every fecond year, till the defired quantity is raifed. 3. By *cuttings* alfo all the forts may be propagated, and this may be done two ways : By *cuttings planted in autumn.* Thefe ſhould be ſtrong ſhoots of the laſt year's wood ; and if the tree to be encreafed is not in fo flouriſhing a ſtate as to make fuch ſhoots, it ſhould be headed the year before, and you will have cuttings proper for your purpofe. The ſtrongeſt ſhoots are the beſt ; and October is the beſt month for the bufinefs. They ſhould be a foot and a half long, and muſt be planted a foot deep, in a ſhady well-ſheltered place, and a moſt foil well worked and fine : By this method many good plants may be raifed. Thefe trees may alfo be encreafed by

cuttings

cuttings planted in the summer. The latter end of June, or the beginning of July, is a proper time for the work, and the management muſt be as follows : Having a ſufficient number of pots ready, the cuttings, or rather ſlips, from the trees, ſhould be gathered, and planted in theſe pots, in any ſort of common garden mould made fine. After this, they ſhould have a good watering, and the pots be plunged up to their rims in the ſtove. Here, if water and ſhade be conſtantly afforded them, they will ſtrike root and become good plants. It may be proper to ob-ſerve farther in this place, that cuttings planted in pots in March, and managed this way, will readily grow. After they have ſtruck root, they may be hardened by degrees to the open air. They ſhould remain under cover in the pots all winter ; for they will be rather tender at firſt, by being ſo nicely nurſed ; but in the ſpring, when all danger of froſt is over, they may be turned out, with the mould, either in nurſery-lines at a foot diſtance and two feet aſunder in the rows, or elſe in the places where they are deſigned to remain ; for they will be hardy enough, after growing openly this ſummer, to be in little danger of ſuffering by almoſt any weather.

M Y R I C A.

LINNEAN Claſs and Order, *Dioecia Tetrandria :* Male flowers containing four ſtamina, and female flowers containing two piſtils ; upon diſtinct plants. There are ſix SPECIES ; two of which are of a fragrant quality, and may be admitted into ſhrubbery quarters.

1. MYRICA *Cerifera :* The CANDLEBERRY-MYRTLE, or WAX-BEARING MYRICK ; *a deciduous ſhrub* ; native of Caro-lina, Virginia and Pennſylvania.

2. MYRICA *Gale :* The GALE, or DUTCH MYRTLE, or SWEET WILLOW ; *a low deciduous ſhrub* ; native of heathy bogs in many parts of England, and alſo of moſt of the northern parts of Europe.

1. CAN-

1. CANDLEBERRY-MYRTLE is a shrub about five feet in growth. Many slender branches are produced from the stalk: They are tough, smooth, and of a yellowish brown, having the older spotted with grey spots. The leaves grow irregularly on them all round; sometimes by pairs, sometimes alternately, but generally at unequal distances. They are of a lanceolated figure; and some are serrated at the top, whilst others have their edges wholly entire. They stand on very short footstalks; having their upper surface smooth, and of a shining green colour, whilst their under is of a more dusky hue. The branches of the old plants shed their leaves in the autumn; but the young plants, raised from seeds, retain them the greatest part of the winter; so as during that season to have the appearance of an evergreen. But this beauty will not be lasting; for they shed their leaves proportionally earlier as the plants get older. There are both male and female trees of this sort: The flowers are small, of a whitish colour, and make no figure; neither does the fruit that succeeds the female, which is a small, dry, blue berry, though produced in clusters, make any show: So that it is from the leaves this tree receives its beauty and value; for these being bruised, as well as the bark of the young shoots, emit the most refreshing and delightful fragrance, that is exceeded by no Myrtle, or any other aromatic shrub.

There is a *variety* of this species, of lower growth, with shorter but broader leaves, and of equal fragrance. This grows commonly in Carolina; where the inhabitants collect, from its berries, a wax, of which they make candles, and which occasions its being called the *Candleberry-Tree*. It delights in a moistish soil.

2. The GALE, or SWEET WILLOW, is a shrub of about the same growth with the other. The branches are tough and slender, and covered with a smooth yellowish-brown bark. The leaves are of the same figure with the other, though not so large: They are placed in the same irregular manner on the branches; and when bruised, like them, emit a delightful and refreshing scent. The flowers will appear in July, and the berries, which succeed them in clusters, make no figure to any except a botanist; so that where that science has no share in

view,

view, it is on account of its fragrance that it is propagated. This sort grows wild upon bogs, in many parts, particularly the northern parts of England; so that when it is designed to be in the shrubbery, the moistest parts must be assigned it.

Both these forts may be PROPAGATED by seeds or layers. 1. The seeds of the Candleberry-Myrtle, and the Spleenwort-leaved Gale, we receive from abroad; those of the Sweet Gale, from the bogs where they grow in England. The best way is to sow them in boxes of earth from a rich pasture, well broken and fine. They should be sown about half an inch deep; and when the hot weather comes on, should be set in the shade. They will often remain until the second year before they come up, especially those seeds that come from abroad. If the boxes are set in the shade, and the plants come up, they will require no other trouble the first summer than keeping clean from weeds; in winter they should be removed to a warm hedge or wall, where they may enjoy the benefit of the sun. In the following spring they will come up in plenty. In the beginning of May they should resume their shady situation; and this summer they will require no other trouble than weeding and watering in dry weather. In the winter they should be removed into a well-sheltered place; and this may be repeated two years; when, in the spring, they should be taken out of the boxes, and planted in the nursery, at about a foot asunder. 2. These forts may be also easily propagated by layers; for this operation being performed on the young wood in the autumn, will occasion them to shoot good roots by the autumn following; many of which will be good plants, fit for any place. 3. These plants may likewise be encreased by suckers; for many of them often throw them out in vast plenty; so that these being taken out, the strongest and best-rooted may be finally set out; whilst the weaker, and those with less root, may be planted in the nursery.

NYSSA

N Y S S A.

LINNEAN Clafs and Order, *Polygamia Dioecia :* Male flowers
containing ten ftamina, and hermaphrodite flowers containing
five males and one female each, upon diftinct plants : There is
only one SPECIES :

NYSSA *Aquatica :* The TUPELO-TREE ; *a deciduous tree or
fhrub ;* native of watery places in America.

The TUPELO comprehends two *varieties :*
> The Entire-leaved Tupelo.
> The Serrated-leaved Tupelo.

The *Entire-leaved Tupelo-Tree,* in its native country, will
grow to be near twenty feet high ; with us, its fize will vary
according to the nature of the foil or fituation. In a moift rich
earth, well fheltered, it will bid fair for twenty feet ; in others,
that are lefs fo, it will make flower progrefs, and will in the end
be proportionally lower. The branches are not very numerous ;
and it rifes with a regular trunk, at the top of which they
chiefly grow. The leaves are of a lanceolated figure, and of a
fine light-green colour. They end in acute points, and are very
ornamental, of a thickifh confiftence, foft, grow alternately on
pretty long footftalks, and often retain their verdure late in the
autumn. The flowers, which are not very ornamental, are
produced from the fides of the branches, growing fometimes
fingly, fometimes many together, on a footftalk. They are of a
greenifh colour ; and, in the countries where they naturally
grow, are fucceeded by oval drupes, inclofing oval, acute, furrowed
nuts. In England, they feldom produce fruit.

The *Serrated-leaved Tupelo-Tree* grows ufually to be near
thirty feet high, and divides into branches near the top like the
other. The leaves are oblong, pointed, of a light-green colour,
and come out without order on long footftalks. The flowers
come out from the wings of the leaves on long footftalks. They
are fmall, of a greenifh colour ; and are fucceeded by oval
drupes, containing fharp-pointed nuts, about the fize of a French
Olive.

The PROPAGATION of thefe forts is from feeds, which we
receive from America. As foon, as they arrive, they fhould be

fown in large pots of light fandy earth, one inch deep. The gardener (who muft not expect to fee any plants come up the firft fpring), after this work is done, fhould plunge his pots up to their rims in the natural ground ; and if it be a moiftifh place, it will be the better. Weeding muft be obſerved all fummer ; and a few furze-bufhes ought to be pricked round the pots in November, which will prevent the ground from freezing, and forward the coming up of the feeds. In the next fpring, the pots fhould be plunged into an hot-bed, and after that the feeds will foon come up. As much air as poffible, and watering, fhould be afforded them ; and they muft be hardened foon, to be fet out. The pots fhould be then plunged to their rims again in the natural mould ; where they may remain until October. Watering muft be given them, and they fhould alfo be fhaded inˢthe heat of the day. In October, they fhould be houfed, with other greenhoufe plants, or elfe fet under a hot-bed-frame, or fome other cover, all winter. The third fpring they fhould be taken out of the larger pots, and each planted in a fmaller, in which their growth may be affifted by a gentle heat in a bed ; but if they are planted up to the rims in a moiftifh place, and fhaded in dry weather, they will grow very well. Though by this time they may have become hardy, yet it will be proper to fhelter them the winter following in bad weather. They will require little more care during their ftay in the pots, which may be either two, three, or more years, if they are large enough ; when in fome fpring they may be turned out, with the mould, into the places where they are to remain, which ought always to be moift and well-fheltered.

O N O N I S.

LINNEAN Clafs and Order, *Diadelphia Decandria :* Each flower contains ten males and one female ; the males being divided at the bafe into two divifions : There are twenty-nine SPECIES ; one of which, being of a fhrubby nature, is proper for our purpofe.

ONONIS

ONONIS *Fruticofa* : The SHRUBBY ONONIS, or PURPLE
SHRUBBY REST-HARROW ; *a low deciduous fhrub* ; native of
the Alps and other mountainous parts of Europe.

The SHRUBBY ONONIS, or REST-HARROW, is a flowering-
fhrub of about a yard in growth. The branches are numerous,
flender, and covered with a purplifh-brown bark, having no
fpines. The leaves are trifoliate, grow irregularly on the
branches, fit clofe, are narrow, fpear-fhaped, and their edges
are ferrated. The flowers come out in panicles from the ends
of the branches : They are of the papilionaceous kind, and
their general charaćters will indicate their ftrućture. They
ftand on lon͜ footftalks, ufually three on one. They are large,
red, appear in May, and are fucceeded by fhort turgid pods,
which will have ripe feeds by July or Auguft.

This fort may be PROPAGATED by the feeds. Common
garden-mould of almoft any foil, made fine, will do for the
purpofe. The beds fhould be made and the feeds fown in March,
and covered about half an inch deep. In May the plants will
appear ; and all the fummer they muft be weeded, and duly
watered in dry weather. In the fpring they fhould be taken
out of the feed-bed, and planted in the nurfery, a foot afunder,
where they may ftand a year or two, and then be planted out
for good. As the feeds of this fort ripen exceedingly well with
us, a few may be fown in different parts of the garden, and
fticks placed for a direćtion. Where there are too many come
up to grow together, they may be drawn, and tranfplanted for
other places, or thrown away, if plenty of feeds can always be
had ; and thus may thefe plants be raifed in their proper
places, without the trouble of removing.

P A S S I F L O R A.

LINNEAN Clafs and Order, *Gynandria Pentandria :* Each
flower contains five males and three females ; the males and
females growing together at the bafe : There are twenty-eight
SPECIES ; one of which is proper for our collećtion :

PASSIFLO

PASSIFLORA *Cærulea :* The PALMATED PASSION-FLOWER, or the COMMON PASSION-FLOWER ; *a ligneous climber* ; native of the Brazils.

The PASSION-FLOWER will climb to a prodigious height ; MILLAR fays, " to forty-feet, with ftalks almoft as large as a man's arm'；" and adds, that it will make fhoots of twelve or fifteen feet long in one fummer. The leaves are palmated, being fhaped like the hand ; each is compofed of five folioles, the middle one of which is, like the fingers of the hand, longer, and the reft are fhorter in the fame proportion. Thefe folioles are fmooth, and have their edges free from ferratures, and all together form a fine leaf. The leaves grow from the joints, on fhort footftalks, from whence alfo the clafpers come out. From the joints, alfo, the flowers are produced, in July, Auguft, and September. They are well known; and in fome countries ferve as monitors to the religious, as fhewing the inftruments of our Bleffed Saviour's Paffion ; for they bring in the leaves of fome of the forts to reprefent fome part of it, and the contorted cirrhi for the flagella with which he was fcourged.

This extraordinary plant is very eafily PROPAGATED ; for it takes freely either by cuttings, layers, or feeds. 1. By cuttings. Thefe fhould be planted in a moift rich foil, at the beginning of March. The beds fhould be immediately hooped, and every day, during the drying March winds and fun, fhould be covered with mats ; and all that time they fhould have frequent waterings in the evening. In moift, hazy, or cloudy weather, they fhould be conftantly uncovered ; and with this management many of them will ftrike root. If, through the heat of fummer, the mats be applied, and evening waterings continued, the plants being thus kept cool and moift, will fhoot to be good ones by the autumn. During the winter, the mats muft be applied in frofty weather ; and in the fpring they may be fet out to ftand. 2. Good plants are obtained by layers ; for thefe being laid in the ground in the fpring, will have ftruck root, and be good plants for removing the fpring following. 3. By feeds. Thefe fhould be fown in pots filled with fine fandy foil, from a rich meadow ; and thefe plunged up to the rims in a fhady border. In thefe pots they will readily come up ; and at the approach of winter fhould be removed into the greenhoufe, or fet under an hot-bed-frame. In the fpring following they
may

may refume their old place; and the fpring after that may be fet out for good.

The after-management will be; if planted to climb up trees in warm well-fheltered places, to take away the dead fhoots in the fpring that have been killed by the frofts; for thefe will not only appear unfightly, but by fhortening the branches it will caufe them to fhoot ftronger and flower better. If planted againft high walls, they muft be conftantly nailed up as they fhoot, and in the fpring following the branches muft be fhort‐ ened, and the others taken away. If they be reduced to about a yard or four feet in length, and all weak fhoots cut out, you will be pretty fure of having plenty of good bloom the fummer after. This fort is fucceeded by a large, oval, yellow fruit, which alfo looks well. As this plant is rather tender, and requires mats to be nailed before it in very hard froft, thefe mats muft be always taken off immediately on the alteration of weather; for otherwife the ftems will grow mouldy, and be deftroyed that way. And as it is ufual to lay ftraw, dung, &c. about the ftems to prevent the froft penetrating the ground, this dung, &c. muft not be laid up to the ftem fo as to touch it, but all round it; for if it is laid up to the ftem, the bark will be de‐ ftroyed, and the tree killed, and alfo very little chance remain of the root's throwing out frefh fhoots, as it often does when the plant is killed down to the ground.

P E R I P L O C A.

LINNEAN Clafs and Order, *Pentandria Digynia :* Each flower contains five males and two females : There are five SPECIES; one of which is fufficiently hardy for this climate.

PERIPLOCA *Græca :* The PERIPLOCA, or VIRGINIA SILK, or CLIMBING DOG'S BANE ; *a ligneous climber*; native of Syria.

The PERIPLOCA is a fine climbing plant, that will wind itfelf with its ligneous branches about whatever tree, hedge, pale, or pole is near it ; and will arife, by the affiftance of fuch fupport, to the height of above thirty feet; and where no tree or fup‐ port is at hand to wind about, it will knit or entangle itfelf together, in a moft complicated manner. The ftalks of the older branches, which are moft woody, are co ered with a dark‐ brown bark, whilft the younger fhoots a e more mottled with

T the

the different colours of brown and grey, and the ends of the youngeſt ſhoots are often of a light-green. The ſtalks are round, and the bark is ſmooth. The leaves are the greateſt ornament to this plant ; for they are tolerably large, and of a good ſhining green colour on their upper ſurface, and cauſe a variety by exhibiting their under ſurface of an hoary caſt. Their figure is oblong, or rather more inclined to the ſhape of a ſpear, as their ends are pointed, and they ſtand oppoſite by pairs on ſhort footſtalks. Their flowers afford pleaſure to the curious examiner of nature. Each of them ſingly has a ſtar-like appearance ; for though it is compoſed of one petal only, yet the rim is divided into ſegments, which expand in ſuch a manner as to form that figure. Their inſide is hairy, as is alſo the nectarium, which ſurrounds the petal. Four or five of the flowers grow together, forming a kind of umbel. They are of a chocolate colour, are ſmall, and will be in blow in July and Auguſt, and ſometimes in September. In the country where this genus grows naturally, they are ſucceeded by a long taper pod, with compreſſed ſeeds, having down to their tops.

The PROPAGATION of this climber is very eaſy; for if the cuttings are planted in a light, moiſt ſoil, in the autumn or in the ſpring, they will readily ſtrike root. Three joints at leaſt ſhould be allowed to each cutting : They ſhould be the bottom of the preceding ſummer's ſhoot ; and two of the joints ſhould be planted deep in the ſoil.

Another, and a never-failing method is by layers ; for if they are laid down in the ground, or a little ſoil only looſely thrown over the young preceding ſummer's ſhoots, they will ſtrike root at the joints, and be good plants for removing the winter following.

PHILADELPHUS.

LINNEAN Claſs and Order, *Icoſandria Monogynia :* Each flower contains about twenty males and one female : There are only two SPECIES :

1. PHILADELPHUS *Coronarius :* The COMMON SYRINGA, or the CUCUMBER-TREE, or the MOCK ORANGE ; *a deciduous ſhrub ;* native place uncertain.

2. PHILADELPHUS *Inodorus:* The CAROLINA SYRINGA, or the SCENTLESS SYRINGA; *a tall deciduous shrub;* native of Carolina.

1. The COMMON SYRINGA admits of three remarkable *varieties:* Common Syringa, Double Syringa, and Dwarf Syringa.

The *Common Syringa,* or *Mock Orange,* is a very beautiful shrub, about six feet in growth: It sends forth numerous branches from the root, which are brittle and full of pith. These also send out others from their sides that are shorter, stand generally oppofite by pairs, and are alternately of contrary directions. These younger shoots are slender, jointed, and covered, some with a smooth pale-brown bark, others with a smooth bark of a darker colour. The leaves are large, and placed oppofite by pairs on short footstalks. They are of an oval, spear-shaped figure, of a strong green colour, and have the flavour of a cucumber. Their edges are irregularly indented, their surface is rough, and they fall off early in the autumn. This shrub, by its flowers, makes a fine figure in May and June; for they are produced in clufters both at the end and from the sides of the branches. They are of a fine white colour, and exceedingly fragrant. The petals of which each is compofed are large, and spread open like thofe of the Orange; and then forming branches, which stand each on its own feparate short footstalk, and being produced in plenty all over the shrub, both at once feast the eye and the smell: The eye, by the pleafing appearance it will then have; the smell, as the air at fome diftance will be replete with the odoriferous particles conftantly emitted from thofe fragrant flowers. Thefe flowers, however, are very improper for chimneys, waterglaffes, &c. in rooms; for in thofe places their scent will be too ftrong; and for the ladies in particular, often too powerful.

The *Double-flowering Syringa* is a low variety of this species, feldom rifing to more than a yard high. The defcription of the other belongs to this fort, except that the leaves and branches are proportionally fmaller and more numerous, and the bark of the shoots of a lighter brown. It is called the Double-flowering Syringa, becaufe it fometimes produces a flower or two with three or four rows of petals; whereas, in general, the flowers, which are very few, and fel-

dom produced, are fingle. They are much fmaller than thofe
of the other; and you will not fee a flower of any kind on
this fhrub oftener perhaps than once in five years. It is hardly
worth propagating on this account; fo that a few plants only
ought to be admitted into a collection, to be ready for obfervation.

The *Dwarf Syringa* is ftill of lower growth than the other,
feldom arifing to more than two feet in height. The defcription of the firft fort ftill agrees with this; only that the branches
and leaves are ftill proportionally fmaller and more numerous,
and the bark is ftill of a lighter brown. It never produces flowers.

2. The CAROLINA SYRINGA is the tallest grower by far of
any fort of the Syringa, and makes the grandeft fhow when
in blow; though the flowers are deftitute of fmell. It will
grow to about fourteen feet in height; the branches are numerous and flender, and the bark on the young fhoots is fmooth
and brown. The leaves alfo are fmooth and entire, and placed
oppofite by pairs on longifh footftalks. The flowers, which are
produced at the ends of the branches, are of a fine white colour,
and, being larger than thofe of the firft fort, have a noble look.

The PROPAGATION of all the forts is very eafy: They are encreafed by layers, cuttings or fuckers. 1. The moft certain method
is by layers; for the young twigs being laid in the earth in the
winter, will be good-rooted plants by the autumn following.
2. Thefe plants may be encreafed by cuttings, which being
planted in October, in a fhady moift border, many of them
will grow; though it will be proper to let thofe of the Carolina fort remain until fpring, and then to plant them in pots,
and help them by a little heat in the bed. By this affiftance,
hardly one cutting will fail. 3. They may be alfo encreafed
by fuckers, for all the forts throw out fuckers, though the Carolina Syringa the leaft of any. Thefe will all ftrike root, and
be fit for the nurfery-ground: Nay, the Double-flowering and
the Dwarf forts are always encreafed this way; for thefe plants
having ftood five or fix years, may be taken up and divided into
feveral fcores. All the plants, however, whether raifed from
layers, cuttings, or fuckers, fhould be planted in the nurfery-
ground to get ftrength, before they are fet out for good. They
fhould be planted a foot afunder, and the diftance in the rows
fhould be two feet. After this, they will require no other
care

care than hoeing the weeds, until they have ftocd about two
years, which will be long enough for them to ftand there.

P H I L L Y R E A.

Linnean Clafs and Order, *Diandria Monygynia :* Each
flower contains two males and one female : There are three
Species :

1. Phillyrea *Media :* The Oval-leaved Phillyrea, or
Mock Privet, or the Medial-leaved Phillyrea ; *a tall
evergreen fhrub ;* native of the South of Europe.

2. Phillyrea *Latifolia :* The Broad-leaved Phillyrea,
or Mock Privet ; *a tall evergreen fhrub ;* native of the South
of Europe.

3. Phillyrea *Anguftifolia :* The Narrow-leaved Phillyrea, or Mock Privet ; *a deciduous fhrub ;* native of Spain
and Italy.

1. The Oval-leaved or Middle Phillyrea has the
following *varieties :* Common Smooth-leaved *Phillyrea*, Privet-
leaved *Phillyrea*, Olive-leaved *Phillyrea*.

The *Common Smooth-leaved Phillyrea* will grow to be twelve
or fourteen feet high, and the branches are many ; the older of
which are covered with a dark-brown bark, but the bark on the
young fhoots is of a fine green colour. They are oval, fpear-
fhaped, and grow oppofite, by pairs, on ftrong fhort footftalks.
The flowers are produced in clufters, from the wings of the
young branches. They are fmall, and of a kind of greenifh-
white colour ; they appear in March, and are fucceeded by
berries, which are firft green, then red, and black in the autumn
when ripe.

Privet-leaved Phillyrea will grow to be ten or twelve feet
high, and the branches are covered with a brown bark. The
leaves a little refemble the Privet ; they are of a fine green
colour, and grow by pairs on the branches. They are of a
lanceolate figure, and their edges are entire, or nearly fo ; for
fome figns of ferratures fometimes appear. The flowers grow

T 3

like

like others, in clusters, in March. They are whitish. and are succeeded by small black berries.

The *Olive-leaved Phillyrea* is the most beautiful of all the sorts. It will grow to be about ten or twelve feet high; and the branches, which are not numerous, spread abroad in a free easy manner, which may not improperly be said to give the tree a fine air. They are long and slender, and are covered with a light-brown bark; and on these the leaves stand opposite by pairs, at proper intervals, on short footstalks. They resemble those of the Olive-tree, and are of so delightful a green as to force esteem. Their surface is exceeding smooth, their edges are entire, and the membrane of a thickish consistence. The flowers are small and white, and like the other sorts make no show. They are succeeded by single roundish berries.

2. The BROAD-LEAVED PHILLYREA will grow to be about twelve feet high. The branches seem to be produced stronger and more upright than those of the former species. The bark is of a grey colour, spotted with white, which has a pretty effect; and the leaves grow opposite by pairs. They are of a heart-shaped oval figure, of a thick consistence, and a strong dark-green colour. Their edges are sharply serrated, and they stand on short strong footstalks. The flowers grow from the wings of the leaves in clusters, in March. They are of a kind of greenish-white colour, make no show, and are succeeded by small round black berries.

The *varieties* of this species are, the *Ilex*-leaved *Phillyrea*, the Prickly *Phillyrea*, the Olive *Phillyrea* with slightly-serrated Edges.

3. The NARROW-LEAVED PHILLYREA is of lower growth, seldom rising higher than eight or ten feet. The branches are few and slender, and they also are beautifully spotted with grey spots. The leaves, like the others, stand opposite by pairs. They are long and narrow, spear-shaped, and undivided, of a deep green colour, and of a thick consistence. Their edges are entire, and they also stand on short footstalks. The flowers, like the others, make no show. They are whitish, and grow in clusters from the wings of the branches, in March; and are succeeded by small round black berries.

The *varieties* of this species are, the Rosemary *Phillyrea*, Lavender *Phillyrea*, Striped *Phillyrea*, &c.

The

The *Phillyreas* are to be PROPAGATED by feeds or layers.
1. By feeds. Thefe ripen in the autumn, and fhould be fown foon
after. The mould muft be made fine; and if is not naturally
fandy, if fome drift fand be added, it will be fo much the
better. The feeds for the moft part remain until the fecond
fpring before they come up; and if they are not fown foon after
they are ripe, fome will come up even the third fpring after.
They muft be fown about an inch deep; and during the following fummer fhould be kept clean from weeds. After they are
come up, the fame care muft be obferved, and alfo watering in
dry weather; and if the beds are hooped, and the plants
fhaded in the hotteft feafon, they will be fo much the better
for it. However, at the approach of winter they muft be
hooped, and the beds covered with mats in the hardeft frofts,
otherwife there will be danger of lofing the whole crop; for
thefe trees, though they are very hardy when grown tolerably
large, are rather tender whilft feedlings. It will be proper to
let them remain in the feed-beds, with this management, for two
fummers; and then, waiting for the firft autumnal rains, whether in September or October (and having prepared a fpot of
ground), they fhould at that juncturebe planted out, and this
will occafion them immediately to ftrike root. The diftance
they fhould be planted from each other need not be more than a
foot, if they are not defigned to remain long in the nurfery: If
there is a probability of their not being wanted for fome years,
they fhould be allowed near double that diftance; and every
winter the ground in the rows fhould be well dug, to break
their roots, and caufe them to put out frefh fibres, otherwife
they will be in danger of being loft, when brought into the
fhrubbery-quarters. 2. By layers they will eafily grow. The
autumn is the beft time for this operation, and the young fhoots
are fit for the purpofe. The beft way of layering them is by
making a flit at the joint; tho' they will often grow well by a
twift being only made. When the gardener choofes the method
of twifting a young branch for the layers, he muft be careful
to twift it about a joint fo as only to break the bark; for if it is
too much twifted, it will die from that time, and his expectations wholly vanifh. But if it be gently twifted with art and
care, it will at the twifted parts be preparing to ftrike root, and

by the autumn following, as well as those layers that had been
flit, will have good roots; the strongest of which will be fit
for planting where they are wanted to remain, whilst the weaker
and worst-rooted layers may be planted in the nursery-ground
like the seedlings, and treated accordingly.

P H L O M I S.

LINNEAN Class and Order, *Didynamia Gymnospermia*:
Each flower contains four males and one female; two of the
males being somewhat longer than the other two; and the seeds
being naked. There are fourteen SPECIES; two of which are
adapted to the shrubbery.

1. PHLOMIS *Fruticosa*: The YELLOW PHLOMIS, or JERU-
SALEM SAGE; *a non-deciduous hoary shrub*; native of Spain and
Sicily.

2. PHLOMIS *Purpurea*: The PURPLE PHLOMIS, or PORTU-
GAL SAGE; *a non-deciduous hoary shrub*; native of Portugal and
Italy.

1. The YELLOW PHLOMIS, or JERUSALEM SAGE, The
varieties of this species are, The Broad-leaved Sage-tree of Jerusalem, The Narrow-leaved Jerusalem Sage-tree, The Cretan
Sage-tree.

The *Broad-leaved Jerusalem Sage-tree* is now become very
common in our gardens, which indeed is no wonder, as its
beauty is great, and its culture easy. It will grow to be about
five feet high, and spreads its branches without order all around.
The older branches are covered with a dirty, greenish, dead,
falling, ill-looking bark; and this is the worst property of this
shrub: But the younger shoots are white and beautiful; they
are four-cornered, woolly, and soft to the touch. The leaves
are roundish and oblong, and moderately large; and these grow
opposite at the joints of the shrub on long footstalks. They
are hoary to a degree of whiteness, and their footstalks also are
woolly, white, tough, and strong. The flowers are produced
in June, July, and August, at the top joints of the young shoots,

in

in large whorled bunches. They are of the labiated kind, each confifting of two lips, the upper end of which is forked, and bends over the other. A finer yellow can hardly be conceived than the colour of which they are poffeffed ; and being large, they exhibit their golden flowers at a great diftance, caufing thereby a handfome fhew.

The *Narrow-leaved Jerufalem Sage-tree* is of lower growth than the other, feldom rifing higher than a yard or four feet. This fhrub is in every refpect like the other ; only the fhoots feem to have a more upright tendency of growth. The leaves alfo, which are narrower, are more inclined to a lanceolate form : They are numerous in both the forts, and hide the deformity of the bark on the older ftems, which renders them lefs exceptionable on that account. In fhort, thefe forts are qualified for fhrubberies of all kinds, or to be fet in borders of flower-gardens, where they will flower, and be exceeded even in that refpect by very few fhrubs.

Cretan Sage-tree is ftill of lower growth than either of the former, feldom arriving to a yard in height. The leaves are of the fame white, hoary nature ; they are very broad, and ftand on long footftalks. The flowers are alfo of a delightful yellow colour, very large, and grow in large whorls, which give the plant great beauty.

2. PURPLE PHLOMIS, or PORTUGAL SAGE. The ftalks of this fpecies are woody, four feet high, and fend forth feveral angular branches, which are covered with a white bark. The leaves are fpear-fhaped, oblong, woolly underneath, crenated, and grow on fhort footftalks. The flowers are produced in whorls, from the joints of the branches. They are of a deep purple colour, and have narrow involucra. They appear in June and July, but are not fucceeded by ripe feeds in England.

There is a *variety* of this fpecies, with iron-coloured flowers ; and another with flowers of a bright purple.

There are fome other fhrubby forts of *Phlomis*, of great beauty ; but thefe not only often lofe their leaves, and even branches, from the firft froft, but are frequently wholly deftroyed, if it happens to be fevere. They are low fhrubs, very beautiful, and look well among perennial flowers, where they will not only clafs as to fize with many of that fort, but, being

rather

rather tender, may with them have fuch extraordinary care as the owner may think proper to allow them.

The PROPAGATION of the above forts is very eafy, either by layers or cuttings. 1. If a little earth be thrown upon the branches, any time in the winter, they will ftrike root, and be good plants by the autumn following, fit for any place. Thus eafy is the culture by that method. 2. The cuttings will alfo grow, if planted any time of the year. Thofe planted in winter fhould be the woody fhoots of the former fummer: Thefe may be fet clofe in a fhady border; and being watered in dry weather, will often grow. This fhrub may be propagated by young flips, alfo, in any of the fummer months. Thefe fhould be planted in a fhady border, like Sage, and well watered. If the border is not naturally fhady, the beds muft be hooped, and covered with matting in hot weather. Watering muft be conftantly afforded them; and with this care and management many of them will grow.

P I N U S.

LINNEAN Clafs and Order, *Monoecia Monadelphia*: Male flowers containing many ftamina joined at the bafe, and female flowers containing one piftil, upon the fame plant; the males being difpofed in fcaly bunches, the females in imbricated cones: There are twelve SPECIES:

1. PINUS *Larix*: The LARCH, or DECIDUOUS PINE; *a tall deciduous tree*; native of Switzerland, the Alps, and fome parts of Italy.

2. PINUS *Sylveftris*: The WILD PINE; *a tall evergreen tree*; native of Scotland and the northern parts of the continent of Europe.

3. PINUS *Strobus*: The WEYMOUTH-PINE, or the WHITE PINE; *a tall evergreen tree*; native of New-England, Virginia, Canada, and Carolina.

4. PINUS *Pinea*: The STONE-PINE; *a tall evergreen tree*; native of Spain and Italy.

5. PINUS

5. PINUS *Cembra*: The CEMBRO, or the CEMBRO-PINE; *an evergreen tree*; native of Switzerland, the Alps, Siberia, and Tartary.

6. PINUS *Tæda*: The SWAMP PINE; *a tall evergreen tree*; native of the fwamps of Virginia and Canada.

7. PINUS *Cedrus*: The CEDAR OF LEBANON; *a tall evergreen tree*; native of Mount Lebanon.

8. PINUS *Picea*: The YEW-LEAVED FIR; *a tall evergreen tree*; native of Scotland, Sweden, and Germany.

9. PINUS *Abies*: The EUROPEAN SPRUCE-FIR; *a tall evergreen tree*; native of the northern parts of Europe and of Afia.

10. PINUS *Canadenfis*: The AMERICAN SPRUCE-FIR, or the NEWFOUNDLAND SPRUCE-FIR; *a tall evergreen tree*; native of Canada, Pennfylvania, and other parts of North America.

11. PINUS *Balfamea*: The HEMLOCK-FIR; *a low evergreen tree*; native of Virginia and Canada.

12. PINUS *Orientalis*: The ORIENTAL FIR; *a low evergreen tree*; native of the Eaſt.

1. The LARCH. This is a lofty tree: its branches are flender, and incline downward: the leaves are of a light green; and, like the Cedar of Lebanon, are bunched together like the pencils or little brufhes of the painter. In fpring, when the leaves and flowers are breaking out, the Larch has a particularly elegant appearance; and in winter, it gives variety to a wooded fcene by a remarkablenefs in its naked branches: It is in good efteem as an *Ornamental*; and its timber is of the more *ufeful* kind: it is fuperior to that of moſt of the *Pinus* tribe. HANBURY fays, "Many encomiums have been beftowed on the timber of the Larch: and we find fuch a favourable account of it in antient authors, as fhould induce us to think it would be proper for almoſt any ufe. Evelyn recites a ſtory of Witfen, a Dutch writer, that a fhip built of this timber and Cyprefs, had been found in the Numidian fea, twelve fathoms under water, found and entire, and reduced to fuch a hardnefs as to refiſt the fharpeſt tool, after it had lain fubmerged above a thoufand four hundred years. Certain it is, this is an excel-

lent

lent wood for ſhip and houſe building. At Venice this wood is frequently uſed in building their houſes, as well as in Switzerland, where theſe trees abound : So that, without all doubt, the Larch excels for maſts for ſhips, or beams for houſes, doors, windows, &c. particularly as it is ſaid to reſiſt the worm.

" In Switzerland, their houſes are covered with boards of this wood, cut out a foot ſquare; and as it emits a reſinous ſubſtance, it ſo diffuſes itſelf into every joint and crevice, and becomes ſo compact and cloſe, as well as ſo hardened by the air, as to render the covering proof againſt all weather. But as ſuch covering for houſes would cauſe great devaſtation in caſe of fire, the buildings are confined to a limited diſtançe, by an order of police from the magiſtrates. The wood, when firſt laid on the houſes, is ſaid to be very white ; but this colour, in two or three years, is changed, by means of the ſun and reſin, to a black, which appears like a ſmooth ſhining varniſh."

Of the *Common Larch* there are ſeveral *varieties.* The flowers which the commoneſt ſort exhibits early in the ſpring are of a delicate *red* colour ; another ſort produces *white* flowers at the ſame ſeaſon, and theſe have a delightful effect among thoſe of the Red ſort ; whilſt another, called the *Black Newfoundland Larix,* encreaſes the variety, tho' by an aſpect little differing from the others. There are alſo Larches with *greeniſh* flowers, *pale-red,* &c. all of which are accidental varieties from ſeeds. Theſe varieties are eaſily diſtinguiſhed, even when out of blow : The young ſhoots of the White-flowering Larch are of the lighteſt green, and the cones when ripe are nearly white. The Red-flowering Larch has its ſhoots of a reddiſh caſt, and the cones are of a brown colour ; whilſt the cones and ſhoots of the Black Newfoundland Larch are in the ſame manner proportionally tinged. The cones, which are a very great ornament to ſeveral ſorts of the Pines, are very little to theſe. Their chief beauty conſiſts in the manner of their growth, the nature and beauty of their pencilled leaves, and fair flowers ; for the cones that ſucceed them are ſmall, of a whitiſh, a reddiſh, or a blackiſh brown colour, and make no figure.

The method of PROPAGATION is from ſeed : The cones may be gathered in November, and ſhould be left in a dry place till the ſpring. Juſt before ſowing, let the cones be opened or torn

into

into four quarters by a knife, the point of which muft be thruft exactly down the center, fo that the feeds in their refpective places may not be damaged. Formerly, great pains were beftowed in getting at the feeds, by cutting off the fcales of the cones fingly, and letting the feeds drop. This occafioned great expence to thofe who wanted a quantity of feeds; fo that it is wholly laid afide now, for the more eafy method of opening them with knives, and then threfhing them. A certain price is generally allowed per thoufand to the poor for opening them. When a fufficient quantity is opened, they fhould be threfhed in a room, which will divide the fcales, and diflodge the feeds, without injuring many of them. Three thoufand cones will generally produce about a pound of good feeds. The cones being fufficiently broken, and the feeds threfhed out, they fhould be winnowed or fieved, to have clear feeds; after which they will be ready for fowing. Let the feminary confift of a fpot of fine light earth; and let the feeds be fowed in beds a quarter of an inch deep. In the fpring, when the plants appear, they fhould be gently refrefhed with water in dry weather, and carefully kept clean of weeds during the whole fummer. By the autumn they will not have fhot more than an inch or two; and in fpring they fhould be pricked out in beds about three inches afunder. The fpring following, they muft be taken out of thefe beds with care, and planted in the nurfery-ground, three feet afunder in the rows, and two feet diftance; and here they may remain until they are fit to be planted out finally, which will be about the fecond or third year after. If they grow well in the nurfery, it is advifeable to plant them where they are to continue after having attained two years ftrength in that place, if the ground can poffibly be prepared for their reception; fince thefe trees always thrive beft when removed fmall from the nurfery, if they are of a fufficient fize not to be injured by the weeds; if they are fmaller, the owner muft keep them clean. The Larch-tree will grow extremely well on almoft any foil, as well in clays as in other forts; it thrives amazingly on the declivities of hills, and fides of high mountains; it is hardy enough to refift the fevereft cold, therefore proper for all expofed places: And, as the timber is fo valuable, and its growth fo quick, it is a tree which may be propagated to the great advantage of the owner.

2. The

2. The WILD PINE. This ſpecies includes two *varieties:* The Scotch Fir and the Pineaſter.

The *Scotch Fir.* This tree is too well known to require any deſcription: and the method of propagating it will be fully treated of under the Article TIMBER GROVES.

The *Pineaſter.* This is a large timber-tree, and naturally throws out very large arms, ſome of which will be nearly horizontal. " Some people think theſe trees are very ornamental on their account; for in the winter eſpecially they appear naked, and are of a yellowiſh colour; and being ſpread abroad thus large, and without order, in the mixture of the more regular ſorts of growing Firs, they make a good contraſt. The Gardener muſt obſerve, that the leaves of this ſort are very large and long, and of a lighter green than thoſe of the Scotch Fir, which is another circumſtance to direct him to its ſituation; and he muſt alſo obſerve, that thoſe long and large leaves which ornament the younger branches only, give the tree a majeſtic air; and as the larger arms appear naked to view, ſo the younger, being thus plentifully furniſhed, have a noble effect, beſides what beauty it receives from its numerous cones." HANBURY.

Its PROPAGATION may be the ſame as that of the Scotch Fir.

3. The WEYMOUTH PINE. This is a princely tree, majeſtic and elegant in the higheſt degree: HANBURY ſays, " it will grow to more than a hundred feet high, and makes ſuch excellent maſts for ſhips, that the Legiſlature in the reign of Queen Anne enacted a law enforcing the encouragement of the growth of theſe trees in America, where they abound." As an *Ornamental* it ſtands firſt of all the Pines, and in the *uſes* of its wood none of them excel it. The bark is ſmooth and ſoft to the touch, and though of a duſky brown colour, on the whole has a delicate look. The leaves are truly ornamental, though their colour is nothing extraordinary; but they are long and ſlender, and are formed into taſſels, which hang in ſo eaſy and elegant a manner, as " to make one in love with the tree."

The PROPAGATION of the Weymouth Pine is not ſo difficult as has been heretofore underſtood: It may be raiſed in common ſeed-beds with ordinary care. HANBURY gives us the following directions " The ſeeds of the Weymouth Pine are larger
than

than thofe of the Scotch Fir; and in order to raife the young plants, it will be proper to fow them in pots or boxes, which may be removed into the fhade after the plants are come up, when the fun's rays are violent. If they are fown in beds of fine light earth, they fhould be hooped, and conftantly covered with mats from the fun's heat, and as carefully uncovered when he fets. In about fix or feven weeks after fowing, the young plants will appear, when they fhould be regularly guarded from birds, otherwife all your feeds, time, and trouble, will be loft; for if the birds take to them at their firft coming up, and are unmolefted, they will not leave a fingle plant. The plants being now above ground, the weeds fhould be conftantly picked out, as they appear, left the fibres of their roots mixing with thofe of the Firs, many of the latter may be drawn out with them. In dry weather they fhould be refrefhed with water: But this muft be done fparingly, and with the utmoft caution; for as the ftems of the young plants are very flender, by over-watering they are frequently thrown afide, which they hardly ever recover. Thus, continues HANBURY: "I have known gentlemen, who, in attempting to raife thefe trees, have feen the young plants go off without perceiving the caufe; and the more watering and pains they have taken, have found the plants perfift in this way more and more, to their great mortification and aftonifhment. In the fpring following thefe plants fhould be pricked out in beds half a foot afunder each way; and here they may ftand two years, when they may be either finally planted out, or removed into the nurfery, at the diftance of one foot afunder, and two feet in the rows. If care has been taken of them in the nurfery, they may be removed at a confiderable height with great affurance of fuccefs: for it is much eafier to make this Pine grow than any of the other forts: So that where they are wanted for ornament in parks, open places, &c. a fhow of them may be made in a little time.

" The foil the Weymouth Pine delights in moft is a fandy loam; but it likes other foils of an inferior nature: and altho' it is not generally to be planted on all lands, like the Scotch Fir, yet I have feen it luxuriant and healthy, making ftrong fhoots, on blue and red clays, and other forts of ftrong ground. On ftony and flaty ground, likewife, I have feen fome very fine

trees:

trees : So that I believe whoever is defirous of having plantations of this pine, need not be curious in the choice of his
ground."

4. The Stone Pine will not grow to the height of the
former ; and the bark is rough, and on fome trees of a reddifh
colour. The leaves are long, very ornamental, and of a fine fea-
green colour. The cones give this tree the grandeft look ;
for they are fometimes near fix inches long, and are large,
thick, and turbinated. The fcales are beautifully arranged,
and the whole cone is large and curious. " The kernels are eatable, and by many preferred to almonds ; in Italy they are ferved
up at table in their defferts ; they are exceeding wholefome,
being good for coughs, colds, confumptions, &c. on which account only this tree deferves to be propagated. Hanbury
continues, It may be very proper here to take notice of a very
great and dangerous miftake Mr. Miller has committed, by
faying, under this article of Stone-Pine, that feeds kept in the
cones will be good, and grow, if they are fown ten or twelve
years after the cones have been gathered from the trees ; whereas
the feeds of this fort, whether kept in the cones or taken out,
are never good after the firft year ; and though fometimes a few
plants will come up from the feeds that are kept in the cones
for two years before, yet this is but feldom ; neither muft a
tenth part of a crop be expected. This caution is the more neceffary, as feveral gentlemen who had cones, upon reading Mr.
Miller's Book, and finding the feeds would take no damage
when kept there, deferred the work for a feafon or two, when
they thought they fhould have more conveniency either of men
or ground for their purpofe ; and were afterwards wholly difappointed, no plants appearing, the feeds being by that time
fpoiled and worth nothing."

The propagation of the Stone Pine is from the feeds, which
may be procured from their large cones by the help of a vife; for
this will fo effectually break the cones, without hurting the feeds,
that they may be taken out with pleafure. The cones fhould
be frefh, not older than a year or two at fartheft, or the feeds
will not be good; for although it has been afferted, that the
feeds of Pines in general will keep in their cones many years,
yet the cones of this fpecies of Pine are an exception, as the
feeds

feeds are rarely found good after the cones are one year old. The feafon for fowing thefe feeds is the middle of March. The weather being fine, and the ground fit for working, they fhould be fown about half an inch deep, in beds of fine light earth. In about feven weeks the plants will appear, which muft be kept clean of weeds, and now and then watered in dry weather till July, by which time they will have made a tolerable fhoot. In the month of July they fhould be taken out of the feed-beds, and pricked in others four inches afunder. Rainy and cloudy weather muft be made choice of for this work; and after they are planted, the beds ought to be hooped, in order to be covered with mats in the heat of the day, which, however, fhould be always uncovered in the night. When they have taken to the ground, farther covering will be needlefs, and here they may remain, with only now and then watering, and keeping them clear of weeds, till the fpring twelvemonth following; when, in the beginning of April, they fhould be planted out in the nurfery, in well prepared ground, a foot afunder, and at two feet diftance in the rows. Here they may ftand two years, and then fhould be finally planted out. But if the trees are defired to be larger before they are brought to the fpot where they are to ftand, they muft be kept conftantly removing every two years in the nurfery; for without this management this is a very difficult tree to be improved.

The Stone-Pine delights in a fandy loam; though, like moft other Pines, it will grow well in almoft any land.

5. The CEMBRO-PINE is a fine tree, though of lower growth than any of the former, and the leaves are very beautiful; for they are of a lighter green than moft of the forts, and are produced five in a fheath. They are pretty long and narrow; and as they clofely ornament the branches all round, they look very beautiful, and render the tree on their account valuable. The cones of thefe trees alfo on their waving heads, have a good effect; for they are larger than thofe of the Pincafter, and the fquamæ are beautifully arranged.

6. The SWAMP-PINE. Of this fpecies there are many varieties: HANBURY gives us the following account of them:

"The Three-leaved American Swamp-Pine is a very large growing tree, if it has the advantage of a moift fituation. The leaves are of a fine green colour, and are exceeding long,

U   flender,

flender, and beautiful; three iffue out of one fheath, and they clofely garnifh the younger branches. This is a tree worthy of propagation, whether we regard its timber, or its fine appearance when growing. Its timber is faid to be equal in value to that of moft forts of the Pine; and befides the beauty it receives from its fine long three-fheathed leaves, its head will be ornamented with very large cones, the good effect of which may be eafily conceived.

" The *Two-leaved American Pine* will grow to be a large tree, and the leaves are long; two only grow in each fheath, which occafions its being fo diftinguifhed. The leaves are of a lighter colour than many of the others. On the whole, it is a fine tree, but will make very little variety, unlefs clofely examined. The cones of this fort are much larger, and the fcales more beautifully arranged, than thofe of the Scotch Fir, though they are not of the fize of the former fort. This Fir alfo likes a moift foil.

" The *Yellow American Pine*, the *Yellow Tough Pine*, and the *Tough Pine of the Plains*, I received by thofe names: There is fome difference in the fize and fhape of the cones, though that feems inconfiderable. Thefe three forts make very little variety among themfelves; for they have nearly the fame manner of growth; and though I have none that are yet grown to any large fize, yet they all feem to have a tendency to throw out large arms, a little like the Pineafter. How valuable the timber may be, I cannot tell; but the younger fhoots of all of them are exceeding tough, and, had we plenty, would make excellent bands for faggoting. The leaves are long, and of a yellowifh green colour; there are three, and fometimes two only in a fheath. If a large quarter of thefe were to be planted, to be feen at a diftance, by any of the darker-coloured forts of Pines, their very different fhade muft have a delightful effect.

" *Baftard Pine* is another fort we receive from America, tho' it differs very little from fome of the other American forts. The leaves are long and flender; fometimes two and fometimes three grow in each fheath. They are generally of a yellowifh colour towards their bafe, though their ends are green. The cones are rather long and flender, and the ends of the fcales are fo pointed, as to occafion its being called by fome the Prickly-coned Pine.

" *Frank-*

" *Frankincense-Pine* is another American fort, which we receive under that name. The leaves of it are long, and of a fine green colour. They are narrow, and three are contained in each fheath. They clofely ornament the younger branches all around. This tree, however, beautiful as it is on their account, makes little variety among the Pines, for many others look like it; but by the cones it makes a ftriking difference; for thefe are exceeding large, even as large as thofe of the Stone-Pine; but their fcales are loofer, and their arrangement is not quite fo beautiful.

" The *Dwarf-Pine*, as its name imports, is the leaft grower of all the forts of Pines. It is an American plant, and the leaves grow two in a fheath; thefe are fhort, and of a pretty good green colour. This fort is coveted by fome, on account of its low growth; but it is the leaft beautiful of any of the Pines, and has naturally a fhabby look. The cones are fmall, and the fcales are pointed. There is very little in the plant to make it defirable.

" There are many other forts of American Pines, which we receive from thence with the like cant names as thofe of the above, which I have chofe to retain, as they will probably be continued to be fent over, and that the Gardener receiving them as fuch may beft know what to do with them. In many of thofe forts I fee at prefent no material difference, fo am induced to think they are the fame, fent over with different names. Some of the forts above-mentioned differ in very few refpects; but I have chofe to mention them, as a perfon may be fupplied with the feeds from Pennfylvania, Jerfey, Virginia, Carolina, &c. where they all grow naturally: and having once obtained the feeds, and from them plants, they will become pleafing objects of his niceft obfervations."

Thefe may all be propagated in the fame manner as the WEYMOUTH.

7. The CEDAR of LEBANON. This, in its native foil, has always been confidered as the moft majeftic tree in nature. The leaves grow in pencils like thofe of the Larch; and the extremities of its branches are likewife declining as thofe of the Deciduous Pine; to which at firft fight it bears a ftrong refemblance; excepting in that it is lefs lofty and more fpreading. There are fome very fine Cedars of Lebanon in Stow Gardens. It ranks

among

among the firſt of the *ornamental* tribe : and the *uſes* of its timber are univerſally acknowledged. HANBURY enumerates the following: " It was greatly uſed in the building of Solomon's temple, which at once convinces us of its ſuperlative excellence. It is ſaid to continue ſound for two thouſand years ; and we arc told, that in the Temple of Apollo at Utica there was found cedar-wood of that age. The magnificent temples of the Pagans, as well as thoſe of the trúe God, were chiefly built of this fa- mous timber. The ſtatue of the Great Goddeſs at Epheſus was made of this material ; and if this tree abounded with us in great plenty, it might have a principal ſhare in our moſt ſuperb edifices. The effluvia conſtantly emitted from its wood are ſaid to purify the air, and make rooms wholeſome. Chapels and places ſet apart for religious duties, being wainſcotted with this wood, inſpire the worſhippers with a more ſolemn awe. It is not obnoxious to worms ; and emits an oil which will preſerve cloth or books from worms or corruption. The ſawduſt will preſerve human bodies from putrefaction, and is therefore ſaid to be plentifully uſed in the rites of embalming, where prac- tiſed."

The method of PROPAGATION is this ; Having procured the cones, whether from the Levant or of our own growth, the ſeeds, a little before ſowing, ſhould be got out in this manner: Let a hole be bored with a gimblet exactly up the center of each cone, from the baſe to the apex ; put them into a tub of water, where they may remain till the next day ; then having a wooden peg, rather bigger than the gimblet, let it be thruſt down the hole, and it will ſo divide the cones, that the different ſcales may be taken away, and the ſeeds picked out. In doing this, great care muſt be taken not to bruiſe and hurt the ſeeds, which will then be very tender. The ſoil in which you ſow theſe ſeeds ſhould be rather of a ſandy nature ; or, for want of this, ſome mould taken from a rich paſture, and ſieved with a little drift ſand, will ſerve the purpoſe. Having the mould and ſeeds ready, in the beginning of March let the latter be ſown in pots or boxes near half an inch deep : In about ſeven or eight weeks the plants will come up, when they ſhould be removed into the ſhade from the heat of the ſun ; where they may ſtand, but not under ſhelter, all the ſummer ; during which time they ſhould

be

be kept clean of weeds, and watered now and then. In the winter-feafon they muft be removed into a warmer fituation ; or, if it is likely to prove very fevere, they fhould be fheltered either by mats, or removed into the greenhoufe, or covered with an hotbed-frame ; for they are fubject to lofe their young tops at firft, by the feverity of frofts. In the beginning of April following, thefe plants may be pricked out in beds four inches afunder ; and if the weather proves dry, they fhould be fhaded and watered till they have taken root ; after which they will want little fhading and lefs watering. Indeed, nothing more is required than keeping them clean from weeds, and covering the ground fo as to keep it moift, and prevent its chapping by the fun's rays. In thefe beds they may remain two years ; when, in the fpring, they fhould be tranfplanted to the nurfery, where they may remain till they are finally planted out. During the time they are in the nurfery, and after planting out, many will frequently have a tendency to droop in their leading-fhoot : As foon, therefore, as this is perceived, an upright ftake muft be driven into the ground, to which the fhoots fhould often be tied with bafs matting to keep them in their upright growth. This, however, will not always effect it ; for fome, after being tied, fo effectually turn the fhoot downwards over the bandage, though loofe, as to appear as if they were beat down on purpofe. The Larch alfo will fometimes rebel in this way : So that it would not be amifs, in both cafes, whenever they firft difcover any figns of fuch a tendency, to lighten the head, by nipping off the extremities of fome few of the largeft branches.

When thefe trees are planted out to remain, they fhould be left to Nature, after being properly fenced : Not a knife nor a hatchet fhould come near them ; lopping even their loweft branches is fo injurious, that it both retards their growth and diminifhes their beauty.

The Cedar of Lebanon will grow well in almoft any fort of foil or fituation. As a proof of this, we need only obferve, that in its native fituation the roots are during part of the year covered with froft and fnow.

8. The YEW-LEAVED FIR. This fpecies includes the Silver Fir and the Balm of Gilead Fir.

The

The *Silver Fir* is a noble upright tree *. The branches are not very numerous, and the bark is smooth and delicate. The leaves grow singly on the branches, and their ends are slightly indented. Their upper surface is of a fine strong green colour, and their under has an ornament of two white lines, running lengthways on each side the midrib, on account of which silvery look this sort is called the Silver Fir. The cones are large, and grow erect; and when the warm weather comes on, they soon shed their seeds; which should be a caution to all who wish to raise this plant, to gather the cones before that happens.

The *Balm of Gilead Fir* has of all the sorts been most coveted, on account of the great fragrance of its leaves; tho' this is not its only good property; for it is a very beautiful tree, naturally of an upright growth, and the branches are so ornamented with their balmy leaves, as to exceed any of the other sorts in beauty. The leaves, which are very closely set on the branches, are broad; and their ends are indented. Their upper surface, when healthy, is of a fine dark-green colour, and their under has white lines on each side the mid-rib length-ways, nearly like those of the Silver Fir. These leaves, when bruised, are very finely scented; and the buds, which swell in the autumn for the next year's shoot, are very ornamental all winter, being turgid, and of a fine brown colour: and from these also exsudes a kind of fine turpentine, of the same kind of (tho' heightened) fragrancy. The tree being wounded in any part, emits plenty of this turpentine; and HANBURY says, "it is supposed by many to be the sort from whence the Balm of Gilead is taken, which occasions this tree being so called. But this is a mistake; for the true Balm of Gilead is taken from a kind of *Terebinthus*; tho I am informed, that what has been collected from this tree has been sent over to England from America (where it grows naturally), and often sold in the shops for the true sort."

These trees are PROPAGATED by sowing the seeds in a shady border, about the middle of March. They will readily come up if the seeds are good; but as this is not often the case, espe-

---

* Mr. MARSHAM says, " The tallest trees I have seen were Spruce and Silver Firs, in the vallies in Switzerland. I saw several Firs in the Dock-yards in Venice 40 yards long; and one of 39 yards was 18 inches diameter at the small end. I was told they came from Switzerland."

cially

cially if they are procured from the feedfmen, they fhould be fown very clofe, otherwife you will be certain of having a very thin crop. The fucceeding fummer the plants will require no trouble, except keeping them clean from weeds ; and the fpring after that they fhould be pricked out in beds at about four inches diftance from each other. Here they may ftand for two years, when they fhould be planted in the nurfery, in rows a foot afunder every way. The year, or at fartheft two years, after they have been fet in the nurfery, they fhould be finally planted out ; for if they are continued longer, many of them will die in the removal, and thofe which grow frequently lofe their leading-fhoot, and meet with fo great a check as to be hardly able to get into a good growing ftate for feveral years.

The Silver Fir is exceedingly hardy, and will grow in any foil or fituation, but always makes the greateft progrefs in a good rich loamy earth.

The latter muft be planted in a deep, rich, good earth ; neither will it live long in any other fort of foil. It matters little whether it be a black mould, or of a fandy nature, provided it be deep, and there is room for the roots to ftrike freely. As thefe trees have hitherto been planted without this precaution, and as fuch a kind of foil does not often fall in the ordinary courfe of Gardening, very few trees that have been planted many years are in a flourifhing ftate ; for if they do not like the foil, or if the roots begin to meet with obftructions, they foon begin to decline, which will be frequently in lefs than feven years ; the firft notice of which is, their leaves, which are naturally of a fine ftrong green colour, lofe their verdure, and appear with a yellow tinge ; and this colour grows upon them daily, until the appearance of the tree is changed. Another fign of this tree being at its *ne plus ultra* is, its producing vaft plenty of cones ; this argues a weaknefs, and they generally die away by degrees foon after. This is always the cafe where the foil does not wholly agree with them ; but where it is deep and good, they will be healthy and flourifhing, and produce cones for feeds.

9. The EUROPEAN SPRUCE FIR. This fpecies includes the Norway Spruce, and the Long-coned Cornifh Fir.

The *Norway Spruce* is a tree of as much beauty while growing,

ing, as its timber is valuable when propagated on that account. Its growth is naturally like the Silver, upright : and the height it will aspire to may be easily conceived, when we say that the white deal, so much coveted by the joiners, &c. is the wood of this tree ; and it may perhaps satisfy the curious reader to know, that from this Fir pitch is drawn. The leaves are of a dark green colour ; they stand singly on the branches, but the younger shoots are very closely garnished with them. They are very narrow, their ends are pointed, and they are possessed of such beauties as to excite admiration. The cones are eight or ten inches long, and hang downwards.

The manner of PROPAGATING this tree is nearly the same as that of the Scotch Fir, only this will more easily grow when of a large size, and consequently will not require removing so often in the nursery. In the middle of March, having got the seeds out of the cones, sow them in a north border ; for when they come up, by being constantly shaded all the summer in such a situation, they will shoot much stronger, and be better to prick out the spring following in the nursery. In about six or seven weeks after sowing, the young plants will appear, when they should be screened with the usual care from the birds, which otherwise would soon destroy them. By the autumn, many of these young plants, if they are kept clean from weeds and watered in dry weather, will have shot three or four inches ; and in spring they should be carefully taken out of their seed-beds, so that the fibres may by no means be broken off or injured. Being thus cautiously taken up, they should be as carefully planted in the nursery-ground, at the distance of one foot asunder each way. Here they may remain, with keeping them free from weeds, for three years, when they should be set out in the places where they are designed to remain. But if larger trees are desired for this purpose, they shoud be taken up and planted in the nursery, a foot and a half asunder, in rows two feet and a half distant, where they may stand, if required, till they are six or eight feet high, without any other removing.

When they are set out finally, they may be planted, with tolerable hopes of success ; for the Spruce-Fir is not so nice or difficult in shifting its quarters as any of the other sorts of Pines. But though these trees may be transplanted at a good height, it

is

is always advifeable to remove them to the places defigned for them with all poffible difpatch, as they are more certain of growing, and will recover the check occafioned in all trees by removal in lefs time.

The better the foil is, the fafter will the Spruce Fir grow, though it will thrive very well in moft of our Englifh lands. In ftrong loamy earth it makes a furprifing progrefs; and it delights in frefh land of all forts, which never has been worn out by ploughing, &c. though it be ever fo poor.

The *Long-coned Cornifh Fir* differs fcarcely in any refpect from the Norway Spruce, except that the leaves and the cones are larger.

10. The AMERICAN SPRUCE FIR. This fpecies includes three varieties: The *White Newfoundland Spruce*: The *Red Newfoundland Spruce*: and the *Black Newfoundland Spruce*. Thefe, however, differ fo little, that one defcription is common to them all. They are of a genteel upright growth, though they do not fhoot fo freely or grow fo faft with us as the Norway Spruce. The leaves are of the fame green, and garnifh the branches in the fame beautiful manner as thofe of that fpecies, only they are narrower, fhorter, and ftand clofer. The greateft difference is obfervable in the cones; for thefe are no more than about an inch in length, and the fcales are clofely placed. In the cones, indeed, confifts the difference of thefe three forts: Thofe of the White fpecies are of a very light-brown colour; thofe of the Red fpecies more of a nut-brown or reddifh colour; and thofe of the Black fpecies of a dark or blackifh colour. Befides this, there is fcarcely any material difference; though it is obfervable, that this trifling variation feems to be pretty conftant in the plants raifed from the like feeds. Thefe forts will often flower, and produce cones when only about five or fix feet high; and indeed look then very beautiful; but this is a fign of weaknefs in the plant, which it does not ofen fairly overget.

11. The HEMLOCK FIR poffeffes as little beauty as any of the Fir tribe; though being rather fcarce in proportion, it is deemed valuable. It is called by fome the Yew-leaved Fir, from the refemblance of the leaves to thofe of the Yew-tree. It is a tree of low growth, with but few branches; and thefe are

long

long and flender and fpread abroad without order. The leaves
do not garnifh the branches fo plentifully as thofe of any
other fort of Fir. The cones are very fmall and rounded;
they are about half an inch long; and the fcales are loofely
arranged. We receive thefe cones from America, by which we
raife the plants; though this caution fhould be given to the
planter, that this tree is fond of moift rich ground, and in
fuch a kind of foil will make the greateft progrefs.

12. The ORIENTAL FIR. This is a low but elegant tree
The leaves are very fhort, and nearly fquare. The fruit is exceeding fmall, and hangs downward; and the whole tree makes
an agreeable variety with the other kinds.

In PROPAGATING the AMERICAN SPRUCE and the HEMLOCK
FIR, the feeds being very fmall, a more than ordinary
care fhould be taken of them, left they be loft. They fhould be
fown in pots or boxes of fine light mould, and covered over
hardly a quarter of an inch. They fhould be then plunged up
to the rims in a fhady place, and netted, to fave them, when
they firft appear, from the birds. If the place in which
they ftand is fhaded, they will need little or no water all fummer, unlefs it proves a very dry one; and being all of a very
hardy nature, they will not require the trouble of covering in
the winter. The beginning of July after that, the Newfoundland Spruce Fir fhould be pricked out in beds at a fmall diftance, though the Hemlock Spruce fhould remain in the pots a
year longer, as they will then be very fmall. After they are
planted, they muft be well watered, and the beds muft be hooped, to be covered with mats for fhade. In hot weather the
mats fhould be put over the beds by nine o'clock in the morning, and conftantly taken off in the evenings, and remain fo in
cloudy and rainy weather. After they have taken root, they
require no further care, until they are planted out; which, fays
HANBURY, "cuftom has taught us to do in the autumn or in
the fpring; but I have by much experience found, that July is a
good month for planting out all the forts of Firs; and if it were
done in a wet time, and the weather fhould continue moift or
cloudy for two or three weeks, it would be by far the beft time
in the whole year. Whoever, then, plants out Firs in July,
unlefs

unlefs fuch weather happens, muft fhade and water them for a month or fix weeks; but as fhade is not to be afforded large trees of this kind, if there be many of them, their removal muft be at the ufual times, left that parching time which often comes in the middle of fummer burn them up before they can have time to take root. On this account, the planting of trees at Midfummer fhould be tenderly enforced; though I muft declare, that I have repeatedly planted Scotch Firs of different fizes, fome, one yard and more, others, fix feet high, in the fcorching heat, and left them to Nature, without giving them any affiftance, and they have for the moft part grown. Let others, if they pleafe, make the experiment with a few, before they venture to plant out quantities at that feafon." This information and the manner in which it is conveyed do Mr. HAN-BURY great credit as a practical man and a writer.

P I S T A C I A.

LINNEAN Clafs and Order, *Dioecia Pentandria*: Male flowers containing five ftamina, and female flowers containing three piftils, upon diftinct plants: There are five SPECIES; four of which will endure our winters, provided they be placed in a warm well-fheltered fituation:

1. PISTACIA *Terebinthus*: The COMMON TURPENTINE. TREE; *a fub-evergreen tree*; native of Italy, Spain, and fome parts of Africa.

2. PISTACIA *Vera*: The COMMON PISTACIA, or PISTACIA-NUT-TREE; *a fub-evergreen tree or fhrub*; native of Perfia, Arabia, Syria and India, from whence we receive the nuts.

3. PISTACIA *Trifolia*: The THREE-LEAVED PISTACIA, or TURPENTINE-TREE; *a low fub-evergreen tree*; native of Sicily.

4. PISTACIA *Narbonenfis*: The LARGE-FRUITED TURPENTINE-TREE; *a fub-evergreen tree*; native of Perfia, Armenia, Mefopotamia, and the fouth of France.

1. The

1. The COMMON TURPENTINE-TREE will grow to the height of about thirty feet. The bark of the trunk is thick, full of cracks, and of a dark-brown colour; whilst that on the young shoots is thin and smooth. The leaves are pinnated and large, of a dark-green colour, and grow alternately on the branches. The folioles of which each leaf is composed are oval, spear-shaped, and consist of three or four pairs, which are placed on the mid-rib, besides the odd one with which they are terminated. There will be male and female flowers on different plants. They exhibit their bloom in April: The male flower is nothing but a katkin, and the females make no figure; so that where philosophy has no view, it is from the desire of having an extensive collection that we procure these trees. In warm countries, the leaves of the *Piſtacia* continue all the year; with us, they fall off when attacked by the frosts. From the trunk flows the true turpentine; in the room of which that taken from some of our Pines is generally substituted.

2. The PISTACIA-NUT-TREE is about twenty feet in height. The trunk of this species also is covered with a dark-brown bark, full of cracks, whilst the young shoots are smooth, and of a light-brown colour. The leaves are likewise pinnated being composed of about two or three pairs of folioles, which do not always stand exactly opposite on the mid-rib, terminated with an odd one. These folioles are large, and nearly of an oval figure: Their edges turn backwards; but have nevertheless a noble look. The male flowers are katkins of a greenish colour; and the female flowers are very small, and produced in clusters from the sides of the branches. April is the month of their flowering; and the female flowers are succeeded by the Piſtacia nuts we eat.

3. The THREE-LEAVED PISTACIA is of about twenty-five feet growth. The bark of the trunk is very rough, and of a dark-brown colour; but that of the young shoots is smooth, and lighter. The leaves of this species are trifoliate. The folioles are of an oval figure, of a very dark green colour, and are greatly ornamental to the plant. Different trees will have male and female flowers: The males are greenish katkins; and the females have no petals, are small, and make no show.

4. LARGER-

4. LARGER-FRUITED TURPENTINE-TREE will grow to be about twenty-five feet high. The bark partakes more of a whitifh colour, and is fmoother, than thofe of the other fpecies. The leaves alfo are pinnated ; but the folioles of which each is compofed are not always of the fame number : Sometimes there are three, fometimes five pair of folioles to form the compound leaf. Thefe are of a paler green than any of the other forts, of a roundifh figure, and ftand on longifh footftalks. The male flower of this fpecies alfo is a katkin ; and the females are fucceeded by nuts, which by many are liked, being eatable, like the Piftachia-nuts. The leaves continue on thefe trees great part of the year, in warm countries.

The PROPAGATION. The feeds, which we receive from abroad, fhould be fown as foon as poffible after their arrival. A compoft fhould be prepared for them, mixed in the following proportions : Six barrows full of earth, from a frefh pafture, taken from thence at leaft a year before, with the green fward, and well turned and rotted ; three barrows of drift or fea fand ; and one barrow of old lime rubbifh, beaten to duft : Thefe fhould be all well mixed together. The feeds fhould be fown about half an inch deep in pots, which may then be fet under a warm wall or hedge, until the hot weather begins to come on, when they fhould be removed into the fhade, and plunged up to the rims in fome mould. At the approach of winter, they may be removed into a warm place, and in fpring a hot-bed muft be prepared for their reception. As thefe plants rarely come up the firft year, this will be a better method than to plunge them in a hot-bed foon after they are fown ; for even with this affiftance, they will be later before they come up, will be very weak and tender plants in the autumn, and will require extraordinary future care to preferve them ; whereas, if they are fuffered to remain unforced for one turn, they will be preparing to vegetate, and of courfe will come up themfelves the fecond fpring ; but an hot-bed will be neceffary, as at that time it will make them fhoot ftronger. But this forcing muft by no means be continued ; a hitch only is to be given them, and they fhould immediately be hardened to the air. Watering and fhade all fummer muft be allowed them ; and they ought to be made as hardy as poffible by the autumn. At the approach of winter, when other plants are to be fet in the green-houfe, thefe fhould

go

go with them, or be placed under an hot-bed frame. They should be set out with them in the spring, and in May the pots must be plunged up to the rim in the shade as before. The next winter they will require the green-house; and in the succeeding spring they will be two years old seedlings; at which time they should be shaken out of the pots, and each planted in a separate pot, in the same sort of compost in which the seeds were sown: This being done, they should be afforded a heat in the bed to set them forward. After they have begun shooting freely, the glasses should be taken off by degrees; and now they will want no more hot-beds. Watering must be given them in dry weather; and in the autumn they must be removed into the green-house, with other plants. And thus they should be treated as a green-house plant for four or five, or if even six years it will be so much the better; observing always, however, in the spring, to shift them into a fresh and larger pot every other year. The plants being now five or six years old, and being become tolerably strong and woody, may be set out in the places where they are to remain. These, as was observed, must be warm well-sheltered places, with a naturally dry soil; and if the two or three succeeding winters should prove mild and favourable, they will by that time be grown to be very hardy, and may bid defiance to almost any weather. The Common Turpentine-tree and the Pistachia Nut-tree, when grown old, resist our severest frosts; and the other sorts, though rather of a more tender nature, even if not old, will droop to none but the most piercing.

P L A T A N U S.

LINNEAN Class and Order, *Monoecia Polyandria:* Male flowers containing many stamina, and female flowers containing several pistils; upon the same plant; the males being collected in a globular katkin, and the females digested in a roundish ball: There are only two SPECIES:

1. PLATANUS *Orientalis:* The ASIATIC or ORIENTAL PLANE; *a tall deciduous tree;* native of Asia.

2. PLA-

2. PLATANUS *Occidentalis:* The AMERICAN or OCCIDEN-
TAL PLANE ; *a tall deciduous tree* ; native of North America.

1. The ORIENTAL PLANE rifes to a very great height, and
in its native foil grows to a prodigious fize : The ftem is covered with a fmooth bark, which falls off annually. The bark of
the young branches is of a dark-brown, inclining to a purple.
The leaves are large and palmated, being deeply cut into five
fegments : their upper fides are of a deep green, and the under
fides pale. The flowers are very minute : they come out at the
fame time as the leaves, which is in June. This is very late,
and is no doubt a blemifh to the beauty of this neverthelefs
highly *ornamental* tree. The antients, we are told, were very
partial to this tree ; which is not to be wondered at when we
confider the extenfive canopy it forms, and the impenetrable
fhade given by the number and fize of its leaves ; and confequently the grateful coolnefs it muft afford in a fultry climate.
EVELYN and HANBURY clafs this and the next fpecies amongft
Foreft or Timber-Trees ; and their wood may rank with that
of the Sycamore, which bears a confiderable refemblance
to this genus of plants, and which in the North of England is
called the *Plane Tree.*

2. The AMERICAN PLANE. This alfo grows to a great
fize ; the ftem not only fwells to an immenfe thicknefs ; but,
rifing erect, fhoots up perfectly ftraight and cylindrical to an
amazing height. The Hon. Paul Dudley, in a Letter to the
Royal Society, fays, " he obferved in New-England one of
thefe Plane-Trees nine yards in girth, which continued its bulk
very high ; containing when felled twenty-two loads of timber."
The bark is fmooth, and, like that of the Afiatic fpecies, falls off
annually. The leaves are broad, with long footftalks, and are
cut into angles at their edges, but not divided nearly fo deep as
thofe of the foregoing fpecies : The upper fide is of a light
green, the under fide paler : The flowers are fmall, and come
out with the leaves about the fame time as thofe of the Oriental
Plane. Altogether, this tree is peculiarly refrefhing to the eye,
and truly *ornamental.*

Befides thefe two diftinct fpecies, there are two *varieties:*

The Maple-leaved Plane.

The Spanifh Plane.

The

The *Maple-leaved Plane*, fays MILLAR, is certainly a feminal variety of the Eaftern Plane. It differs from the two forts
before-mentioned in having its leaves not fo deeply cut as thofe
of the Eaftern Plane, but much more deeply than thofe of
the Occidental Plane. The footftalks of the leaves are much
larger than thofe of either of the former, and the upper furface
of the leaves is rougher.

The *Spanifh Plane* has larger leaves than either of the other
forts. They are divided in a fimilar manner to thofe of the
Maple-leaved Plane. Some of them are cut into five and others
into three lobes : Thefe are fharply indented on their edges,
and are of a light-green. This is by fome called the *Middle
Plane*, from its leaves being fhaped between thofe of the two
firft forts. This too is probably a variety of the Oriental
Plane.

The method of PROPAGATING the ORIENTAL PLANE is
from feeds, when they can be eafily procured ; but whoever
enjoys not this convenience muft have recourfe to layers. 1. The
ground proper for the feminary fhould be moift and fhady,
well dug, and raked till the mould is fine; then in the autumn,
foon after the feeds are ripe, let them be fcattered over this
ground, and the feeds raked in, in the fame manner as turnep-
feeds. In the fpring, many of the young plants will come up,
though you muft not expect the general crop till the fecond
year ; the fpring after which they may be taken out of the feminary, and planted in the nurfery in rows one yard afunder, and
at one foot and a half diftance in the rows. Here they may remain, with the ufual care of digging between the rows and keeping them clean, till they are of fufficient fize to be planted out.
2. Where the feeds of thefe trees cannot be procured, layering
muft be the method of propagation. For this purpofe, a
fufficient number muft be planted out for ftools, on a fpot of
earth double dug. After they have ftood one year, they fhould
be cut down, in order to make them throw out young wood
for layering. The autumn following, thefe fhould be laid in
the ground, with a little nick at the joint ; and by the fame
time twelve months after, they will be trees of a yard high,
with a good root, ready to be planted out in the nurfery,
where they may be managed as the feedlings ; and as the ftools
will

will have shot up fresh young wood for a second operation, this treatment may be continued at pleasure.

The AMERICAN PLANE is PROPAGATED by cuttings; which, if they be taken from strong young wood, and planted early in the autumn, in a moist good mould, will hardly fail of succeeding. They are generally planted thick, and then removed into the nursery-ground, as the layers of the other sort: But if a large piece of ground was ready, the cuttings might be placed at such a distance as not to approach too close before they were of a sufficient size to be planted out to stand; and this would save the expence and trouble of a removal. The Oriental Plane-tree will grow from cuttings, but not so certainly as this; and whoever has not the convenience of proper ground for the cuttings, must have recourse to layers, which, indeed, is for either sort the most effectual and sure method.

Plane-trees delight in a moist situation, especially the Occidental sort. Where the land is inclined to be dry, and Plane-trees are desired, the others are to be preferred. But in moist places, by the sides of rivulets, ponds, &c. the Occidental makes such surprising progress, that it might be ranked among the Aquatics.

P O P U L U S.

LINNEAN Class and Order, *Dioecia Octandria* : Male flowers containing eight stamina, and female flowers containing one pistil; upon distinct plants; the males and females being similarly situated in long, loose katkins: There are five SPECIES :

1. POPULUS *Alba:* The ARBEEL, or WHITE POPLAR ; *a deciduous aquatic tree*; growing common in England and most parts of Europe.

2. POPULUS *Nigra:* The BLACK POPLAR ; *a deciduous aquatic tree*; this also grows common in England and most parts of Europe.

3. POPULUS *Tremula* : The ASPEN ; *a deciduous tree*; native of England and the colder parts of Europe.

X

4. POPULUS *Balfamifera* : The CAROLINA POPLAR ; *a deciduous tree* ; native of Carolina and many parts of North America.

5. POPULUS *Heterophylla* : The VIRGINIA POPLAR ; *a deciduous tree* ; native of Virginia.

1. The WHITE POPLAR. This is a tall fpreading tree, one of the largeft of the aquatic tribe. The trunk is covered with a fmooth whitifh bark. The leaves are about three inches long, and ftand upon footftalks about an inch in length : they are indented at the edges ; and are of a dark green on the upper furface, but white and woolly underneath. The Arbeel and Poplars in general, whilft young, are elegant and *ornamental* ; but the litter which is made by their katkins renders them difagreeable neighbours to kept walks and fhrubberies. The *ufes* of its wood are not many : it makes good boards, which have one peculiar good quality for flooring : they can with difficulty be made to burn, and will never blaze out like thofe of other wood : it would be needlefs to add, that the wood of the Poplar is the worft of fuel. It is a quick-growing tree, and may frequently be made ufe of as a fkreen to hide fwamps or other deformities.

2. The BLACK POPLAR. This tree will alfo grow to a large fize. Its leaves are not fo large as thofe of the former. Their colour is a pleafant green ; they are heart-fhaped, and appear about the middle of April.

The *Lombardy Poplar*, or the *Po Poplar*, feems to be a *variety* of the Black Poplar : the chief difference is, the Black Poplar throws out a large fpreading head, whilft the Lombardy afpires with a remarkably clofe one, rifing like an obelifk. There is a peculiar elegance in this plant when young, and in a moift rich fituation it flourifhes extraordinarily ; but it will not thrive in a dry barren foil, foon growing ftunted and moffy.

3. The ASPEN. This tree will grow to a great height, and takes a good outline. Its leaves are fmaller than thofe of the Black Poplar ; they ftand on long flender *flat* footftalks, which render it of all the other forts the moft tremulous ; they are roundifh, and fmooth on both fides ; but do not make their appearance before the beginning of May : This, together with their inelegant fhape and their want of brilliancy, render the Afpen the leaft *ornamental* of the Poplar tribe.

The

The PROPAGATION of thete three fpeçies of Poplar is very eafy : they will grow from cuttings, fetts, truncheons, &c. ; but, fays HANBURY, " I by no means approve of the planting of truncheons, as has been often practifed on boggy places; becaufe I have always obferved, that plantations of thefe luxuriant trees, attempted to be raifed in this manner, have been frequently ftunted, and very unpromifing ; and that the moft promifing trees have never equalled, in goodnefs or beauty, thofe planted with regular trees raifed in the nurfery. In order, therefore, to obtain a quantity of Poplars, proper to be planted in avenues or clumps, by the fides of rivulets, bogs, or any other places where they are defired, you muft get a piece of ground double dug for the nurfery. If the trees wanted are to be planted for good in a watery fituation, this nurfery-ground fhould be pretty near it; but if they are defigned for pafture-grounds, fields, or fuch as have no more than a common degree of moifture, the foil of the nurfery fhould be proportionably drier. The latter end of October is the beft feafon for planting the cuttings; though they will grow if planted in any of the winter months. They fhould be all of thofe laft year's fhoots which have been vigorous, or at leaft not older than two years wood. Thefe cuttings fhould be one foot and a half in length ; and muft be planted in the nurfery-ground in rows, a yard afunder, and at a foot and a half diftance from one another. They fhould be planted a foot in the ground, while the other half muft remain to fend forth the leading-fhoot. Now in order to have one leading-fhoot only, in fummer thefe plants fhould be carefully looked over, and all young fide-branches nipped off, in order to encourage the leading-branch. After this, no farther care need be taken of them than keeping them clean of weeds, and digging between the rows in the winter, till they have attained a proper fize to be planted out.

4. The CAROLINA POPLAR will grow to be a large timber-tree, and has a majefty both enchanting and peculiar. It is an exceeding fwift grower, infomuch that it has been known to fhoot ten feet in the fpace of one fummer, and to be in thicknefs, neareft the bafe, an inch in diameter. The bark is fmooth, and of a whitifh colour ; though that on the young fhoots is of a fine green. The young fhoots are cornered, having five angles ; and the bark of which thefe are compofed, being extended by the future growth, leaves only the traces on the older

X 2
branches

branches of thefe angles. "This (continues HANBURY) gives the tree in winter a particular look; for at the bafe of each bud they curve over and meet. Thus there will be between every bud formed by the bark, figures like niches, as it were, of public buildings, tho' with an upright in the middle, at the top of each of which, like an ornament, is feated the bud, for the future fhoot or leaf. Thefe buds are only to be found on the younger branches; but the figure is retained on the bark of the older without thofe ornaments. But of all the trees in a collection, none more agreeably by its leaves entertains us than this, whether we confider their colour, figure, or fize. The colour is a light fhining green, which is heightened in the autumn by the ftrong mid-rib, and the large veins that iffue from it, turning to a red colour; the leffer veins alfo being in fome degree affected, occafions upon the fame leaf a fweet contraft. Their figure nearly refembles that of an heart, and they are notched at their edges. But the chief majefty this tree receives is from the fize of the leaves: I have meafured fome of the younger trees, and found the leaves ten inches long and eight broad, with a ftrong footftalk of four inches in length. Thefe majeftic leaves are placed alternately on the branches; though, as the tree advances in height, they diminifh in fize. This fpecies fhoots late in the autumn; and thefe young fhoots have their ends often killed in hard winters; which is an imperfection, as it caufes the tree to have a very bad look in the fpring, before and when the leaves are putting out: However, thefe laft will not fail afterwards to make ample amends for the former defect. The flowers afford no pleafure to the gardener: They are only katkins, like other Poplars, and fit only for the curious Botanift's infpection."

5. The VIRGINIA POPLAR grows to be a large timber-tree. The branches are numerous, veined, and angular. The leaves are heart-fhaped, broad, flightly ferrated, and downy on their firft appearance. The flowers come out in loofe katkins, and make little fhow: They appear early in the fpring; and are fucceeded by numerous downy feeds, which are difperfed all about to a confiderable diftance.

Thefe two fpecies are PROPAGATED, 1. By cuttings. In order o obtain proper cuttings for the purpofe, the plants fhould be eaded the year before, and a foot and a half of the thickeft part

of

of the former fummer's fhoots fhould be taken. The month of October is the feafon; and thefe cuttings fhould be planted in a moift fhady foil, one foot deep, with the other half foot above ground. Many of them will grow; though it is generally allowed to be a good crop if half fucceed. 2. By layers. Thefe muft be of the laft fummer's fhoots; and the operation ought to be performed in the autumn, before they have done growing; for the fap being then in motion, they may readily be brought down; whereas, if it is deferred until winter, the young fhoots are then fo exceedingly brittle, that though all poffible care be taken, many of them, in attempting to bring them down, will be broken. A fmall flit with the knife muft be given to each; and after the operation is performed, fome furze-bufhes fhould be ftuck round each ftool, to break the keen edge of the black frofts, and preferve the ends of the layers from being killed. In the fpring they fhould be cut down to within one eye of the ground; and by the autumn they will have ftruck root, and be good plants, either for the nurfery-ground, or where they are intended to be fet out to ftand.

P R I N O S.

Linnean Clafs and Order, *Hexandria Monogynia* : Each flower contains fix males and one female: There are only two Species :

1. Prinos *Verticillatus:* The Virginia Winterberry, or Deciduous Prinos; *a deciduous fhrub*; native of Virginia, Pennfylvania, and other parts of North-America.

2. Prinos *Glaber:* The Canada Winterberry, or Ever-green Prinos; *an evergreen fhrub*; native of Canada.

1. The Deciduous Winterberry is a fhrub of about fix or eight feet in growth, fending forth many branches from the bottom to the top, which are covered with a brownifh bark. The leaves are fpear-fhaped, pretty large, of a ftrong green colour, lengthways ferrated, and placed alternately on flender footftalks on the branches. The flowers are produced at the fides

of the branches, growing one or two together at the joints ; but make no fhow. They appear in July ; and are fucceeded by purple-coloured berries, which remain on the trees all winter, and look well.

The beft way of PROPAGATING this plant is from its feeds. Thefe fhould be fown, foon after they are ripe, in beds of fine fandy earth ; and if the garden does not naturally afford fuch, a few barrows full of drift fand muft be brought to mix with the common mould. The beds being thus prepared, and made ready for fowing, the feeds fhould be fown about three quarters of an inch deep. It is very feldom that any of the feeds come up the firft fpring after ; if any do, there will be but few ; fo that all the fummer they muft be kept clean from weeds. The fpring following the plants will come up ; though many will lie until the third fpring before they make their ap- pearance. After they are come up, weeding and watering muft be afforded them in the fummer ; and with this care they may remain in the feed bed two years. In March, being then two years-old feedlings, they fhould be taken up, and planted in the nurfery, at very fmall diftances ; and here they may remain, with the ufual nurfery-care, until they are fet out.

2. The EVERGREEN WINTERBERRY grows to about eight or ten feet high, fends forth many branches from the bottom to the top, and the whole plant affumes the appearance of an *Ala- ternus*. The leaves are oblong, fpear-fhaped, acute, ferrated, of a ftrong green colour, and placed alternately on the branches. The flowers come out from the wings of the leaves, two or three together on a footftalk. They are fmall, white, appear in July, and are fucceeded by red or purple berries, which remain on the trees all winter.

The PROPAGATION of this fhrub is exactly the fame as that of the deciduous. fpecies, except that this fpecies is of a more tender nature ; and inftead of fetting out the feedlings in the nurfery-ground, each fhould be fet in a feparate pot, to be placed under fhelter in winter for a few years, until they are grown ftrong plants, and after that to be turned out, with the mould at the roots, into the places where they are defigned to remain, which ought always to be in a dry fandy foil, and a well-fheltered fituation.

P R U,

P R U N U S.

Linnean Clafs and Order, *Icofandria Monogynia :* Each
flower contains about twenty males and one female : There are
fifteen Species ; twelve of which are here treated of ; moft of
them well-known fpecies ; including a numerous tribe of orchard,
garden, and fhrubbery plants.

1. Prunus *Padus:* The Padus, or Common Bird-Cherry ;
a deciduous fhrub or tree ; native of England and moft parts of
Europe.

2. Prunus *Virginiana :* The Virginia Padus ; *a low deciduous tree* ; native of Virginia, Pennfylvania, and Carolina.

3. Prunus *Canadenfis :* The Canadian Padus ; *a deciduous fhrub or tree* ; native of Canada and many other parts of
America.

4. Prunus *Mahaleb :* The Mahaleb, or Perfumed
Cherry ; *a tall deciduous fhrub* ; native of Switzerland and the
north of Europe.

5. Prunus *Armeniaca :* The Apricot ; *a low deciduous
tree* ; whofe native country is unafcertained.

6. Prunus *Cerafus :* The Cherry, or the Cultivated
Cherry ; *a deciduous tree* ; native of England and moft parts
of Europe.

7. Prunus *Avium :* The Wild Cherry ; *a tall deciduous
tree* ; native of England and the north of Europe.

8. Prunus *Domeflica :* The Plum : *a deciduous tree* ; native
of many parts of Europe.

9. Prunus *Infititia :* The Bullace ; *a deciduous tree or
fhrub* ; native of England and Germany.

10. Prunus *Spinofa :* The Sloe-Thorn, or Black Thorn ;
a deciduous fhrub or tree ; native of England and moft countries
of Europe.

11. Prunus *Lauro-Cerafus:* The Laurel, or the Common
Laurel ; *an evergreen fhrub or tree* ; native of Trebifond,
near the Black Sea.

12. Prunus

12. PRUNUS *Lufitanica* : The PORTUGAL LAUREL ; *an ever-green fhrub* ; native of Portugal ; alfo of Pennfylvania, and other parts of America.

1. The COMMON PADUS, or BIRD-CHERRY, is a tree of about twenty feet growth ; oftentimes it rifes higher. It grows with an upright ftem, and makes a handfome appearance. The bark of the older fhoots is of a dark-brown, inclined to a purple colour, and is befprinkled with a few greyifh fpots ; while the preceding fummer's fhoots are fmoother, and of a reddifh caft. The buds early in the winter will begin to fwell, for the future fhoots. The leaves are large, and grow alternately on the branches. Their figure is nearly oblong. They are rough, and have their edges ferrated. Their under furface is of a lighter colour than their upper, and they have two glandules at their bafe. The flowers are white, and produced in May, in long bunches. A kind of fpike of white flowers grows from the fides of the branches ; and thefe waving about on every fide, in a loofe and eafy manner, have a genteel and pleafing effeft. The flowers of which thefe fpikes are compofed, ftand each on their own proper pedicles, and are all arranged alternately along the main ftalk, which is tolerably long. Thefe flowers are fucceeded by fruit, which is a fmall berry, that ripens in Auguft, at which period it will be black ; but befides this, it will undergo the changes of being firft green and afterwards red. When thefe berries are ripe, they are of a fweet, difagreeable tafte, but fo liked by the birds (which will flock from all parts to feed on them) as to occafion its being called the Bird-Cherry ; and for their fake purely many perfons plant a more than common quantity of thefe trees, that they may have thefe feathered fongfters in greater plenty.

There is a *variety* of this tree, called the *Cornifh Bird-Cherry*, which differs from it in fome refpefts ; but thefe differences are inconfiderable.

2. The VIRGINIA PADUS will grow to thirty or forty feet high, and is faid to afford wood of great value. The bark is of a dark-brown, inclined to a purple colour, and fpotted irregularly with fome greyifh blotches. The young fhoots are of a lighter colour, and very fmooth ; and the whole tree is more ramofe than the former fort. The leaves are oval, and of a fhining green

green colour. Their edges are ferrated, and placed alternately on the branches. They ftand on fhort footftalks, and continue on the trees late in the autumn. Their flowers are white, and produced in May, in the fame fort of long bunches as the other; and are fucceeded by black berries, which are equally coveted by the birds, for whofe fake only this fpecies alfo is frequently planted. HANBURY claffes this tree amongft his Foreft-Trees, and fays, " The wood is very valuable ; is much ufed by the cabinet-makers; will polifh very fmooth, and difplay beautiful veins, both black and white." MILLER tells us, " that the *Padus Virginiana* will grow to be a large tree when it is planted in a moift foil, but in dry ground it rarely rifes more than twenty feet high." He has alfo a *Padus Caroliniana* (probably a *variety* of this fpecies), the feeds of which he fays " were fent from Carolina by the title of BASTARD MAHOGANY, from the colour of the wood, which is fomewhat like Mahogany." He adds, however, that " this feems to be little more than a fhrub, if we may judge from the growth here."

3. The CANADA PADUS is of much lower growth than the former forts. The branches are fmooth. The leaves are broad, fpear-fhaped, rough, downy, and deftitute of glands, like thofe of the former fpecies. The flowers grow in long, branching bunches: Their colour is white ; they come out in May, and are fucceeded by fmall, round, black berries, which will be ripe in the autumn.

4. The PERFUMED CHERRY feldom grows to be more than ten or twelve feet high. The branches are covered with a fmooth, whitifh-grey bark. The leaves are fmall, of a lucid green colour, of an oval figure, and ftand alternately on the branches. The flowers are white, produced in May in roundifh clufters, and are fucceeded by berries, of which the birds alfo are very fond. The wood of all thefe forts is much efteemed by the cabinet-makers, particularly amongft the French, as it always emits a very agreeable odour.

5. The APRICOT-TREE is often planted as a flowering fhrub ; for though it will grow to be thirty feet high, it may neverthelefs be kept down to what height the owner defires.

This

" This tree, fays HANBURY, as well as moſt ſorts of fruit-
trees, is exceeded by few in ornament ; for being permitted to
grow in its natural ſtate to twenty or thirty feet high, with all
its luxuriancy of branches, covered with their delightful heart-
ſhaped leaves, what a glorious figure will it preſent! But when
we reflect on the fine appearance ſuch a tree muſt make, early
in the ſpring, when covered all over with the bloom of ſuch
fine flowers as thoſe of the Apricot are known to be, this enhances the value ; and either of theſe motives is ſufficient for
introducing theſe trees into plantations of this kind. Add to
this, ſome of the ſorts, in warm well-ſheltered ſituations, will
produce fruit when growing in this manner, as well as if
planted and trained againſt walls ; ſo that additional returns
will be made by the fruit to the curious planter of theſe
trees."

6. The CHERRY-TREE of our orchards is too well known,
with all its varieties, to need any deſcription. HANBURY obſerves, " were the tree ſcarce, and with much difficulty propagated, every man, tho' poſſeſſed of a ſingle tree only, would look
upon it as a treaſure. For beſides the charming appearance theſe
trees have, when beſnowed, as it were, all over with bloom in
the ſpring, can any tree in the vegetable tribe be conceived
more beautiful, ſtriking, and grand, than a well-grown and
healthy Cherry-tree, at that period when the fruit is ripe ?"

The many kinds of Cherry-trees afford an almoſt endleſs *variety* ; all differing, in ſome reſpect, in their manner of ſhooting, leaves, flowers, or fruit : two in particular demand admiſſion into the pleaſure-garden ; the Double-bloſſomed and the
Red-flowering.

The *Double-bloſſomed Cherry.* The pleaſing ſhow the common Cherry-tree makes when in blow is known to all ; but that
of the Double-bloſſomed is much more enchanting. It bloſſoms,
like the other, in May ; the flowers are produced in large and
noble cluſters ; for each ſeparate flower is as double as a roſe,
is very large, and placed on long and ſlender footſtalks, ſo as
to occaſion the branches to have an air of eaſe and freedom.
They are of a pure white ; and the trees will be ſo profuſely
covered with them, as to charm the imagination. Standards of
theſe trees, when viewed at a diſtance, have been compared to

balls

balls of fnow; and the nearer we approach, the greater plea-fure we receive. Thefe trees may be kept as dwarfs, or trained up to ftandards, fo that there is no garden or plantation to which they will not be fuitable. By the multiplicity of the petals the organs of generation are deftroyed; fo that thofe flowers which are really full are never fucceeded by any fruit.

The *Red-flowering Cherry-tree* differs in no refpeɛt from the Common Cherry-tree, only that the flowers are of a pale-red colour, and by many are efteemed on that account.

Befides the *ornament* and utility afforded us by the flowers and fruit of the Cherry, its *timber* is a further inducement for propagating it; more efpecially that of the fmall *Black Wilding* fort; which may perhaps with propriety be confidered as the genuine fpecies, and a native of this ifland. Be this as it may, it will grow, in a foil and fituation it affeɛts, to be a large timber tree, which, if taken in its prime before it become tainted at the heart, will turn out perhaps not lefs than a ton of valua-ble materials, peculiarly adapted to the purpofes of furniture. The grain is fine, and the colour nearly approaching to that of mahogany, to which valuable wood it comes nearer than any other which this country produces.

7. The WILD RED CHERRY is a very large-growing tree, and may like the Black Wilding be an objeɛt for timber. The leaves are oval, fpear-fhaped, and downy underneath. The flowers come out from the fides of the branches in feffile um-bels. They appear rather later than the cultivated forts; and are fucceeded by fmall red fruit, which ripens late in the autumn. This is often called the Wild Northern Englifh Cherry.

8. The PLUM-TREE, with all its varieties, is fo well known as to require no defcription. No one need be told, that the Plum-Tree is a large-growing tree, and that it has a beautiful appearance in fpring when in blow. The fruit that fucceeds the bloffom is of many colours, fhapes, and fizes; and the trees of the variety of forts will be fo adorned with them in the autumn, as have a noble and delightful effeɛt, being hardly exceeded by the Cherry itfelf. Thefe are feldom planted any

where

where except in orchards ; but let them be set where they will, they never fail to repay the owner with pleasure and profit.

The *varieties* which are principally eligible for ornamental plantations are, The Cherry Plum-Tree, the Double-blossomed, the Stoneless, the Gold-striped, and the Silver-striped Plum.

The *Cherry Plum-Tree* is always planted among flowering-shrubs, on account of its early flowering. It may be kept down to any height ; and the flowers will be produced in March, in such plenty, and so close, as almost to cover the branches. It is admired by all for the early appearance of its flowers, which are succeeded, after a mild spring, by a round reddish plum, on a long slender footstalk, that has the resemblance of a Cherry. Unless there is little or no frost after these trees have been in blow, it rarely happens that any fruit succeeds the flowers.

The *Double-blossomed Plum-Tree* is another variety. The flowers of this sort are exceedingly double, and the twigs will be richly furnished with them in the month of May. Their petals, like those of the Cherry, are of a pure white, though amongst these some filaments with darkish antheræ appear. As soon as the show of flowers is over, we are not to give up all expectations from this tree ; for many of them will be succeeded by fruit, which is of the same colour, shape, and taste, with the common Damascene, though smaller, and is liked by many.

The *Stoneless Plum*. This is a variety that should be admitted on no other account than because the pulp surrounds a kernel, without having any stone. It is a small blue plum ; and those people who have it in possession, take a pleasure in shewing it as a curiosity.

The *two Striped sorts* make a variety by their variegated leaves ; on which account they are frequently sought after by the curious.

9. The BULLACE-TREE is sometimes planted in shrubbery quarters, for the sake of the fruit ; which by many persons is deemed very agreeable, being possessed of a fine acid. It ought to be pulled and eaten immediately from the tree.

The *varieties* of this species are, The *Black*, the *White*, and the *Red Bullace*.

10. The SLOE-TREE. The Sloe-bush is, without all doubt, a species distinct from either Plum or Bullace. And indeed it is such a species, that, were it not for its commonness, it would

be

be thought inferior in beauty to none of our ſhrubs. The commonneſs of this tree, however, cauſes its beauties to be unnoticed, and forbids us to admit too many into our collection.

11. The COMMON LAUREL will grow to about thirty feet,
and the leaves, which are ſometimes five or ſix inches long
and three broad, being likewiſe of a firm ſtructure even at the
edges, garniſh the branches of the tree in ſuch a manner as
would excite our admiration, did not the frequency of this noble
plant diminiſh our reſpect. The Laurel, however, will ever,
we apprehend, preſerve its rank as a *Stock Plant* in ſhrubberies
and other *ornamental* grounds. EVELYN places the Laurel
among Foreſt Trees; and HANBURY ſpeaks of the *uſes* of its
timber to the turners and cabinet-makers. It ſeems peculiarly
adapted to the purpoſe of ORNAMENTAL UNDERWOOD; as it
is of quick growth, and will flouriſh under the drip and ſhade
of other trees.

The Common Laurel affords two *varieties:* the *Gold-ſtriped*
and *Silver-ſtriped* Laurel.

12. The PORTUGAL LAUREL is a lower-growing tree than
the former; and though its leaves, flowers, &c. are proportionally ſmaller, it is thought by many to be much the moſt
beautiful; the commonneſs of the one, and ſcarcity of the
other, may perhaps not a little contribute to this opinion. The
Portugal Laurel will grow to be ſix, eight, or ten feet high, accordingly as the ſoil in which it is placed contributes to its encreaſe. The branches are produced in an agreeable manner,
being chiefly inclined to an upright poſture; and the young
ſhoots are clothed with a ſmooth reddiſh bark. The leaves are
ſmooth, and of a fine ſtrong green colour, though their under
ſurface is rather paler than the upper. They are much ſmaller
than thoſe of the Common Laurel, are of an oval figure, and
have their edges ſerrated; they are of a thick conſiſtence, and
juſtly entitle the tree to the appellation of a fine Evergreen.
The flowers are produced in the ſame manner as thoſe of the
Common Laurel, but are ſmaller. They are white, appear in
June, and are ſucceeded by berries, which when ripe are black;
though before that they will undergo the different changes of
being firſt green, and then red.

The

The PROPAGATION of the feveral tribes of *Prunus* varies with the refpective fpecies.

The fpecies and varieties of PADUS, or BIRD-CHERRY, may he raifed, 1. From feeds, fown in autumn, in beds of light earth, about half an inch deep. The plants will appear the firft fpring, and the fpring following they may be planted out into the nurfery at the diftance of two feet by one ; in which fituation they may remain till wanted for planting out. 2. This clafs may alfo be propagated by Layers ; the young twigs, being fimply laid in the ground, will without any other trouble ftrike root in one year, and may be taken up and tranfplanted into the nurfery, or be planted where they are to remain, as circumftances may fuit. 3. Thefe trees will alfo grow from cuttings, planted in October, in a moift fituation : if the fpring and fummer prove dry, they will require to be watered.

This clafs of *Prunus* affects a moift fituation.

The PERFUMED CHERRY is PROPAGATED by grafting, or by budding upon any of our Cherry-ftocks.

The APRICOT-TREE is PROPAGATED by budding it upon the Plum-ftock.

The FRUIT-BEARING, DOUBLE-BLOSSOMED, and RED-FLOWERING CHERRY-TREES are PROPAGATED by grafting upon ftocks raifed from the ftones of the Black-Cherry-tree ; though it may be proper to obferve here, that when the Double-bloffomed Cherry is wanted to be kept very low, in its dwarf ftate, the Common Bird-Cherry will be a much more proper ftock to work it upon, as that fort is naturally of much lower growth than the Black Cherry-tree.

The PLUM-TREE, in all its varieties, and the Bullace-tree, the Cherry Plum, the Double-bloffomed Plum, and the Stonelefs Plum, are PROPAGATED by grafting upon Plum-ftocks raifed from feeds ; though it is obfervable, that fuckers of the Bullaces will grow to be trees, and produce plenty of good fruit ; but thefe will not be fo good as thofe grafted on the Plum-ftocks.

The SLOE-BUSH may be obtained from the places where they grow ; for from thence a fucker or two may be taken, and
planted

planted for the conveniency of obſervation; but theſe will not be ſo good as thoſe raiſed from the ſtones.

The COMMON LAUREL is PROPAGATED either from ſeeds or from cuttings. 1. If the former method is practiſed, the ſeeds muſt be gathered from the trees when they are full ripe: This will be known by their being quite black, which is generally about the beginning of October. Theſe ſeeds ſhould be ſown directly in beds of light earth, half an inch deep, which muſt be afterwards hooped over, to be covered in very ſevere froſts. A hedge of furze-buſhes alſo ſhould be made around them, to break the force of the freezing black winds, and ſecure the ſeeds, together with the mats, from being deſtroyed. This is a much ſafer method than covering the beds with litter, which, if neglected to be taken off when the froſt is over, will retain the rains which generally ſucceed ſuch weather, ſodden the beds, and make them ſo wet as frequently to deſtroy the whole of the expected crop. The ſeeds being ſown, and preſerved with the above care, will appear in the ſpring. During the ſummer they ſhould be kept clear of weeds, as well as watered in dry weather; and all the enſuing winter they muſt remain untouched in their beds, the furze-hedge ſtill ſtanding till the froſty weather is paſt; for if theſe young ſeedlings are planted out in the autumn, the major part of them will be in danger, before the winter be expired, of being thrown out of the ground by the froſt; and not only ſo, but of being really killed by it, as they are not very hardy at one year old. In the ſpring, therefore, when the bad weather is ceaſed, let them be planted out in the nurſery-ground, in rows two feet aſunder, and the plants a foot and a half diſtant in the rows; where they may ſtand till they be finally planted out. 2. Trees raiſed from ſeeds generally grow more upright, and ſeldom throw out ſo many lateral branches as thoſe reared from cuttings; nevertheleſs, as the expectation of a crop from ſeeds has ſo often failed, notwithſtanding great care has been uſed; and as the difficulty of procuring the ſeeds, and preſerving them from the birds, has been very great; the moſt certain and expeditious method of raiſing quantities of theſe trees is by cuttings, and is as follows:

lows : In the month of Auguſt the cuttings ſhould be gathered, about a foot and a half in length.   They will thrive the better for having a bit of the laſt year's wood at the end, though without this they will grow exceeding well.   The under leaves ſhould be cut off a foot from the thick end of the cuttings, which muſt all be planted about a foot deep in the ground ; the other half foot, with its leaves, being above it.   No diſtance need be obſerved in planting theſe cuttings, which may be ſet as thick as you pleaſe, though the ground for raiſing them ſhould be ſheltered, leſt the winds, which are frequently high at this time of the year, or ſoon after, looſen the plants juſt when they are going to ſtrike root, if not wholly blow them out. The weather when the cuttings are to be planted ſhould be either rainy or cloudy ; and if no ſhowers ſhould fall in Auguſt, the work muſt be deferred till they do ; for if cuttings are planted in Auguſt, when the weather is parching and dry, they will be burnt up, without great care and trouble in ſhading and watering.   Neither is cloudy or rainy weather only to be recommended in planting theſe cuttings, but a ſhady ſituation alſo, either under a north wall, or in beds which are covered the greateſt part of the day with the umbrage of large trees.   This ſhady ſituation is very neceſſary for them ; ſince, though the weather be rainy and cloudy when they are planted, yet ſhould it prove fair afterwards, the ſun will ſoon dry up the moiſture at that ſeaſon, and endanger the plants, if they are not conſtantly watered and protected with a ſhade ; which at once ſhews the expediency of pitching on a ſpot where ſuch a conveniency is natural.   If theſe cuttings are planted in Auguſt, they will have taken root before winter, eſpecially if they have ſhade, and water in dry weather : but they ſhould remain undiſturbed till the ſpring twelvemonth following, in order to acquire ſtrength to be planted in the nurſery.   During the ſummer, they will require no other trouble than watering in dry weather, and being kept clean from weeds ; and by the autumn they will have made a ſhoot of perhaps a foot or more in length.   In the beds, neverthelefs, they may remain till the ſpring, when they ſhould be all carefully taken out, and planted in the nurſery, as was directed for the ſeedlings.

The

The Portugal Laurel is to be raifed the fame way as the Common Laurel, by *Seeds* and *Cuttings*; but the cuttings of the Portugal Laurel do not take fo freely as thofe of the Common fort; and the young practitioner, out of a good bed of cuttings, muft expect to fee but a few real plants fucceed. If they are planted in July or Auguft, they muft be fhaded, and kept moift during the hot weather; and that will be the moft probable way to enfure fuccefs. If a perfon has the conveniency of a good ftove, the beft method is not to plant them until the fpring; and then many cuttings may be planted in one pot, and afterwards plunged into the bark-bed; and by this means numerous plants may eafily be obtained.

P T E L E A.

Linnean Clafs and Order, *Tetrandria Monogynia:* Each flower contains four males and one female: There are only two Species: one of them introduced into our fhrubberies; the other a late difcovery.

Ptelea *Trifoliata:* The Three-leaved Ptelea, or the Trefoil Shrub; *a deciduous fhrub*; native of Virginia and Carolina.

The Ptelea will grow to the height of ten feet. The branches are not very numerous;—when broken they emit a ftrong fcent: They are brittle, full of pith, and covered with a fmooth purplifh bark. The leaves are trifoliate, and grow irregularly on the branches, on a long footftalk. The folioles are oval, fpear-fhaped, of a delightful ftrong green colour on their upper fide, lighter underneath, fmooth, and pretty large when they are fully out, which will not be before part of the fummer is elapfed; for they put out late in the fpring. The flowers are produced in bunches at the end of the branches: Their colour is a greenifh-white. They come out in June; and are fucceeded by roundifh bordered capfules; but the feeds feldom ripen in England.

This fhrub may be propagated either by feeds, layers, or cuttings. 1. By feeds. Thefe fhould be fown in a warm border, in the spring, in common garden mould made fine; and if the seeds are good, they will grow, and come up the first summer. We generally receive the seeds from abroad; though they will in some warm seasons ripen here with us. When the young plants begin to come up, which will be, if the seeds are good, by the end of May, they should be shaded, and every second evening duly watered; and this, together with constant weeding, will be all the care they will require until the autumn. At the approach of winter, it will be proper to prick some furze-bushes round the bed, to break the keen edge of the black frosts. They will then require no other trouble until the second spring after they are come up; when they should be all taken out of the seed-bed, and planted in the nursery, a foot asunder; and in two or three years they will be fit to be finally planted out. 2 .By layers. For this purpose a number of plants must be planted for stools; and, after they have stood a year or two, these should be cut down pretty near the ground. By the autumn they will have made shoots, some of which will be five or six feet, or more, in length; and these are the shoots for layering. October is the best month for the work; and the operation is to be performed by cutting the twig half through, and making a slit half an inch long. Any thing may be put into this slit, to keep it open; and after the mould is levelled all round, the longest ends should be taken off. By this method they will generally have good roots by the autumn following: and the stools will have shot out fresh wood for a second layering. At this time they should be taken up, and the weakest planted in the nursery, to get strength; whilst the stronger layers will be good plants to set out to stand. After this, the operation may be again repeated, and so continued annually, at pleasure. 3. By cuttings. In order to obtain plenty of good cuttings, the plants should be headed as for layering. In October the young shoots should be taken off, and cut into lengths of a little more than a foot, two-thirds of which should be set in the ground. Some of these cuttings will grow; though, says HANBURY, I ever found this way very uncertain, and not worth the practising: But if the cuttings are planted in pots, and assisted by artificial heat, they will grow readily. This, however, is not a good method; for they will be tender the first winter, as well as require to be protected in the greenhouse, or

under

under some cover, which will occasion more trouble than if they
had been layered. By layers and seeds, therefore, are the best
and most eligible methods of encreasing these trees.

P Y R U S.

LINNEAN Class and Order, *Icosandria Pentagynia*: Each
flower contains about twenty males and five females: There
are eleven SPECIES: four of which are as follow:

1. PYRUS *Communis*: The PEAR; a well-known *deciduous
tree*; native of most parts of Europe.

2. PYRUS *Malus*: The APPLE; an equally well-known *deciduous tree*; native also of most parts of Europe.

3. PYRUS *Coronaria*: The SWEET-SCENTED CRAB; *a deciduous tree*; native of Virginia.

4. PYRUS *Cydonia*: The QUINCE; *a deciduous shrub or tree*;
native of. the Banks of the Danube.

1. The PEAR. Of the numerous *Varieties* of this species of
Pyrus there are two admissible into ornamental grounds:
       The Double-blossomed Pear,
       The Twice-flowering Pear.
The *Double-blossomed Pear* differs from the other sorts only
in that the flowers are double. The leaves, indeed, are not so
much serrated as some of the other Pears; nay, scarcely any
serratures appear, excepting on the oldest leaves; for the
younger are perfectly entire and downy. The multiplicity of
the petals of this flower is' not sufficient to entitle it to the
appellation of a full flower; for it consists only of a double row
of petals; but as these are' all large produced in clusters, and
of a pure white, they entitle the tree to be called a flowering-
tree, with greater propriety than the ordinary Pears can be so
styled. The planter of this species is rewarded in a double
respect; for as the petals are not multiplied in so great a degree as to destroy the stamina, the flowers are succeeded by a
good

good fruit, whofe properties are fuch as entitle it to the rank
of a good baking Pear.

The *Twice-flowering Pear.* This fpecies is fufficiently defcribed by the title; it being a Pear that often produces flowers
in the autumn, when the fruit that fucceeded thofe of the fpring
are near ripe. This tree deferves to be planted both for its
beauty and fingularity; for it fometimes happens, though by
no means conftantly, that it is covered over in September
with bloom and fruit. This autumnal bloom falls away, and
the chilling cold often prevents its coming to any embryo
fruit.

2. The APPLE. This fpecies likewife affords us two ornamental *varieties:*

　　　The Paradife Apple,
　　　The Fig Apple.

The *Paradife Apple* is rather a fhrub than a tree. There are
two forts of it, which gardeners diftinguifh by the names of the
French and the Dutch Paradife Apple. They are both low-
growing trees; and the only difference between them is, that
the Dutch fort is rather the ftrongeft fhooter. They are chiefly
ufed for ftocks to graft apples upon, in order to make them
more dwarfifh; fo that a plant or two in a collection, for the
fake of variety, will be fufficient.

Fig Apple has a place here for no other reafon than its being
deftitute of the moft beautiful parts of which the flowers are
compofed; viz. the petals: They have all the ftamina, &c.
but no petals, which is a fingular imperfection; tho' by many
they are coveted on that account. As the ftamina and other
parts are all perfect, the flowers are fucceeded by a tolerable
good eating Apple; for the fake of which this tree deferves to
be propagated.

3. The SWEET-SCENTED CRAB of VIRGINIA differs from
our Crab in the leaves, flowers, and fruit. The leaves are
angular, fmooth, of a fine green colour, and have a look
entirely different from any of our Crabs or Apples. The flowers
ftand on larger footftalks than thofe of the generality of our
Crabs, and are remarkable for their great fragrance. This tree
is feldom in full blow before the beginning of June. The
flowers, when they firft open, are of a pale-red, though the
petals foon after alter to a white colour. They are fucceeded

by

by a little round Crab, which, of all others, is the foureſt, rougheſt, and moſt diſagreeable, that can be put into the mouth.

There is a *ſub-evergreen Crab* of America, ſuppoſed to be a *variety* of this ſpecies. Its natural growth ſeems to be not more than twelve feet; and the branches are covered with the ſame kind of ſmooth brown bark as our common Crab-tree. The leaves are long and narrow, and will often be found of different figures; for though ſome will be angular, others again are oblong, or of a lanceolate figure. They are fine, ſmooth, of a ſtrong dark green colour, and have their edges regularly ſerrated. They will remain until late in the ſpring, which rather entitles this ſhrub to a place here; though in an expoſed ſituation, the ends of the branches will be often ſtripped of thoſe ornaments, after a few ruffian attacks of the piercing northern blaſts: So that this tree, when confidered as an evergreen, ſhould always be planted in a well-ſheltered place, where it will retain its leaves, and look very well all winter.

4. The QUINCE. There are many *varieties* of the Quince tree, which are chiefly raiſed for the fruit. The Quince-tree ſeldom grows to be higher than eight or ten feet; and the bark on the branches is often of a kind of iron colour. The leaves are large and oval: Their upper ſurface is of a pleaſant green colour, though often poſſeſſed of a looſe downy matter, and their under ſide is hoary to a great degree. The flowers are produced in May, all along the branches: They grow upon young ſhoots of the ſame ſpring, and are very large and beautiful; for although each is compoſed of about five petals only, yet theſe are often an inch long, are broad and concave, and of a fine pale-red as they firſt open, though they afterwards alter to a white; and thoſe flowers being produced the whole length of the branches, and befpangling the whole tree in a natural and eaſy manner, juſtly entitle this ſpecies to no mean place among the flowering kinds. They are ſucceeded by that fine large yellow fruit which is ſo well known, and which at a diſtance, on the tree, appears like a ball of gold. Indeed, theſe trees ſhould always be planted at a diſtance from much-frequented places; for the fruit, valuable as it is when properly prepared for uſe, has a ſtrong diſagreeable ſcent, that

Y 3   will

will fill the air all around with its odour, which to moſt people is offenſive.

PROPAGATION. HANBURY ſays, all theſe ſorts will take by grafting or budding upon one another, notwithſtanding what MILLAR has alledged to the contrary. He continues, "I have a tree that bears excellent Apples grafted upon a Pear-ſtock ; and Pears grafted upon Crab-ſtocks that have not yet borne." The uſual way is to graft the PEARS on ſtocks raiſed from the kernels of Pears, and the APPLES on Crab-ſtocks. Theſe ſhould be ſown, ſoon after the fruit is ripe, in beds half ar inch deep, and carefully guarded from mice, which will ſoon deſtroy the whole ſeminary, if once found out. In the ſpring the plants will come up ; and in the winter following they ſhould be planted out in the nurſery, in rows two feet aſunder. In a year or two after this, they will be fit for working ; and by this method all the ſorts of Pears and Apples are propagated.

The PARADISE APPLE is generally raiſed by layers or cuttings ; and all the ſorts of QUINCES grow readily by cuttings, planted any time in the winter ; though the early part of that ſeaſon is to be preferred.

The *Evergreen Crab* will take by grafting or budding on the common Crab or Apple-ſtock : but great care and nicety of execution is requiſite in performing the operation : Budding towards the latter end of July, HANBURY ſays, he has always found to be the moſt certain method.

QUERCUS.

LINNEAN Claſs and Order, *Monoecia Polyandria :* Male flowers containing many ſtamina, and female flowers containing one piſtil ; upon the ſame plant : There are thirteen SPECIES :

1. QUERCUS *Robur :* The ENGLISH OAK : a well-known *tall deciduous tree ;* native of England ; and is found in moſt parts of Europe.

2. QUERCUS *Phellos :* The WILLOW-LEAVED OAK ; *a deciduous tree ;* native of moſt parts of North America.

3. QUERCUS

3. QUERCUS *Prinus :* The CHESNUT-LEAVED OAK ; *a deciduous tree ;* native of moft parts of North America.

4. QUERCUS *Nigra :* The BLACK OAK ; *a low deciduous tree ;* native of North America.

5. QUERCUS *Rubra :* The RED OAK ; *a tall deciduous tree :* native of Virginia and Carolina.

6. QUERCUS *Alba :* The WHITE OAK ; *a deciduous tree ;* native of Virginia.

7. QUERCUS *Efculus :* The ITALIAN OAK , or the CUT-LEAVED ITALIAN OAK ; *a low deciduous tree ;* native of Italy, Spain, and the South of France.

8. QUERCUS *Ægilops :* The SPANISH OAK, or OAK WITH LARGE ACORNS AND PRICKLY CUPS ; *a tall deciduous tree ;* a native of Spain.

9. QUERCUS *Cerris :* The AUSTRIAN OAK, or the OAK WITH PRICKLY CUPS AND SMALLER ACORNS ; native of Auftria and Spain.

10. QUERCUS *Suber :* The CORK-TREE ; *an evergreen tree ;* native of the fouthern parts of Europe.

11. QUERCUS *Ilex :* The ILEX, or COMMON EVERGREEN OAK ; *an evergreen tree ;* native of Spain and Portugal.

12. QUERCUS *Coccifera :* The KERMES OAK ; *a tall evergreen fhrub ;* native of France and Spain.

13. QUERCUS *Molucca :* The LIVE OAK : *an evergreen tree ;* native of America.

1. The ENGLISH OAK will grow to great ftature and live to a great age. EVELYN, whofe learning and induftry are evident in every page of his elaborate work, fatigues us with a tedious account of large trees, which either were growing in his time, or which he found in the mouth of tradition, or in the pages of learning and hiftory. We would rather however refer our readers to his detail than either copy or abridge it ; confining ourfelves to a few individuals of our own time, which now are (or were very lately) actually ftanding in this kingdom. The COWTHORP-OAK, now growing at Cowthorp, near Wetherby in Yorkfhire, has been held out as the father of the foreft. Dr. Hunter of York, in his brilliant edition of Mr. Evelyn's book, has favoured us with an engraving of this

tree ;

tree; the dimenſions of which, as he juſtly obſerves, " are
almoſt incredible." Within three feet of the ſurface, the
Doctor tells us, " it meaſures ſixteen yards, and cloſe to the
ground, twenty-ſix yards. Its height in its preſent ruinous ſtate
(1776) is about eighty-five feet, and its principal limb extends
ſixteen yards from the bole. Throughout the whole tree the
foliage is extremely thin. ſo that the anatomy of the antient
branches may be diſtinctly ſeen in the height of ſummer. When
compared to this, all other trees (the Doctor is pleaſed to ſay) are
but *children* of the foreſt." If indeed the above admeaſurement
might be taken as the dimenſion of the *real ſtem*, its ſize would
be trulv enormous, and far exceed that of any other Oak in
the kingdom. But the Cowthorp Oak has a ſhort ſtem, as moſt
very large trees it is obſervable have, ſpreading wide at the
baſe, the roots riſing above the ground like ſo many buttreſſes
to the trunk. which is not like that of a tall-ſtemmed tree, a
cylinder, or nearly a cylinder, but the fruſtum of a cone. Mr.
MARSHAM gives us a plain and accurate account of this tree: He
ſays, " I found it in 1768, at four feet, forty feet ſix inches;
at five feet, thirty-ſix feet ſix-inches; and at ſix feet, thirty-two
feet one inch." Therefore in the principal dimenſion, *the ſize
of the ſtem*, it is exceeded by the BENTLEY OAK; of which the
ſame candid obſerver gives the following account: " In 1759, the
Oak in Holt-Foreſt, near Bentley, was, at ſeven feet, thirty-
four feet. There is a large excreſcence at five and ſix feet that
would render the meaſure unfair. In 1778, this tree was in-
creaſed half an inch, in nineteen years. It does not appear to
be hollow, but by the trifling increaſe I conclude it not
found." Extraordinary, however, as theſe dimenſions may appear, they are exceeded by thoſe of the BODDINGTON OAK; a
tree which we believe does not appear any where upon record,
except it be alluded to in Mr. EVELYN's Liſt. This Oak grows
in a piece of rich graſs land, called the Old-Orchard Ground,
belonging to Boddington Manor Farm, lying near the turnpike
road between Cheltenham and Tewkſbury, in the Vale of
Gloceſter. The ſtem is remarkably collected and ſnug at the
root, the ſides of its trunk being more upright than thoſe of
large trees in general; nevertheleſs its circumference at the
ground, as near to it as one can walk, is twenty paces: meaſuring

ing with a two-foot rule, it is fomewhat more than eighteen yards. At three feet high it meafures forty two feet, and at its fmalleft dimenfions, namely, from five to fix feet high, it is thirty-fix feet. At about fix feet it begins to fwell out larger; forming an enormous head, which heretofore has been furnifhed with huge, and in all probability extenfive arms. But age and ruffian winds have robbed it of a principal part of its grandeur ; and the greateft extent of arm at prefent (1783) is eight yards, from the ftem. From the ground to the top of the crown of the trunk is about twelve feet ; and the greateft height of the branches, by eftimation, forty-five feet. The ftem is quite hollow ; being, near the ground, a perfect fhell ; forming a capacious well-fized room ; which at the floor meafures, one way, more than fixteen feet in diameter. The hollownefs, however, contracts upwards, and forms itfelf into a natural dome, fo that no light is admitted except at the door, and at an aperture or window in the fide. It is ftill perfectly alive and fruitful, having this year a fine crop of acorns upon it. It is obfervable in this (as we believe it is in moft old trees), that its leaves are remarkably fmall ; not larger, in general, than the leaves of the Hawthorn.

In contemplating thefe wonderful productions of nature we are led to conjecture the period of their exiftence : Mr. MARSHAM in his Paper publifhed in the Firft Volume of the Tranfactions of the Bath Agriculture Society, has given us fome very ingenious calculations on the age of trees; and concludes that the Tortworth Chefnut is not lefs than eleven hundred years old. We have however fhewn under the Article CHESNUT, that Mr. MARSHAM is miftaken in the dimenfions of that tree. Neverthelefs, if it ftood in the days of King John, fix centuries ago, and was then called the Great Chefnut *, we may venture to fuppofe it not much lefs than one thoufand years of age ; and further, if we confider the quick growth of the Chefnut compared with that of the Oak, and at the fame time the inferior bulk of the Tortworth Chefnut to the Cowthorp, the Bentley, and the Boddington Oaks ; may we not

* As Tradition fays it was.

venture

venture to infer, that the exiſtence of theſe truly venerable trees commenced ſome centuries prior to the era of Chriſtianity ?

The root of the Oak ſtrikes deep, eſpecially the middle or tap-root, which has been traced to a depth nearly equal to the height of the tree itſelf: nor do the lateral roots run ſo ſhallow and horizontal as thoſe of the Aſh and other trees ; but perhaps the roots of very few trees range wider than thoſe of the Oak. The ſtem of the Oak is naturally ſhort, and if left to itſelt, in an open ſituation, it will generally feather to the ground. It has not that upright tendency as the Aſh, the Eſcu-lus, and the Pine-tribe : neverthelefs, by judicious pruning, or by planting in cloſe order, the Oak will acquire a great length of ſtem : in this caſe, however, it rarely ſwells to any conſider-able girt. Mr. MARSHAM indeed mentions one in the Earl of Powys's Park near Ludlow, which in 1757 meaſured, at five feet, ſixteen feet three inches, and which ran quite ſtraight and clear of arms near or full ſixty feet. But, as has before been obſerved, Oaks which endure for ages have generally ſhort ſtems ; throwing out, at ſix, eight, ten, or twelve feet high, large horizontal arms ; thickly ſet with crooked branches ; terminating in clubbed abrupt twigs ; and cloſely covered with ſmooth gloſſy leaves ; forming the richeſt foliage, irregularly ſwelling into the boldeſt outline we know of in nature. The Pine-tribe and the Eſculus may be called elegant or beautiful ; but the general aſſemblage of a lofty full-furniſhed Oak is truly ſublime.

It is ſomewhat extraordinary, that the moſt *ornamental* tree in nature ſhould, at the ſame time, be the moſt *uſeful* to man-kind. Its very leaves have been lately found to be of eſſential uſe to the gardener ; the huſbandman is well acquainted with the value of its acorns ; and every Engliſhman experiences daily the uſeful effects of its bark. It is wholly unneceſſary to mention the value of its timber : it is known to the whole world. The Oak raiſed us *once* to the ſummit of national glory : and *now* we ought to hold in remembrance that our exiſtence as a nation depends upon the Oak. If therefore our fore-fathers, merely from the magnitude and majeſty of its appearance, the veneration due to its age, and gratitude perhaps for ſome few economical uſes they might apply it to, paid divine honours to this tree ; how much more behoves it us, circumſtanced as we are, to pay

due

due homage to this our national faviour! How could our Kings be invefted with the enfigns of royalty, or our Creator receive at ftated times the gratitude and praife which we owe to him, with greater propriety than under the fhadow of this facred tree? Acts like thefe would ftamp it with that refpectability and veneration which is due to it : and to corroborate thefe ideas, as well as to inftitute fuch laws as might be found neceffary, the ftate of the growth of Oak in Great Britain ought to be a ftanding enquiry of the Britifh Legiflature. It is far from being impracticable to have annual returns of Oak fit for fhip-building in every parifh in the kingdom ; with the dif_ tance it ftands from water-carriage. It avails but little our making laws of police, or forming foreign alliances, unlefs we take care to fecure in perpetuity the defence of our own coaft. It is idle to think of handing down to pofterity a national independency, if we do not at the fame time furnifh them with the means of preferving it.

The PROPAGATION of the ENGLISH OAK. We do not purpofe in this place to give directions for raifing woods or plantations of Oak : this we referve until we come to treat of plantations in general, under the title WOODLANDS ; for by collecting the more ufeful trees into one point of view, we fhall be better able to judge of their comparative value ; and the methods of raifing the feveral fpecies for the purpofe of timber (fhip-timber excepted) being nearly the fame, we fhall be enabled to give our directions more fully, yet upon the whole much more concifely, than we could have done, had we retailed them feparately under each article : therefore, we mean to abide by the fame rule under the prefent head that we have obferved throughout this part of our work ; namely, to treat of the plant under confideration merely as a *nurfery plant.*—There are various opinions about the *choice of acorns*; *authors* in general recommend thofe of " fair, ftraight, large and fhining trees ;" but *nurferymen*, we believe, pay little attention as to the tree from which the acorns are gathered. And indeed, when we confider that the feeds of the diftinct *varieties* of any individual fpecies of plants produce one and the fame feedling-ftock, or a fimilar variety of feedling plants, we muft conclude that little attention is due. If however it be true that the feeds of fome *varieties* pro-

duce

duce *more* of its own kind than thofe of other *varieties* of the fame fpecies, it may be worth the trouble, when only a fmall quantity of feed is wanted, to gather it from the moft valuable tree. The *prefervation of Acorns* is extremely difficult : if we fow them in autumn, they become obnoxious to vermin and birds : if we keep them above-ground, it is very difficult to pre-vent their fprouting, and at the fame time preferve their vege-tating power, Upon the whole, the fall of the Acorn feems the propereft *time of fowing.* For fpring fowing February and March are the proper months. The ufual *method of fowing* is either in drills, or promifcuoufly in beds, covering them about two inches deep. But we would rather recommend placing them in beds in the quincunx manner, from four to fix inches apart, covering them one half to two and a half inches deep, ac-cording to the ftiffnefs or lightnefs of the foil. Sowing feeds in drills renders them peculiarly obnoxious to mice and rooks ; and by fcattering them promifcuoufly the plants are liable to come up double and irregularly, and the ufe of the hoe is pre-cluded. The oakling rifes the firft fpring after fowing. The feedling plants, having ftood two years in the feed-bed, fhould be removed into the nurfery, placing them in rows from two and a half to three feet afunder, and the plants from nine to twelve inches in the rows ; the tap-root and all long fprawling fibres having been firft taken off, and the top trimmed to a fwitch, if tolerably ftraight, or, if deformed or maimed, cut down within two or three inches of the ground ; remembering to fort the plants as directed in the introductory part of our work. Hav-ing remained two or three years in the nurfery, they will be ready to be planted out into fenced plantations. Such as are wanted to be trained for ftandards, may be removed into fome vacant ground ; firft pruning them in the conoidic manner, and afterwards remembering from time to time to pay proper at-tention to their leaders.

The Englifh Oak admits of fome *varieties :* indeed, if we attend minutely to particulars, we fhall find them almoft infinite. There is one variegation under the name of the *Stripe-leaved Oak :* But the moft interefting variety of the Englifh Oak is the *Lucombe* or *Devonfhire Oak.* In the Sixty-fecond Volume of the Philofophical Tranfactions, a particular account is given of

this

this Oak ; fetting forth that Mr. LUCOMBE, a gentleman of Devonfhire, having, about the year 1765, fowed a parcel of acorns faved from a tree of his own growth, and obferving that one of the feedling plants preferved its leaves through the winter, he paid particular attention to it, and propagated, by grafting, fome thoufands from it. Its being an evergreen is not the only peculiarity of this variety ; it has a fomewhat more upright tendency, and feems to be of a quicker growth, than Oaks in general. The plants however, which we have feen, do not anfwer altogether the defcription given in the account abovementioned ; but as they are now in the hands of almoft every Nurferyman, we forbear faying any thing further refpecting them.

2. The WILLOW-LEAVED OAK will grow to be a large timber-tree. It receives its name from its leaves refembling very much thofe of the Common Willow. Thefe long narrow leaves have their furface fmooth, and their edges entire ; and their acorns will be almoft covered with their large cups.

There are feveral *varieties* of this fort ; fome having fhorter leaves, others broader, and hollowed on the fides ; fome large acorns, others fmaller, &c. all of which are included under the appellation of Willow-leaved Oaks.

3. The CHESNUT-LEAVED OAK. This alfo will grow to be a large timber-tree ; and in North America, where it grows naturally, the wood is of great fervice to the inhabitants. It is fo called, becaufe the leaves greatly refemble thofe of the Spanifh Chefnut-tree. They are about the fame fize, fmooth, and of a fine green colour.

There are two or three *varieties* of this fort ; but the leaves of all prove that they are of the fpecies called the Chefnut-leaved Oak ; fo that nothing more need be obferved, than that the leaves of fome forts are larger than thofe of others ; that the acorns alfo differ in fize, and grow like thofe of our Englifh Oak, on long or fhort footftalks as it fhall happen.

4. The BLACK OAK is a tree of lower growth, it feldom rifing to more than thirty feet high. The bark of this tree is of a very dark colour, which occafioned its being named the Black Oak. The leaves are fmooth, very large, narrow at their bafe, but broad at their top, being in fhape like a wedge : They have
indentures

indentures at the top, fo as to occafion its having an angular
look ; they are of a fhining green colour, and grow on fhort
footftalks on the branches.

There is a *variety* or two of this fort, particularly one with
trifid leaves, and another flightly trilobate, called the *Black Oak
of the Plains*, the leaves and cups of all which are fmall.

5. RED OAK. The Red Oak will grow to be a timber-tree
of fixty or feventy feet high, and the branches are covered with
a very dark-coloured bark. It is called the Red Oak from the
colour of its leaves, which in the autumn die to a deep red colour.

There are feveral *varieties* of this fpecies, the leaves of which
differ in fize and figure ; but thofe of the larger fort are finely
veined and exceeding large, being often found ten inches long
and five or fix broad : They are obtufely finuated, have angles,
and are of a fine green colour in the firft part of the fummer,
but afterwards change by degrees to red, which is mark
enough to know thefe trees to be of this fpecies. There are
feveral varieties of this tree, which exhibit a manifeft difference
in the fize of the leaves, acorns, and cups. That is the beft
which is commonly called the *Virginian Scarlet Oak* ; and the
bark is preferred for the tanners ufe before that of all the other
forts.

6. The WHITE OAK. The White Oak will not grow to the
fize of the former, it feldom being found higher than forty feet
even in Virginia, where it grows naturally. But though the
timber is not fo large, yet it is more durable, and confequently
of greater value for building to the inhabitants of America, than
any of the other forts. The branches of this tree are covered
with a whitifh bark ; the leaves alfo are of a light colour. They
are pretty large, being about fix inches long and four broad.
They have feveral obtufe finufes and angles, and are placed on
fhort footftalks.

There is a *variety* or two of this fpecies ; and the acorns are
like thofe of our Common Oak.

7. The ITALIAN OAK will grow to about the height of thirty
feet. The branches are covered with a dark-purplifh bark. The
leaves are fmooth, and fo deeply finuated as to have fome refemblance

femblance of pinnated leaves; and each has a very fhort footftalk. The fruit of this fpecies fits clofe to the branches. The
cups are in fome degree prickly and rough, and each contains
a long flender acorn, that is eatable. This (fays HANBURY) is
the true *Phagus* of the Greeks, and the *Efculus* of Pliny; in
the places where thefe trees grow naturally the acorns are, in
times of fcarcity, ground into flour, and made into bread.

8. The SPANISH OAK will grow to be as large a tree as our
Common Oak, and is no way inferior to it in ftatelinefs and
grandeur; for the branches will be far extended all around,
caufing, with the leaves, a delightful fhade. Though the bark of
thefe branches is of a whitifh colour, yet they are neverthelefs
fpotted with brownifh fpots. The leaves are of an oblong oval
figure, but not very long, feldom being longer than three
inches, and two broad. They are fmooth, and have their edges
deeply ferrated: Thefe ferratures are acute, and chiefly turn
backwards. Their upper furface is of a fine light green colour,
and their under of an hoary caft; and with thefe beautiful leaves
each branch is plentifully ornamented all over the tree. The
cups are moft peculiar and fingular; for they are very large,
and compofed of feveral rough, black, large fcales, that lap
over one another like the fcales of a fifh. They almoft cover
the acorn, though they are pretty large, narrow at the bottom,
but broader higher, and have their tops flat. The Greeks
call the acorns *Velani*, and the tree itfelf *Velanida*. The acorns
are ufed in dyeing.

9. The AUSTRIAN OAK is of lower growth than the preceding fpecies, it feldom rifing to more than forty feet high. The
leaves are of two colours; their upper furface being of a fine
green colour, and their under downy. Their figure is oblong;
but they are fo indented about the middle as to make them have
the refemblance of a lyre. They are wing-pointed, tranfverfely
jagged, and ftand on flender footftalks on the branches. The
cups of this fort alfo are fmaller and prickly, and the acorns
alfo proportionally fmaller than thofe of the preceding
fpecies.

All thefe foreign deciduous forts may be PROPAGATED from
the acorns, which muft be procured from the places where the
trees naturally grow. They fhould be fown as foon as poffible
after

after they arrive; and if any of them have fprouted, great care muſt be uſed in taking them out of the boxes in which they were conveyed. Any fort of our common garden-mould, made fine, will fuit them ; and they ſhould be fown in drills, in beds an inch deep. The firſt ſpring after fowing, the plants will come up; they ſhould be always kept clean from weeds, and if they are watered in dry weather, it will be the better. They will want no prefervation in winter ; for they are all very hardy, even when young. In March they ſhould be all taken out of the feed-bed, have their tap-roots ſhortened, and be planted in the nurfery-ground a foot afunder, and two feet diſtant in the rows, where they may ſland, with the ufual nurfery care, until they are to be planted out.

The STRIPED-LEAVED OAK is ufually PROPAGATED by in-arching into the Common Oak ; but it is beſt increafed by grafting. In the fame manner, alfo, any particular variety belonging to the other fpecies may be continued and multiplied.

EVELYN fays the Oak " will endure the LAYING, but never to advantage of bulk or ſtature."

10. The CORK-TREE admits of two *Varieties :*
The Broad-leaved Cork-Tree,
The Narrow-leaved Cork-Tree.

The *Broad-leaved Cork-Tree* is a timber-tree in Portugal and Spain, and other fouthern parts of Europe, where it grows naturally. In our prefent plantations, it ſhould be placed near the middle of our largeſt quarters, among others of about forty feet growth ; and a few alfo ſhould be planted fingly in opens, that its fungous bark may be in view : not that there is any great beauty merely in the fight, but with us it is a curiofity ; being the true Cork, and is of the fame nature with what comes from abroad, and we ufe for bottles, &c. Thus rough and fpongy is the bark on the trunk and main branches ; but the bark on the young ſhoots is fmooth and grey, and that on the youngeſt white and downy. The leaves are of an oblong, oval figure, with fawed edges. Their upper furface is fmooth, and of a ſtrong green colour, but their under is downy. They grow alternately on the branches, on very ſhort though ſtrong footſtalks, and indeed differ in appearance very little from many forts of the *Ilex.* As the flowers of the *Quercus* make no ſhow, we
ſhall

fhall proceed to the next fort, after obferving, that the acorns of the Cork-tree are longifh, fmooth, and brown when ripe, and of the fize and fhape of fome of our common acorns, to which they are fo much alike, as not to be diftinguifhed, if mixed together.

The *Narrow-leaved Cork-tree* is a variety only of the common and moft general fort ; fo that as this article requires nothing more than obferving that the leaves are fmaller, and as fuch make a variety in plantations, it may not be amifs to fay fomething of the Cork, which we receive from abroad, and which is collected from thefe trees. The beft cork, then, is taken from the oldeft trees, the bark on the young trees being too porous for ufe. They are, neverthelefs, barked before they are twenty years old; and this barking is neceffary, to make way for a better to fucceed ; and it is obfervable, that after every ftripping the fucceeding bark will encreafe in value. They are generally peeled once in ten years, with an inftrument for the purpofe ; and this is fo far from injuring the trees, that it is neceffary, and contributes to their being healthy ; for without it they thrive but flowly ; nay, in a few years they will begin to decay, and in lefs than a century a whole plantation will die of age ; whereas thofe trees that have been regularly peeled will laft upwards of two hundred years. " Wonderful, then, is the wifdom and goodnefs of Almighty God, and calls for our profoundeft admiration, that he fhould not only provide for us his creatures fuch variety of things for ufe, but caufe, as in this inftance, what would be death to one tree, to be refrefhment to another, for the fupply of our neceffaries ; and in the formation of this tree, not only caufing the cork to grow, but providing alfo an interior bark fufficient to nourifh the tree, and even in a manner exhilarate it, as the loaded wool is fhorn from the fleecy kind. To make our gardening to the utmoft degree ufeful, we fhould be always exercifed in thefe confiderations, and this will infpire us with acts of gratitude and obedience. " HANBURY.

11. The ILEX is a well-known Evergreen, of which there are many *varieties,* all of which add great beauty to the large quarters of Evergreen-trees. The bark of all thefe forts is entire, and that of the younger forts fmooth ; but the leaves are of different

Z   fhapes

shapes and composition, according to the nature of their variety. Some of them are nearly like those of both sorts of the Cork-tree; others again are nearly round and prickly; some are long, smooth, and narrow, with few indentures; whilst others are broad and much serrated: All these *varieties* will often proceed from acorns gathered of the same tree; nay, the leaves of the same tree will not be always alike, being often found very different on the same plant; so that a quantity of plants of this species raised from seeds, will of themselves afford considerable variety. The acorns of all these sorts are of different sizes, though their shape is nearly the same, which is like that of some sorts of our Common Oak, but smaller. The most striking *variety* of the *Ilex* is the *Holly-leaved Oak*, which differs from the other sorts only that the leaves are shaped like those of the Holly-tree. They are of an oblong, oval figure, sinuated, prickly, and downy underneath; but many sorts raised from seeds of the *Ilex* will have such kind of leaves; and it constitutes no further a *variety*, than what may reasonably be expected from a quantity of the acorns of the *Ilex* sown.

12. KERMES OAK. This is a low-growing tree, and a fine Evergreen: It seldom grows to be twenty feet high, and it may be kept down to what height is required. It has the appearance of some of the sorts of the *Ilex*, from which it looks to be a variety only, though doubtless this is of itself a distinct species. The leaves are smooth, and of an oval figure. They are of a thickish consistence, and larger than most sorts of the *Ilex*. Their verge is indented, and many of them are possessed of small spines; and they are placed on short strong footstalks on the branches. The acorns of this sort are small, though there are to be found in our woods acorns of about the same size and shape.

MILLAR says, "this is the Oak from which the Kermes or Scarlet Grain is collected, which is an insect that harbours on this tree."

13. The LIVE OAK is common in America, where it grows to timber. The leaves are large, spear-shaped, oval, of a fine dark-green colour, entire, and placed on short footstalks on the branches. The acorns of this sort are small, though they grow in cups with footstalks like the other sorts. The wood of this

tree

tree is very ufeful to the inhabitants of Carolina, Pennfylvania, and Virginia, where it grows naturally, being very tough and hard, and ferves for many purpofes that require fuch a fort. The acorns ferve for food for the meaneft people, who not only eat them as fuch, but, being of a very fweet nature, they are liked by perfons of all ranks. From thefe acorns a fweet oil alfo is extracted, which is very good.

There are many other *varieties* of the different fpecies of Evergreen Oaks, which it will not be fo neceffary to fearch for here, as the forts mentioned are the bulk of the tribe, and of themfelves afford much variety ; and indeed, if much coft and trouble were beftowed in procuring others, the variety would be little heightened, particularly as the pleafure received from the variation arifes principally from the different forms of the leaves ; for none of thefe trees produce flowers for ornament, and the acorns afford too minute a variety to require dwelling long on here.

All the fpecies of Evergreen Oaks are to be raifed from *acorns*, in the manner which has been directed for the foreign deciduous forts. The beft acorns we receive from abroad ; for they feldom ripen well with us.——The acorns which come from abroad, and which are by far the fineft, often fprout in the paffage ; fo that care muft be ufed in taking them out of what they are enclofed in, and they fhould be put into the mould as foon as convenience will permit. Traps for mice, &c. muft be fet; and after they come up, they will want nothing but weeding for at leaft three years ; for I would not have them taken out of the feed-beds fooner ; efpecially the forts of the *Ilex*; for when thefe have been pricked out of the feed beds at one year old, they have feldom grown ; and though fometimes fome of them will be green, and have the appearance of growing, during one fummer, they will ofteneft turn brown, and gradually go off afterwards. " After thefe plants have ftood to be two or three feet high, I always found them more fure of growing when moved. I have tranfplanted fuch plants at moft times of the year with fuccefs ; in the fpring, in the depth of winter, and in the autumn, and have had them grow well when moved in July; and indeed I am pretty well perfuaded there is no

month in the year more proper than that for the removing of
moſt ſorts of Evergreens, provided the weather be rainy or
hazy at their planting, and ſhade can be afforded them for ſome
time after." HANBURY.

Theſe trees may be alſo encreaſed by *inarching*, for they
will grow very readily this way on ſtocks of our Common
Oak ; ſo that having a tree or two of any of the ſorts, if young
Oaks are planted round each of them, after they have grown a
ſummer or two, they will be ready to embrace the young ſhoot.
After they are well joined, they may be cut off from the mother
tree, and tranſplanted into the nurſery-ground, or where they
are to remain, and freſh Oaklings planted round the trees to
be multiplied : and the continuance of the repetition of this
may be at pleaſure. In removing of the inarched plants, the
time ſhould be obſerved as in removing young plants of our
Common Oak, the roots ſtill remaining of that kind and
nature.

Theſe trees will take by *grafting* on the young ſtocks of our
Common Oak. The ſtocks ſhould be young and healthy, the
cuttings ſtrong and good, and great care muſt be taken in properly joining and claying them, or they will not grow ; which
makes the inarching more neceſſary, as by that practice no
cutting is in danger of being loſt.

R H A M N U S.

LINNEAN Claſs and Order, *Pentandria Monogynia :* Each
flower contains five males and one female : There are twenty-
ſeven SPECIES ; ſeven of which may be admitted into our
collection.

1. RHAMNUS *Catharticus :* The COMMON BUCKTHORN ; *a
tall deciduous ſhrub* ; native of England, and (one of its varieties)
of Spain, Italy and France.

2. RHAMNUS *Frangula :* The FRANGULA, or the BERRY-
BEARING ALDER ; *a deciduous ſhrub* ; native of England and
moſt of the northern parts of Europe.

3. RHAMNUS

3. RHAMNUS *Alpinus* : The ALPINE RHAMNUS, or the ROUGH-LEAVED FRANGULA ; *a deciduous shrub* ; native of the Alps.

4. RHAMNUS *Paliurus* : The PALIURUS, or THORN OF CHRIST, or CHRISTI THORN ; *a deciduous shrub or tree* ; native of Paleſtine, alſo of Spain, Portugal and Italy.

5. RHAMNUS *Alaternus* : The COMMON ALATERNUS ; *an evergreen tree* ; native of the South of Europe.

6. RHAMNUS *infectorius* : The NARROW-LEAVED EVERGREEN BUCKTHORN; *an evergreen shrub or tree* ; native of Spain.

7. RHAMNUS *Oleoides* : The OLIVE-LEAVED EVERGREEN BUCKTHORN ; *an evergreen shrub* ; native of Spain.

1. The COMMON BUCKTHORN. Of this ſpecies there are the following *varieties* : Dwarf Buckthorn, Long-leaved Dwarf Buckthorn, and the Common Buckthorn of our hedges. Variety is the ſole motive for admitting theſe ſorts into a collection. The flowers have no beauty to catch the attention ; though their berries, their manner of growing, the colour of their bark in winter, and verdure of their leaves in ſummer, court us to admit a few of them.

Dwarf Buckthorn is a ſhrub of about a yard high. The branches grow irregular, and are covered with a blackiſh-coloured bark. The leaves are nearly oval, though they end in a point. They are ſcarcely an inch long, about half that breadth, and ſtand oppoſite by pairs for the moſt part. The flowers grow on ſhort footſtalks, on ſpurs, by the ſides of the branches. They are of a greeniſh colour, and make little ſhow.

Long-leaved Dwarf Buckthorn differs little from the other, only that it grows to be rather a larger ſhrub, and the leaves are longer. The flowers are about the ſame colour as the Dwarf ſort ; but neither of theſe ſcarcely ever produce berries : This makes them much leſs valuable than our Common Buckthorn, which will exhibit its black berries in plenty in the autumn, either for ſhow or uſe.

Common Buckthorn is well known in England. Where it does not grow common about a habitation, a few ot theſe ſhrubs ſhould be admitted ; for it is a well-looking tree, either in winter or fummer, and its black berries in the autumn are no fmall ornament. The Common Buckthorn will grow to be near fixteen feet high, and will fend forth numerous branches on all fides. Thefe are fmooth ; and the bark is of a blueifh colour Many ftrong fharp fpines come out from the fides and ends of the branches. The leaves are oval, fpear fhaped, about two inches long, and one broad. Their under furface is of a lighter green than the upper. They have ferrated edges, and ftand, fometimes by pairs, fometimes fingly, on longifh footftalks on the branches. The flowers are pro- duced in clufters from the fides of the branches, in June. Their colour is green : and they are fucceeded by black berries, each containing four feeds. Syrup of Buckthorn is made of thefe berries, and is well known as a cathartic. From the juice of thefe berries alfo an admirable green colour is prepared, which is in great requeft with miniature-painters.

All the forts of Buckthorn are eafily PROPAGATED, either by feeds or cuttings. The feeds of the Purging Buckthorn may be gathered in plenty in moft parts of England ; but the feeds of the Dwarf forts muft be procured from abroad, where they grow naturally, for they produce no feeds with us. They fhould be fown as foon as poffible after they are ripe, in almoft any kind of garden-mould made fine. They will not always come up the firft fpring ; fo that the beds muft remain un- difturbed and weeded during the fummer. After they are come up, and have ftood in the feed-bed a year or two, they may be planted out in the nurfery-way, at fmall diftances. Thefe plants are alfo to be raifed by cuttings, which fhould be planted int he autumn ; and if they are not planted very clofe, they will want no removing until they are finally fet out. If a large quantity of thefe plants is wanted, and little ground is prepared for the cuttings, they may be fet very clofe, and in the winter following taken up, and planted in the nurfery-way, like the feedlings. In two or three years, they may be planted out to ftand.

2. FRANGULA, or BERRY-BEARING ALDER. This fpecies affords us the following varieties : Common Black-berry-bear- ing Alder, Dwarf Berry-bearing Alder, and the American Smooth-leaved Berry-bearing Alder.

The

The *Common Black-berry-bearing Alder* will grow to the height of about ten feet. It will afpire with an upright ftem, and produce numerous branches on all fides. The bark is fmooth, of a blueifh colour, and is all over fpotted with white fpots, which make it refemble a blueifh-grey. The leaves are oval, fpear-fhaped, and grow irregularly on the branches. They are about two inches long and one broad. Their upper furface is fmooth and of a fhining green, and their under furface is poffeffed of many ftrong veins that run from the mid-rib to the edges. The flowers are produced in bunches in June, each having a feparate footftalk. They are of a greenifh colour, and make no fhow; but they are fucceeded by berries, which are firft red, afterwards (when ripe) black, and are a great ornament to the tree.

Dwarf Berry-bearing-Alder is of very low growth. It feldom rifes higher than two feet. The branches are of a blueifh-brown, and the leaves are nearly round. They are placed on fhort footftalks, and many ftrong veins run from the mid rib to the border. It makes no fhow, either in the flowers or fruit; the firft being fmall, and the latter rarely happen ing.

American Smooth-leaved Berry-bearing Alder will arrive at the height of our common fort; and hardly in any refpect differs from it, either in leaves, flowers, or fruit.

3. ROUGH-LEAVED ALPINE FRANGULA, or Berry-bearing Alder, differs in no refpect alfo from the common fort, only that it is unarmed with thorns, will grow to be rather taller, and the leaves are tough, larger, and doubly laciniated.

There is a *variety* of this fpecies, with fmooth leaves and of rather lower growth, called the *Smooth-leaved Alpine Frangula.*

The method of PROPAGATING thefe forts of the Berry-bearing Alder is exactly the fame as that laid down for the Buckthorn; and if thofe rules are obferved, any defired quantity may be raifed.

4. PALIURUS, or *Chrifti Thorn.* The PALIURUS will grow to be a tree of near fourteen feet high, and may be trained to an upright ftem, which will fend forth numerous flender branches on all fides. Thefe are armed with fharp thorns,

Z 4

two

two of which are at each joint. One of thefe thorns is about
half an inch long, ftraight and upright ; the other is fcarcely
half that length, and bent backward. Between thefe is the bud
for the next year's fhoot. The bark on thefe twigs is fmooth,
and of a purplifh colour, and the fpines themfelves are of a reddifh caft. The joints alternately go in and out, forming at each
bud an obtufe angle. The leaves are nearly of an oval figure,
of a pale green colour, and ftand on very fhort footftalks. They
are fmall, being fcarcely an inch in length, have three longitudinal veins, and are placed alternately on the branches. The
flowers are produced in clufters from the fides of the young
fhoots. They are of a yellow colour; and though each fingle
flower is fmall, yet they will be produced in fuch plenty all
over the plant, that they may make a very good fhow. June is
the time of flowering ; and they are fucceeded by a fmall fruit,
that is furrounded by a membrane.

The plant under confideration, fays HANBURY, " is undoubtedly the fort of which the crown of thorns for Our Bleffed Saviour was compofed. The branches are very pliant, and the
fpines of it are at every joint ftrong and fharp. It grows naturally about Jerufalem, as well as in many parts of Judæa ;
and there is no doubt that the barbarous Jews would make choice
of it for their cruel purpofe. But what farther confirms the
truth of thefe thorns being then ufed, are the antient pictures of
Our Bleffed Saviour's crucifixion. The thorns of the crown on his
head exactly anfwer to thofe of this tree ; and there is great
reafon to fuppofe thefe were taken from the earlieft paintings of
the Lord of Life ; and even now our modern painters copy from
them, and reprefent the crown as compofed of thefe thorns.
Thefe plants, therefore, fhould principally have a fhare in thofe
parts of the plantation that are more peculiarly defigned for religious retirement ; for they will prove excellent monitors, and
conduce to due reflection on and gratitude to *Him who hath loved
us, and has wafhed us from our fins,*" &c.

Thefe deciduous forts may be PROPAGATED by feeds and
layers. The foil for the feed fhould be that taken from a frefh
pafture, with the fward ; and having lain a year to rot, and
been turned three or four times, to this a fourth part of drift
fand fhould be added ; the whole being well mixed, the feeds
fhould

fhould be fown half an inch deep. They rarely come up before the fpring twelve-month after fowing; fo that the beds muft be undifturbed all the fummer, and kept free from weeds. After the plants are come up, they may ftand a year or two in the feed-bed, and be then planted out in the nurfery, at the ufual diftance: In about three years they will be fit to be finally planted out. Thefe plants may alfo be propagated by layers; but this is not always a very eafy tafk, and it is feldom that plants can be obtained under two years. Nicking them like carnations is a very uncertain method to be practifed on thefe twigs: for the end of the nick where the root is expected to ftrike will fwell, and be covered with a clofe watery fubftance, without fending out any fibres; and the branch growing in the ground will in two or three years grow this out, and thus all hopes of a root will be loft. By twifting them, alfo, is an uncertain method (though many plants may be raifed this way); for if the twifting be too great, you kill the twig defigned for the layer; and if it is too little, you may look at the end of two or three years, and find no roots at your layers. However, by a gentle twift, juft breaking the bark, plants may be raifed. HAN-BURY continues, " Finding thefe methods precarious and uncertain, I had recourfe to another, by which I obtained numbers of plants. With a fharp knife I made a gentle nick or two the depth of the bark, about the bud and thorns which are at a joint. Having done this in two or three places in every fhoot, and having laid them in the ground, every twig had ftruck root, and were become good plants by that time two years; many of which were fit to plant out for good, and the fmaller proper for the nurfery-gound to gain ftrength."

5. The ALATERNUS. The *varieties* of this Species are:

The Common Alaternus,
The Broad-leaved Alaternus,
The Jagged-leaved Alaternus.

The *Common Alaternus* is again *variegated:* There are of it, the Gold-ftriped, the Silver-ftriped, the Blotch-leaved, the large and the fmaller growing *Alaternus*; and whoever is for having them in plantations of the prefent kind, will ftill encreafe the variety. This is indeed objected to by fome, as, they fay, they cannot be Evergreens; others again think they are moft proper, as they retain their leaves, and appear amongft others, of different

ferent colours, like flowers in fummer. The branches of thefe
forts of *Alaternus* are numerous ; and the younger branches are
covered with a fmooth green bark. In winter, indeed, they
will be brown, and fome of a reddifh colour ; others will have
their fides next the fun red, and the oppofite green. The leaves
are oval, of a lucid green in the common forts, and look very
beautiful. Their edges are crenated, and they grow alternately
on the branches. The flowers are produced in April, from the
wings of the leaves, in little clufters : They are of a greenifh
colou , but make no fhow ; and are fucceeded by berries,
which are very grateful to blackbirds, thrufhes, and the like
kinds of birds.

The *Broad-leaved Alaternus* is the grandeft looking tree of all
the forts : It will grow to the greateft height, if permitted to
fhoot freely, though it may be kept down to any height wanted.
The leaves are the longeft of any of the forts, and their edges
are lightly crenated. They differ a little in figure from the
preceding fort, being more heart-fhaped. They are of a fine
fhining ftrong green colour, both in winter and fummer ; and
this tree produces flowers and feeds like the other.

The *Jagged-leaved Alaternus* has as different a look from the
other as any two Evergreens whatever. It is a well-looking
upright tree, and the branches are covered with a fmooth fine
bark, which in winter is of a reddifh colour. The leaves, like
thofe of all the forts, grow alternately. They are long and
narrow, and are fo jagged as to caufe them to have a particular
look. Their furface is fmooth and fhining, and their figure
lanceolate ; and this, together with the nature of their ferratures, caufes in the tree a beautiful as well as fingular look. The
flowers are produced in the fame manner as the others ; and are
fucceeded by berries, which are ufed by painters in compofing
fome of their yellows. There are variegated forts of the Jagged-leaved *Alaternus* in both filver and gold ftripes, which are
indeed very beautiful ; but they are very apt to turn green, if
planted in a rich foil ; fo that to continue the ftripes in perfection, the worft fort of hungry land fhould be allotted them.

There are more varieties of the *Alaternus*, but their differences
are fo inconfiderable as fcarcely to be worth enumerating. All
the forts have been confounded by the unfkilful with thofe of
Phillyrea, which have indifcriminately paffed one for the other :

That

That the Gardener, therefore, may be guarded from running again into thefe errors, he muft obferve, that the leaves of all the forts of *Phillyrea* grow always oppofite by pairs, whereas thofe of the *Alaternus* grow fingly and alternately on the branches, which firft gave occafion to the fhrub's being fo called. The Botanift will fee a more material difference, when, upon examining the flowers, he finds they belong to diftinct claffes.

6. The NARROW LEAVED BUCKTHORN grows to be a tree of ten or twelve feet high, fending forth feveral branches from the fides from the bottom to the top. They are covered with a blackifh or dark-coloured bark, and each of them is terminated by a long fharp thorn. The leaves are very narrow, flefhy aftringent, of a ftrong green colour, and grow together in bunches on the fides of the branches. The flowers come out from the fides of the branches in fmall bunches: They are of an herbaceous colour, appear early in the fpring, and are fucceeded by large round berries, like thofe of the Sloe-bufh, which are harfh and four to the tafte, and of a fine black colour when ripe The fruit of this fort continues on the trees all winter, making a beautiful appearance among the narrow cluftered leaves at that feafon.

7. OLIVE-LEAVED BUCKTHORN will grow to be eight or ten feet high, fending forth numerous branches, each of which is terminated by a long fharp fpine. The leaves are fmall, oblong, obtufe, undivided, veined, fmooth, of a thickifh confiftence, and grow two or three together on their own feparate footftalks. The flowers come out from the fides of the branches in the fpring. They are fmall, of a whitifh green colour; and are fucceeded by round black berries, about the fize and colour of thofe of the Common Purging Buckthorn.

Thefe Evergreen forts are to be PROPAGATED, 1 By layers. This bufinefs muft be done in the autumn, when the laft fummer's fhoots fhould be laid in the ground. Thefe will often ftrike root at almoft every joint; though they have been found in fome ftrong foils, upon examining them in the autumn, after being layered a whole year, without any roots; fo that it would be proper to give the layer a flit at the joint, and bend it fo in the ground as to keep it open; and it will have plenty of root by the autumn. Another thing to be obferved is, that in order

to

to obtain good layers, the plants defigned to be encreafed fhould be headed the year before, and this will caufe them to fhoot vigoroufly ; and from thefe fhoots the ftrongeft and beft layers may be expected ; many of which will be good plants, to fet out where they are to remain, while the weakeft may be planted in the ufual nurfery-way, to gain ftrength. 2. Thefe plants may be raifed by feeds, the variegated ones excepted, for they muft always be encreafed by layers. The feeds will be ripe in September, or the beginning of October, when they fhould be guarded from the birds, or they will foon eat them all. Soon after they are ripe they fhould be fown, for even then they will often remain two years before they come up. The beds fhould be compofed of fine light mould, and they fhould be fown an inch deep. If few or no plants appear in the fpring, you muft wait, and weed the beds with patience, until the fpring following, when you may expect a plentiful crop. Let them ftand two years in the feed-bed, with conftant weeding, and frequent watering in dry weather ; and in March let them be planted out in the nurfery, where they will be afterwards ready for removing when wanted. As thefe trees produce plenty of good feeds, by this means a prodigious quantity of plants may be foon raifed ; and thofe from feeds are always obferved to grow ftraighter, and to a greater height than thofe raifed from layers ; fo that where many of thefe trees are wanted for large plantations, the raifing them from feeds is the moft eligible method.

All the forts of *Alaternus* are very hardy, and may be planted in almoft any foil or fituation ; but the Narrow and Olive-leaved Buckthorn fhould be ftationed in a dry, warm, well-fheltered place.

R H O D O D E N D R O N.

Linnean Clafs and Order, *Decandria Monogynia* : Each flower contains ten males and one female ; There are feven Species : fix of which are here treated of :

                                         1. Rho-

RHODODENDRON *Ferrugineum* : The FERRUGINEOUS DWARF ROSE-BAY ; *a low deciduous fhrub* ; native of the Alps, Apeninnes, and other mountains of Europe.

2. RHODODENDRON *Hirfutum* : The HAIRY DWARF ROSE-BAY ; *a low deciduous fhrub* ; native of the Alps and many mountains of Switzerland and Auftria.

3. RHODODENDRON *Chamæciftus* : The CHAMÆCISTUS, or CILIATED-LEAVED DWARF ROSE-BAY ; *a low deciduous fhrub* ; native of Mount Baldus, and near Saltzburg in Germany.

4. RHODODENDRON *Dauricum* : The DAURIAN DWARF ROSE-BAY ; *a low deciduous fhrub* ; native of Dauria.

5. RHODODENDRON *Maximum* : The AMERICAN MOUNTAIN LAUREL ; *an evergreen fhrub* ; native of Virginia.

6. RHODODENDRON *Ponticum* : The PONTIC DWARF ROSE-BAY ; *an evergreen fhrub* ; native of the Eaft, and of 'moft fhady places near Gibraltar.

1. The FERRUGINEOUS DWARF ROSE-BAY is a fhrub of about two or three feet in growth. The branches are numerous, irregular, and covered with a dark-brown bark, having a tinge of purple. The leaves are of two very different colours ; the upper furface is of a fine green, but the under is of an iron colour. There will be numbers of thefe on every twig ; and they grow in a pleafing irregular manner : They are of a lanceolated figure, have their furfaces fmooth, and are little more than an inch long. Their edges are reflexed ; but they have no ferratures, and, on the whole, conftitute a great beauty when in leaf only. The flowers grow at the ends of the branches, in round bunches. Their petals are funnel-fhaped, of a pale rofe colour, appear in June, and are rarely fucceeded by feeds in England.

2. HAIRY DWARF ROSE-BAY is a fhrub of about the fame, or rather of a lower growth. The branches of this fpecies alfo are numerous, and the bark with which thev are covered is of a lightifh brown colour. They are ornamented with plenty of leaves, in an irregular manner. They are not fo large as thofe of the former fort ; but are of the fame figure, only a little more inclined to an oval. They fit clofe to the branches, and have no ferratures, but hairs on their edges like the eye-lafhes. Their under furface alfo is poffeffed of the fame fort of hairs, which

are

are all of an iron colour. The flowers will be produced at the
ends of the branches, in bunches, in May. These are also
funnel-shaped, of a light-red colour, make a good show, and
are succeeded by oval capsules, containing ripe seeds, in
August.

3. CHAMÆCISTUS, or CILIATED DWARF ROSE-BAY, will
grow to be about a yard high. The branches are numerous,
produced irregularly, and covered with a purplish bark. The
leaves are produced in great plenty, and without order, on the
branches. They are oval, spear-shaped, small, and their under
surface is of the colour of iron. The edges also are possessed of
many iron-coloured hairs, which are placed like those on the
eyelids. The flowers are produced at the ends of the branches,
in bunches. They are of a wheel-shaped figure, pretty large,
of a fine crimson colour, and make a handsome show. They
appear in June, and are succeeded by oval capsules, containing
ripe seeds, in September.

4. DAURIAN DWARF ROSE-BAY is a low shrub, sending
forth many branches covered with a brownish bark. The leaves
are broad, naked, smooth, and come out without order on
short footstalks. The flowers are wheel shaped, large, and of a
beautiful rose colour: They appear in May ; and are succeeded
by oval capsules full of seeds, which do not always ripen in
England.

All these deciduous sorts are PROPAGATED best by the seeds,
and as they grow naturally on the Alps, Apennines, and other
snowy and cold mountains, and are seldom made to grow and
flourish fair in gardens, it will be the best way for a Gentleman
who has extended his plantation, and has any part of it mountainous, hilly, or rocky, on the north side, to get some spots
well cleared of all roots and weeds ; and these being made fine
and level, let the seeds be sown therein. They will want no
covering ; a gentle patting down with the spade will be sufficient ; for the seeds are so exceeding small, that they will be
washed into the ground deep enough by the first shower of rain
that follows. Whoever is not content with sowing seeds, and
covering them no more than what they will get by being
patted down, must only lightly dust some earth over them ;
for if they are covered half an inch, the general depth for most
seeds, you must expect no crop. After the young plants come

up,

up, they muft be watered in dry weather, weeded, and in the winter protected from the frofts, which will deftroy them. And here one thing is to be obferved, that though the north fide, at the foot of or on a hill, is thought moft proper for their growth, as being moft fuitable to their nature, yet a place muft be chofen for them that has trees and hedges to fhelter them from the northern black frofts ; for thefe trees, hardy as they are, will be liable to be deftroyed by them, for want of fnow, as in other places, to cover them and keep them warm in the winter feafon. After thefe plants are come up, they fhould be thinned; and leaving only a proper number in each refpective place, and being protected for the firft two or three winters, either by mats or hand-glaffes, in the fevereft weather, they will be afterwards ftrong enough to be left to themfelves, efpecially if the places are tolerably fheltered. If a gardener has no other ground than his feminary for raifing plants, his beft method will be to prepare a compoft for thefe feeds in the following manner : Take four bufhels of earth from fome neighbouring hill,which if rocky, that neareft the furface, on which the fheep have been ufed to lie and dung, will be the beft ; but if it be of any other nature, the mould neareft the furface, mixed with the following, will do very well : Take fix bufhels of maiden earth, from a rich loamy pafture, that has been dug up with the fward, and by frequent turning is well rotted and mixed, and four bufhels of drift or fea-fand. Let thefe be well mixed together, and of this let the bed be made. The bed being made level and fine, the feeds fown, and gently patted down with the fpade, or at fartheft no other covering than being gently dufted over with the fineft mould, may be left to nature. This bed fhould be in a fhady well-fheltered place ; and the plants after they are come up fhould be weeded and watered in the fummer, and protected from frofts by mats in the winter. In the fpring they may be pricked out in beds in the nurfery-ground, at a very fmall diftance, that they may be hooped and matted if the following winter fhould prove very fevere. The fecond winter they will require no other trouble than pricking furze-bufhes round the bed for their defence ; and after that they may be fet out to ftand.

5. The AMERICAN MOUNTAIN LAUREL is a plant fo diftinguifhed becaufe, in America, it grows naturally upon the highest

highest mountains, and on the edges of cliffs, precipices, &c.
There it will grow to be a moderate-sized tree; with us it seldom rises higher than six feet. The branches are not numerous,
neither are they produced in any order. The leaves are large
and beautiful, of an oval spear-shaped figure, and a little resemble those of our Common Laurel. They are of a shining
strong green on their upper surface, though paler underneath ;
but they lose this delicacy as they grow older, altering to a
kind of iron-colour. Their edges are acutely reflexed, and they
grow irregularly on short footstalks on the branches. The
flowers are produced at the ends of the branches about Midsummer, though sometimes sooner ; before which time the buds
will be large and turgid ; and indeed, as they begin to swell
early in the autumn before, these have a good effect, and look
well all winter. When the shrub is in blow, the flowers appear close to the branches, in roundish bunches. Each is composed of one petal, which is divided at the rim into five parts,
one of which is dotted in a pretty manner. They are very beautiful, and alter their colour as they grow older ; for at first the
petal is of a very pale blush colour, which dies away to a white ;
but the outside, which is a peach colour, is not subject in so high
a degree to this alteration. They will continue, by succession,
sometimes more than two months ; and are succeeded by oval
capsules, full of seeds.

6. Pontic Rose-Bay grows to about four or five feet high,
sending forth several branches without order from the sides. The
leaves are spear-shaped, glossy on both sides, acute, and placed
on short footstalks on the branches. The flowers are produced
in clusters from the ends of the branches ; each of them is
bell-shaped, and of a fine purple colour. They appear in
July ; and are succeeded by oval capsules containing the seeds,
which seldom ripen in England.

The PROPAGATION of these Evergreen sorts must be from seeds,
which we receive from the places where they grow naturally. The
best way is to sow them very thin in the places where they are
designed to remain ; and if these places be naturally rocky,
sandy, and shady, it will be so much the better (especially for
the first sort ; the second requires a moistish soil, in a warm
shady place) ; if not, a quantity of drift-sand must be added to
the natural soil, and all made fine and level. Some spots for
the

the reception of the feeds are to be pitched on   A few feeds
fhould be put in each, and covered about half an inch deep,
and then fome fticks ftuck round them to direct to the
true places, that they may not be difturbed by hoeing the
weeds, but that thefe may be all carefully plucked up by the
hand as often as they appear ; for it will be a whole year, and
fometimes two or more, before the plants come up   This careful weeding muft always be repeated; and after the plants come
up, thofe that grow too clofe may be drawn the fpring following, and each fet in a feparate pot, and then plunged into a
hotbed, to fet them growing. The plants that remain without
removing will be the ftrongeft and beft, and will be more likely
to produce flowers than any other ; though this feems to be a
plant that will bear tranfplanting very well, efpecially if it is
not to be carried at too great a diftance for the roots to dry, and
a ball of earth be preferved to them. Whenever they are not
to be raifed and remain in the places, the beft way is to fow
them in pots filled with fandy earth, or fuch as is made fo by
at leaft a third part of fand being added. After the plants come
up, they may be planted in feparate pots the fpring following,
and then fet forward by a plunge in the bed ; and afterwards
they may be any time turned out into the places where they
are to remain, which ought to be in a naturally fandy fituation,
otherwife there will be little hopes of feeing them in any
degree of perfection.

R H U S

Linnean Clafs and Order, *Pentandria Trigynia:* Each flower
contains five males and three females. There are twenty-
four Species ; eight of which are fufficiently hardy to ftand
this climate.

1. Rhus *Coriaria :* The Tanner's Sumach, or the Elm
leaved Sumach; *a tall deciduous fhrub* ; native of Turkey,
Paleftine, Syria, Italy and Spain.

2. Rhus *Typhynum* : The Virginia Sumach ; *a deciduous shrub* ; native of Virginia.

3. Rhus *Glabrum* : The Smooth Sumach ; *a tall deciduous shrub* ; native of North-America.

4. Rhus *Coppallinum* : The Lentiscus-leaved Sumach ; *a deciduous shrub* ; native of North-America.

5. Rhus *Vernix* : The Varnish Tree, or Poison Ash Tree ; *a deciduous shrub* ; native of North-America, also of Japan.

6. Rhus *Toxicodendron* : The Toxicodendron, or Poison Oak ; *a low deciduous shrub* ; native of North-America.

7. Rhus *Radicans* : The Radicant Toxicodendron ; *a deciduous shrub* ; native of Virginia and Canada.

8. Rhus *Cotinus* : The Venetian Sumach, or Coccygria ; *a deciduous shrub* ; native of Italy, Spain, and many parts of Europe.

1. The Tanner's Sumach will grow to be about twelve feet high ; and the branches are covered with a brownish hairy bark. It is said that this bark is equal to that of the English Oak for tanning of leather, and that the leather from Turkey is chiefly tanned with it. The leaves of this shrub, which are placed alternately on the branches, have a grand look. They are pinnated, and each ends with an odd foliole. The midrib of each is garnished with about eight pairs of folioles, which all terminate with an odd one. The folioles of which the compound leaf is composed are oval, and not large, being scarcely two inches long, and three-fourths of an inch broad ; but the whole leaf makes a fine show. Their colour is a light-green ; their under surface is hairy, and they are sawed at their edges. The flowers, which are produced in large bunches at the ends of the branches, are of a whitish colour, with a tinge of green. Each is composed of many spikes, on which the flowers sit close. They come out in July ; but are not succeeded by ripe seeds in England, like some of the subsequent sorts. The leaves and seeds are possessed of many excellent virtues.

2. Virginia Sumach. Of this species there are several *varieties* ; such as, the Common Stag's Horn, Large Virginian, and Dwarf Sumach.

The

The *Stag's Horn Sumach* is so called from the younger
branches much resembling a stag's horn, called the Velvet
Horn. It will grow to be about ten feet high, and the older
branches are covered with a smooth brownish bark, in some
places of a greyish colour, whilst the younger ones are covered
with a hairy down, which much resembles the velvet horn of a
stag. The leaves have a noble look ; for they are large and
pinnated. The folioles are oblong, and larger than those of
the preceding sort : about seven pairs are stationed along the
mid-rib, which are terminated by an odd one. Their under
surface is hairy, and they die to a purplish scarlet in the
autumn. The flowers are produced in June, at the ends of the
branches : they will be in large tufts, but make no show ;
though some admire them when succeeded by seeds in the
autumn ; for at the end of that season, even after the leaves are
fallen, there will be large tufts of seeds, of a scarlet colour, left
at the ends of the branches, which have an uncommon appearance.

The *Large Virginian Sumach* differs in no respect from the
preceding species, only that it shoots stronger, and grows to be
larger, even sixteen or eighteen feet high, and is a more regular
tree. The young shoots also are of a more reddish colour ; and
though possessed of the like hairy down, on the whole do not so
much resemble those of the velvet stag's horn as the other.

Dwarf Sumach differs in no respect from the Common Stag's
Horn, except that it is of a very low growth, seldom rising
higher than three feet.

3. SMOOTH SUMACH. This includes many notable *varieties*,
commonly called New-England, Smooth Carolina, and Canada
Sumach.

New England-Sumach will grow to about sixteen feet high,
sending forth many strong shoots from the root and the sides,
covered with a smooth downy bark. The radical shoots will
often be near an inch in diameter in one summer's growth.
The young branches also from the sides will be large : they
are smooth, though a little downy in the summer ; and the
bark in the winter is of a light-brown colour. The leaves of
this sort are the largest of any, being composed of ten or

       more

more pairs of folioles, proportionally large, and which are terminated by an odd one. The flowers are produced at the ends of the branches, in large loofe panicles : They are of a greenifh-yellow colour, and come out in June, but are not fucceeded by feeds with us.

The *Carolina Sumach* 'feldom rifes to more than ten feet high. The branches are fmooth, of a fine purplifh colour, and dufted over with a whitifh powder. The leaves are pinnated like the other, and the flowers are produced in panicles at the ends of the branches. They are of a fine fcarlet colour, appear in July, and are fucceeded by bunches of feeds, which in autumn are of a very beautiful red, though they never ripen in England.

The *Canada Sumach* grows to about ten feet in height, and the branches, which are fmooth and of a purplifh colour, are dufted over, like the former, with a kind of whitifh powder. The leaves are pinnated like the other, and the folioles are on both fides fmooth ; but their furfaces are of two colours, the upper being of a fhining green, whilft the under is hoary. The flowers are red, and produced in July, in large panicles, at the ends of the branches. They appear as if a whitifh powder had been dufted in among them, which attracts notice ; but their feeds do not ripen in England.

4. LENTISCUS-LEAVED SUMACH. The chief *varieties* of this fpecies are, the True Lentifcus-leaved, and the Canada Lentifcus-leaved Sumach.

The *True Lentifcus-leaved Sumach* feldom rifes to more than four feet in height, and the branches are covered with a fmooth brown bark. The leaves alfo are pinnated, and are the moft beautiful of all the forts ; for the folioles, though fmall, are of a fhining green. There are about four or five pairs on the mid rib, which are beautifully arranged, having a membrane or wing on each fide running from pair to pair : they are terminated by an odd one, refemble in appearance thofe of the Lentifcus, and are the greateft ornaments of this fhrub. The flowers are produced in July, at the ends of the branches. They are of a greenifh colour ; and though produced in large loofe panicles, make no great figure ; neither do the feeds ripen with us.

Canada Lentifcus-leaved Sumach grows to be ten feet high. The leaves have chiefly the properties of the former, but are larger, lefs delicate, and dufted or pounced over with a whitifh matter. The flowers are produced in the fame manner as the other: they are greenifh, and fucceeded by feeds in England.

5. The Poison-Ash. This is called the Poifon-Tree becaufe it abounds with a milky poifonous juice, and is diftinguifhed by the title Poifon-Afh, becaufe the leaves fomewhat refemble thofe of the Afh-tree. It is called alfo by fome the Varnifh-tree, being the fhrub from which the true varnifh is collected. The Poifon-Afh, with us, will grow to the height of about eight feet; and the branches, which are not very numerous, are covered with a fmooth light-brown bark, tinged with red. The leaves are pinnated, and the folioles of which each is compofed confift of about three or four pairs, with an odd one. Thefe are of an oblong pointed figure, of a fine green colour, and have their edges entire. In the autumn, they die to a red or purple colour, and at that time their leaves, juft before they fall, make a charming appearance, fome being red, others purple, others between both; the colours of the footftalks and mid-ribs will alfo be various, thereby in the fame tree affording a variety of fhades. The flowers are fmall, and make no fhow: they are whitifh, and produced in May, from the wings of the branches. There will be male and female flowers on different plants; and the females are fucceeded by fmall roundifh fruit, which feldom ripens in England.

6. The Poison-Oak is a lower fhrub, feldom growing to be more than four feet high. The branches are fmooth, and of a light-brown colour. It will coft the gardener fome trouble to keep thefe plants properly, as upright fhrubs; for they will fend out fhoots from the bottom, which will naturally trail on the ground, and ftrike root. But thefe muft be conftantly taken off; for were they to be neglected a few years, a fingle plant would have fpread itfelf to fuch a diftance as to occupy a great fpace of ground, in a manner not becoming a well-or-dered fhrubbery or wildernefs. The leaves of this fhrub are trifoliate. Each foliole has a fhort pedicle to itfelf, and the common footftalk of the whole three is very long. They are

A a 3 of

of a shining green, smooth, and have their edges sometimes sinuated, though generally entire. They are roundish, angular, large, and on the whole make a good show. The flowers are of a whitish colour, are produced from the sides of the branches, in July, and are succeeded by cream-coloured berries, which growing in the autumn, and even in the winter, after the leaves are fallen, in a kind of panicles, are by many taken notice of.

There are several *varieties* of this species; some with hairy leaves, some with leaves very downy, others of fine upright growth. In other respects their difference is inconsiderable.

7. RADICANT TOXICODENDRON. Of this species there are several *varieties*; some of which are of upright growth, though the stalks of all have, more or less, a tendency to lie on the ground, and strike root at the joints. The leaves of all the sorts are trifoliate, of an oval figure, smooth, and entire. The flowers are greenish, appear in June and July, and are succeeded by roundish yellow berries, which rarely ripen in England.

8. The VENETIAN SUMACH is a shrub of about ten feet growth, and has many valuable properties to recommend it. The bark on the older branches is of a light-brown colour, whilst that on the young shoots is smooth, and of a purple hue. The leaves are nearly of an oval figure, and stand singly upon long footstalks on the branches. From these the tree receives great beauty : they are of a delightful green, are smooth, and when bruised emit a strong scent, which by many is thought very grateful; and on that account only makes this shrub desirable. The flowers are produced at the ends of the branches, in July, in a singular manner: The end of the last year's shoot about that time will divide itself, and produce hair like bunches of purplish flowers, so as to cover the tree; and in the autumn, though they do not perfect their seeds with us, these tufts will still remain, be of a darker colour, and almost cover it; on account of which singular oddness this shrub is valued by some persons. The bark is used by the tanners; whilst the wood and leaves are sought after by the dyers; the former being said to dye a yellow, and the latter, together with the young branches, to dye a good black.

The PROPAGATION of the Sumach is not very difficult; for the second, third, and fourth sorts, with their varieties, produce

suckers

fuckers in fuch plenty as to over-run, if not taken off, all that is near them. Thefe fuckers when taken up will be each a good plant; nay, their very roots will grow; and though they be thrown upon a bed and dug carelefsly in, even then many young plants will fpring from them.

The POISON-OAK and RADICANT TOXICODENDRON alfo PROPAGATE themfelves very faft by their trailing branches, which ftrike root as they go, and each of which will be a plant.

The VENETIAN SUMACH is eafily encreafed by layers; for the young fhoots being flit and layered in the autumn, by the autumn following will be good plants, either for the nurfery-ground, or where they are to be planted out to ftand.

The ELM-LEAVED SUMACH and the POISON-ASH, however, do not throw out fuckers in this manner; and thefe are to be PROPAGATED from the feeds, which we receive from the places where they naturally grow. An eaft border of garden-mould (made fine) fhould be prepared; and in this the feeds fhould be fown as foon as poffible after we receive them. The depth they will require will be about half an inch. After being fown, and the border dreffed up, nothing more need be done till the weeds begin to come up, which will be before the plants: as often as thefe appear, they muft be plucked up; and when the hot parching weather comes on, the border muft be fhaded in the heat of the day, and, every evening, fhould be gently fprinkled over with water. In the beginning of June many of the plants will come up; though they frequently remain, at leaft the greateft part of them, until the fecond fpring before they make their appearance. After the plants are come up, they will want no other care than fhading, weeding, and now and then a watering during the firft fummer; and if the winter fhould be fevere, they fhould be matted, efpecially the Elm-leaved fort, which is rather the moft tender whilft young. After this they will require no other care than weeding until they are two-years-old feedlings; when, in the fpring, they fhould be taken up and planted in the nurfery-ground, and in two or three years more will be fit to fet out for good. And here it muft not be omitted to obferve, that the other forts before-mentioned, which propagate themfelves fo faft by fuckers, may be raifed this way if the feeds can be obtained; and, indeed,

whoever

whoever has not the conveniency of procuring a few plants of each, and can have the feeds, muft practife this method with them, by which he will foon procure plenty.

R O B I N I A.

LINNEAN Clafs and Order, *Diadelphia Decandria :* Each flower contains ten males and one female ; the males being divided into two fets at the bafe : There are nine SPECIES ; four of which will bear the open air of this country.

1. ROBINIA *Pfeudacacia :* The FALSE ACACIA ; *a deciduous tree* ; native of moft parts of North America.

2. ROBINIA *Caragana :* The CARAGANA ; *a deciduous fhrub* ; native of Siberia.

3. ROBINIA *Frutefcens :* The SHRUBBY ASPALATHUS ; *a deciduous fhrub* ; native of Siberia and Tartary.

4. ROBINIA *Pygmæa* The DWARF ASPALATHUS ; *a low deciduous fhrub* ; native of Siberia.

1. The FALSE ACACIA will grow to the height of thirty-five or more feet. The branches are covered with a fmooth purplifh-coloured bark, and armed with ftrong fpines, which are placed at the buds. Each bud, efpecially of the young vigorous fhoots, will be generally guarded by two of thefe fpines, one of which will be on one fide, while the other will occupy the oppofite place. The branches are very brittle, and in fummer, when the leaves are on, are often broke by the high winds. The leaves come out late in the fpring ; but for this they make ample amends by the beautiful foliage they will difplay foon after. They are pinnated leaves, the moft beautiful of all the compound forts. The folioles of which each is compofed are of a fine green ; and as there are no lefs than nine or ten pair of them placed along the mid-rib, with an odd one, the whole leaf appears very large ; and all the tree being thus ornamented has a noble look, even at that time. But this fhrub will be in its greateft beauty when in flower ; for thefe will be
pro.

produced in long pendulous bunches, in June. They are of the papilionaceous kind ; their colour is white ; and when the tree blows freely, its head will be enchantingly covered with them ; for they will hang all over it in a free and eafy manner ; fome bunches appearing wholly in view, others again half hid by the waving leaves, that will fometimes alternately hide and fhew them ; at which time alfo, when there is a current of air, the flowers themfelves receive frefh beauty from being thus agi‧ tated. But this is not all : nature has granted them a fmell, which is very grateful ; fo that in an evening, or after a fhower, they will perfume the circumambient air to fome diftance : Thus they will prove a feaft to all thofe who will attend at thofe times, as they will never fail of regaling one of the fenfes by their grateful and profufe fragrance. Thefe flowers, it is to be lamented, are of fhort duration ; and are fucceeded by pods, which in fome feafons will perfect their feeds with us

The principal *varieties* of this fpecies are, the *Scentlefs*, *Prickly-podded*, *Rofe-coloured*, *Scarlet*, *Smooth-podded*, &c. *Acacia*.

The *Falfe Acacia with Rofe-coloured flowers* appears to have the moft material difference. Is is of lower growth ; the young branches, and the footftalks and very cups of the flowers, are armed with prickly hairs or fpines. The flowers are produced rather earlier than thofe of the other forts ; they are large, and of a moft beautiful rofe-colour. They have no odour like the others ; but make a moft noble fhow when in blow." They are fucceeded by flat pods ; and the variety is moft beautifully en-creafed by this fort. The difference of the others is chiefly pointed out by the names which are in common ufed to exprefs them.

2. CARAGANA rifes, with a fhrubby ftalk, to the height of about eight or ten feet, fending forth feveral branches, which are covered with a greenifh-yellow bark. The leaves are ab-ruptly pinnated ; the folioles are oval, fpear fhaped, pointed, and confift of about five or fix pair arranged along the mid-rib. The flowers come out from the fides of the branches, on fingle footftalks : they are fmall, of a yellowifh colour, appear in May, and are fucceeded by fmooth comprefled pods containing the feeds, which will be ripe in September.

3. The

3. The Shrubby Aspalathus is a beautiful flowering shrub. Its growth will be seven or eight feet; and the branches naturally grow upright. The bark is smooth, and of a yellowish colour; but that of the youngest twigs partakes more of a purplish colour on one side, and is on the other often of a light green with a yellow tinge. The leaves are each composed of about four folioles, which are oval and pointed. The flowers are produced in May, from the joints of the branches, upon single footstalks: they are of a fine yellow colour, and of the butterfly make; and so adorn the tree when in blow, as to render it inferior to few of the flowering-shrubs. These flowers are succeeded by pods, containing ripe seeds, in the autumn.

4. Dwarf Aspalathus is a pretty little shrub, sending forth several slender branches, which are covered with a golden bark. The leaves are quaternate, wedge shaped, obtuse, have no footstalks, and, unless very severe weather happens, continue on the plant the greatest part of the winter. The flowers come out from the sides of the branches, on single footstalks: they are small, of a yellow colour, appear in May, and are succeeded by ripe seeds in the autumn.

The Propagation of all these sorts is very easy, and may be done, 1. By seeds. If these are sown the beginning of March, half an inch deep, in a bed of any common garden-mould, plants will come up in May, which will want no other care than weeding all the first summer, and no protection of any kind in the winter; for they are all hardy enough. In the following spring they should be planted out in the nursery-ground, a foot asunder, and two feet distant in the rows; and here (the three first sorts) they should not stand longer than two or three years before they are set out to stand, as they will grow exceeding fast, and by that time will be perhaps six feet in height. The fourth sort being of lower growth, the plants may be pricked in beds, a foot asunder, which will be room enough for them to grow in, before they be finally set out. It may not be amiss to observe also, that the seeds of this sort often remain until the second spring before they come up; so that when they do not appear the first after sowing, the beds must be kept weeded all summer; and, if the seeds were good, there will be no fear of a crop the following spring. 2. These sorts are easily propagated by cuttings, which if planted in October, in a
moistish

moiftifh fhady border, many of them will grow. Here they fhould ftand two years, when they will be proper plants to be planted out ; though we muft obferve, that the fourth fort may remain longer before they are fet out ; and as the cuttings of that fort have often failed growing, the moft certain method, and what is generally practifed when there are no feeds, is to encreafe it by layers. 3. The firft forts will encreafe themfelves by fuckers, in fufficient plenty ; for the old plants will fpawn at a confiderable diftance, and afford fuch a quantity of free-fhooting fuckers, that they will be all good plants, fit to be fet out for continuance.

R O S A.

Linnean Clafs and Order, *Icofandria Polygynia:* Each flower contains about twenty males and many females : There are eighteen Species ; thirteen of which we here enumerate :

1. Rosa *Canina :* The Dog Rose, or Hep Tree ; *a deciduous fhrub ;* common in our hedges, and moft parts of Europe.

2. Rosa *Pimpinellifolia :* The Burnet-Rose, or Cat-Whin ; *a deciduous fhrub ;* natural to England and moft parts of Europe.

3. Rosa *Spinofiffima :* The Scotch Rose ; *a deciduous fhrub ;* native of Scotland, England, and moft parts of Europe.

4. Rosa *Alpina :* The Alpine Rose ; *a deciduous fhrub ;* native of the Alps of Switzerland.

5. Rosa *Eglanteria :* The Eglantine, or the Sweet-Briar ; *a deciduous fhrub ;* native of England and Switzerland.

6. Rosa *Cinnamomia :* The Cinnamon Rose ; *a deciduous fhrub ;* grows in the fouthern parts of Europe.

7. Rosa *Carolina :* Carolina Rose ; *a deciduous fhrub ;* native of North America.

8. Rosa *Villofa :* The Apple Rose ; *a deciduous fhrub ;* native of moft parts of Europe.

9. Rosa

348

ROS

9. ROSA *Centifolia* : The HUNDRED LEAVED ROSE ; *a decid-uous shrub* : it is not known where this Rose grows na-turally.

10. ROSA *Gallica* : The GALLICAN ROSE ; *a deciduous tree* ; grows naturally in most parts of Europe.

11. ROSA *Sempervirens* : The EVER-GREEN ROSE, or MUSK ROSE ; *an ever-green shrub* ; native of Germany.

12. ROSA *Pendulina* : The LONG-FRUITED ROSE ; *a deci-duous shrub* ; native of Europe.

13. ROSA *Alba* : The WHITE ROSE ; *a deciduous shrub* ; native of Europe.

1. The DOG-ROSE grows all over England, and is seldom cultivated in gardens. It is, nevertheless, possessed of many beauties, if observed with due attention ; and, if it was not so very common, would deserve a place in the choicest Collection.

The *varieties* of this species are, the Hep-tree with Red Flowers, the White-flowered Hep-tree.

2. BURNET-ROSE is a small-growing tree, seldom rising higher than one yard. The flowers are single, and make no great figure ; but what renders this Rose valuable is, that the leaves are pinnated in such a manner as to resemble those of the burnet, which occasions its being so called, and by which it constitutes. an agreeable variety among the leafy tribe.

The *varieties* of it are, Red-flowered Burnet-leaved Rose, Black Burnet-leaved Rose, White Burnet-leaved Rose.

3. SCOTCH ROSE. The *varieties* of this species are all of low growth, and known by the respective names of,
   Dwarf Scotch with a White Flower,
   Dwarf Scotch with a Red Flower,
   Dwarf Scotch with a Striped Flower,
   Dwarf Scotch with a Marbled Flower.
They are all beautiful flowering shrubs. The White flowering sort will grow to the highest size, as it will commonly grow to be three feet, whilst the others seldom rise to above two feet in height. The branches are upright and numerous, and smartly set off by their beautiful pinnated leaves ; for the leaves of these sorts excel those of all other Roses in delicacy, the folioles be-ing small, of a good green colour, and arranged along the mid-
rib

rib in the manner of thofe of the burnet. The flowers will be produced from the branches in vaſt profuſion ; and though they are all ſingle, they make a ſhow inferior to few ſhrubs. In winter they will be full of heps that have the appearance of black-berries ; and if the weather be mild, the young buds will ſwell early, and appear like ſo many little red eyes all over the ſhrub, which is a promiſe of the reviving ſeaſon. The young branches of all theſe forts are exceeding full of prickles.

4. ALPINE ROSE. This is uſually called the Roſe without Thorns, the branches being perfectly free from all kinds of prickles. They are exceedingly ſmooth, of a reddiſh colour, and look well in winter. The flowers are ſingle, and of a deep-red colour. They come out in May, before any of the other forts ; and the plant is valued by ſome people on that account. They are ſucceeded by long narrow heps, which look ſingular, and, together with the early appearance of their flowers, and their beautiful twigs, that are wholly free from the armature of the other forts, cauſe this ſpecies to be much admired.

5. EGLANTINE, or Sweet-Briar. The *varieties* of this ſpecies are, Common Sweet-Briar, Semi-double Sweet-Briar, Double Red Sweet-Briar, Maiden Bluſh Double Sweet-Briar, Sweet-Briar with Yellow Flowers.

The *Common Sweet-Briar* is well known all over England. The branches, which are of a reddiſh caſt, are all over cloſely armed with prickles ; the flowers are ſingle, and of a pale-red colour, like thoſe of the Common Wild-Briar. The leaves conſtitute the value of this plant ; for they are poſſeſſed of ſo grateful an odour, as to claim admittance for this fort into the firſt claſs of aromatic plants : the odoriferous particles they emit are ſweet and inoffenſive ; and they beſtow them in ſuch profuſion, eſpecially in evenings or after a ſhower, as to perfume the circumambient air to a conſiderable diſtance. For this reaſon, plenty of Sweet-Briars ſhould be planted near much-frequented walks ; or if the borders of theſe are deſigned for more elegant flowering ſhrubs or plants, they may be ſtationed at a diſtance, out of view, and then they will ſecretly liberally beſtow their ſweets, to the refreſhment of all. For noſegays, alſo, there is nothing more proper than ſprigs of the Sweet-Briar, when diveſted of its prickles ; for they will not only have a good look-

as

as a fine green in the center of a pofy, but will improve its odour, let the other flowers of which it is compofed be what they will.

Semi-double Sweet-Briar differs in no refpect from the Common, only that the flowers confist of a double feries of petals that furround the stamina. The leaves are poffeffed of the fame fragrance ; but this fort is thought more valuable on account of the flowers, which, being poffeffed of more petals, make a better figure.

Double Sweet-Briar. The number of petals are fo multiplied in this fort as to form a full flower ; and it feems to differ in no other refpect from the other Sweet-Briars. The flowers are red, and fo large and double as to be equal in beauty to many of the other forts of rofes. As by the fragrance of their leaves they afford us a continual treat during the fummer months, as well as by their fair flowers at the time of blowing, all who pretend to make a Collection are careful of procuring plenty of this fort.

Double Blush Sweet-Briar is a moft valuable, and at prefent a very fcarce, plant. It feems to have a tendency not to grow fo high as the other forts of Sweet-Briars. The branches are green, and clofely armed with ftrong prickles. The flowers are of a pale-red or blufh colour, and every whit as double as the Cabbage-Provence-Rofe : it cabbages in the fame manner, and is very fragrant. No one need be told the value of a Rofe which has every perfection and charm, to the higheft degree, both in the leaves and flowers, to recommend it.

Sweet-Briar with Yellow Flowers. The flowers of this fort are fingle ; the petals are of a bright-yellow colour ; but it differs in no other refpect from the Common Sweet-Briar.

6. CINNAMON-ROSE. The *varieties* of this fpecies are, Single Cinnamon-Rofe, Double Cinnamon-Rofe.

The *Single Cinnamon-Rofe* is a much ftronger fhooter than the Double fort, which is better known. It will grow to be ten or twelve feet in height. The young branches are of a reddifh colour. The flowers are fingle, and have the fame hue as thofe of the Double. It is rather a fcarce plant at prefent ; on which account chiefly it is thought valuable.

The

The *Double Cinnamon-Rose* will grow to about fix or feven feet high, and the branches are many and flender. The prickles are pretty numerous, and the young fhoots in winter are of a red colour, with a purplifh tinge. This fort, which ufhers in the flowery tribe of Double Rofes, will be in blow fometimes pretty early in May. The flowers are fmall, but very double : they are of a purplifh red, very fweet, and have a little of the fmell of cinnamon, which occafions this Rofe to be fo called ; and on that account only, not to mention their early appearance, this fort is defirable.

7. CAROLINA ROSE. The *varieties* of this fpecies are ufually called, Wild Virginian Rofe, Pennfylvania Rofe, Pale-red American Rofe.

The *Wild Virginian Rose* will grow to be nine or ten feet high. The branches are covered with a fmooth red bark, and guarded by a very few prickles. It produces its flowers in Auguft, when moft of the other forts are out of blow, and is by many valued for that reafon. The flowers are fingle, of a red colour, are produced in clufters, and will continue blowing from the beginning of Auguft until October. Neither is this the fole beauty this fort affords us ; for the flowers will be fucceeded by heps, which in winter appear like fo many red berries all over the fhrub. Thefe heps ferve as food for birds, and are therefore much frequented by thrufhes and others of the whiftling tribe, who will be ready to ufher in, by their fweet warbles, the earlieft dawn of fpring. This tree grows wild in Virginia, and many parts of North-America, from whence we receive the feeds, and propagate it not only on fome of the above accounts, but becaufe it is naturally an upright well growing tree, and makes a good figure in winter by its red and beautiful fhoots.

The *Pennfylvania Rofe* feems to differ in nothing from the former, except its fize, it feeming to be a plant of lower growth ; and the Pale-red fort occafions variety only from the lobes of the flowers.

8. APPLE-ROSE. This fpecies is a curiofity, not fo much from the fingularity of the fhoots, leaves, or flowers, as fruit. The fhoots, indeed, will be ftrong and bold, and in winter diftinguifh the tree from others by a degree of eminence. They

are

are then covered with a smooth reddish bark ; and the prickles which guard them are thinly placed, though those are very strong and sharp. Many think this tree has a good look in winter, and value it much on that account. As to the leaves, they are nearly the same as the other sorts of Roses ; but are large, and very hairy and downy underneath. The flowers are single, of a red colour, and are succeeded by heps as large as little apples. To their account the value chiefly of this sort is to be placed ; for being thus large, they occasion a singular look ; and this is heightened by being all over beset with soft prickles. For use as well as beauty this sort is propagated by some ; for these heps or fruit, when preserved, make a sweetmeat greatly esteemed.

9. HUNDRED-LEAVED ROSE. This is a very extensive species, and includes all *varieties* whose stalks are hispid, prickly, and have leaves growing on footstalks which are not armed with prickles ; and whose flowers have oval, hispid germina and footstalks. Of this kind are, the Deep Red Provence, the Pale-red Provence, the Large Cabbage Provence, the Dutch Provence, the Childing Provence, the Moss Provence, the Great Royal Rose, the Blush hundred-leaved Rose, the Dutch hundred-leaved Rose,

The *Provence-Roses* are all well known. The Red and the Pale Provence sorts differ, in that one is a deep, the other a pale red ; the petals are larger and looser than the Cabbage-Provence, and make varieties. The Cabbage-Provence is the best of all the sorts ; and, if its commonness does not detract from its value, is inferior to no Rose. The Dutch Provence has a tendency to cabbage, and is of a deeper red than the Common Provence. The Childing is of lower growth than any of the other sorts, seldom growing to be more than four feet ; it is naturally of upright growth, and the bark is brown and prickly. The flowers at first are globular, though they will afterwards open at top, and display their petals folded a little like those of the Belgic. All these are beautiful roses, and greatly ornamental either to shrubberies or gardens.

The *Moss-Provence* is a sort that has been sought after of late more than any of the others. Its branches are of a dusky brown, and they are all over closely beset with prickles. The flowers
are

are like thofe of the Common Provence; though they have a
ftronger footftalk, and grow more upright. About the calyx of
the flower grows a kind of mofs, which is of a yellowifh-green
colour, and by which it will be wholly furrounded. This Rofe
has not been many years known in England, and from whence
it was firft brought is uncertain. It feems to owe its excellence
to the moffy fubftance growing about the footftalk and calyx
of the flower; but were this as common as the other forts of
Provence-Rofes, that would be looked upon as an imperfection;
for though this flower naturally is poffeffed of the fame agreeable
fragrance as the other Provence-Rofes, yet this moffy fubftance
has a ftrong difagreeable fcent, and is poffeffed of a clammy
matter.

Great Royal Rofe is one of the largeft, though not the
compacteft, Rofes we have. It will grow to be eight or nine feet
high. The branches are brown, and have a number of prickles.
The flowers are red, and poffeffed of a very grateful odour, and
the petals very large. Upon the whole, this is a fort very much
coveted, and is one of the beft Rofes in England.

The *Blufh and Dutch Hundred-leaved Rofes* differ in no refpect,
only that the flowers of one are of a paler red than thofe of the
other; and both thefe forts may contend for the prize of beauty
with any of the Rofe tribe. They feldom grow more than four
feet high. The branches are green and upright, and have very
few fpines. The flowers are large, and exceedingly double:
Each is compofed of numerous fhort petals, which are arranged
in fo regular a manner as to form a complete flower; and it is
on account of the extraordinary number of thefe petals that this
Rofe takes the name of Hundred-leaved Rofe. We feem to do
injuftice to this Rofe, when we do not pronounce it the faireft of
the whole lift; but when we reflect on the furpaffing delicacy
and beauty of many other forts, we are obliged to give the
preference to none.

10. GALLICAN ROSE. Under this title are arranged all
thofe Rofes whofe branches and footftalks of the leaves are hifpid
and prickly, and whofe flowers have oval, hifpid germina, and
grow on hifpid footftalks. Of this kind are,

The Semi-double Red Rofe,

The Old Double Red Rofe,

The *Rofa Mundi*, or Variegated Rofe,

B b The

The York and Lancafter Rofe,
The Semi-double Velvet Rofe,
The Full-double Velvet Rofe,
The Blufh Belgic Rofe,
The Red Belgic,
The Blufh Monthly,
The Red Monthly,
The White Monthly,
The Striped Monthly,
The Red Damafk,
The White Damafk,
The Blufh Damafk,
The Doubled Virgin,
The Marbled,
The Great Spanifh,
The Yellow Auftrian Rofe,
The Copper-coloured Rofe,
The Double Yellow,
The Franckfort Rofe.

11. The Musk-Rose. The Ever-green fort is naturally a climbing plant, but if planted fingly will form itfelf into a bufh of five or fix feet high: its flowers are fingle, white, and fragrant.

Befides the Evergreen, there are two deciduous *varieties* of this fpecies, called, the Single Mufk-Rofe, and the Double or Semi-double Mufk-Rofe.

Single Mufk, or White Clufter, is a fcarce and valuable Rofe. The young fhoots are covered with a fmooth green bark, and are not poffeffed of many fpines; thofe few they have are very ftrong, and of a dark-brown colour. This fort produces its flowers in Auguft, in very large clufters; they are of a pure white; and the tree will continue to exhibit its fucceffion of flowers until the froft puts a period to the blowing. The ends of the branches are frequently killed by the frofts in the winter; fo that early in the fpring they fhould be gone over with the knife, and all dead wood taken off, which would have an ill look, amongft the healthy leaves and young fhoots.

The *Semi-double and Double Mufk,* or White Clufter-Rofes,

are

are late-flowering forts. They will begin blowing in Auguſt, and continue ſo till the froſt puts an end to the glories of that ſeaſon. The ſtalks are covered with a ſmooth green bark, which will be armed with a few very ſtrong, brown, crooked ſpines. The flowers are of a pure white, and produced in large cluſters, at the ends of the branches. Theſe at preſent are not common, and are much coveted by the curious.

12. *Pendulous-fruited Roſe* grows only to about five or ſix feet high, ſending forth ſeveral hiſpid branches from the bottom to the top. The leaves are compoſed of many oval folioles, arranged along the mid-rib, and their footſtalks have few or no prickles. The flowers have oval, ſmooth germina, grow on hiſpid footſtalks, and are ſucceeded by long pendulent fruit, full of ſeeds.

13. The WHITE ROSE. The characteriſtics of this ſpecies are, the ſtalks and footſtalks of the leaves are prickly, the flowers have oval ſmooth germina, and grow on hiſpid footſtalks. Of this kind are,

The Double White Roſe,
The Semi-double White,
The Dwarf White,
The Maiden's Bluſh Roſe.

All the ſorts of Roſes are to be PROPAGATED, 1. By layers. For this purpoſe, in order to obtain plenty of them, a ſufficient number ſhould be planted for ſtools; and after theſe have been planted a year or two, they ſhould be headed near the ground, which will make them throw out plenty of young ſhoots. In the autumn, theſe ſhould be layered in the ground. The beſt way to do it is by a ſlit at the joint, though a gentle twiſt will often do as well, particularly for all the ſorts of Monthly Roſes, Damaſk-Roſes, and Sweet-Briar, which will readily take if the bark be juſt broke, and will often ſend forth roots at every joint by the autumn following. Moſt of the other ſorts do not ſtrike root ſo freely; ſo that amongſt them, by the autumn, after layering, few will be found ſtrong enough, and with root ſufficient, to be planted out to continue. However, in general, they will have roots, and oftentimes very good ones. In the autumn every layer muſt be taken up, the ſtools neated up, and a freſh operation performed on the young ſhoots that may have ſhot the preceding ſummer. The layers that have

been taken up fhould be planted in the nurfery, at no very
great diftance, and the forts fhould be kept feparate and
booked, number-fticks being made to the feparate forts, that
they may be diftinctly known. The Mofs-Provence and the
Mufk-Rofes do not ftrike root fo freely by layers ; neither does
the Apple-bearing Rofe ; fo that for all thefe forts you
muft often wait two years before you take off the layers from
the ftools, and fometimes longer ; which is the reafon of thefe
plants being rather fcarce, they not being to be expeditioufly
propagated in plenty. 2. Thefe trees may be propagated by
fuckers, which moft of the forts have a natural tendency to
throw out ; and thefe may be taken up, and the ftrongeft and
beft rooted fet out to ftand, whilft the weakeft may be planted
in the nurfery for a year or two, to gain ftrength. But here
we muft obferve, that the Mofs-Provence, Mufk and Apple-
bearing Rofes feldom throw out fuckers; fo that we muft not
wait for them from thefe forts, but muft get forward with our
layering. 3. The Common Sweet-Briar is to be propagated
by feeds. Thefe fhould be fown as foon as they are ripe, in a
bed of common garden-mould made fine. They generally remain until the fecond fpring before they come up, and afterwards will require no other care than weeding until the fpring
following, when they may be taken up, and planted in the
nurfery at fmall diftances; and in two or three years time
they will be good plants for the fhrubbery, wildernefs, or
hedges. And indeed as great quantities of thefe odoriferous
plants are often wanted, this is the eafieft and moft expeditious
way of raifing them in plenty.

By feeds alfo the Burnet-leaved, Apple-bearing, and Red
or White Scotch Rofes may be raifed ; which are doubtlefs diftinct fpecies, and will preferve the forts by feeds.

R U B U S.

Linnean Clafs and Order, *Icofandria Polygynia :* Each
flower contains about twenty males and many females : There
are

are eighteen SPECIES; four of which are applicable to our purpofe :

1. RUBUS *Fruticofus:* The COMMON BRAMBLE ; *a well-known trailing plant* ; common in moft countries in Europe.

2. RUBUS *Hifpidus:* The CANADA BRAMBLE ; *a trailing plant* ; native of Canada.

3. RUBUS *Cæfius:* The DEWBERRY or CÆSIUS ; *a trailer* ; native of moift places in moft parts of England and Europe in general.

4. RUBUS *Odoratus :* The VIRGINIA RASPBERRY ; *a deciduous fhrub* ; native of Virginia and Canada.

1. The COMMON BRAMBLE admits of the following *varieties :*

The Double-bloffomed Bramble, the Bramble without Thorns, the Bramble with White Fruit, the Cut-leaved Bramble, the Variegated Bramble.

The *Double-bloffomed Bramble* differs in no refpeſt from the Common Bramble, only that the flowers are very double. The ftalks, like that, are clofely armed on all fides by ftrong crooked prickles, that turn backwards. They are, like that, channelled; and in the winter have fome of a reddiſh-purple colour, others green, fome red on one fide and green on the other. The leaves alfo are fhaped like the hands, and are compofed fometimes of three, fometimes of five lobes. They have their upper furface fmooth, and of a fine green colour, whilft their under is of a whitiſh colour. The footftalks that fupport them are prickly, and a feries of prickles ate arranged all along the midrib of each lobe. They continue on the plants moſt part of the winter, at the beginning of which they are green ; but after Chriftmas they turn brown, and feldom look well after. This is the defcription of the Common Bramble, and of the Double fort alfo, which differs in no other refpeſt than in the doublenefs of the flower. They are produced in the fame manner at the ends of the fhoots, each of which is exceeding double. The petals are whiter ; and as a profufion of thefe ornament the ends of moft of the fhoots in the fame manner as the flowers of the Common fort, they make a fhow, and are beautiful beyond expreffion. It may be

kept down and confined, to have the appearance of a flowering-
fhrub. The flowers are fucceeded by no fruit. It will thrive
and flower exceedingly well under the drip of trees; fo that
for old plantations, this is a ufeful plant for the under fhrubs,
as it will flourifh where hardly any thing elfe will grow.

Bramble without Thorns, is not near fo ftrong a fhooter as the
Common Bramble, the fhoots being more trailing and flender,
perfectly fmooth, and of a blueifh colour; and on this account
it is that this plant is held as a curiofity. A curiofity, indeed,
it is; and many have expreffed their agreeable furprize to find
a Bramble that they could familiarly handle without hurt. The
leaves of this fort have a blueifh tinge, and the footftalks and
mid-rib are entirely free from prickles. It flowers in the
fame manner as the Common Bramble, though the flowers are
rather fmaller; and are fucceeded by black berries, on which
the infects do not feem to fwarm in fuch plenty as they do on
the other fort.

Bramble with White Fruit is deemed curious only on that account, and has (fays HANBURY) often given occafion to a hearty
laugh, by a bull which has been made by many on their firft
feeing this fruit, who have cried out with furprize, "Here is
a Bramble that bears white black-berries." It is, therefore,
the colour of the fruit that makes this fort coveted, though the
leaves are of a lighter green than any of the other forts, and
on that account make a variety among the leafy tribe.

Bramble with Cut Leaves differs from the Common only in
that the leaves are cut in an elegant and beautiful manner. It
affords a variety in no other refpect; and thofe that are fond
of fuch, are fure of meeting one in this, whofe leaves being
thin and elegantly cut, make the plant have a different look
from the other forts.

Variegated Bramble differs in no refpect from the Common
Bramble, only it is a weaker plant. The leaves are ftriped;
and it is valuable only to thofe who are fond of variegated
fhrubs.

2, AMERICAN BRAMBLE. The fhoots of this fpecies are
long, ligneous, procumbent, rough, and hairy. The leaves
are trifoliate, naked, cut at the edges, ferrated, and grow on
hifpid

hifpid footftalks. The footftalks of the flowers alfo are hifpid. They come out from the ends and fides of the branches, in July and Auguft ; and are fucceeded by round reddifh fruit in the autumn.

3. CÆSIUS, Small Bramble, or Dewberry-bush. The ftalks of this fort are weak, flender, prickly, and trailing. The leaves are trifoliate, large, and ufually of a dufky-green co-lour. The flowers are whitifh, come out from the ends and fides of the branches, in July and Auguft, and are fucceeded by large blue fruit, which will be ripe in the autumn, and of which an excellent wine is made.

All thefe forts may be PROPAGATED by cuttings. They fhould be planted in the autumn, in a fhady border, and by the autumn following they will be fit to remove. But as a crop from cuttings often fails, the beft way will be to throw fome mould over the fhoots, as they ftrike in the fpring; and when they have fhot two or three feet farther, cover them afrefh, and fo on all fummer. By this means, thofe parts that were firft covered will have either ftruck root, or they, together with all the others, will be preparing to ftrike root : fo that being cut into lengths, and the parts before covered planted again in earth, and about three or four inches of the uncovered part being above ground, almoft every one of the cuttings of this nature being thus prepared will grow, and thus plenty of plants may be foon obtained.

4. The VIRGINIA RASPBERRY. All the forts of Rafp-berries are fpecies of *Rubus*, and are propagated for their fruit; but this fort is cultivated folely to mix with our flowering fhrubs. It rifes from the ground like the Common Rafpberries, though it will naturally grow higher ; but its growth is either higher or lower in proportion to the nature of the land or fituation, as it will grow higher by two or three feet in a deep, rich, moift foil, than it will in a foil of the oppofite nature. The ftalks are of a brown colour, and wholly without prickles ; and the ftrongeft will divide into feveral fmaller branches. The leaves are exceedingly large for a fhrub of that height ; from whence the plant derives no fmall beauty. They are broader than they are long, and of a fine green on both fides, the upper being of a dark, the under of a lighter colour. Each is divided into an

B b 4.   uncertain

uncertain number of lobes, which are ferrated, and end in
acute points. Thefe leaves grow alternately, on footftalks that
are of a proportionable length and ftrength to the fize of the
leaves, they being often eight or nine inches broad, and feven
or eight in length. The flowers are produced in July, in plenty,
at the end of the ftalks; and the fucceffion will be continued
for often more than two months; though they are always the
moft beautiful on their firft appearance. They are of a purplifh
red, a colour which is very defirable at that time, when moft
of the other fhrubs that are in blow will have yellow flowers.
Each ftands on a long footftalk; and many of them being collected into a kind of loofe bunch, they make a tolerable
figure. They are feldom fucceeded by any fruit with us; and
when this happens, it is of no flavour, and on that account of no
value.

It is eafily PROPAGATED from the fuckers, which it fends forth
in fuch abundance, that from a few plants, in a few years,
almoft any defired quantity may be obtained: nay, fo faft do
they creep and fend forth ftalks on all fides, that, unlefs they
are conftantly taken up as they grow, they will foon overfpread
and choak all fmaller plants that grow near them. The beft
time for taking off the fuckers is the autumn; though they
will grow very well if planted either in the winter or fpring.

R U S C U S.

LINNEAN Clafs and Order, *Diœcia Syngenefia*: Male flowers
containing three ftamens, and female flowers containing one
piftil; upon diftinct plants: There are four SPECIES:

1. RUSCUS *Aculeatus*: The COMMON BUTCHER's BROOM;
an evergreen fhrub; native of England, Italy and France.

2. RUSCUS *Hypophyllum*: The BROAD-LEAVED BUTCHER's
BROOM; *an evergreen fhrub*; native of Italy.

3. RUSCUS

3. Ruscus *Hypoglossum* : The Hypoglossum ; *an evergreen shrub* ; native of Italy and Hungary.

4. Ruscus *Racemosus:* The Alexandrian Laurel ; *an evergreen shrub* ; native place not known.

1. The Common Butcher's Broom will rife with tough, ligneous, ftreaked, green, fpreading ftalks, to about a yard in height. Thefe proceed from a large, white, tender, creeping root, which will, if the plant has remained long, be found very deep in the ground. The leaves are of an oblong figure, of a dark dufky-green colour, and grow alternately on the ftalks. Their edges are entire ; they are of a thick ftiff confiftence ; and their points are prickly, and as fharp as needles. The flowers grow on the middle of the upper furface of the leaves, and will be ripe in June. They are fmall and greenifh ; and the females are fucceeded by large beautiful red berries, of a fweetifh tafte. This plant is of great ufe to the butchers, who gather it to make different befoms, both for fweeping of their fhops and cleaning of their blocks ; from whence it has the appellation of Butcher's Broom. The young tender fhoots of this fhrub, in the fpring, may be eaten like hop-tops or afparagus, and fome people are very fond of them. The feeds and roots are much ufed in medicine.

2. The Broad-leaved Butcher's Broom has large white roots, with long thick fibres, and from thefe rife pliable ftalks, which will grow to be near a yard high. Thefe ftalks are of a very fine green colour, and are very tough and nu.merous. They produce their leaves in an alternate manner, are of a very fine fhining-green colour, and of a thick confiftence. They are longer and broader than the other fort ; their figure is oval, and they end in acute points. The flowers of this fort grow on the under furface of the leaves, near the middle. Thefe are fmall, and of a greenifh white. They are produced in July ; and the feeds that fucceed them are fmall and red, and will be ripe in winter.

3. The Hypoglossum is the loweft of all the forts, as the ftalks feldom get to above a foot high, and has very few pretenfions, indeed, to be called a fhrub ; neverthelefs, it may juftly claim a place at the edge at leaft of all evergreen fhrubberies. The roots are nearly of the fame nature with the other forts, and the

ftalks

ftalks are numerous and pithy. They are of a dull-green colour, and ftriated ; and they produce their leaves in an irregular manner, being fometimes alternate, whilft others again
may be feen ftanding oppofite by pairs. Thefe leaves are of a
lanceolated figure, and are of the fame dull-green colour with
thofe of the ftalks. They are from three to four inches long,
and about one broad. They grow without any footftalks,
being narrow at both ends, and their edges naturally turn towards the center of the upper furface. They are free from
ferratures ; and from the ftalk or bafe of the leaves run feveral
veins the whole length, which gradually diverge from the
middle, but approach again in the fame manner until they all
end in the point of the leaf. Each of thefe leaves produces
another fmall leaf of the fame fhape, from the middle of its
upper furface; and from the bottom of thefe fmall leaves are
produced the flowers. Thefe will be ripe in July, are fmall
and yellowifh ; and the fruit that fucceeds them is large and
red, and will be ripe in winter.

4. The ALEXANDRIAN LAUREL has the fame kind of
white fcaly roots with long thick fibres as the others, and the
branches are very numerous and pliable. They are fmooth
and round, of a fhining-green colour, and produce others
fmaller, alternately from the bottom to the top. They will
grow to be four or five feet high, and their pliable branches are
neverthelefs brittle near the bottom. The leaves grow chiefly
on the fmaller fide-fhoots, and on thefe they are placed alternately. They fit clofe to the branches, are fmooth, of a delightful fhining-green colour, and have feveral fmall veins running the whole length, diverging from the middle, but approaching again to end at the point. They are from two to
three inches long, and about one broad, are of an oblong
lanceolated figure, and end in very acute points. The flowers
are produced in long bunches, at the ends of the branches.
Each of them is fmall, and of a yellowifh colour ; and they
are fucceeded by large red berries, which will be ripe in
winter.

There is a *variety* of this fort with red flowers. " This fpecies
of *Rufcus*," fays HANBURY, " is fuppofed to be the Laurel
which compofed the wreaths worn by the ancient victors and
poets ; and indeed with good reafon, not only on account of
its pliablenefs, by which it might be eafily wrought for fuch

pur-

purpofes, but the wreaths on the ancient bufts, &c. feem to figure to us the leaves and flender branches of the plant we are treating of."

There is another fort of *Rufcus*, which has oval acute-pointed leaves, growing by threes round the ftalks, and which produce the flowers and fruit from the mid-rib on the under furface ; alfo another fort, with oval acute-pointed leaves which produces the flowers from the mid rib, on the upper furface. But as thefe are only varieties of the above forts, have the fame kind of roots, produce the fame kind of flender pliable branches, and have their flowers fucceeded by nearly the like kind of berries, nothing more need be faid of them.

All thefe forts may be eafily PROPAGATED. 1. After having obtained a plant or two of each, their roots will encreafe fo faft, and will proportionably fend forth fuch a quantity of ftalks, that each of them will foon form itfelf into a little thicket: thefe, then, are to be taken up and divided ; and from one original root or off-fet many will be foon produced. The beft time for this work is early in the autumn ; though they will grow very well if divided and removed in the fpring, or any time in the winter. 2. Thefe plants are alfo to be encreafed by feeds. This, however, is a flow way ; but muft, neverthelefs, be practifed, when the plants cannot be obtained. The beds for their reception muft be made fine, and cleared of the roots of all weeds. They will require no other compoft than that of good common garden mould. They fhould be fown an inch and a half or two inches deep, and the beds fhould be neated up to lie undifturbed, for they will not come up before the fecond, and fometimes the main crop the third, fpring after fowing. All the fummer they fhould be kept clean of weeds ; and if the beds wear away fo as to endanger the feeds being laid bare, a little fine mould fhould be riddled over them, to fupply what may be loft by wear in weeding, fettling, &c. After they are come up, they will require no other care than weeding, for they are very hardy ; and when they come too thick in the fpring after the frofts are over, the ftrongeft fhould be drawn out and planted in beds fix inches afunder. This will make room for the others to flourifh ; and though mention is made of removing thefe plants after the frofts are over, it is not becaufe they are tender and fubject to be deftroyed by it, but if they are removed in the autumn, or early in the winter, being

then

then small, the frosts generally throw them out of the ground, to the great danger, if not entire lofs, of the whole stock of the new-removed feedlings. This, however, is confidered by few Gardeners who have not paid dear for their experience, and is what is chiefly recommended by our modern authors, to tranf-plant feedlings of moft forts from the beds in October; which indeed would be an excellent month, were no frofts to enfue. But good thought and experience, by fatal práctice, have taught the Gardener now, to defer the removing his fmall feedlings until the fpring, when they will not be liable to be turned out of their warm beds when they fhould leaft like it, by the ri-gours of the winter. But to return: After the feedlings are two or three years old, whether they have been removed or not, they will by that time be good ftrong plants, fit for removing, and may be then taken up and planted out.

S A L I X.

Linnean Clafs and Order, *Diœcia Decandria*: Male flowers containing two ftamens, and female flowers containing one piftil; upon diftinct plants: There are feveral Species; fifteen of which are cultivated in this country:

1. Salix *Alba*: The Common White Willow; *a deci-duous tree*; common about towns and villages in moft parts of Europe.

2. Salix *Vitellina*: The Golden Willow; *a decidu-ous tree*; native of England and moft parts of Europe.

3. Salix *Purpurea*: The Purple Willow; *a deci-duous tree*; native of England and the fouth of Europe.

4. Salix *Pentandria*: The Sweet Willow; *a decidyous tree*; native of mountainous and marfhy fwampy grounds in moft parts of Europe.

5. Salix *Babylonica*: The Weeping Willow; *a deci-duous tree*; native of the Eaft,

6. SALIX *Hermaphroditica:* The SHINING WILLOW; *a deciduous tree;* grows about Afton in Cumberland, and alfo Upfal in Sweden.

7. SALIX *Triandria:* The TRIANDROUS WILLOW; *a deciduous tree;* native of Switzerland and Siberia.

8. SALIX *Phylicifolia:* The PHYLICA-LEAVED WILLOW; *a deciduous tree;* native of the north of Sweden.

9. SALIX *Amygdalina:* The ALMOND-LEAVED WILLOW; *a deciduous tree;* native of England and moft parts of Europe.

10. SALIX *Haftata:* The HASTATED WILLOW; *a deciduous tree;* native of Lapland and Switzerland.

11. SALIX *Fragilis:* The CRACK WILLOW; *a deciduous tree;* native of England and the north of Europe.

12. SALIX *Helix:* The ROSE WILLOW; *a low deciduous tree;* native (though not common) of England and the fouthern parts of Europe.

13. SALIX *Caprea:* The SALLOW; a well-known *low deciduous tree;* native of England and moft parts of Europe.

14. SALIX *Viminalis:* The OZIER; *a low deciduous tree;* native of England and moft parts of Europe.

15. SALIX *Glauca:* The GLAUCUS WILLOW, or ALPINE SALLOW; *a deciduous fhrub or tree;* native of the Alps of Lapland and the Pyrenees.

1. The WHITE WILLOW. This is a tall-growing tree, and being univerfally known needs no defcription. The filvery elegance of its leaves would render it very *ornamental,* were it not for its too great commonnefs : it is a quick grower, and its wood is *ufeful* when lightnefs and a cleannefs of grain is required *.

2. The

---

* HANBURY, fpeaking of Aquatic Foreft-Trees, fays, " The forts ufed for plantations of thefe trees have hitherto been our common white and red Willow. Thefe, however, feem now to give place to more forts, which have been lately introduced. A few years ago I faw in the public papers an advertifement of a Willow which would grow large enough for mafts of fhips, &c. in twenty or thirty years; and in another paper there was an account, that thefe trees might be feen in full maturity at one Squire Angel's, about three miles from Weftminfter-Bridge. I went to examine them, but when I came found them the Common White Willows, which, having liked the fituation, had grown to a great fize and beauty. I enquired out the author
thor

2. The GOLDEN WILLOW may be admitted into ornamental plantations, not for any extraordinary figure thefe trees will make in fummer, but from the fhow they make in winter; for their bark is fmooth, and of a clear yellow; and in that feafon they have a fingular and ftriking effeẗ among other trees. This will not grow to near the fize of the other fort.

3. The RED WILLOW is a free fhooter, and will grow to a fize almoft as large as the Common White Willow. A few of thefe only fhould be admitted into our plantations; for they have no fingular look in fummer; but in winter their bark appears of a red colour, which makes a pretty variety among other trees at that feafon; but it is, neverthelefs, not near fo ftriking as the yellow fort.

4. The SWEET-SCENTED WILLOW. This will grow to be a large timber-tree, and the branches are covered with a fmooth brown bark. The leaves of this fort refemble thofe of the Bay-tree, and are by far the broadeft of any of the forts of Willows. They are fmooth, and have their upper furface of fhining green; but their under furface is paler, and they are ferrated at their edges. They emit, efpecially when bruifed, a grateful odour; fo that as an aromatic it claims a place in thefe plantations among others of its own growth. Indeed it deferves it; for air will frequently be perfumed by the fragrance of its leaves after a fhower to a confiderable

thor of the advertifement, but found he knew nothing of the nature of thefe Willows, and that he had his account from a bafket-maker near Weft-minfter-Bridge. Upon applying to the bafket-maker, he difavowed knowing any thing of the trees growing by Mr. Angel's, but faid he had two forts of Willows, which would anfwer in every refpeẗ to the firft advertifement; that they were of all others the freeft fhooters; that they were not fo fubjeẗ to rot in the fides as the large White Willow Tree: but that they would grow found to timber fit for mafts of fhips, &c. in lefs than thirty years. He added, that he had cuttings many years ago brought him from the coaft of France, by a Captain whofe name I have forgot. I immediately procured fome cuttings of thefe forts, which grow to a miracle, and feem as if they would anfwer the promifed expeẗation; fo that thefe now are the trees of which our future timber-plantations fhould confift: nay, whether they are defigned for the bafket-makers or for hurdles, they ought to have their fhare; and fhould always be preferred to be planted out for ftandards for lopping, by the fides of rivers, rills, ditches, &c. The cuttings of thefe two forts have been difperfed into almoft every quarter of England; fo that there is no doubt but that in a few years the planting of them alone for timber will become general, as they may be encreafed at pleafure, by every flip or twig"

distance;

diftance ; fo that it will readily join with other aromatics in perfuming the air with their fpicy odours.

5. The Weeping Willow of Babylon will grow to be a large tree ; and no tree is more proper to be planted by rivers, ponds, over fprings, &c. than this ; for its flender branches are very long and pendulous ; the leaves, alfo, are long and narrow ; and when any mift or dew falls, a drop of water will hang at the end of each of thefe leaves, which, together with the pendulous branches and leaves, caufe a moft pleafing appearance. Lovers garlands are faid to have been made of the wreaths of this Willow, the branches of which are very flender and pliable ; and the plant itfelf has always been fought after for ornamental plantations, either to mix with others of the like growth in the largeft quarters, or to be planted out fingly over fprings, or in large opens, for the peculiar variety they will occafion by the elegance of their outline.

6. Shining Willow is a large-growing tree, fending forth feveral flender branches, which hang down, and are covered with a pale-brown bark. The leaves are fmooth, glandulous, ferrated, and of a yellowifh-green colour. The flowers are numerous hairy katkins, and the male flowers have two ftamina only. They appear early in the fpring ; and the females are fucceeded by downy feeds, like the Common Willow.

7. Triandrous Willow is a large-growing trees, fending forth numerous, erect, flexible branches, which are covered with a greyifh bark. The leaves are oval fmooth, fpear-fhaped, acute-pointed, ferrated, green on both fides, and eared at their bafe. The katkins are long, narrow, loofe, and appear early in the fpring. This fort is planted by the basket-makers, to mix with other kinds for their different forts of work.

8. Phylica-leaved Willow. This is a tree of rather lower growth than the former. The branches are numerous, flexible, tough, and ferviceable for feveral articles in the bafket way. The leaves are fpear-fhaped, fmooth, ferrated, and waved on their edges. The flowers are long katkins, which come out early in the fpring from the fides of the branches ; and they foon afford a large quantity of down,

which

which is wafted about with the winds to a confiderable diftance.

There is a *variety* of this with broad leaves.

9. ALMOND-LEAVED WILLOW. This is a Willow of the middle fize, fending forth numerous flexible tough branches, covered with a light-green bark. The leaves are fpear-fhaped, fmooth, ferrated, acute, eared at their bafe, and of a light-green colour on both fides. The flowers are oblong katkins, which turn to a light down in the fummer.

There are feveral forts of this fpecies, that are of inferior value to this, which is generally diftinguifhed from the others by the name of the *Old Almond-leaved Willow*. The branches are very tough and flexible, and when planted in the Ozier-way, and grown to be one year's fhoots from the ftools, are very ftrong, and highly ferviceable for the different purpofes of bafket-making.

10. HASTATED WILLOW. This is a middle-fized tree for the Willow kind, fending forth feveral long, green fhoots from the ftools, which are full of pith, but neverthelefs tough and ferviceable to the bafket-maker. The leaves are nearly oval, acute, fmooth, ferrated, fit clofe to the branches, and have broad appendices at their bafe. The flowers, are an oblong, yellow katkin, and come out in the fpring from the fides of the young fhoots, almoft their whole length.

11. CRACK WILLOW is another middle-fized tree for the Willow kind. The branches are very brittle, and covered with a brownifh bark. The leaves are oval, fpear-fhaped, long, fmooth, ferrated, green on both fides, and have glandulous footftalks. The katkins are long, flender, and the fcales are loofely difpofed.

There is a *variety* of this fpecies, with a yellow bark, which it cafts every year, called the *Almond-leaved Crack Willow*. Both forts are unfit for the bafket-makers ufe, being very brittle ; on which account this fpecies gained the appellation of Crack Willow.

12. ROSE WILLOW. This is of much lower growth than the former. The body of the tree is covered with a rough, yellow bark. The branches are upright, tough, and
of

of a reddifh colour. The leaves are fpear-fhaped, narrow, fmooth, of a blueifh-green colour, and, towards the upper part of the branches, are nearly oppofite to each other. The flowers come out from the fides of the branches, and numbers of them are joined together in a rofe-like manner. They are of a greenifh-white colour, and have a fingular and beautiful look.

There are two or three *varieties* of this fpecies. The leaves of one are downy underneath; the ftalks of another are brittle, and the leaves green on both fides; whilft another has its leaves of a light-green on the upper furface, and glaucous underneath: They are all low-growing plants, and feldom cultivated for ufe.

13. SALLOW. The Sallow is well known all over England, and delights in a dry rather than a moift foil. It is a tree rather below the middle growth. The branches are numerous, fmooth, of a dark-green colour, and their chief ufe is for hurdle-wood and the fire; though the trunk, or old wood, is admirable for feveral ufes in the turnery way. The leaves are oval, rough, waved, indented at the top, and woolly underneath. The katkins are very large, yellow, appear early in the fpring, and are much reforted to by the bees, on their firft coming out of their hives at that early feafon.

There is a *variety* of this fpecies with long leaves, which end in acute points; and another with fmooth leaves, beautifully ftriped with white, called the *Striped Sallow*.

14. OZIER is a tree of rather low growth, though the fhoots grow amazingly long and ftrong in one year from the ftools. The leaves are fpear-fhaped, narrow, long, acute, almoft entire, of a blueifh-green on their upper fide and hoary underneath, and grow on very fhort footftalks. This is the moft propagated of all the kinds for bafket-making: it admits of feveral forts of different value, but all are neverthelefs ufeful to the bafket-maker.

The *varieties* ufually go by the names of the *Green Ozier*, the *Old Bafket Ozier*, *Welfh Wicker*, &c. &c. &c.

15. GLAUCOUS WILLOW. This is a low Alpine Willow, of little ufe for ornament or profit. The leaves are oval, oblong, entire, of a glaucous colour, and poffeffed of fine hairs on their under fide. The katkins are large, oval, of a white colour, and appear about the time of thofe of the Common Sallow.

All the SALICES may be PROPAGATED by planting the cuttings, which may be done at all times of the year, for they will
grow if it is in fummer; though the beft feafon is the winter, or
early in the fpring, juft before they begin to fhoot. The cuttings fhould be of the laft year's wood, fhould be in height in
proportion to their thicknefs, and always ought to be planted
in an upright pofition.

The method of planting an OZIER GROUND will be given
under WOODLANDS.

S A L S O L A.

LINNEAN Clafs and Order, *Pentandria Digynia*: Each
flower contains five males and two females: There are fixteen
SPECIES; one only of which is adapted to our Collection:

SALSOLA *Fruticofa*: The SHRUBBY GLASSWORT, or the
STONECROP-TREE; *an evergreen fhrub*; native of the fea-
coafts of England, France, Spain, and Perfia.

The STONECROP-TREE is a fhrub of about four or five feet
growth. It will fhoot rather higher, if permitted; but is never more beautiful than when about a yard high. The branches
are numerous, naturally grow upright, are covered with a grey
bark, and are verv brittle. As to the leaves, they are very
much like the Common Stonecrop of our walls, which is
well known, being narrow, taper, and flefhy like them. They
are of the fame light pleafant green, and the branches are
ftored with them in plenty. The flowers make no fhow; neither is there any thing that is defirabie to the Gardener that
fucceeds them. This is a very hardy fhrub; but, as we have
introduced it as an Evergreen fhrub, it may not be improper
to give a hint or two for its being properly ftationed. It
fhould be fet in a well fheltered place; for although the leaves
remain on all winter, yet our fevere black frofts fuddenly coming
en them, when in an open expofed place, deftroy them, and
caufe them to turn black; and although the fhrub will fhoot
out again early in the fpring, yet the black deftroyed leaves
will

will look very difagreeable all winter, and be as blots among others that are lefs fubject to thefe difafters. One hint more may be neceffary; and that is, whenever this fhrub is planted, either in fmall or large gardens, among deciduous or evergreen trees, not to circumfcribe the tree, with ftrings or bafs mattings, in order to confine the branches and keep them clofer : this will effectually deftroy all the branches and leaves, if not the whole plant; for being thus clofely confined, the free admiffion of the air will be excluded, which will caufe thefe fucculent leaves to rot and decay. This precaution is the more neceffary, as their upright branches being heavy laden with fuch plenty of fucculent leaves, are fubject to be blown down from the bottom by the high winds : and as they then muft of courfe look irregular, and may probably overfpread fome little plant that grows near them, it is a common thing to tie them up again to the other branches. This cuftom, however, ought never to be practifed; but when any of them happen to be blown down in that manner, they fhould be taken off and thrown away.

Nothing is more eafy than the PROPAGATION of the Stonecrop-tree; for it is encreafed by layers, cuttings, and fuckers. In fhort, if fome of thefe fhrubs are planted, they will foon fend forth many ftalks from the roots ; and if the whole be then taken up, thefe, without any other trouble, may be divided, and will each of them be a good plant ; and thus in a few years, from a plant or two of this fhrub, numbers may be obtained.

S A M B U C U S.

LINNEAN Clafs and Order, *Pentandria Trigynia :* Each flower contains five males and three females : There are four SPECIES; three of which are here treated of; the fourth, *Sambucus Ebulus,* or Dwarf Elder, is an herbaceous plant.

1. SAMBUCUS *Nigra :* The COMMON ELDER ; *a deciduous fhrub or tree* ; common in moft parts of England, but is faid to be originally a native of Germany.

2. SAMBUCUS *Canadenfis :* The AMERICAN ELDER ; *a deci-duous'.Jhrub* ; native of Canada, Pennfylvania, and Virginia.

3. SAMBUCUS *Racemofa :* The MOUNTAIN ELDER, or the MOUNTAIN RED-BERRIED ELDER ; *a tall deciduous Jhrub* ; native of the mountainous parts of the South of Europe.

1. The COMMON ELDER admits of many *varieties :*
The Black Elder.
The White-berried Elder.
The Green-berried Elder.
The Parfley-leaved Elder.
The Gold-ftriped Elder.
The Silver-ftriped Elder.
The Silver-dufted Elder.

The *Common Black Elder* is too well known to require any defcription. It will grow to thirty feet high, with a large trunk ; and in this cafe its wood is very valuable. The leaves and flowers have a ftrong and difagreeable fmell, which renders it improper to be planted near buildings or walks which are much frequented ; but if they could be planted fingly, or a fmall clump of them, at a diftance from any place of refort, there is no tree in the world will make a grander figure, or be more ftriking when in blow ; for at that time they will be covered all over with large bunches of white flowers, which will affume an air of majefty at that diftance, equal to any of the flowery tribe. " Neither may a few of them only be ftationed in this manner ; but any acute corner of the plantation, that fhews itfelf at a diftance, may end with one of thefe trees ; for there it will difplay its gaudy pride when in blow, and the eyes of all be feafted by its delicious appearance, whilft the fenfe of fmelling is no way incommoded by its ftrong difagreeable fcent."

The *White-berried Elder* differs from the former, in that the berries are whiter ; the bark, alfo, of the young fhoots is whiter ; the buds, likewife, at their firft appearance, are in-clined to a whiter colour ; the leaves, too, are of a paler green ; and the plant in general has not fuch a ftrong difagreeable fcent, though it neverthelefs has a proportionable fhare. A plant or two only of this fort is to be admitted, merely for variety ; though where they are required for the fake of the berries to

make

make wine, an hedge of them may be planted, in a place that is little frequented, and they will plentifully furnish the owner with berries for his purpose.

Green-berried Elder differs, in that the berries are green; the bark, also, of the young shoots is of a darker grey than that of the White; and the buds at their first appearance have nearly as dark a colour as that of the Common Elder. We must have only a plant or two of this sort for variety; and where the berries are wanted for wine, an hedge of them may be planted in some distant place, in the same manner as those of the White sort.

The *Parsley-leaved Elder* varies in no respect from the Common sort, except in the nature of the leaves; which are laciniated in such a manner as to resemble the leaves of some sorts of parsley. These leaves occasion a wonderful variety in shrubbery-quarters among the leafy tribe, and on their account the plant is deemed worthy of a place in any Collection; though the flowers possess the same nature with the Common sort, and emit the same disagreeable scent.

The *striped* sorts are distinguished by their different-coloured stripes; whilst the Silver-dusted kind is remarkable for leaves finely powdered or dusted over, in a pounce-like manner, causing thereby a very beautiful and striking appearance.

2. The AMERICAN ELDER is of a lower growth than any of the above sorts, seldom rising higher than eight or ten feet. The young shoots are of a reddish colour. The leaves on the lower part of the plant are trifoliate; others are composed of about two or three pairs of folioles, terminated by an odd one. These folioles are serrated, and of a pleasant green colour; neither do they emit so strong a scent as any of the other sorts. The flowers are produced in the same manner as the folioles; and are succeeded by berries of a reddish colour. Though these berries have not quite such a strong disagreeable taste as the Common Elder berries, yet they have a kind of physical flavour: nevertheless they are liked by some persons, who are as fond of them as they are of some sorts of fruit. What was said of the first sort, recommending its being planted singly, or in small clumps at a distance, will hold good in all these sorts, which when in blow will equally have the same noble appearance as that, except the American, which is of lower growth,

and

and confequently of lefs figure than the others, and as fuch lefs proper for the purpofe.

3. The MOUNTAIN ELDER will grow to about ten or twelve feet high, and is a tree that is with great juftice univerfally admired. The bark of the young fhoots is of a reddifh colour, and the buds in winter will be very large and turgid, and of a ftill deeper red. The leaves are pinnated with an odd one ; their folioles are ferrated ; they are placed at a good diftance on the mid-rib, which is pretty long ; and they die to a reddifh colour in the autumn. The reddifh-coloured branches, with their large turgid buds, have a fingular and noble look in winter amongft other trees ; and in the fpring, as flowering-fhrubs, thefe trees feem to attempt to vie with any of the flowering tribe ; for in April, and the beginning of May, they will produce their bunches of flowers at the ends of every joint of the laft year's fhoots. Thefe bunches of flowers are of an oval figure ; a figure in which compound flowers are not commonly produced. They are not, however, of fo clear a white as any of the other forts, being tinged with green ; and although the tree will be covered with them, they have not the fame ftriking appearance ; but this defect is made amends for by the peculiar form which they affume, and the fcarcity of the plant itfelf. Were there nothing but the above recited properties to recommend this fhrub, it might juftly claim admiffion in plenty into our choiceft plantations : but thefe are not all its beauties ; what remains is much more ftriking and engaging ; for thefe oval bunches of flowers are fucceeded by oval bunches of berries, that are of a deep fcarlet colour. A crop, indeed, does not always enfue ; but when it does, no tree is more fingularly beautiful than this is rendered by them, chiefly occafioned by their colour and form, which any one muft conceive to be delightful.

All the forts of Elder are PROPAGATED by cuttings. Thefe fhould be of the laft year's fhoot, and each cutting fhould confift of three joints ; two of which muft be in the ground, whilft the third is left above, to make the fhoot. October is the beft month for this bufinefs ; and almoft any foil will do, though the moifter it is the better. Thefe cuttings may be either planted very clofe, and removed the autumn following into the nurfery-ground ; or they may be planted a foot or more afunder,

der,

der, and then they will be of a fufficient diftance until they are
finally taken up, which may be any time after two years.
Thus eafy is the culture of thefe plants when known.

S M I L A X.

Linnean Clafs and Order, *Diœcia Hexandria:* Male
flowers containing fix ftamina, and female flowers containing
three piftils, upon diftinct plants. There are fourteen Species;
eight of which are as follow :

1. Smilax *Afpera :* The Italian Smilax, or Common Rough Bindweed, or Prickly Bindweed; *a low
climber* ; a native of Italy, Spain, Sicily, and France.

2. Smilax *Exceffa .* The Oriental Smilax; *a lofty
climber* ; native of many parts of the Eaft.

3. Smilax *Sarfaparilla :* The Peruvian Smilax, or
Sarsaparilla ; *a climber* ; native of Peru, Mexico, and
Virginia.

4. Smilax *Rotundifolia :* The Canada Smilax ; *a climber* ;
a native of Canada.

5. Smilax *Laurifolia :* The Laurel-leaved Smilax ;
a climber ; native of Virginia and Carolina.

6. Smilax *Tamoides :* The Briony-leaved Smilax ;
a climber ; native of Carolina, Virginia, and Pennfylvania.

7. Smilax *Lanceolata :* The Lance-leaved Smilax ; *a
climber* ; native of Virginia.

8. Smilax *Herbacea :* The Ivy-leaved Smilax ; *a
climber* ; native of Virginia and Maryland.

1. The Italian Smilax is poffeffed of a long, creeping,
white, flefhy root, which fends forth many flender, angular
ftalks, armed with ftrong, fhort, crooked fpines, and having
clafpers. If any thing is near for it to climb on, it will, by
fuch affiftance, arrive at the height of ten or twelve feet. The
leaves are cordated, end in acute points, are of a fine dark-
green colour, indented, have nine longitudinal veins, have

C c 4   their

their edges befet with fome fhort fpines, and are placed on to lerable long tough footftalks. The flowers make no figure: They are white, and are produced from the wings of the ftalks, in fmall bunches, in June or July; and the female flowers will be fucceeded by round red berries.

There is a *variety* of this fpecies, which produces black berries; and from which it differs in no other refpect; and which occafions its being called by Gardeners the *Black-fruited Rough Bindweed.* There is alfo another fort, with brown fruit.

2. ORIENTAL SMILAX is a lofty climber; for being planted near pretty tall-growing trees, it will afcend to their very tops, and proudly, by fuch affiftance, fhew itfelf to a great diftance. The roots are thick, white, and flefhy; and the ftalks are angular, and armed with fpines. The leaves are of a pleafant green colour, and are nearly of a fagittated figure. They are poffeffed of no fpines, have longitudinal veins, and their footftalks are tolerably long and tough. Their flowers are white, and are produced in fmall bunches, in June and July; and the females are fucceeded by round red fruit in their own countries, but not with us.

3. PERUVIAN SMILAX, or SARSAPARILLA, has alfo white, thick, flefhy roots. Thefe fend out angular ftalks, that are armed with fharp fpines; but they will not climb up trees to near the height of the former. The leaves are fmooth, being unarmed with fpines. They are retufe, oval, cordated, of a ftrong green colour, have three nerves, and grow on ftrong rough footftalks. The flowers are produced in fmall bunches, from the fides of the branches. They are of little figure, and the females are fucceeded by a fmall, round, red fruit, where they grow naturally.

4. CANADA SMILAX has long creeping roots, which fend forth round flender ftalks, that are thinly guarded with fharp ftraight fpines. The leaves are reniform, cordated, and have no fpines. They are broader than they are long, have five ftrong nerves, and fhort footftalks, from each of which grow two flender clafpers. The flowers are produced in fmall bunches, in June and July. They will be fucceeded by a fmall berry, which will not come to perfection here.

5. LAUREL-LEAVED SMILAX has round taper ftalks, that are befet with fpines. The leaves are of a ftrong green colour,

and

and a thick confiftence. They have no fpines, have three nerves, are of an oval lanceolate figure, and are about the fize of thofe of our Common Bay-tree. The flowers are produced in fmall round bunches, in June and July, from the wings of the ftalks ; and thefe are fucceeded by fmall black berries in the autumn. This fort is rather of a tender nature, and unlefs the foil be naturally dry and warm, and the fituation well fheltered, they will be pretty fure of being killed in the winter.

6. BRIONY-LEAVED SMILAX has large, flefhy, white roots, which fend forth round, taper, prickly ftalks. The leaves are oblong, heart-fhaped, have no fpines, but have many veins running lengthways. Their upper furface is of a fine ftrong green colour, and, being tolerably large, they make a goodly fhow. The flowers are produced in July, in fmall loofe bunches, and are fucceeded by black berries.

7. LANCE-LEAVED SMILAX. The ftalks are flender, taper, and free from prickles. The leaves are fpear-fhaped, pointed, and unarmed with fpines. The flowers come out in fmall clufters, and are fuccceded by red berries.

8. The IVY-LEAVED SMILAX. The ftalks are angular, herbaceous, unarmed with fpines, but poffeffed of clafpers, by which they lay hold of any thing near them for fupport. The leaves are oval, free from fpines, feven-nerved, and grow on footftalks. The flowers of this genus make no fhow, being poffeffed of no ornament except the fegments of the calyx. Thofe of this fpecies are very fmall, and are collected in fmall umbels. They appear in June ; and are fucceeded by roundifh berries, which feldom ripen in England.

Thefe forts are all eafily PROPAGATED : indeed they will propagate themfelves, if a plant or two of each fort can be obtained ; for they are poffeffed of long creeping roots, which run under the furface of the ground, and will, both near the main plant and far off, fend up young ones ; which being taken up in the autumn or fpring, or in any time of the winter, will be good plants for ufe. Thus will thefe plants by nature furnifh you foon with plants enough for your purpofe, if one or two of each can be firft procured, and planted in a light good foil, in proper beds prepared for the purpofe, under warm hedges, or amongft trees in well-fheltered places.

S O L

S O L A N U M.

LINNEAN Clafs and Order, *Pentandria Monogynia* : Each
flower contains five males and one female : There are forty
SPECIES ; one of which, though common, claims our attention.

SOLANUM *Dulcamara:* The WOODY NIGHTSHADE, or the
BITTER-SWEET ; *a ligneous climber;* native of England and
moft parts of Europe.

THE WOODY NIGHTSHADE. Were it not for the common-
nefs of this plant, it would defervedly claim a principal place in
our efteem, as one of thofe forts that require fupports to fet
them off; for befides the flowers, which are of an exquifite fine
purple, and grow in bunches, it has many beauties to recommend it to our obfervation and care. The leaves ftand on large
footftalks, and the upper ones are of an haftated figure. Their
beautiful purple flowers will be produced in fmall clufters, in
June and July ; and they are fucceeded by oblong red berries,
which will be ripe in autumn. This is the Common fort,
which is of all the moft beautiful, though hardly ever propagated. The *varieties* of it, however, are in great efteem with
moft people, and of thefe there are, 1. A variety with *white
flowers,* which is much coveted on that account ; and although
thefe flowers are not fo beautiful as the purple ones, yet the
fort being a rare plant, makes it defirable ; and this is the
fort that is cultivated, and which differs in no refpect from the
purple, only in its white flowers, thereby pleafing the fpectator
by the variety it affords. 2. The next remarkable variety of the
Woody Nightfhade is that with beautifully *variegated leaves.*
Thefe plants are feduloufly propagated for the fake of their
finely-ftriped leaves ; fo that there is fcarcely a nurferyman who
does not raife plenty of them for fale amongft other fhrubs ;
and they are fo generally liked, that his difpofing of them will
be pretty certain. This plant, as has been obferved, is only
the Common Woody Nightfhade with the leaves delightfully
variegated, its flowers being of the fame fine purple, and the
fruit that fucceeds them exactly the fame. 3. Another variety
has *thick leaves,* which are *very hairy.* This fort grows chiefly
in Africa, and muft have a warm fituation to live through our
winters.

winters. It is, however, a very fine plant, and where such a
situation is not found, ought to be treated as a greenhouse-
plant.

All thefe forts are eafily PROPAGATED by cuttings; for they
will grow, if planted in any of the winter months, in almoft
any foil or fituation, and will be good plants for removing by
the autumn following. If the owner has only a plant or two
of thefe, which he is defirous of multiplying with certainty,
let him lay the young ftalks upon the ground, and draw over
them a little foil, and they will effectually be good plants by
the next autumn; and this will be the fureft way, as cuttings
of moft forts, though they will for the moft part take very
well, are often attended with much hazard. The variegated
fort muft be planted upon a poor foil, or it will be in danger
of running away from its colours.

S O R B U S.

LINNEAN Clafs and Order, *Icofandria Trigynia*: Each flower
contains about twenty males and three females: There are three
SPECIES:

1. SORBUS *Aucuparia*: The WILD SORB, or MOUNTAIN
ASH, or QUICK-BEAM, or QUICKEN-TREE, or ROAN-TREE;
a deciduous tree; native of England and moft of the northern
parts of Europe.

2. SORBUS *Domeftica*: The CULTIVATED SORB, or SWEET
SERVICE; *a deciduous tree*; native of the fouth of Europe.

3. SORBUS *Hybrida*: The MONGREL SORB, or SEMI-PIN-
NATED SERVICE; *a deciduous tree*; native of Gotlandia.

1. The WILD SORB, or MOUNTAIN ASH. Although
we generally fee this fpecies in a fhrubby underwood ftate, it
will neverthelefs, if properly trained, grow to a tree of the
middle fize. It has no claim to the appellation of Mountain
Afh, except fome diftant refemblance of the common Afh in
the formation of its leaves. The flowers of this tree have a
pleafing effect in the fpring, and its berries, in autumn and
winter,

winter, render it highly *ornamental*. Evelyn enumerates its *uses:* he says, " befides the ufe of it for the hufbandman's
" tools, goads, &c. the wheel-wright commends it for being
" all heart. If the tree be large, and fo well grown as fome
" there are, it will faw out into planks, boards, and timber.
" Our fletchers commend it for bows next to yew, which we
" ought not to pafs over, for the glory of our once Englifh
" anceftors: In a ftatute of Henry VIII. you have it men-
" tioned. It is excellent fuel ; but I have not yet obferved
" any other ufe." Hanbury follows him, and places the
Mountain Afh among his foreft-trees. Its wood is undoubtedly
pliable and tough. Its fhoots, from the ftool, are generally
numerous, ftraight, and long. In the north of England the
hufbandmen ufe them for whip-ftocks to drive their teams
with. For ftakes, edders, and perhaps for hoops, no wood is
better adapted than the Wild Sorb ; and as an Underwood it
feems well worth the planter's notice.

The propagation of the native Sorb is from feeds or
by layering. Having procured a fufficient quantity of berries,
they fhould be fowed, foon after they are ripe, in the feminary,
about half an inch deep, in beds made as has been before directed. They frequently lie till the fecond fpring before they
make their appearance ; and, in the fpring following, may be
planted out in the nurfery. We need not repeat that the feminary
fhould be kept clear of weeds, and that the young plants in dry
weather now and then ought to be refrefhed with water ; neither
need the gardener be reminded, that after they are planted in
the nurfery way, digging the ground in the rows muft be obferved every winter, taking off all fhoots alfo which would
make the tree forked, and keeping the weeds hoed in the rows,
til they are of fufficient fize to plant out where they are intended to remain.

This fpecies will take very well from *layers* ; fo that whoever
cannot procure the berries, and has a few of thefe trees, may
cut them down clofe to the ground, when they will throw out
many ftools ; and if the year following thefe are laid in the
ground in the fame manner as carnations, they will have taken
good root in one year. But trees cultivated this way will not
grow fo ftraight and handfome, neither will they arrive at fo
great a magnitude as thofe raifed from the feeds.

The

The Quicken-tree will grow upon almoſt any ſoil, either ſtrong or light, moiſt or dry. It flouriſhes both on the mountains and in the woods; it is never affected by the ſeverity of the weather, being extremely hardy; and if even planted on bleak and expoſed places, it grows exceedingly well.

2. The CULTIVATED SORB, or SWEET SERVICE, is ſo diſtinguiſhed from the other, becauſe it produces eatable fruit, which in France, Italy, and other parts, is ſerved up in deſerts; and the tree is cultivated there ſolely on that account. It will grow to be larger than the Quicken-tree; and in many reſpects is ſuperior in beauty to moſt trees. It will grow with an upright ſtem; and the young ſhoots in the ſummer are ſo downy as to appear covered with meal. In the winter they are inclined to a purpliſh colour, and are ſpotted all over with whitiſh ſpots; the buds at the ends of them will be turgid, preparing for the next year's ſhoot. The leaves reſemble thoſe of the Quicken-tree; they are finely pinnated, and compoſed of ſeven or eight pair of lobes, which are terminated by an odd one. They are broader than thoſe of the Quicken-tree, ſerrated in a deeper and more irregular manner, and their under ſurface is of a much more downy nature. The flowers are white, grow in umbels, come out in May, and are ſucceeded by an agreeable fruit, which is large, fleſhy, and of various ſhapes in the different varieties.

3. The MONGREL SORB. This ſeems to be a mongrel, between the Quicken-tree and *Aria*. It is an upright-growing tree, and the young branches are of a whitiſh colour. The leaves are very downy, and pinnated at the baſe; but the upper lobes join together, thereby forming a half-pinnated leaf. The flowers are white, grow in umbels, and are ſucceeded by bunches of roundiſh berries, which will be ripe in the autumn.

The PROPAGATION of the Wild Sorb has already been given; and that culture will ſerve for all the ſorts: but in order to have good fruit of the Sweet-Service-tree, the beſt ſorts ſhould be grafted or budded upon pear or quince-ſtocks.

Theſe trees are very hardy, for they will grow in almoſt any ſoil; though they make the ſwifteſt progreſs, and arrive at the greateſt height, in a moiſt ſituation.

S P A R T I U M.

LINNEAN Clafs and Order, *Diadelphia Decandria:* Each flower contains ten males and one female ; the males being divided into two fets at the bafe : There are fixteen SPECIES ; feven of which are proper for our collection :

1. SPARTIUM *Scoparium:* The COMMON ENGLISH BROOM, *a deciduous fhrub* ; native of England, and the fouthern parts of Europe.

2. SPARTIUM *Junceum:* The SPANISH BROOM ; *a tall deciduous fhrub* ; native of Spain, Portugal, and fome parts of Italy and Sicily.

3. SPARTIUM *Radiatum:* The STARRY BROOM ; *a low deciduous fhrub* ; native of Italy.

4. SPARTIUM *Monofpermum :* The SINGLE-SEEDED BROOM ; *a deciduous fhrub* ; native of the barren parts of Spain.

5. SPARTIUM *Angulatum :* The EASTERN BROOM ; *a deciduous fhrub* ; native of the Eaft.

6. SPARTIUM *Spinofum:* The THORNY BROOM, or PRICKLY CYTISUS ; *a tender deciduous fhrub* ; native of the fea-coafts of the fouthern parts of Europe.

7. SPARTIUM *Scorpius:* The PRICKLY BROOM ; *a deciduous fhrub* ; native of Spain and the fouth of France *.

1. The ENGLISH BROOM will grow to be about fix feet high. The branches are very flexible and numerous ; they are angular ; and the bark with which they are covered is of a delightful green. The leaves are both trifoliate and fingle, the lower part of the branches producing the former, the upper part the latter. The flowers are large, and produced in May, all along the fides of the laft year's fhoots, from the bottom to the top. They ftand upon fhort footftalks, and fo ornament each twig of which the whole fhrub is compofed, that they have a look grand beyond moft of thofe of the flowery tribe. Thefe flowers are fucceeded by compreffed pods, containing kidney-fhaped feeds, all of which are very well known.

* For another Clafs of Brooms fee GENISTA.

2. Spanish Broom. There are two notable *varieties* of
this species: Common Spanish Broom, Double Spanish
Broom.

Common Spanish Broom is a fine plant, and has been much
fought after as a flowering-shrub. It will grow to be ten feet
high. The branches are taper, placed opposite, and covered
with a smooth green bark. The leaves, which are not very
numerous, are of a spear-shaped figure, and, like the twigs, of
a fine green colour. The flowers are produced at the ends of
the branches, in loose spikes, in July; and there will be a succession still kept up, at the end of each spike, often until the
frost puts a period to their blowing. The flowers of this sort,
also, are succeeded by compressed pods, which contain kidney-
shaped seeds, which often ripen in the autumn.

The *Double Spanish Broom* differs in no respect from the
other, except that the leaves are very double. The manner of
growing, colour of the shoot, and nature of the leaves, are
exactly the same; and it produces very full double flowers;
but these flowers do not come out so early as the single sort, it
being often September before any of them will be in blow; and
the succession will be continued so slowly, that sometimes not
more than two or three flowers on a spike will be fully out before the frosts nip them from any further blow. This sort is succeeded by no seeds.

3. Starry-Broom is a low plant, seldom growing more
than a yard high, even when it has the advantage of culture; in
the places of its natural growth, two feet it seldom aspires to.
Notwithstanding the low growth of this shrub, however, it will
occupy a large space of ground in proportion to its size, for it
extends its flexible branches all around to some distance. The
branches of which it is composed are very narrow, angular, and
grow opposite by pairs. The leaves are trifoliate, grew opposite to each other, and the folioles are awl-shaped, placed opposite, and spread out in such a manner as to resemble the rays
of a star, which occasions its being so called. The flowers are
produced in June and July, at the ends of the branches: they
will be in a kind of small clusters or spikes, are of a bright-
yellow colour, and of the same figure with the former,.but proportionally smaller. They are succeeded by short hairy pods,

in

in which are a few kidney-fhaped feeds, which will be ripe in Auguft or September.

4. SINGLE-SEEDED BROOM. The *varieties* of this fpecies are, the Common Yellow, and the White-flowered.

The *Common Single-feeded Broom* is a plant about fix feet in growth. Its branches, which are very numerous and tough, are angular; and the leaves, which are not very many, are of a lanceolated figure. The flowers are produced in bunches from the fides of the branches. Thefe bunches are fmall ; but being of a fine deep-yellow colour, and alfo being in plenty all over the fhrub, give it a beautiful look. This fhrub blows in July ; and the flowers are fucceeded by fhort pods, each of which contains a fingle feed only, which feldom ripens in England.

White-flowered Single-feeded Broom, which is alfo called the *White Spanifh Broom*, is of a more tender nature than the former fort ; yet not fo tender but that it will thrive abroad with us, in any dry foil and well-fheltered fituation, if the winters are not too fevere. After this precaution, we would advife the gardener not to venture his whole ftock of thefe plants abroad, left a fevere winter fhould take them all off ; but to have a few planted in pots, and fet under fhelter, that, in cafe the others fhould be killed, a fhare of thefe may fupply their places. The White Spanifh Broom, then, will grow to about eight feet high ; and the branches are numerous, flender, and tough. Their bark is of a whitifh colour, and they are taper, almoft like a rufh. The leaves. which are not many, are of a lanceolated figure. The flowers are white, come out in clufters from the fides of the branches in July, and are fucceeded by fhort pods, each of which contains one fingle feed only.

5. The EASTERN BROOM will grow to about fix or eight feet high. The branches of this, alfo, are numerous, flender, and tough. They are rather of a fingular ftructure, each of them affording fix angles. The leaves, which are few, are of dif-ferent figures, fome being found fingle only, whilft others are trifoliate. The flowers are produced in July, at the ends of the branches, in a kind of fpikes. They are of a paler yellow than moft of the other forts, and are rarely fucceeded by feeds with us.

6. PRICKLY

6. PRICKLY CYTISUS, or THORNY BROOM, has fcarcely any bufinefs in this place, being generally reared as a greenhoufe plant ; but as it will bear our moderately mild winters in a warm foil and fituation, with this caution it may be introduced. It is about fix feet in growth ; and the branches are numerous, flender, tough, angular, and armed with long fpines. The leaves are trifoliate; and the flowers are produced in clufters, in June, at the ends of the branches. They ftand on long footftalks, are of a bright yellow, and make a good figure. They are fucceeded by fhort hard pods, which contain a few feeds of the fame figure with the others.

7. PRICKLY BROOM. The ftalk of this fpecies is woody, and fends forth feveral flender, prickly branches, which fpread themfelves every way. The leaves are oval, fmooth, and in fome varieties hairy. The flowers are moderately large ; and fome are of a deep-yellow colour, whilft others are pale. They appear in July; and are fucceeded by fhort pods, containing the feeds, which feldom ripen in Ehgland.

All thefe forts of Broom, the Double-bloffomed excepted, are to be PROPAGATED from feeds; and one method may be obferved for all the forts. The forts that ripen their feeds in England are fuppofed to be ready at hand ; the feeds of the others muft be procured from the places where they grow naturally. The firft week in April is the beft time for fowing the feeds; and this fhould be either in drills, or on beds, half an inch deep. It will not be long before the plants appear ; and as the hot weather comes on, they fhould be fhaded from nine o'clock in the morning till within an hour of funfet. Watering and conftant weeding muft be given them ; and this is all the trouble they will require in fummer. The reader will perceive our Common Broom to want none of this care; neither will the Common Spanifh Broom need much of it ; it is to be afforded thofe only which are lefs common, that we may be more certain of a plentiful ftrong crop. In the fpring all thefe feedlings are to be taken up, and pricked out in the nurfery-ground, a foot afunder, and two feet diftant in the rows. This work muft be done when they are one year-old feedlings; becaufe they naturally fend down a ftrong tap-root,

D d

which,

which, if deferred longer, will be grown fo big as to endanger the growth of the plant. After they have ftood in the
nurfery-ground two years, they will be good plants for fetting
out where they are to remain. Thus may all the forts of *Spartium* be raifed by feeds ; though it will be highly proper to
have particular regard to the fituation of the tenderer forts;
fuch as the White Spanifh Broom, the Oriental, and the forts
called the Prickly *Cytifus* and Prickly Broom. Thefe may be
raifed the fame way ; but the foil and fituation muft be naturally warm and well-fheltered, and the beds fhould be hooped
to be covered with mats in frofty weather ; otherwife the whole
crop will be in danger of being loft the firft winter. In the
fpring they may be planted, fome in pots (to preferve the
forts), others in the warmeft places of the fhrubbery. Another
method will not be improper to be followed in raifing the tenderer forts ; namely, by fowing them in pots in April, and
plunging them in a fhady border up to the rim. At the approach of the firft froft, they may be removed into the greenhoufe, or placed under fome fhelter, when they will be effectually preferved until the fpring ; then they fhould be turned
out, and planted in feparate pots, which fhould be plunged in
a fhady border, and removed under cover in the winter. By
thus protecting them for a winter or two, they will get ftronger,
and be able to refift the cold ; and then a fhare may be planted
out in the warmeft fituation, whilft the others may be removed
into larger pots, to be kept, if wanted, as greenhoufe plants.

S P I R Æ A.

Linnean Clafs and Order, *Icofandria Pentagynia*: Each
flower contains about twenty males and five females. There
are nineteen Species : fix of the ligneous kind are here treated
of :

1. Spiræa *falicifolia*: The Common Spiræa Frutex, or
the Willow-leaved Spiræa ; *a low deciduous fhrub*; native of Siberia and Tartary.

2. Spiræa

2. SPIRÆA *Tomentofa:* The RED-FLOWERING SPIRÆA; *a low deciduous fhrub* ; native of Philadelphia.

3. SPIRÆA *Hypericifolia:* The HYPERICUM - LEAVED SPIRÆA, or HYPERICUM FRUTEX; *a deciduous fhrub*; native of Canada.

4. SPIRÆA *Crenata:* The SPANISH SPIRÆA ; *a low deciduous fhrub* ; native of Spain and Siberia.

5. SPIRÆA *Opulifolia :* The GELDER-ROSE SPIRÆA; *a deciduous fhrub*; native of Virginia, Canada, and Carolina.

6. SPIRÆA *Sorbifolia:* The SERVICE-LEAVED SPIRÆA; *a low deciduous fhrub* ; native of moift land in Siberia.

1. The COMMON SPIRÆA FRUTEX rifes to about four feet high. The root is fpreading ; fo that befides the common ftalks which fend forth branches, others are produced from the roots called Suckers, which by the autumn will be as high or higher than any of the whole plant. The bark on all thefe is fmooth, and of different colours; that on the old ftalks is red, though for the moft part clouded with a dufky matter: The young fhoots that grow from thefe ftalks are lighter, though neverthelefs of a reddifh tinge ; whilft the bark on the fummer fhoots, that fprung from the root, are nearly white. The leaves of this fpecies are of a fine green, and grow without order on the branches. They are fpear-fhaped, obtufe, naked, and their edges are ferrated. The flowers are produced in June, at the ends of the branches that grow from the main ftalk ; and before thefe have done blowing, the fuckers that arife from the roots will exhibit their flower-buds at the ends. Thefe are generally larger and fairer than thofe that were before in blow ; and by thefe fuckers a fucceffion of flowers is often continued even until late in the autumn. The flowers are produced in double-branching fpikes, which are larger downwards, diminifh gradually, and end with an obtufe fpike at the top. They are of a pale-red colour; and though feparately each flower is fmall, yet being produced in thefe thick fpikes, four or five inches long, they have a good look. Thefe flowers, with us, are fucceeded by no ornamental feeds. The vigorous fhoots

of this ſhrub that ariſe from the roots are very tough, pliable, and taper, and make good riding-ſwitches.

2. The RED-FLOWERING SPIRÆA will grow to the height
of about four feet; and the branches are covered with a purple
bark. The leaves grow on theſe without order : they are of
an oval, lanceolated figure, and unequaily ſerrated. Their
upper ſurface is of a fine green colour, but their under is
downy : the ſtalks, alſo, are poſſeſſed of a good ſhare of this
meally kind of matter. The flowers are produced in July, at
the ends of the branches, in double-branching ſpikes, like the
former ; and being of a bright-red colour, make a fine appearance.

There is a *variety* of this ſpecies, with white flowers.

3. The HYPERICUM-LEAVED SPIRÆA, or HYPERICUM
FRUTEX, will grow to the height of about five or ſix feet, and
has beauty and elegance beyond deſcription ; not ſo much from
its natural form of growth, or the colour of the bark or leaves,
as from the flowers ; for the branches are produced irregularly.
The older ſhoots are covered with a dark-brown bark; the
younger ſhoots are ſmooth and lighter, and are tinged with
red. The leaves are ſmall, though of a pleaſing dark-green
colour; they are produced irregularly on the ſhrub, and have
their edges entire. The flowers are produced in May, almoſt
the whole length of the branches: they are of a white colour;
and though each flower is ſeparately ſmall, yet they are collected in umbels that ſit cloſe to the branches, which being thus
ornamented their whole length, ſcarcely any thing but flowers,
beſides the main ſtalks, are to be ſeen ; ſo that the ſhrub has
the appearance of one continued flower, branched out into
as many different diviſions as there are twigs ; for every twig at
a little diſtance will look like a long narrow ſpike of flowers ;
and theſe being all over the ſhrub, of a pure white, the ſhow
they then make is delightful.

4. SPANISH SPIRÆA will grow to be about four feet high ;
and the branches, which are produced irregularly, are covered
with a dark-brown bark. The leaves are ſmall, of a pleaſant
green

green colour, and ferrated at their ends. The flowers are produced from the fides of the branches, in May; they grow in
roundifh bunches, are of a whiter colour than, and being produced nearly the whole length of, the branches, make a
charming fhow, like the preceding fort; from which this appears very little to differ, without being ftrictly examined.

5. GELDER-ROSE SPIRÆA. Of this fpecies there are two
varieties, called, Virginian Gelder-rofe, and Carolina Gelder-rofe.

Virginian Gelder-rofe will grow to be feven or eight feet
high. The branches are covered with a dark-brown bark,
which peels off in the winter, and difcovers an inner, which
is fmooth, and of a lighter colour; fo that in winter this fhrub
has a very ragged look. The leaves refemble thofe of the
common currant-bufh, which has occafioned its being called by
fome the Currant-leaved Gelder-rofe. They are for the moft
part lobed like them; though all the leaves will not be alike,
fome being divided into more than three lobes, whilft others
are fcarcely divided at all. They are ferrated at their edges,
are of a palifh-green colour, and placed irregularly on the
branches, on long green footftalks. The flowers are produced
in June, at the ends of the branches: they are white at their
firft opening, and afterwards receive a reddifh tinge, which is
ftill heightened before they die off. Each flower feparately is
rather fmall; but many of them grow together, each having
its feparate footftalks, in large umbels. The beauty of the
Common Hawthorn is known to all; and it may not be amifs
here, as the fimile is juft, and that the Reader may have a true
idea of the flowers, to mention, that each flower feparately
has the appearance of a fingle flower of the Hawthorn, and
that they are produced in bunches. Thefe flowers are fucceeded by the fame kind of bunches of reddifh, cornered fruit,
which caufes a pretty variety in the autumn.

Carolina Gelder-rofe differs very little from the former
fort. The branches are covered with the fame kind of falling
bark; though the leaves are not lobated in the fame manner;
for thefe will be of different fhapes; yet moft of them are nearly

D d 3 oval,

oval, but end in points, and are all unequally ferrated round their edges. The flowers of this fort, also, are white, but grow in rounder and fmaller bunches than the other. They are fucceeded by the like kind of cornered fruit, which is of a reddifh colour in the autumn.

6. SERVICE-LEAVED SPIRÆA is a fhrub of very low growth; a yard is the higheft we ever yet know it arrive to. The young branches are covered with a purplifh bark. The leaves are beautifully pinnated, fo as nearly to refemble thofe of the Service-tree. The folioles are oblong, and generally about four pair in number: they are uniformly ferrated, and exceedingly ornamental to the fhrub. The flowers are white, and produced at the ends of the branches, in July, in panicles. They are feldom fucceeded by feeds in England.

The PROPAGATION of all the forts is very eafy. It may be done by cuttings; for if the ftrongeft parts of the fhoots of the laft fummer's growth be planted in October, in a fhady border, moft of them will grow, and become good plants by the autumn; fo that by the autumn after that, they will be very proper plants to be fet out to ftand. But if a perfon has only a plant or two of a fort, from which he can get but a very few cuttings, the beft way is to layer them, and not hazard their growing this way; for although they will take freely, yet (fays HANBURY) by fome unfeafonable weather, I have known whole crops of *cuttings of all forts* to fail. Thus, of the many thoufand cuttings of all forts I planted in the winter preceding the dry fummer in 1762, very few grew; for although they were fhaded and watered, and others planted in fhady borders, yet fuch large cracks and chafms would open among them (as they did almoft all over my plantations) as to caufe watering to be of no fervice; nay, the more I watered them, the harder the mould fet, and the chafms became greater; and notwithftanding many of the cuttings were planted in parts that were poffeffed of a natural moifture, yet the crevices there were larger, and the ground harder; and all attempts to prevent it feemed to be in vain. Though this is the nature of the foil of few nurferies, I mention this to fhew, that there is an hazard in planting of cuttings, unlefs the feafon fhould prove good; for this turn

I had

I had fcarcely any grew : fo that whatever trees will grow by cuttings, if a gentleman has only a plant or two, and wants to have them encreafed, the beft way is to do it by layers; and hence, of all *the forts before-mentioned*, if the twigs be but laid in the ground in the autumn, they will have good roots by the autumn following, many of which will be plants ftrong enough to be planted in the fhrubbery, whilft the weaker may be fet in the nurfery-ground for a year or two, to gain ftrength Some of thefe forts will throw out fuckers, which will be good plants when taken up : nay, the firft fort will propagate itfelf faft enough this way; for after it has ftood a year or two, it will throw them out fo vigoroufly, as has been before obferved, that in one fummer they will grow to be as high as the whole plant, and will have fair flowers at their ends in the autumn. And here the Gardener muft obferve, that after this fort is planted in the fhrubbery, the fuckers muft be conftanty cleared off the old plants every winter, otherwife they will foon be fo numerous and clofe, as to lofe that beauty which always attends plants that arife with fingle or with few ftems.

STAPHYLÆA.

LINNEAN Clafs and Order, *Pentandria Trigynia* : Each Flower contains five males and three females : There are two SPECIES :

1. STAPHYLÆA *Pinnata* : The COMMON STAPHYLÆA, or BLADDER-NUT ; a *deciduous fhrub* ; native of many parts of Europe.

2. STAPHYLÆA *Trifolia* : The TRIFOLIATE STAPHYLÆA, or BLADDER-NUT ; a *deciduous fhrub* ; native of Virginia.

1. The COMMON STAPHYLÆA will grow to be eight or ten feet high. The older branches are covered with a brown bark ;

that on the younger shoots is of a much lighter colour. The bark is exceeding smooth; the twigs are very pithy, and when broken have a very strong scent. - The buds will be turgid and large early in winter, as if ready to burst out of their stipulæ, and begin their shoots; this causes the plant at that season to have an air of health and verdure, which of course must then be very pleasing. The leaves are pinnated, of a light-green colour, and, like all others of that nature, are very ornamental. They consist of two pair of folioles, that are terminated with an odd one; which occasions this sort being frequently called the Five-leaved Bladder-Nut. These folioles are tolerably large, oblong, pointed, and stand on pretty long footstalks. The flowers are produced in long pendulous bunches, from the wings of the leaves; and are white. The buds appear in the spring, almost at the first dividing of the stipulæ, though they will not be in full blow until May. These flowers are succeeded by large inflated bladders, in which the seeds are contained, and have a very striking and singular look in the autumn. The nuts of this tree are smooth, and said to be eaten as food by the poor people in some countries. They are also used by the Catholics, who compose some of their rosaries of them.

2. The TRIFOLIATE STAPHYLÆA grows to about the same height with the former. The elder branches will be besprinkled, as it were, all over with greyish spots. The bark on the younger branches is perfectly smooth, and of a yellowish colour. The buds will be swelled early in the winter, though they will not be so large and turgid as those of the former sort. The leaves are trifoliate, and grow by threes on a footstalk; which has occasioned this plant being distinguished by the name of Three-leaved Bladder-Nut. They are of a light green colour; and the folioles are generally pretty large, oval, pointed, and serrated at their edges. The flower-buds appear at the first beginning of the buds to open in the spring; which has been known to be sometimes so early as January; though the flowers will not be in full blow until May. These flowers, like the former, are produced from the sides of the branches, in long pendulous bunches: their colour is white; and they are succeeded by large inflated bladders, in which the seeds are contained. The seeds of both species ripen well in England.

These

Thefe fpecies may be PROPAGATED by feeds, layers, or cuttings. 1. The feeds fhould be fown, foon after they are ripe, in the autumn, three quarters of an inch deep, in almoft any fort of common garden-mould made fine. In the fpring fome fhare of the plants will appear ; though you muft not expect the whole crop until the fecond fpring following : nay, if the fowing of the feeds is deferred until the fpring, fcarcely any of them will come up until the fpring after. All the fummer the beds muft be kept clear of weeds ; and if it fhould prove dry, a gentle watering fhould be given the young plants, which will encreafe their growth. The fpring after the remainder of the crop will come up ; and the bufinefs of weeding muft be continued that fummer. In the autumn the two years-old plants fhould be drawn out and planted in the nurfery, a foot afunder, and two feet diftant in the rows ; and in the beginning of March the one-year-old feedlings fhould be taken up, and planted in the fame manner. The reafon of deferring the planting out of the younger feedlings is, that, being fmall when planted out in autumn, they are often thrown out of the ground by the froft, and many of them loft ; whereas, of larger plants there will be little danger. After they have ftood two or three years in the nurfery, they will be good plants for any places where they are wanted. 2. Thefe fhrubs may alfo be propagated by layers ; and this muft be performed in the autumn, on the fhoots of the preceding fummer, by flitting them at a joint, and laying them in the ground. The making of this flit will be neceffary, or at leaft the well breaking of the bark ; otherwife they will not ftrike root ; and if this be done with judgment, they will have good roots by the autumn following, many of which will be good plants, and fit for the fhrubbery ; whilft the weaker may be planted in the nurfery-ground for a year or two, to gain ftrength. One caution is to be obferved : If the layering is to be performed by twifting the young fhoots fo as to break the bark, be careful not to over-do this ; for being very pithy, it will kill them to be much twifted ; and if the bark is not well broke, they will not ftrike root this way. 3. Thefe trees are to be encreafed alfo by cuttings ; from which they will grow very well. The cuttings muft be the bottom part of the laft fummer's fhoot, which fhould be planted

in

in October, in a shady border of light earth. If the spring
should prove dry, give them some watering, and there will be
little fear but that most of them will grow.

_____

S T E W A R T I A.

LINNEAN Class and Order, *Monadelphia Polyandria* : Each
flower contains many males, and five females ; the males being
joined in one set at the base : There is only one SPECIES :

STEWARTIA *Malacodendron* : The STEWARTIA ; *a deciduous
shrub* ; native of Virginia.

The STEWARTIA is a shrub of about eight or ten feet
growth with us, and the branches, which are produced irregularly from the sides of the main stem, are covered with a brown
bark. The leaves are placed alternately on the branches, and
are of much the size and make of those of the Cherry-tree.
Their upper surface is of a fine green, though they are lighter
and hairy underneath, and have their edges most acutely serrated. In the beginning of June this tree will be in blow.
The flowers are produced from the sides of the branches : they
are white, and seem to be composed of five large oval petals;
but upon examining them to the bottom, we find them joined
at the base. The flowers have a genteel look, are possessed of
an air of delicacy; and this being at present a very scarce
plant, makes it more valuable. It was named *Stewartia* in
honour of the right hon. the Earl of Bute, as a compliment to
his great skill in the science of botany.

This Plant is PROPAGATED by layers and seeds. 1. The
young shoots should be layered in autumn, by making a slit at
the joint, as is practised for Carnations. In the spring, a tall
hedge of some kind should be made on the south side of them,
bending also a little towards the east and west, that they may
be shaded all the summer. In dry weather they should be watered; and then they should remain until the March following,
when they should be examined to see if they have struck root ;
for sometimes they will strike root pretty freely, if so shaded and

watered ;

watered; and sometimes they have disappointed our expectations after waiting two years; though cuttings will sometimes grow. In March, however, a sufficient quantity
of pots must be provided, filled with good garden-mould,
mixed with a share of drift-sand; and the layers should be taken
up, whether they have struck root or not, and planted in these
pots, which must be plunged up to their rims in a bark-bed.
Those layers that have no roots will have the parts ready for
striking, and this assistance will set them all forward; so that
in a very little time they will become good plants. They must
be hardened as soon as possible to the open air. For this purpose the pots should be taken out of the beds, and plunged up
to the rims in a shady place; and though these are hardy trees,
it will be proper to take the pots up, and remove them into the
greenhouse, or under some shelter, for the first winter. At the
latter end of March they may be turned out of the pots, with
their mould, into the places where they are wanted to stand.
2. Another method of propagating these plants is from seeds,
which we receive from abroad. These should be sown in pots of
light earth, about half an inch deep, and the pots should be
plunged up to the rims in a bark-bed; where all the advantages
of heat, water, and shade, must be afforded them; for without these requisites, it is not often that they will grow.

S T Y R A X.

LINNEAN Class and Order, *Decandria Monogynia:* Each
flower contains about ten or twelve males, and one female:
There is only one SPECIES:

STYRAX *Officinalis:* The STORAX-TREE; *a tall deciduous
shrub*; native of Italy, Palestine, and Syria.

The STORAX, in its native places of growth, will arrive to
be more than twenty feet high; with us, twelve or fourteen
feet is the height we may expect it to grow to. The branches
are covered with a smooth greyish bark; and the younger shoots
are of a reddish colour. The very wood of this tree is finely
scented; and in Turkey and other places where it naturally

grows, that fragrant refin called Storax exfudes from its trunk, an incifion being firft made. The virtues of this refin are well known, and the tree is rendered valuable on that account. The leaves which ornament the flender branches, that are produced without order all around, are of a moderate fize, and of an oval, pointed figure. Their edges are a little waved, though free from ferratures. They grow on fhort footftalks, without any order, being fometimes by pairs, fometimes fingly, producing a pleafing irregularity. They a little refemble the leaves of the quince-tree, and are of two colours; their upper furface is of a lucid green, but their under is hoary; and this difference of colours makes a good contraft, efpecially when waving with the wind, on this charming fweet-fcented tree. The flowers are produced in June, from the fides of the branches, in bunches; feven or eight flowers will conftitute a tuft. Their form and colour fomewhat refemble thofe of the orange-tree, and their odours are diffufed all around. Thefe flowers are fucceeded by no fruit with us; fo that the height of its beauty is when it is in full blow.

The PROPAGATION is from feeds, which we receive from abroad. Thefe muft be fown an inch deep, in pots of light fandy earth, which pots fhould be plunged in a fhady well-fheltered place, there to remain until the fecond fpring after fowing. In March the feeds will be ready to fprout; and to affift them, it will be neceffary to take up the pots, and fet them up to the rims in a hot-bed. When the plants come up, all convenient air muft be given them; often water; and they fhould be hardened foon to the open air. They fhould be then fet abroad in the fhade, and in the winter fhould be removed into the greenhoufe, and placed under fhelter. In the fpring it will not be neceffary to force them a fecond time in the hot-bed; for if the pots are fet in a fhady place up to the rims, and now and then a little watering afforded them, the plants will grow very well, and make good fhoots that fummer. Like greenhoufe-plants, at the approach of winter, they muft be removed into fhelter; and in fpring they muft be fhook out of thefe larger pots, and each planted in a feparate fmaller pot; and being well watered, if they are plunged into a hot-bed, it will fet them growing finely. After they have had help this

way,

way, they muſt be ſoon hardened, and the pots taken up, and ſet up to the rims in mould in a ſhady place. In winter they ſhould be placed in the greenhouſe as before; and this method muſt be continued for ſix or eight years, treating them exactly as hardy greenhouſe-plants, and ſhifting them into freſh pots, as their encreaſe of ſize by growth requires. By this time they will be woody and ſtrong; and may then, the beginning of April, be turned out of the pots, with the mould, into the places where they are deſigned to remain. If the ſoil be naturally dry and warm, and the place well ſheltered, nothing but very ſevere froſts will injure them, eſpecially after having ſtood a winter or two.

S Y R I N G A.

LINNEAN Claſs and Order, *Diandria Monogynia :* Each flower contains two males and one female: There are only two SPECIES:

1. SYRINGA *Vulgaris:* The COMMON LILAC; *a tall deciduous ſhrub ;* native of Egypt.

2. SYRINGA *Perſica :* The PERSIAN LILAC; *a deciduous ſhrub ;* native of Perſia.

1. The COMMON, or EGYPTIAN LILAC. The *varieties* of this Species are,

The Purple Lilac,

The Blue Lilac.

The White Lilac,

The *Purple Lilac* generally riſes to the higheſt ſize of any of the three ſorts, though the height of all of them is either greater or leſs, according to the ſoil in which they are planted. The Purple, in good, light, rich earth, will grow to be ſixteen or twenty feet high ; and the others, in the ſame ſort of mould,

nearly

nearly as high. The Purple Lilac is naturally of an upright growth, though it foon divides into branches; and thefe alfo, as the tree grows older, into others, all of which are covered with a fmooth brownifh bark. All winter the plant has a bold and healthy look, occafioned by the large and turgid purplifh buds, which will have begun to fwell early the preceding fummer, and which will burft forth into leaf foon in the fpring following. The leaves are large and fmooth, and of a pleafant dark green colour. They are of an oval, cordated figure, end in acute points, and grow oppofite by pairs on the branches. The flowers will be produced in May, at the end of the fame fpring's fhoot, in very large and almoft conical bunches. They are of a purplifh colour, are clofely placed, and the number of which each bunch is compofed is very great. " I have meafured a bunch of them, fays HANBURY, a foot long ; and can any thing be thought to excel fuch a profufion of flowers, in its aggregate ftate, of which each clufter is compofed ! But many of thefe flowers appear all over the tree, mixed in an eafy manner among the delightful leaves ! fome peeping as it were above them, and feveral reclining their tops, to make the appearance ftill more free and eafy. The value of thefe flowers is ftill heightened by their delightful fragrance ; and when their blow is over, which it will be in a fortnight or three weeks, they have paid us their tribute, except what they afford from their leaves and manner of growth ; for they are fucceeded by feed veffels, of fuch a colour and nature as none but the curious botanift can find any pleafure in obferving."

The *Blue Lilac* differs in no refpect from the Purple, except that the branches are rather more flender and lefs erect, and that it feldom rifes higher than twelve or fourteen feet. The branches are covered with a fmooth brownifh bark ; and the buds in the winter will be turgid like the former, though fmaller ; and they, as well as the young fhoots, will have a blueifh tinge. The leaves are exactly like the preceding fort, though they will have a caft of blue. The flowers are produced in May, in not quite fuch large bunches as the former fort ; the bunches will be alfo loofe. They are of a fine blue colour, and admirably fcented ; and the preference is to be given with juftice to neither of thefe trees.

The

The *White Lilac* feems rather a ftiffer plant than the Blue, and the branches grow more erect than any of the forts. The young branches are covered with a fmooth light-coloured bark; and in winter the buds, which will be large and turgid, are of an herbaceous yellow colour, by which this fort at that feafon may be eafily diftinguifhed from the others. The leaves are of the fame figure and nature, though their colour is lighter, thereby making a variety. The flowers are of a fine white colour; and are produced in the fame kind of large clofe panicles as the others, which ftand upright. They are very fair, and, in the bunches, are fet very clofely together, which caufes them to be more erect than either of the two former forts. Thus may any perfon who has never feen thefe trees form an idea of their beauty when in blow; which will be very early, when the plants are fmall; for they will begin flowering at the height of four or five feet, and will every year after afford greater plenty of flowers as they advance in growth. The bunches generally grow by pairs, two at the end of the fame fpring-fhoot, though of unequal fize, the one being generally much larger than the other.

2. The PERSIAN LILAC. The *varieties* of this fpecies are:

The Common Perfian Lilac (or Perfian Jafmine),
The White Perfian Lilac,
The Blue Perfian Lilac,
The Cut-leaved Perfian Lilac.

The *Common Perfian Lilac* feldom grows higher than five feet, and is deemed a moft delightful flowering-fhrub. The branches are long, flender, flexible, and covered with a fmooth brownifh bark, with a blueifh tinge, on which are often feveral yellowifh punctules. The buds will be large and turgid in winter; and the leaves and flower-buds will come out early in fpring. The leaves are of a lanceolated figure, of a fine green colour, and grow oppofite by pairs on the branches. The flowers will be in full blow before the end of May. They are of a blueifh colour, and are produced in the fame kind of panicles as the other forts, though they will be fmaller and loofer. Their odour is more heightened than that of the others; and the fhrub, on the whole, is very valuable, though now pretty common. The long
flexible

flexible branches have a natural tendency to hang downwards; and when in blow their bunches of flowers will greatly encreafe this tendency; on which account it will be proper to place a few fticks to fupport them, which may be difpofed in fuch a manner as to efcape notice, unlefs by the nicelt examiner; and this will be proper, as the feeing the branches tied to fticks in full view, would fhew a degree of ftiffnefs which would not look well.

White Perfian Lilac will grow to the fame height with the former. The leaves, buds, and fhoots are of a lighter colour, It produces its flowers at the end of May, in the fame kind of panicles as the other (though thefe are of a white colour) and poffeffed of the fame heightened odour.

Blue Perfian Lilac differs from the preceding, in that the flowers are of a deep blue colour, thereby caufing a pleafing variety on that account.

Cut-leaved Perfian Lilac affords the greateft variety by its leaves; though the bark is rather darker, and the twigs feem flenderer, and are ftill more pendulous than the other forts. The leaves of this fort are divided, almoft to the mid-rib, into an un-certain number of fegments; and as this occafions them to have a different, an unfrequent and a fingular look, the value of the plant is much heightened on their account; particularly as it is in no refpect diminifhed in the elegance and fragrance of its flowers.

The beft way of PROPAGATING all thefe forts is by layers; for if this work be performed in autumn, on the young fhoots, they will be good plants by the autumn following. This me-thod is particularly to be preferred in the three firft forts of Li-lacs, as they naturally throw out fuch plenty of fuckers as to weaken, unlefs conftantly taken off, and diminifh the beauty of the mother-plants. Plants raifed by layering will be lefs liable to throw out fuckers, and confequently will be more valuable. The common way, indeed, is to take up the fuckers, and plant them in the nurfery for a year or two, and then fet them out fi-nally; but thefe plants will not be fo valuable as the others, as they will be more liable to produce fuckers, which to the gar-dener, when he has got a fufficient ftock of plants, become very troublefome.

The

The Perfian forts being lefs liable to put up fuckers, may not only be encreafed by layers; but when they do throw out any, the fuckers may be taken up, and deemed good plants. Cuttings of thefe forts, alfo, planted in Auguft, in a fhady moift place, will often grow.

The Perfian Lilacs never produce feeds with us, but the firft three forts do; and by thefe the plants may be encreafed; which alfo is a good method. The feeds ripen in the autumn; and in October they fhould be fown. They are rather fmall; and therefore the mould of the beds fhould be very fine, and they fhould be covered over lightly. In the fpring they will come up, and will want no other care than weeding. In the fpring following they may be planted in the nurfery, a foot afunder, and two feet diftant in the rows; and here they may ftand two or three years, when they will be of a proper fize to be finally planted out, and will flower in a year or two after. The differences of all thefe three forts are generally permanent from feeds; fo that a perfon may fow them with reafonable hopes of obtaining the forts the feeds were gathered from.

T A M A R I X.

LINNEAN Clafs and Order, *Pentandria Trigynia:* Each flower contains five males and three females; There are only two SPECIES:

1. TAMARIX *Gallica:* The FRENCH TAMARISK; *a tall deciduous fhrub*; native of France, Italy, and Spain.

2. TAMARIX *Germanica:* The GERMAN TAMARISK; *a deciduous fhrub*; native of low overflowed places in Germany.

1. The FRENCH TAMARISK will grow to the height of about fourteen feet. The branches are few, and fpread abroad in an irregular manner, fome being upright, others horizontal, whilft others decline with their ends towards the earth. The bark is fmooth, and of a deep-red or purplifh colour next the

E e   fun;

fun; but on the oppofite fide of the branch of a pale brown.
The leaves are rather of a pale green, and very beautiful. They
are very narrow; and upon examining them we find them fcaly
in a fine degree. The flowers will be produced in plenty at the
ends of the branches: They grow in feemingly very large loofe
panicles; but on examining them, we find that each is compofed
of numerous compleat flowers, which grow in fpikes, and are
produced near the extremities of the branches on the flender
twigs all around. Each of thefe fpikes feparately is but fmall;
and they are of a pale-red colour. The flowers of each fpike
are exceeding fmall; and the number of ftamina is five, which
differs from the other fpecies by only having half the number.
This fort flowers in July, and we have known it in full blow in
September, and fometimes in October, and even November,
when the weather has been all along mild. Nothing ornamental
fucceeds the blow.

2. The GERMAN TAMARISK is of lower growth, feldom afpiring higher than eight or ten feet. It is a more regular tree
than the former, as the branches all naturally grow in an upright
pofition. They are very brittle, are fcented, and covered with
a fmooth yellowifh bark. The leaves have a fcaly appearance,
and ftand much clofer together than thofe of the other fort:
They are of an exceeding light-green colour, and very ornamental. The flowers are produced in July, at the ends of the
branches, in long loofe fpikes. Each feparate flower is fmall,
though much larger than the other fort, and is poffeffed of ten
ftamina, which are alternately fhorter. Thefe fpikes attract the
attention when in blow, and are acknowledged by all to have
a fine look; neither is the noble appearance loft when the
flowers are faded; but it is continued in the fpikes even
until the feeds are ripe, which then feem to diffolve into a
fhattered down and fcales.

The PROPAGATION of thefe forts is very eafy: Every cutting will grow that is fet in winter, and will be a good plant
by the autumn following. The encreafing of thefe forts by
layers has been recommended; but this is bad advice, not only
as being unneceffary trouble, when they will grow fo freely by
cuttings, but becaufe layers of this tree very often will not ftrike
root at all. We have layered them, and found them, after lying
two years, without any roots; and the wound being grown up,

differed

differed from the other branches only in that the mould had a little altered the colour of the bark; which fhould warn all perfons who want a ftock of thefe plants to beware of lay-ering: and this, no doubt, they will do when we affure them, the cuttings will ftrike root as freely as thofe of the common willow. The beft time for the work is October, though any time of the winter will do. The cuttings fhould be of the laft fummer's fhoot; and a moift part of the garden is moft eligible for them to be planted in. In two years they will be good plants for the wildernefs or fhrubbery, and may then be planted out in almoft any foil, though they beft like a light moift earth, efpecially the German fort; as in other countries, where it grows naturally, it is generally found in low watery grounds.

T A M U S.

LINNEAN Clafs and Order, *Diœcia Hexandria:* Male flowers containing fix parts, and female flowers containing one part; upon diftinct plants. There are only two SPECIES:

1. TAMUS *Communis:* The COMMON BLACK BRIONY; *a climber;* native of England, fouth of Europe, and the Eaft.

2. TAMUS *Cretica:* The CRETAN BLACK BRIONY; *a climber;* native of Crete.

1. The COMMON BLACK BRIONY. This has a very thick flefhy root, full of a vifcous juice, blackifh without, white within, and from which iffue numerous flender twining ftalks, which wind about themfelves, or any thing that is near them, and will mount, if fupported, to about twelve feet high. The leaves are heart-fhaped, fmooth, undivided, of a fhining-green colour, and grow alternately on the ftalks. The flowers come out from the fides of the ftalks, in long bunches. They are fmall, of a whitifh colour, appear in June and July, and the females are fucceeded by round red berries, which ripen in the autumn.

E e 2

There

There is a *variety* of this with brown, and another with black berries.

2. CRETAN BLACK BRIONY. This has a large, fleshy root, from which issue many slender twining branches, which, if supported, will rise to about the height of the former. The leaves are trifid, or divided into three lobes. They are of a good green colour, smooth, and grow alternately on the branches. The flowers come out in bunches, from the sides of the branches. They appear about the same time as the former; and are succeeded by the like kind of red berries.

The PROPAGATION of both these sorts is very easy : It is, effected by parting the roots, or sowing the seeds. 1. The best time of parting the roots is early in the autumn, that they may be established in their new situation before the frosts come on. 2. The seeds also should be sown in the autumn, soon after they are ripe, otherwise they will often lie until the second spring before they make their appearance. A very few of these plants in the Shrubbery-quarters, will be sufficient. The best way is to well-dig the ground under the trees or bushes where you chuse they should grow; then put five or six berries in a place, covering them over about half an inch depth of mould. They will readily come up, will twist about the trees, and shew themselves to greater advantage than when directed by art in their course.

T A X U S.

LINNEAN Class and Order, *Diœcia Monadelphia*: Male flowers containing many stamina joined in one set at the base, and female flowers containing one pistillum; upon distinct plants : There are two SPECIES : TAXUS *Nucifera*; and

TAXUS *Baccata* : THE YEW; *an evergreen tree*; said to be a native of Britain, and most parts of Europe. It grows also in Canada.

The

The Yew will grow to a great fize. Evelyn mentions fome very large ones in his time; and Mr. Pennant, in a Tour in Scotland, took notice of a Yew in Fotheringall church-yard, the ruins of which girted fifty-fix feet and a half. The leaves of this tree form perfect feathers: the young leaflets are of a pale yellowifh hue; but the old leaves are of a darker green. Having been accuftomed to fee this tree, either as a fubject of torture or a companion of the dead, and gene-rally in an old declining and frequently difeafed ftate, we are either wholly unacquainted with its natural beauties, or over-look them. If, however, the Yew, ftanding fingly, be fuffered to form its own head, it becomes *ornamental* in a fuperior de-gree; it throws out its lower branches to a great extent; and, fhooting upwards, takes a ftrikingly conical outline; puting on a loofe genteel appearance. The timber of the Yew is tough and durable. Evelyn enumerates its *ufes:* " Befides the ufes of the wood for bows (for which the clofe and more deeply dyed is beft) the artifts in Box, cabinet-makers, and inlayers, moft gladly employ it; and in Germany they ufe to wainfcot their ftoves with boards of this material: alfo for the cogs of mills, pofts to be fet in moift grounds, and everlafting axle-trees, there is none to be compared with it: likewife for the bodies of lutes, theorboes, bowls, wheels, and pins for pulleys; yea, and for tankards to drink out of." He mentions whole woods of thefe trees diverfe miles in circuit, growing in the neighbourhood of Box-hill, in Surrey. Thefe woods, or rather, we apprehend, *plan-tations,* have lately been taken down (a few ftandards excepted); and the timber of fuch trees as were found were fold to the cabinet-makers at very high prices, for inlaying: one tree in particular was valued at a.hundred pounds, and half of it was actually fold for fifty. The leaft valuable were cut up into gate-pofts; which are expected to laft for ages: even ftakes made from the tops of Yew have been known to ftand for a number of years. We do not mention thefe circumftances as an induce-ment for making plantations of Yew, fo much as hints to thofe who may have Yew-trees in their poffeffion. Indeed, ornamental and ufeful as the Yew-tree undoubtedly is, there is one great objection to planting it: we mean its poifonous effects upon cattle. It is obfervable, however, that in the extenfive Yew-plantations above mentioned cattle were admitted with impunity, and ſtill range amongſt the ſtragglers that are left, without any evil conſequence. They are brouzed to the very bole: ſheep are particularly fond of the leaves, and, when the ground is covered with ſnow, will ſtand upon their hind legs, and de-vour them as high as they can reach. HANBURY, ſpeaking of this matter, ſays, " It has been thought dangerous to turn cattle into fields where Yew-trees grow; but, I believe, no beaſts will touch them, unleſs compelled by extreme hunger. It is true, ſeveral have loſt both horſes and cows by their eating the leaves of this tree; but this accident muſt be attributed either to the above-mentioned cauſe, or to the gardener having thrown the clippings careleſsly in places where cattle come; who, par-ticularly cows, will eat them when about half dry as greedily as new hay. By ſuch an accident, viz. a gardener's having thrown the clippings of a Yew-tree over the wall, a neigh-bouring farmer of mine loſt ſeven or eight of his beſt cattle; which ought to be a caution to all gardeners, whenever theſe trees are cut, to be careful that the clippings be either carried in for the fire or buried."

The PROPAGATION of this tree is from ſeeds: In autumn, when the ſeeds are ripe, a ſufficient quantity ſhould be gathered; and being firſt cleared of their mucilage, let them be ſown neatly, in beds about half an inch deep. By being thus expe-ditious in planting them, many will come up the next ſpring; whereas, if the ſeeds are kept out of the ground till February, the plants will not appear until the ſpring after. During the ſummer the beds muſt be kept clean from weeds; and if the weather ſhould prove very dry, now and then watered. This will promote the growth of the hidden ſeeds, and at the ſpring may be expected a general crop. The plants being come up, no other care will be neceſſary, for two years, than keeping the beds weeded, and refreſhing them in dry weather with gentle watering. After they have gained ſtrength in theſe beds by ſtand-ing two years, a piece of ground muſt be prepared for them, in which they ſhould be planted at a foot aſunder. Here they may ſtand for three or four years, and may be then planted where they are deſigned to remain, or ſet out in the nurſery in rows two feet aſunder, and three feet diſtance in the rows, in order to be trained for hedges, or raiſed to a good ſize to be planted out for ſtandards.

There

There are two *varieties* ; one with very *ſhort leaves*, and another with *ſtriped leaves*. Theſe are encreaſed by *layers*, and the ſtriped ſort, HANBURY tells us, muſt be ſet in a very barren ſoil, or it will ſoon become plain.

T H U Y A.

LINNEAN Claſs and Order, *Monœcia Monadelphia :* Male flowers containing four parts joined at the baſe, and female flowers containing one part ; upon the ſame plant : There are four SPECIES ; Two of them as follow :

1. THUYA *Occidentalis :* The COMMON ARBOR VITÆ ; *an evergreen tree ;* native of the moiſt ſwampy parts of America and Siberia.

2. THUYA *Orientalis :* The CHINESE ARBOR VITÆ ; *an ever-green tree ;* native of China.

1. The COMMON ARBOR VITÆ will riſe to thirty or forty feet high. The leaves of this tree are peculiarly formed, the leaflets being broad, and, in an advanced ſtate of the tree, thinly ſcattered : when bruiſed they emit a ſtrong, and to moſt people, very diſagreeable ſcent. In a youthful ſhrub-like ſtate, the Thuya nevertheleſs gives no unpleaſing variety, and may be admitted amongſt *ornamentals.* EVELYN and HANBURY arrange it amongſt Foreſt Trees ; and in Canada, the Indians, we are told, apply it to many *uſes.* HANBURY tells us " the wood is reddiſh, firm and refinous ; ſo that we may eaſily judge of its value for curioſities of moſt ſorts when worked up by the reſpective artificers of turnery, joiners, cabinet-makers, &c." He ſeems however to ſpeak from theory rather than from experience.

The Arbor Vitæ is to be PROPAGATED either from ſeeds, layers, or cuttings, the former of which produce the beſt trees, though the two latter methods of propagating are more generally practiſed. 1. In order to propagate this tree from ſeeds, theſe laſt ſhould be gathered as ſoon as they are quite ripe, which will be by the beginning of October. They muſt be ſown in pots or boxes of light fine earth, being covered about a quarter of an

E e 4
inch

inch deep. The boxes fhould immediately after be put in a well-fheltered place, fo that the feeds in them, whilft they are preparing to difclofe, may not be deftroyed by violent frofts. Being thus protected till the month of February, they muft be brought out, and fet along a South wall, that the fun warming the mould may fet the powers of vegetation at work; and whenever fevere weather is expected, they fhould be removed into their fhelter, but muft be brought out again when the fine fpringing weather returns. With this care, the plants will come up in the fpring; whilft, without it, they frequently lie till the fecond fpring before they make their appearance, by which neglect one year is loft. When the young plants are up, and all danger of the froft is ceafed, they fhould be fet in the fhade where they can have the free air; and in this place they may remain all fummer. During that feafon, little water fhould be given them; keeping them clean from weeds is the principal trouble they will caufe. By the autumn they will have made a poor fhoot; for this reafon they fhould continue in their pots or boxes, which muft be placed in the fame fheltered fituation they had at firft, where they may remain all winter. In the fpring they may be brought out into the fun again, to reap the benefit of his influence at that feafon; and if they are fet in the fhade at the beginning of May to remain there all fummer, it will forward their growth. The fpring following, being then two-years-old feedlings, they fhould be taken out of the boxes, and planted in beds nine inches afunder. Here they may ftand two years, before they are fet in the nurfery. When they are taken from thefe beds with this intent, a moift feafon ought always to be made choice of, and they fhould be planted a foot and a half afunder, and two feet and a half diftance in the rows, where they may ftand till they are fet out to ftand. 2. In order to propagate this tree from layers, the ground fhould be dug, and made light round about the ftools, and the branches laid down fo deep as that the top eyes may but juft peep above the ground, all being of the young wood. But if it fhould fo happen, that a few of the laft year's fhoots on the branches fhould have fhot out vigoroufly, and that there are many healthy twigs which would make good layers, that are not fo long; in order to have the greater plenty of layers, and that the fhorter fhoots may not be buried, it will be proper to fhorten the longeft, fo that, being

all

all laid in the ground, their nofes may juft appear above the surface. This will be a means of preferving every twig, and confequently of propagating the greater number of plants from the fame ftool. When thefe plants are layered, the fhoots ought to have a gentle twift or a fmall nick; for without this they will not always ftrike root: nay, if the land is ftrong and heavy, it is great odds but you find them without root, as you laid them, only grown bigger. Thus will one year be loft, which fhews the neceflity of obferving thefe precautions. Being layered in this manner in the autumn, by the autumn following they will have taken root; and in the fpring, when the fevere frofts are paft, they may be taken from the ftools, and planted in the nurfery, at the diftance directed for the feedlings. 3. In order to propagate thefe trees from cuttings, young fhoots fhould be taken from the trees in Auguft, if rain has fallen; if not, the bufinefs muft be deferred till it does; for work of this kind fhould never be performed till the early autumnal rains have fallen upon the earth, and made it cool and moift. All thefe cuttings ought to be of the laft year's fhoot; and if a bit of the old wood be left at the end of each, it will encreafe the certainty of fuccefs. The fituation thefe cuttings fhould have ought to be fhady and well fheltered; and the foil in which they are planted, to enfure the greater fuccefs, fhould be a red loam. They may be planted almoft as thick as you pleafe; not more, however, than four or five inches afunder, in rows; the rows may be a foot and a half diftance from each other; and after they are planted, a little litter may be laid between the rows, to keep the froft out of the ground in winter, and the fun from over-drying them in the fummer. This litter will not only keep down the weeds, but will fave the trouble of watering, which will be much better for the plants; for thefe young plants, juft ftriking root, do not much like watering, at leaft not in great plenty, as it often caufes the tender fibres to rot at firft ftriking, and fo deftroys the young plant. In one year thefe cuttings will have good roots; fo that the litter may be taken away, and the furface of the earth turned over in the fpring, which will cherifh the plants and prepare them to fhoot vigoroufly the fucceeding fummer. In the autumn, being then two years old, they may be taken up, and planted in the nurfery, at the diftance directed for the feedlings and layers. After they are

planted

planted in the nurfery, they will require nothing more than the ufual care of keeping them clean from weeds, and digging between the rows in winter, till they are planted where they are to remain.

October is the beft month for planting out thefe trees, tho' any of the winter or fpring months will anfwer. When they are planted, they fhould be fet a yard afunder, and thinned and managed as has been all along directed for others which are nearly of the fame growth.

There is a *variety* of this tree difcovered by Mr. HANBURY, which he has named The *American Sweet-fcented Arbor Vitæ:* this feems to remove a principal objection to the Common fort ; namely, the difagreeablenefs of its fmell. He fays, " It came up from fome fcattered feeds at the bottom of a box I had from Pennfylvania. It has the fame dufky look in winter as the Common fort, though it is better furnifhed with branches ; neither are they produced fo horizontally, or hang down in the manner of the Common fort. What makes this fort moft valuable is the property of its leaves ; for, being bruifed, they emit a moft refrefhing odour, which is by many fuppofed to be as fine an aromatic as any we have; whereas the leaves of the other forts being bruifed, to moft people are foetid and difagreeable. Whether this property will be continued by feeds, I have not yet experienced."

2. The CHINESE ARBOR VITÆ is a much more beautiful plant than the Common fpecies ; for its branches are more numerous, and grow in a more picturefque erect manner, and the leaves are of a fine pleafant light-green colour ; whereas thofe of the other in winter are of a dark difagreeable green, inclined to a dufky-brown, which is the worft property of this tree in the winter-feafon. The branches of the Common Arbor Vitæ are of a dark-brown colour, and the bark on the young branches is fmooth ; the bark of the Chinefe is alfo fmooth, and of a light-brown. The leaves of this fort, like the others, are imbricated, that is, they grow over each other; but they are more numerous and fmaller, and grow clofer together; and being of fo fine a green, which continues all winter, makes this fort the moft valuable, though not to the rejection of the others, even in pleafurable plantations ; for thofe caufe good variety by their manner of growth, as well as the colour of their leaves. The flowers

of

of none of the forts have any beauty : they have males and females diftinct; and the females of the Common Arbor Vitæ
are fucceeded by fmooth cones, whereas the cones of the Chinefe
fort are rugged. They are larger than the Common fort, and
are of a fine grey colour.

This fpecies, as well as the Sweet-fcented fort, may be PRO-
PAGATED by layers and cuttings, as has been directed above, for
the Common fort.

T I L I A.

LINNEAN Clafs and Order, *Polyandria Monogynia :* Each
flower contains many males and one female : There are only
two SPECIES :

1. TILIA *Europæa :* The EUROPEAN LIME, or the LINDEN-
TREE ; *a tall deciduous tree* ; native, it is faid, of England, and
moft parts of Europe.

2. TILIA *Americana :* The AMERICAN LIME ; *a deciduous
tree* ; native of Virginia and Canada.

1. The EUROPEAN LIME will grow to eighty or ninety feet
high, and from twenty to thirty feet in circumference. The
foliage is peculiarly foft and delicate, and its flowers fweet in the
extreme. It naturally forms a moft perfectly elliptical head ; and
even in winter its general appearance is rendered pleafing, by
the elegance of its long flender twigs. As ftandards, efpecially
in a rich deep foil, Limes are peculiarly eligible ; they are, in
fuch fituations, of very quick growth, and except the Oak and
the Efculus, few or no trees exceed them in point of *ornament.*
The *wood* of the Lime is light, foft, and peculiarly fine-grained :
it ranks with that of the Sycamore and the Poplar, and may ferve
upon many occafions as a fubftitute for the Beech : indeed, in one
point of view, it feems to exceed any of thofe woods, and ftands
upon its own bafis ; namely, for the purpofe of the carver : we
cannot, however, upon the whole, recommend it in general terms
to the planter as a timber-tree : land fuch as this tree requires
to render it of quick growth, ought rather to be applied to the
more ufeful purpofe of hufbandry, or, if convenient or neceffary

to

to be planted, fhould be occupied by the more valuable Oak or Afh ; for which neceffary woods, a certain and perpetual market may be expected.

The European Species affords feveral *varieties* ; as,

The Narrow-leaved Mountain Lime,
The Broader-leaved Mountain Lime,
The Elm-leaved Lime,
The Green-twigged Lime,
The Red-twigged Lime.

All thefe are very inconfiderable differences ; and though, if nicely obferved, they caufe fome variety, yet that is fo fmall as not to deferve much pains to procure them, except the Red-twigged fort, which of all others is the moft beautiful ; becaufe, when divefted of their leaves, its young branches exhibit their fine, fmooth, red bark all winter, which has a pleafing effect in all places ; though in the younger plants this effect will be more ftriking and delightful, as the bark only is red of the laft year's fhoots ; and the fmaller the plants are, the more of thefe and the lefs of older wood the compofition of the tree will be ; whereas, when the trees get older, the twigs will be fhorter and lefs vifible ; and though ftill of a red colour, yet not of fo delicate a red as the young plants wear on their bark at firft. Sometimes thefe trees will run away from their colour, and grow with green branches ; but as this is not common, the Red-twigged fort muft be ftill allowed to be preferable to all others ; and the feeds of this muft always be fown for the raifing of forts.

The PROPAGATION of the EUROPEAN LIME is from feeds, cuttings, and layers : HANBURY, however, fays, " That trees from layers or cuttings never grow fo handfome nor fo faft as thofe from feeds. Thefe fhould be gathered from thriving healthy trees of the true Red-twigged kind ; and then by far the greateft part of the young plants will be of that fort. The feeds will be ripe in October ; and let a dry day be made choice of for gathering them. As the feeds grow at the extremity of the branches, and as it would be tedious to gather them with the hand, they may be beaten down by a long pole, having a large winnowing fheet, or fome fuch thing, fpread under the tree to receive them.

When

When you have got a fufficient quantity, fpread them in a dry place, for a few days, and then fow them. The manner of fowing them is in beds of rich mould, about an inch deep, and about an inch afunder all over the bed. The plants will appear the firft fpring, and fhould ftand in the feminary two years, when they fhould be removed to the nurfery, planting them in rows, about two feet and a half afunder, and a foot and a half in the rows ; and here they may remain till wanted for ufe.

2. The AMERICAN LIME. Of this fpecies alfo there are a variety or two, which indeed differ very little in appearance from any of the Common European forts; for the leaves are heart-fhaped like theirs. There are a larger and a fmaller leaved fort. Their edges are finely ferrated, and end in acute points. Thefe beautifully cordated leaves, that thus run into acute points, have their under furface of a paler green than their up-per. The larger-leaved kind is by far the fineft fort ; and the branches vary from all others of this genus, in that they are covered with a dark-brown'bark. The flowers excite no atten-tion in the Gardener ; but the Botanift is delighted when he finds they are furnifhed with nectaria, whereas the flowers of our Common Lime-tree have none. The flowers are produced in bunches, like our Common fort, but make no better figure. They are very fragrant ; and are fucceeded by coriaceous cap-fules, containing the feeds.

The PROPAGATION of this fpecies is the fame as that of the European fort, if feeds can be procured from abroad; if not, a few plants muft be obtained. Thefe fhould be planted in a light rich foil, if fuch can be had, for in fuch they fhoot the ftrong-eft ; though almoft any other will do. After thefe plants have ftood a year or two, they fhould be headed near the ground, for ftools. They will then fhoot out many young branches from thefe, which may be layered in the autumn ; though if they ftand two years there will be greater plenty of young twigs for layering ; for every fhoot of the firft fummer will the year fol-lowing divide into feveral. When the layering of thefe is to be performed, which ought to be in the autumn, the ftrong two-years fhoots muft be brought down ; and if they are ftiff and do not bend readily, they muft have a gentle plafh with the knife

near

near the bottom ; a flit fhould be made at the joint for every one
of the youngeft twigs, and their ends bent backwards that the
flit may keep open. This being done, the mould muft be levelled among the layers, and the ends of them taken off to within one eye of the ground. The bufinefs is then done ; and
the autumn following they will have all good roots, many of
which will be ftrong, and fit to plant out to ftand, whilft the
weakeft may be removed into the nurfery-ground, in rows, to gain
ftrength. All the forts of Lime-trees will alfo grow from
cuttings ; but this is found to be an uncertain method ; and if
it was more certain, plants raifed either by them or layers are
not near fo good as thofe raifed from feeds, which way ought
always to be practifed where they can be obtained. Where that
is not to be done, any art muft be ufed to obtain fome few
plants ; and if the gardener fhould happen to procure a cutting
or two of the American forts, fet them in pots, and plunge
them in the bark-bed ; let him water and fhade them, and they
will be fure to grow ; and thefe he may afterwards encreafe at
pleafure.

V I B U R N U M.

LINNEAN Clafs and Order, *Pentandria Trigynia :* Each
flower contains five males, and three females : There are eleven
SPECIES ; eight of which are proper for our collection :

1. VIBURNUM *Lantana :* The COMMON VIBURNUM ; or
WAYFARING-TREE, or PLIANT-MEALLY-TREE ; *a deciduous
fhrub or tree ;* native of England, and moft of the northern parts
of Europe.

2. VIBURNUM *Dentatum :* The SAW-LEAVED VIBURNUM ;
a deciduous fhrub ; native of Virginia.

3. VIBURNUM *Nudum :* The ENTIRE-LEAVED VIBURNUM ;
a deciduous fhrub ; native of Virginia.

4. VI-

4. VIBURNUM *Prunifolium:* The PLUM-LEAVED VIBUR-
NUM, or the BLACK HAW; *a deciduous shrub;* native of Virginia and Canada.

5. VIBURNUM *Opulus:* The MARSH ELDER; *a tall deciduous
shrub;* native of moist grounds in England, and most parts of
Europe.

6. VIBURNUM *Acerrifolium:* The MAPLE-LEAVED VIBUR-
NUM; *a deciduous shrub;* native of Virginia.

7. VIBURNUM *Caffinoides:* The BASTARD CASSINE; or
CASSIOBERRY, or SOUTH-SEA THEA; *a deciduous shrub;* native ·
of Virginia.

8. VIBURNUM *Tinus:* The LAURUSTINUS; *an ever-green
shrub;* native of Italy and Spain.

1. The COMMON VIBURNUM will grow to be twenty or more
feet high, though it may be kept down to any height desired;
and in such gardens as are at a distance from the places
where it grows common, and in which it has not been before observed; in such gardens, it is enquired after, and
attracts the attention of those who walk therein, almost as
much as any shrub in the whole collection. The branches are
not very numerous, and in winter they are covered with a
smooth greyish bark, inclined to a brown colour, especially near
the bottom of the shoots. The younger, as they shoot, are
white and downy, and the ends, especially in winter, feel
soft and woolly. The branches are long, and exceeding tough.
They will often shoot near six feet from the bottom in a
year; and make the best bands for faggoting. The leaves
are very large, heart-shaped, very full of large veins, and have
their edges serrated. Their upper surface is of a dark-green
colour, but their under is white, and like cotton; and they
are placed opposite by pairs on the branches. The flowers are
produced at the ends of the branches: the buds will be formed
the preceding summer, which continue to get larger in the au·
tumn; all winter they will be in a state of increase, and at that
season they terminate the ends of the branches like so many
rough buttons. The flowers, when out, will be in large umbels, to form which these buds encrease in size all spring,
but shew little of what may be expected from them until about

May,

May, when they begin to divide, and shew that they are
growing to be bunches of flowers. In June, they will be wholly
out, and formed into large umbels; they are of a white colour,
and have a most noble look. These flowers are succeeded by
berries, which are also ornamental, and cause variety; for they
will be first of a fine red colour, and afterwards of a deep
black.

There is a *variety* of this sort, with more oval leaves; but
the differences are very inconsiderable in all respects. There is
also the *striped leaved Viburnum*, which is coveted by those who
are fond of variegated plants.

2. The Saw-leaved Viburnum is so called, because the
leaves are more beautifully serrated than any of the sorts. It is at
present not very common. Its branches, leaves, and flowers, are
not so large as the former, but they are of a more genteel
growth. It will grow to the height of about ten feet. The
bark is smooth, and of a light colour; and the leaves are of a
fine light green. They are tolerably large, though nothing like
those of the other sorts, and stand on longish footstalks, which
give them a fine air. They are strongly veined, and have their
edges finely serrated. They are of a roundish oval figure,
and are placed opposite by pairs on the branches. The flowers
are produced in June, at the ends of the branches, in very large
round bunches: Their colour is white; they appear in June;
and are seldom succeeded by any berries in England.

3. Entire-leaved Viburnum. The sorts of Lauruftinus
are evergreens, and have all entire leaves; but this species of
Viburnum agrees in every respect in description with two sorts,
one of which sheds its leaves in winter, whilst the other retains
its verdure during that season. The deciduous kind grows to
about ten feet high: the younger branches are covered with
a smooth, deep-red bark; whilst that of the older, tho' smooth,
is of a dark brown colour. The leaves are pretty large, and
of a delightful shining green on their upper surface; but their
under is paler, and much veined: they are of a lanceolated,
oval figure, though their ends are rounded; their edges are entire, and they stand opposite by pairs on the branches. The
flowers are produced in July at the ends of the branches in large
umbels; their colour is white; and they have much the resemblance of those of the Common Lauruftinus, though they
are

are rather fmaller. They have a genteel look ; and are fucceeded by berries, which never ripen with us.

4. PLUM-LEAVED VIBURNUM, or BLACK HAW. This fpe.
cies, for the moft part, goes by the name of Black Haw, becaufe the fruit a little refembles that of the Haw, though of a
black colour. It will grow to be about ten feet high ; and the
branches are covered with a fmooth reddifh bark. The leaves
are oval, and not fo large as any of the other forts, being feldom more than two inches long, and proportionally broad.
They are of a light pleafant green colour, and have their edges
finely ferrated. Their footftalks are pretty fhort, and they grow
for the moft part oppofite by pairs on the branches. The
flowers are produced in June, at the ends of the branches, in
large umbels. Their colour is white; but they are feldom fucceeded by berries in England.

5. MARSH ELDER. Of this fpecies there are two notable varieties. Marfh Elder with Flat Flowers. Gelder-Rofe.

The *Marfh Elder with flat flowers* will grow to be a tree
near twenty feet high. The young branches are covered with
a fmooth and almoft white bark. They are often produced oppofite by pairs ; though in general they are of an irregular
growth. The young fhoots will be cornered ; and this is
more perfect in the more vigorous ones, being compofed of
five or fix flat fides. The leaves are large and ornamental, of
a fine green colour and a foft contexture, compofed of three
large lobes, which are jagged at their edges, and grow on
glandulous footftalks. In autumn thefe leaves have exquifite
beauty ; for they die to fo fine a red, as to have a ftriking effect at that feafon. The flowers are produced in large umbels, the beginning of June, all over the tree, and have a grand
look. Each umbel is compofed of very many hermaphrodite
flowers, which of themfelves make no great figure ; but they
are furrounded by a border of male flowers, which are white,
and are fo ornamental to each bufh as to throw a luftre over the
whole tree. Neither does this fhrub ceafe to exhibit its beauties when the flowers are over ; for befides what it affords by
its leaves, which are inferior to few other trees, both in fummer and autumn, the hermaphrodite flowers will be fucceeded
by fine fcarlet berries, which will grow in fuch large bunches
and be produced in fuch plenty all over the fhrub, as to give it

F f

an

an appearance fuperior to almoft any thing of the berry kind and were it not for its commonnefs, this would, on their account only, be ranked amongft trees of the firft value.

The *Gelder-Rofe*, or *Snow-ball-Tree*, is a variety only of the preceding fort; its origin was accidental, and it is kept up and continued by culture in our gardens. The nature of the fhoots and fize of the tree, together with the colour of the bark, differ in no refpect from the former. The leaves alfo are of the fame form, are produced in the fame manner, and die away to the fame delightful red in the autumn. The variety this fort occafions, then, is by the flowers; and by thefe this variety is fo great, as to be exceeded by fcarcely any two diftinct fpecies whatfoever. They are produced in the beginning of June, all over the tree, in large globular bunches. Each bunch is compofed of numerous male flowers, of the fame nature with thofe that furround the hermaphrodite flowers of the former fort. Their colour is white, like thofe; but being produced in large globular heads, and in great plenty, have a much finer appearance. HANBURY adds, "It is delightful to fee this tree ufher in the month of June, as it were, with its glorious flowers, which will then at a diftance have the appearance of balls of fnow, lodged in a pleafing manner all over its head."

6. MAPLE-LEAVED VIBURNUM. This is a middle-fized fhrub, fending forth feveral branches, which are rough, and full of pith. The leaves are compofed of three principal lobes, like thofe of the Maple-tree, and grow on fmooth footftalks. The flowers come out from the fides of the branches, in umbels: Their colour is white, they appear in June, and are rarely fucceeded by feeds in England.

7. BASTARD CASSINE, Caffioberry-bufh, or South-Sea Thea, is rather tender, will grow to about ten feet in height, forming itfelf into a bufh by rifing with three or four ftems, and fending forth numerous branches from the bottom to the top. The leaves are of an oblong, lanceolated figure, ferrated, grow oppofite by pairs, and continue on the trees until the nipping frofts come on; infomuch that in the early part of a mild winter, they have been taken for an evergreen. Thefe leaves are of an exceeding bitter nature, if chewed; and it is faid that an infufion of them proves efficacious in removing

pain,

pain, bracing a relaxed ftomach, and reftoring a loft appetite. The flowers are produced in bunches from the fides of the branches. Their colour is white; they appear at the end of· July; and are fucceeded by red berries in the autumn. Whenever this plant is to have a fhare in a Collection, a naturally warm and dry foil, that is well fheltered, muft be fought 'for, otherwife there is a chance of lofing it by frofts; or if the plant is not wholly deftroyed, the young branches will be killed, and the tree fo haggled, as to have rather a bad appearance with others in the fpring.

The firft fix forts are very eafily PROPAGATED, either by feeds, layers, or cuttings. No particular art need be ufed for the feeds, whether they be of the forts of our own ripening, or of thofe we receive from abroad. A border of common garden-mould, made fine, will be fufficient; though it may be proper to obferve, that many of them will lie until the fecond fpring before they appear. The beds, before and after the plants are come up, will want nothing except weeding; and when they are a year or two old, they may be planted in the nurfery, at fmall diftances; and in two or three years more they will be fit to be finally planted out. 2. They are all eafily propagated by layers alfo; for if branches are pegged down, and the mould any-how thrown on them, they will have plenty of roots by the next autumn; and moft of them will be good plants for almoft any place. This freedom, however, fhould be given to none but thofe of our own country; for the American forts, as being ftrangers, demand mcre care and neatrefs in the performance. 3. They are alfo eafily propagated by cuttings; for the young fhoots of thefe trees cut into lengths, and planted in a moift garden foil, in the autumn, will any of them grow; and this is our common method of propagating them: However, if a perfon has only a few plants of the American kinds, the beft way is to make fure of encreafing them by layers.

If a large quantity is wanted, the beft way to PROPAGATE the MARSH-ELDER is by feeds. As the GELDER-ROSE is a male flowering variety, and never produces any feeds, it muft always be propagated by layers or cuttings, by which the variety will always be preferved.

The BASTARD CASSINE is PROPAGATED by layers. The young shoots are fit for this purpose; and when they have taken root, if they are planted in pots, and protected for two or three winters, until they are grown strong plants, either in a greenhouse, or under a hotbed-frame or some cover, there will be less danger of losing them than by planting them immediately in the nursery, or where they are to remain for continuance. However, a person who has not these conveniences, must fix on the warmest and best-sheltered spot he can find; and having prepared the ground, let the layers be taken from the old plants in the spring; if the weather be moist, it will be so much the better; and let him plant them in the nursery, row by row, at two feet asunder. In the summer, they should be watered in dry weather, and when the winter frosts begin to come on, the ground should be covered with pease-straw almost rotten, old thatch, or tanners bark, to keep them from penetrating the roots. By this means many of the plants will be preserved; and this care may be repeated every winter until they are planted out to stand. But this is not so good or so safe a method as potting them, and managing them as before directed; for they may be then turned out of their pots, when wanted, mould and all together, without feeling the effect of a removal.

8. The LAURUSTINUS is one of the greatest ornaments of our gardens in the winter months, not only as it is a fine evergreen, but because, during that season, it will either be in full blow, or else exhibit its flowers and buds in large bunches ready to burst open, in spite of all weather that may happen; and the boldness of these buds, at a time when other flowers and trees shrink under oppressive cold, is matter of wonder and pleasure. There are many VARIETIES of *Laurustinus*; but those most remarkable are, The Narrow-leaved *Laurustinus.* The Broad-leaved *Laurustinus.* The Hairy-leaved *Laurustinus.* The Shining-leaved *Laurustinus.* The Silver-striped *Laurustinus.* The Gold-striped *Laurustinus.*

The *Narrow-leaved Laurustinus* is so called because, of all the sorts, the leaves of this are smallest. It is generally planted among the low shrubs; tho' we have known it trained up against a wall to fourteen or sixteen feet high. It produces

its

its branches irregularly, which will grow fo thick and clofe as to form a bufh ; for it hath that appearance when planted fingly in open quarters. The bark in fummer is green, and often a little hairy and glandulous ; in winter it is frequently of a dark-brown colour. The leaves grow by pairs, ftanding oppofite, on ftrong and very tough footftalks. They are of an oval figure, and their edges are entire. Their upper furface is fmooth, and of a ftrong green colour ; but their under is lighter, and a little hairy ; and they are at all feafons ery ornamental. The flowers are produced in large umbels, and are well known. It generally will be in full blow in January, February, March, and April ; during which time it will be covered with bloom, caufing a delightful effect.

The *Broad-leaved Lauruftinus* differs from the former fort, in that the leaves are broader, and the roots proportionably ftronger. It will arrive to a greater height than the other forts, and the umbels of the flowers are larger, though they will not be produced in fuch plenty ; it, neverthelefs, makes an excellent figure.

The *Hairy-leaved Lauruftinus* is as free a fhooter as the other, and the leaves are frequently as large, and differs from that in fcarcely any thing but that the leaves are hairy ; the young fhoots alfo are hairy to a great degree. In this refpect it makes a fmall variety. It flowers like the other forts ; but blows rather later than thofe.

The *Shining-leaved Lauruftinus* is ftill of about the fame growth, and the leaves are large and fair. They are of an oval figure ; and their upper and under furfaces are both fhining, though their under is veined, and of a paler green. It differs only in that the leaves and young fhoots are fmooth, fhining, and free from hairs ; and being of this lucid green, force efteem. It generally flowers later than the firft two forts.

The *two variegated forts* are only one or other of the above forts, ftriped with white or yellow ; though the forts ftriped with Silver we have met with have been the Broad-leaved kinds ; but the Gold-ftriped forts have always been the firft, or Narrow-leaved kind, with leaves ftriped or blotched with yellow ; and on thefe accounts, thofe who are fond of variegated plants covet them in their collection.

All

All thefe forts are eafily PROPAGATED; for if in winter a little mould be any-how thrown amongft the young branches, they will ftrike root, and be good plants by the next autumn. Notwithftanding thefe plants, however carelefsly the mould be thrown, will grow, it is not here recommended to the Gardener to practife that cuftom; it is expected that he be always neat in all his work; it is mentioned here only to fhew what may be done; but let him gently lay the branches down, ftrip off fome of the lower leaves, and with his hand draw the mould amongft the young fhoots, and leave them neated up, as if a workman had been there; and thefe will be all good plants by the autumn, the ftrongeft of which may be fet out to remain, whilft the youngeft may be planted out in the nurfery, at fmall diftances, to gain ftrength.

V I N C A.

LINNEAN Clafs and Order, *Pentandria Monogynia*: Each flower contains five males and one female: There are five SPECIES; three of which will bear our open air.

1. VINCA *Major*: The LARGE-LEAVED PERVINCA or PERIWINKLE; a *fhrub* or *climber*; native of England, France, and Spain.

2. VINCA *Minor*: The SMALL-LEAVED PERVINCA or PERIWINKLE, or the COMMON PERIWINKLE; a *climber*; native of Germany, France, and England,

3. VINCA *Lutea*: The YELLOW PERVINCA, or PERI-WINKLE; a *climber*; native of Carolina.

1. The LARGE GREEN PERIWINLKE has fmooth ftalks of a pale green colour, which, if fupported, will arife to about four or five feet high; but, unfupported, the tops turn again at about two feet high, and thus at a diftance form the appearance

pearance of a round evergreen fhrub, of that low fize; and when they are defigned for this, the fuckers muft be always taken off, otherwife they will foon form themfelves into a pretty large bed; for they will fend out thefe at fome diftance from the rotten plant, and the very tops bending to the ground will often take root, which, unlefs taken away or prevented, will foon fpread abroad, and take off the fhrub-like appearance of the plant. The leaves are of a delightful evergreen, and ftand oppofite by pairs on ftrong footftalks. Their edges are entire, and they are of an oval heart-fhaped figure. They are fmooth and fhining, and very ornamental in the winter months. The flowers are produced from the wings of the ftalks, almoft all the year round; are blue; but there will be fometimes white ones feen amongft them. They are compofed of one petal, ftanding fingly on upright footftalks. The tube is narrow, and nearly of a funnel fhape; but their brim is large and fpreading, fo as to form a pretty large well-looking flower.

2. The COMMON GREEN PERIWINKLE has fmooth green ftalks, like the former, though they are much more weak and flender, and will trail along the ground, and ftrike root at almoft every joint: fo that they will foon run a great way their general furface putting on a kind of rock-work-like appearance; though if they are planted near other fhrubs, they will rife to two or three feet high, and will caufe a pretty look amongft them this way. The leaves are fmooth, and of a fine fhining green colour. They are of an oval figure, their edges are entire, and they ftand oppofite by pairs on ftrong fhort footftalks. The flowers are compofed of one petal. They fpread open at the rims, and grow from the wings of the ftal's in the fame manner as the former, though they are much fmaller; and as they are not fo fubject to flower in winter, that is another reafon for their being held lefs valuable.

The *varieties* of this fpecies are,

The Green Periwinkle with Blue Flowers.

The Green Periwinkle with White Flowers.

The Green Periwinkle with Double Blue Flowers.

The Green Periwinkle with Double White Flowers.

        The

The Green Periwinkle with Double Purple Flowers.

The Gold-ftriped Periwinkle with White, Blue, and Double Flowers.

The Silver-ftriped Periwinkle with White, Blue, and Double Flowers.

All thefe forts are *varieties* of the Common Periwinkle ; tho' they may differ in the colour or properties of the flowers, or the variegation of the leaves. The White-flowering Periwinkle is this very fort, only the flowers are white ; the Double Periwinkle is the fame fort, only the flowers are double and of a reddifh colour ; the Gold-ftriped Periwinkle is alfo this fort, only the leaves are beautifully variegated with a gold colour ; and the Silver-ftriped with that of filver : the variegations are fo completely done, and their ftripes fo little fubject to vary or run away, that they are highly efteemed amongft the variegated tribe. There are Double Blue and Double White flowers belonging to both thefe forts ; and thefe are all the hardy varieties Nature affords us from this genus.

3. YELLOW PERIWINKLE has a twining flender ftalk, which twifts about whatever is near it. The leaves are oblong, and not much unlike thofe of fome of our willows. The flowers are both fingle and double ; and thus continue in fucceffion from June to the end of fummer. This fpecies muft have a warm light foil, and a well-fheltered fituation.

The PROPAGATION of thefe forts may be eafily feen to be not very difficult. With regard to the firft fort, the fuckers it naturally fends out may be taken up and multiplied at pleafure ; and the ends of the fhoots that turn again, and ftrike root into the ground, will be good plants when taken off : Nay, the very cuttings will grow ; fo that any defired number of thefe plants, be it ever fo great, may be foon obtained. With regard to the other forts, there is no end of their multiplying ; for as they will ftrike root, if permitted to lie on the ground, at every joint, one good plant of each fort will produce a hundred of the like in a feafon or two.

All thefe forts are very hardy, and will grow under the drip of trees, and flourifh in all foils and fituations ; no plants are more proper to be fet among low or larger fhrubs, either in the evergreen or deciduous quarters, to form tufts or beds

in

in the refemblance of rock-work, or to be placed near other
fhrubs, by whofe affiftance their flender ftalks may be fupported to the height Nature will admit them to rife.

V I S C U M.

LINNEAN Clafs and Order, *Diœcia Tetrandria:* Male flowers
containing four parts, and female flowers containing one
part; upon diftinct plants: There are nine SPECIES; one of
them common in many parts of this ifland:

VISCUM *Album:* The MISLETOE; a *parafitical plant;*
native of England and moft parts of Europe.

The MISLETOE is a fingular plant. It will grow upon trees
only; more efpecially upon the Crab, the Hawthorn and the
Maple: It is not unfrequent upon the Afh; but feldom, very
feldom indeed, is feen upon the Oak; and but rarely upon the
Willow. It has a thick flefhy leaf ftanding ftiff upon the
twigs, which are green and forked. The whole of the plant
is of a green colour, and of the fhrubby, bufhy kind, rifing in
numerous ftems; dividing into forked branches; and thefe
again into forked twigs, thickfet with leaves. This thickens
the general furface of the plant, and forces it into a fpherical
or more generally a hemifpherical form. A tree thickly fcattered with this plant, has fomewhat the appearance at a diftance as if overgrown with Ivy. The Mifletoe, however, is of
a lighter green than the Ivy; efpecially when full of berries,
which are of a light tranfparent Pea-green colour, and about
the fize of the common Field Pea; but when full ripe they
become paler, taking the appearance of white currants. The
pulp is vifcid in the extreme, being of the confiftence of thick
gum-water. Each berry inclofes one vetch-like feed. In the
cyder counties the Mifletoe is a mifchievous intruder upon the
Apple-tree; fo much, that were not the Farmers to cut it

out

out every three or four years, or as often as neceſſary, it would deſtroy the tree. It is very common to ſee Crab-trees, eſpecially in or near woods, entirely killed by this truly para-ſitical plant. This is a curious faſt in Nature, and affords ample ſubjeſt for reflection. The Miſletoe may be ſaid to be a ſuperior order of Plants ; for like the animal creation it feeds not upon the juices of the earth, but upon thoſe of vege-tables. This, added to its ſuppoſed medicinal qualities, aſſiſted, probably, in rendering it ſacred among the antient Bri-tons ; eſpecially when found growing upon the Oak ; which tree they alſo held ſacred.

The PROPAGATION of this Plant is ſuppoſed to be, na-turally, by the Miſletoe Thruſhes, which delight in its glu-tinous berries, and which in autumn, the ſeaſon of their be-coming ripe, repair in flights to the places where the Miſletoe abounds. It ſeems to remain unaſcertained whether the ſeed be conveyed in the fœces of the bird, or whether, ſticking to its beak amongſt the glutinous matter, the bird in cleaning its beak wipes it off upon the branch of the tree it happens to perch upon. This laſt is the more probable ſuppoſition ; as it has been found, that by ſtriking the ſeeds upon the clean ſmooth part of the bark of ſome or all of the trees abovementioned, this plant may be artificially propagated. We do not learn, however, that the attempt has yet been ſucceſsful upon the Oak or the Willow. It ſeems probable that the Bird, in wiping its beak acroſs the branch, ripples the cuticle or outer rind ; and this ought perhaps to be copied in attempting artificial propagation. In places where this plant is un-known, the cultivation of it would add a ſtriking variety to ſhrubbery quarters.

V I T E X.

V I T E X.

Linnean Clafs and Order, *Didynamia, Angiofpermia.* Each flower contains four males and one female ; two of the males being longer than the other two ; and the feeds being covered. There are eight Species; one only of which is proper for our Collection :

Vitex *Agnufcaftus :* The Agnuscastus, or the Tree of Chastity ; a *deciduous fhrub* ; native of marfhy, moift places in fome parts of France, Spain, and Italy.

The Tree of Chastity (being held by the antients as conducive to that amiable virtue) affords two *varieties* :

The Broad-leaved Chafte-tree.

The Narrow-leaved Chafte-tree.

One defcription will nearly ferve for both forts ; though it has been obferved, that the Narrow-leaved fort will grow to be the talleft. The branches are produced from the bottom and fides of the ftalk. They are very pliable, and the joints are long. It is difficult to exprefs the colour of the bark. To fay it is grey is not proper ; and to fay it is brown is not true ; it is of a colour between both, though, in different foils, the bark of fome trees will be of a darker colour than others. The leaves are digitated, being compofed of feveral folioles, which fo unite at their bafe in one common footftalk as to refemble an open hand. Thefe folioles are of a dark-green colour ; and their number is uncertain ; being five, fix, feven, and fometimes eight. They are narrow, and the longeft grow always in the middle, whilft the fhorter occupy the outfides. This character is common to both the forts ; though it is obfervable, that the folioles of the Broad-leaved fort are both fhorter and broader, which occafions its being fo called. Their edges are alfo ferrated, whilft thofe of the Narrow-leaved are intire ; and in this the moft important difference of thefe plants confifts. The flowers of both forts are produced at the ends of the branches, in whorled fpikes. Thefe fpikes are pretty long, and their colour is that of a bluifh pur-ple.

ple. They appear in September and October; and are not
fucceeded by feeds in England. Each individual flower is
inconfiderable; but the whole fpike makes a good fhow; and
the circumftances of the flowers being produced late, even
often when moft other flowers are over, as well as being alfo
very fragrant, greatly heighten their value. The early frofts
often deftroy the beauty of thefe fpikes, before and when they
are in full blow; fo that it is no wonder their ornamental fruit
feldom, if ever, fucceeds them.

There is a *variety* of each kind with white flowers.

The PROPAGATION of thefe forts is eafily done, either by
layers or cuttings. 1. The young fhoots being layered, any
time in the winter, will have roots by the autumn following;
though it will be proper not to take them up until the fpring,
as they fhoot late in the autumn, and have often their ends
deftroyed by the frofts. When this work is deferred till the
fpring, all the killed ends may be taken off; and all danger from
fevere frofts being over, they will meet with no check in their
preparing to fhoot. The removing of thefe trees in the fpring,
however, is not abfolutely neceffary; for it may be done any
time in the winter, though the cutting off the dead ends fhould
be deferred until the latter end of March, when they fhould be
gone over with the knife, and cut down to within an eye or
two of the ground, whether planted in nurfery-lines, or finally
fet out to ftand. 2. Plenty of plants may be foon raifed by
cuttings. About the middle of March is the beft time for
planting them; and they fhould be fet in a fhady border of
good light garden-mould. Nothing but weeding, and now
and then watering, will be required all fummer; though if the
place is not naturally well fheltered, they muft be defended
from black frofts by fticking plenty of furze-bufhes all around
them. If this be judicioufly done, it will take off the keen
edge of frofty winds, fufficiently, and will occafion much lefs
trouble and expence than reed-hedges, &c. All thefe plants
are very hardy; but they require this protection, to preferve
the young fhoots. Here they may grow until they are fully
planted out; and if it be a moift, light, rich foil, and a
well-fheltered fituation, they will like it the better,

VITIS.

V I T I S.

Linnean Clafs and Order, *Pentandria Monogynia:* Each flower contains five males and one female: There are eight Species; four of which are adapted to ornamental plantations:

1. Vitis *Labrufea:* The Wild Virginia Grape; a *climber;* native of many parts of North America.

2. Vitis *Vulpina:* The Fox Grape; a *climber;* native of Virginia.

3. Vitis *Laciniofa:* The Parsley-leaved Grape; a *lofty climber;* native of Canada.

4. Vitis *Arborea:* The Pepper-Tree; a *fhrub* or *climber;* native of Virginia and Carolina.

1. The Wild Virginia Grape, if defired for its climbing property, fhould be planted among pretty large trees or fhrubs; for, by the affiltance of its well-holding tendrils, it will arrive to a great height; and if the fhrubs that grow near it be low-growing ones, it will entirely over-top them; and in fummer, its leaves being large, almoft conceal them from the fight. Thefe large ornamental leaves have their edges indented, and are nearly divided into three lobes, though they are of a heart-fhaped appearance; and downy on their under fide. The flowers are produced in bunches, like the other fpecies of the Vine; and they are fucceeded by round, rough-flavoured, black fruit.

2. The Fox-Grape. The name of this fpecies naturally brings the fable of the fox and grapes to the memory; and it is very common for thofe who are not fkilled in the hiftory and nature of plants, to afk if this fpecies is not poffeffed of more excellent properties, or produces more defirable fruit, than moft of the other forts of the vine; whereas, alas! this fort is called the Fox-Grape from the ill flavour of its fruit, which is like the fcent of a fox, and which name the inhabitants of Virginia, where it grows naturally, have given it on that account. It muft, like the former, be planted among largifh trees; for it
will

will over-top the fmall ones. The leaves are large, fmooth on both fides, of an heart-fhaped figure, and their edges are indented. The flowers are produced in the Vine-like bunches; and they are fucceeded by black fruit of the above-named difagreeable flavour.

3. The PARSLY-LEAVED GRAPE. The leaves of this fort are finely divided, and at a diftance refemble thofe of parfly, though larger. The ftem is very thick, and the fhoots are ftrong; fo that when it is planted for a climber, the talleft trees muft be appropriated for its fupport; otherwife it will be too powerful for trees of lower growth.

4. The PEPPER-TREE is a weaker fhooting plant than any of the others, and affords fingular beauty from its leaves. Their upper furface is of a fine fhining green colour; their under is paler, and they are compofed of a multitude of folioles of the moft elegant and delicate texture. The fhoots will arrive to a tolerable height by their tendrils, if they have trees near for their fupport; but they are very liable to be killed down very low in fevere winters; on which account the plant fhould be ftationed at firft in a well-fheltered place. Every fpring the Gardener fhould carefully cut off not only the dead fhoots, but fhorten them within an eye or two of the old wood, which will make them fhoot ftronger, and the leaves will be larger and finer. The flowers are white, and are produced in bunches from the wings of the ftalks; but we have never yet perceived any fruit to fucceed them. The name Pepper-tree is a cant name, and was given it without any meaning by the inhabitants where it grows naturally.

All thefe forts are PROPAGATED by cuttings, layers, or fuckers. 1. The cutting muft be the bottom of the laft year's fhoot; and if there be a bit of the old wood to it, it will be the better. 2. When raifed from layers, the young branches fhould be pegged down, and a little foil drawn over them. They will ftrike root, and become good plant by the feafon following. 3. Suckers may be taken from thefe plants, and immediately planted; or may be fet in the nurfery for a year to gain ftrength before they are fet out.

U L E X.

U L E X.

LINNEAN Clafs and Order, *Diadelphia Decandria:* Each flower contains ten males and one female ; the males being joined at the bafe in two fets : There are two SPECIES; one of which is a green-houfe plant; the other is,

ULEX *Europæus:* The FURZE, WHIN, or GORSE ; an *ever-green fhrub*; native of England, France, and Brabant.

The FURZE is fo extremely common in this country, that how ornamental foever it may be in nature, it cannot with much propriety be admitted into our. *ornamental* plantations. Its *ufes* however are many ; as a fuel where wood and cou's are fcarce ; and as hedge-wood upon light barren land : its ufe as horfe provender too feems to be fully proved, though not yet eftablifhed.

HANBURY enumerates the following *varieties*; fome of which, if properly trained, may add a kind of fecondary orna-ment to our grounds and fhrubberies.

The White-flowered Furze.
The long Narrow-fpined Furze.
The Short-fpined Furze.
The large French Furze.
The Round-podded Furze.
The Dwarf Furze.

The FURZE is PROPAGATED from feeds fown very fhallow, in February or March. See the Article HEDGES.

U L M U S.

LINNEAN Clafs and Order, *Pentandria Digynia:* Each flower contains five males and two females.—LINNEUS makes only three SPECIES of ULMUS:

1. ULMUS

1. ULMUS *Campestris:* Leaves double-sawed; unequal at the base.

2. ULMUS *Americana:* Leaves equally sawed; unequal at the base.

3. ULMUS *Pumila:* Leaves equally sawed; equal at the base.

MILLER enumerates six SPECIES:

" 1. ULMUS *Campestris:* Elm with oblong acute-pointed leaves, which are doubly sawed in their edges, and unequal at their base; called The COMMON ROUGH or BROAD-LEAVED WITCH ELM.

2. ULMUS *Scaber:* Elm with oblong oval leaves, which are unequally sawed, and have leafy empalements to the flowers; called The WITCH HAZEL, or VERY BROAD-LEAVED ELM; by some unskilful persons called The ENGLISH ELM.

3. ULMUS *Sativa:* Elm with oval acute-pointed leaves, which are double sawed, and unequal at the base; called The SMALL-LEAVED or ENGLISH ELM.

4. ULMUS *Glaber:* Elm with oval smooth leaves, which are sharply sawed on their edges; called The SMOOTH-LEAVED WITCH ELM.

5. ULMUS *Hollandica:* Elm with oval acute-pointed rough leaves, which are unequally sawed, and a fungous bark; called The DUTCH ELM.

6. ULMUS *Minor:* Elm with oblong smooth acute-pointed leaves, which are doubly sawed; called The SMOOTH NARROW-LEAVED ELM, and by some The UPRIGHT ELM."

These six species of MILLER are all of them comprehended in the ULMUS *Campestris* of LINNEUS; so that MILLER is silent as to the second and third species of LINNEUS; and so is HANBURY, who only treats botanically of one species; namely, the ULMUS *Campestris* of LINNEUS: He nevertheless enumerates seven SORTS:

1. The true English Elm.
2. The Narrow-leaved Cornish Elm.
3. The Dutch Elm.
4. The Black Worcestershire Elm.
5. The Narrow-leaved Wych Elm.

6. The

6. The Broad-leaved Wych Elm.

7. The Upright Wych Elm.

In another part of his work he says, " It would be endless, as well as needless, to enumerate the forts of Elms : I have counted in my time more than twenty, in woods, hedges, &c. that have fell in my way when in queft of plants." The fact is, no genus of plants whatever is more incomprehenfible to the Botanift than the Ulmus ; for although we fee among the cultivated Elms of this country, individuals as different from each other as are fome individuals of diftinct genera, yet every man who has attended clofely to the feveral kinds of Elms grow-ing in different parts of the kingdom, muft have obferved fuch a chain of intermediate kinds as renders claffification extremely difficult ; and muft frequently have met with an individual, which he was puzzled to find a name for. LINNEUS, no doubt, having experienced this, lumped the whole mafs of cultivated Elms in one fpecies,— The *Ulmus Campeftris :*— and as a BOTANIST he may be right : as PLANTERS, however, we muft beg leave to attempt a feparation ; and yet we are obliged to confefs, that we cannot *defcribe* more than TWO *obvioufly diftinct* VARIETIES. With refpect to the fecond and third fpecies of LINNEUS, we take it for granted they have not yet been intro-duced, or are but little known, in this country (if we are wrong, we beg to be fet right) : We therefore proceed to

ULMUS *Campeftris :* The CULTIVATED ELM ; *a tall deci-duous tree* ; found growing more or lefs, in one or other of its varieties, in hedges about villages, in moft parts of Europe.

The CULTIVATED ELM. Notwithftanding the chain of varieties abovementioned, if we examine the two extremities, we fhall find two plants very different from each other in their general appearance ; and fufficiently diftinct in the analyfis to be confidered, in a work of this nature, as diftinct fpecies : The leaf of the one is nearly oval, with an obtufe lance-like point ; that of the other nearly circular, faving a narrow flender point, growing as it were out of the periphery of the circle. The membrane of the one is grofs and rigid, of the other comparatively thin, delicate, and fupple : When held againft a ftrong light, the former appears opake ; the latter, comparatively tranfparent. The nerves of *that* are ftronger, fet clofer, and run more parallel ; of *this,* more flender, fewer

G g   in

in number, and divide more into branches. *That*, in its general appearance, bears fome refemblance to the leaf of the Chefnut *;* *this*, a very ftrong one to that of the Hazel. The branches of the Coarfe-leaved fort are clean, ftraight, and flender, with a filvery bark; thofe of the Fine-leaved kind more divided, run fhorter lengths, and are covered with a lefs delicate bark. The general tendency of the latter is more upright, being eafily trained to a great length of ftem; that of the former is to divide into fpreading arms, and when attempted to be trained up with a tall ftraight ftem, generally grows ftooping with a nodding head like the panicles of an oat. The Coarfe-leaved kind matures its feed in this ifland, and is probably a native; the Fine-leaved fort feldom if ever perfects its feed with us, and is probably an exotic.

We therefore proceed to treat feparately of thefe two forts; confidering the intermediate kinds as fubordinate varieties of thefe two:

1. The COARSE-LEAVED ELM, or the Chefnut-leaved Elm, or the Broad-leaved Elm, or the Wych Elm, or the North-country Elm.

2. The FINE-LEAVED ELM, or the Hazel-leaved Elm, or the Narrow-leaved Elm, or the South-country Elm.

1. The COARSE-LEAVED ELM will grow to a very great fize. Mr. Marfham mentions a Wych Elm by Bradley Church in Suffolk, which in 1754 meafured (at five feet high) twenty-five feet five inches and a half, and in 1767 twenty-fix feet three inches. The leaves of this fpecies of Elm have been already defcribed to be of an oval figure, with a thick membrane and ftrong nerves; their fize varies with the individuals they grow upon, fome trees of this fpecies bearing leaves confiderably fmaller and *much narrower* than thofe of fome individuals of the Hazel-leaved fort: The common diftinction therefore of thefe two kinds of Elms into *Broad-leaved* and *Narrow-leaved* is altogether improper—their *figures* forbid it: It would be equally proper to diftinguifh an oval from a circle, by calling the former broad, and the latter narrow.

The outline or general appearance of this tree is fometimes ftrongly featured, coming near to that of the Oak: In general, however, it is liable to be ragged, rather than irregular, and in point of *ornament* is frequently exceeded by the Lime, the

Beech,

Beech, and its fifter tree the Fine-leaved Elm. Its *ufes* are many. The whole tribe of Elms have a peculiar excellency by which they ftand alone, and are rendered in a great meafure independent of other woods. The Oak is pre-eminent for durability, the Afh for toughnefs, the Beech for clofenefs of texture and cleannefs of grain, and the Elm for its tenacity or adhefive quality, being lefs liable to be fplit than other woods: This renders it fingularly ufeful for many important purpofes. The keels of fhips are now almoft univerfally laid with Elm, and fometimes the gunwales, efpecially of fhips of war, are made of this wood; it being lefs liable to fplinter off in action even than Englifh Oak; as keels made of this wood are lefs apt to fplit in taking the ground. Another very important ufe of Elm is for naves of wheels of carriages of every kind, whether of ufe or of pleafure. There is a fort in Yorkfhire peculiarly adapted to this purpofe, which goes by the name of the *Nave-Elm*; it is of the Coarfe-leaved kind.

The Coarfe-leaved Elm may be PROPAGATED from feeds, or by layering. HANBURY fays, " In order to propagate them by layers, proper ftools for the purpofe muft be firft obtained; to procure which, let a piece of good ground be double dug, and plant Elms of about four or five feet high over it, at the diftance of about ten feet: If they make good fhoots in the firft year after, they may be cut down early in the fpring following; if not, they fhould remain two years before they are headed for ftools; which fhould be by cutting them down to within half a foot of the ground. After they are cut down, they fhould be fuffered to grow undifturbed for two years: The ground between the ftools muft be dug in the winter, and conftantly hoed as the weeds arife in the fummer; and at the end of that time, that is two years, the branches growing from thefe ftools will be fit for layering; which may be performed thus: Excavate a piece of ground wide enough to receive a whole branch, and let the hollow be about half a foot deep; then fplafh the branch with a knife, near the body of the ftool, that its head may be more readily brought into the prepared place: Next, thruft an hooked ftick into the ground, to hold it clofe; take off all the fuperfluous branches, which crofs and would otherwife incommode thofe that are to be continued. After this, cut all the remaining young branches acrofs half through

G g 2 with

with the knife; turn the edge towards the end, flitting it about half an inch. When this is done on all the young branches, the mould should be gently put amongst them, and every one of them should have their ends bent towards the stool, that the slit may be open. Lastly, having the whole vacuity filled with its own mould, smooth and even, take the end of each twig off that peeps above the ground, down to one eye, and the branch is layed, and will afford you as many plants as there are buds peeping out of the ground. Proceed in like manner to the other branches of the same stool, then to the next stool in order, and so on until the whole business of layering is finished. By the autumn following, these layers will have taken root, and many of them will have made a shoot of near a yard in length. It is now necessary to take them from their stools, and plant them in some double-dug ground in the nursery. They should be set in rows three feet asunder, and the distance allowed them from each other in the rows ought to be a foot and a half. Here they may stand till they are planted out where they are to remain, with no farther trouble than digging the ground between the rows every winter, and in the summer carefully watching those which shoot out two branches at the head, and nipping the weakest of them off. After the layers are taken up, the stools must have all the wounded parts, occasioned by the former splashing, taken away; the old branches also should be cut off, pretty close to the stem; and in the spring they will begin to shoot out fresh branches again, for a second layering, which will likewise be ready to have the same operation performed the second year after: and thus may this layering be performed on these stools every other year. But nurserymen who would raise great quantities of trees this way, should be provided with two quarters of stools, to come in alternately, so that from one or other of them they may annually receive a crop." We have given Mr. HANBURY's method in his own words, in order to convey to our readers in the fullest manner *his method of layering.*

His method of PROPAGATING the Elm from seeds, we also give at length; for the Elm standing next to the Oak at the beginning of his book, he has treated more fully of that article than any other (the Oak only excepted), and frequently refers to it in the courfe of his work. He fays, " Let the feeds be gathered the beginning of June, it being the time when they are full ripe. When gathered, fpread them three or four days to dry; for if they were to be fown immediately after they were gathered, they would rot. Having been fpread about that time, and the mould, which ought to be frefh and good, being in readinefs for their reception, mark out your beds four feet wide, and let the alleys between them be a foot and a half or two feet broad. Rake the mould out of the beds until they are about an inch deep; riddle that which came out of the beds into them again, until the bottom of each bed is raifed half an inch (*i. e.* half filled) with riddled mould; then gently prefs the mould down with the back of the fpade, and fow the feeds thinly all over it with an even hand, covering them down with fine earth about half an inch deep. When the feeds are all fown this way, the beds fhould be hooped, and covered with mats, to be fhaded in that hot feafon of the year; and they fhould alfo fometimes be refrefhed with water: Part of the young plants will come up in about a month, or fooner; the others not till the fpring following. From the time the feeds are fown to their appearance above ground, whenever rain falls, be careful to uncover the beds, and as ready to cover them again when the fcorching beams of the fun break out. About the end of Auguft, the mats fhould be wholly taken away, that the plants may be hardened againft winter: The fpring following, a frefh breed will prefent themfelves among thofe that came up the fummer before. All the fummer following they fhould be conftantly kept free from weeds, and watered as often as dry weather fhall render it neceffary; and in October or fpring they may be planted out in the nurfery, at the diftance before prefcribed for the layers, and afterwards fhould be managed like them."

2. The FINE-LEAVED ELM will alfo grow to a great height and confiderable bulk: We do not however find any tree of this kind upon record. The largeft Elms we have feen, of the Fine-leaved fort, grow in the Vale of Gloucefter. There are feveral in the parifh of Church-down which girt, at five feet high, from ten to twelve feet. But the fineft Elm in the Vale ftands in the road between Cheltenham and Tewkefbury,—within a few hundred yards of the Boddington Oak (See QUERCUS). It is known by the name of PIFFE'S ELM; and the turnpike-gate,

the fence belonging to which is faftened at one end to this tree, takes its name from it; being called " Piffe's Elm 'Pike." The fmalleft girt of this tree, which falls about five feet high, is at prefent (1783) exactly fixteen feet. At ten feet high it throws out large arms, which have formerly been lopped, but which now are furnifhed with tree-like fhoots, rifing, by eftima- tion, to feventy or eighty feet high, with an extent proporti- onable, exhibiting all together the grandeft tree we have feen; — not fo much from its prefent fize, as from that fullnefs of growth and vigour which it now wears. There is an Elm of the Small-leaved fort in Hyde Park whofe ftem is larger than this; but it is hollow, its head much impaired, and is a mere dotard compared with Piffe's Elm; which we mention the rather as it may be a tree in ages to come, and, ftanding as it does in a well-foiled country, may fwell out to twice its prefent fize.

The leaves of this fpecies of Elm have been already fully defcribed; it remains only to obferve, that notwithftanding we are accuftomed to fee trees of this fort trimmed up to mere maypoles, or at beft with clofe afpiring heads, yet, if planted fingly and fuffered to form their own head, they will take an outline equal to that of the Beech or the Linden; and where an immediate object or fkreen is wanted, the Elm has two material advantages: it may be removed when of a great fize, and its growth is quicker than that of any other tree which is equally ornamental. The *ufes* of this fpecies of Elm are fimilar to thofe of the Coarfe-leaved kind; and in places where bricks are rendered dear by a want of a proper fupply of fuel, as in Surry and Kent, great quantities of this Elm are cut up for ftuds and weather-boarding for the fides of barns, ftables, and even dwelling-houfes; and in the fouthern counties in general, it is much ufed in carpenter's work of all kinds as a fubftitute for Oak,

The propriety of planting the Elm depends entirely upon the foil: It is the height of folly to plant it upon light fandy land. There is not, generally fpeaking, a good Elm in the whole county of Norfolk. By the time they arrive at the fize of a man's waift they begin to decay at the heart, and if not taken at the critical time, they prefently become ufelefs as timber. This is the cafe in all light foils; it is in ftiff ftrong

land

land which the Elm delights.   It is obfervable, however,
that here it grows comparatively flow.   In light land, efpecially if it be rich, its growth is very rapid ; but its wood is
light, porous, and of little value, compared with that grown
upon ftrong land ; which is of a clofer ftronger texture, and, at
the heart, will have the colour and almoft the heavinefs and the
hardnefs of iron :   On fuch foils the Elm becomes profitable,
and is one of the four *Cardinal Trees* which ought in preference
to all others to engage the Planter's attention.

The method of PROPAGATING this fpecies of Elm is principally by layering (in the manner already defcribed) ; the
feeds not coming to perfection here.   HANBURY recommends
in very ftrong terms the grafting what he calls the True Englifh Elm upon the Wych Elm ; which he fays has a ftronger
and more porous root, and will thrive upon poorer land.   His
reafoning, however, feems to flow from a theory perhaps ill
grounded, rather than from practice.   Neverthelefs, as he
feems to have taken particular pains in drawing up directions
for this operation we will, for reafons already affigned, tranfcribe them at length.   " The ftocks for the purpofe fhould
be the Broad-leaved Wych Elm, which muft be raifed from
the feed, and planted out as before.   When they have grown
two years in the nurfery, they will be of proper fize to receive the graft; and the laft week in January is the beft
time for the work.   If a large quantity of Elm ftocks are to be
grafted, procure fix men in readinefs for the purpofe.   The
bufinefs of the firft man is to take the mould from the ftem of
the ftocks, with a fpade, down to the root, laying the top of
the root bare ; the next man is to follow him with a fharp
pruning-knife, cutting off the heads of the ftocks, and leaving
the ftumps to be grafted only about two inches above the
root; the third man is the grafter himfelf, who having his
grafts cut about four or five inches in length, all of the young
wood, and fuch as has never bore lateral branches, in a difh,
takes out one of them, and holding it in his left hand, the
taper end being from him, with the knife that is in his right
he takes off a flope about an inch and half or two inches long ;
and if the grafter be an artift, it will be cut as true as if
wrought by a plane.   This done, he makes a fmall cut acrofs,

G g 4   nearly

nearly at the top of the flope, and then proceeds to prepare
the ftock to receive it, which is effected by floping off a fide of
it, of the fame length with the floped graft, that the parts
may fit as near as poffible. He then makes a cut nearly at the
top of the ftock downward, to receive the tongue he had made
in the graft; and having properly joined them, he proceeds
to the next. After the grafter, follows a perfon with bafs
matting, cut into proper lengths; and with thefe he ties the
grafts pretty clofe to the ftock. The fifth man brings the clay,
which fhould have been prepared a week or longer before, and
well worked and beaten over, mixed with a fourth part of
horfe-dung, and fome chopped hay, in order to make it hang
the better together: with this he furrounds the graft and the
ftock. Laftly, the fixth man comes and clofes the clay, fo
that there may be no probability of its being wafhed off. Two
or three rows being grafted, let an additional hand or two be
employed, either in drawing the earth up above the clay, fo
that it may be wholly covered, or digging the ground between
the rows, and levelling it fo that nothing of the performed
work may appear, except the tops of the grafts, above ground.
The danger of froft renders this precaution highly neceffary;
for if it fhould be delayed a night or two, and fharp frofts
fhould happen, the clay will moft of it fall off, and thus the
work will require to be repeated; whereas, when it is lapped
warm in the manner directed, there will be no danger of fuch
an accident. A good workman, with the above-mentioned
neceffary affiftance, will graft about fifteen hundred ftocks in a
day. In the fpring, the buds will fwell, difclofe, and fhoot
forth nearly as foon as thofe of the tree from which they were
taken. By the latter end of June, they will be fhot a foot
and half, when they fhould be freed from the clay; the matting fhould be alfo taken off, and themfelves left to fport at eafe
with all the vegetative powers. At this time, of thofe which
have put forth two fhoots, the weakeft fhould be taken up, to
ftrengthen the other, and to lighten the head, which would
otherwife be fubject to be broken off by high winds. By
autumn the fhoot will have grown about a yard in length; and
in the winter dig the ground between the rows. In this place
they may remain till they are of a fize to be planted out for
con-

continuance, with no other trouble than what was directed for the layers; namely, keeping them clear of weeds, digging between the rows in the winter; at the same time taking off all very large side-branches; and in the summer pinching off such young shoots, in the head, as may have a tendency to make the tree become forked. This practice of grafting will be found a valuable improvement of the English Elm, if we confider the nature of the Wych Elm, on which it is grafted. First, the Wych Elm will not only grow to the largest fize of all the forts, but will grow the fastest. However, this is not to be wondered at, if we examine the root, which we shall find more fibrous, and the pores larger and in greater numbers than in any of the other Elms. Now, as all roots are of a spongy nature, to receive the juices of the earth for the nourishment and growth of the tree, that tree must necessarily grow the fastest whose root is most spongy and porous; and therefore the true English Elm, being set upon the root of the Wych, a greater quantity of nutriment is received from the earth for its encrease, in proportion as the root of the Wych Elm is more spongy and porous than that of its own fort. Thus the English Elm, on this basis, will arrive at timber many years sooner than those raised by layers, and be also forced to a greater fize. If we consider too that the roots of the Wych Elm will imbibe such juices as are proper for the growth of its own forts, timber thus raised must be better, as the wood of the Wych Elm is so excellent in its kind as to answer the purposes of all the other kinds."

See more of the ELM under WOODLAND.

ZANTHOXYLUM.

LINNEAN Clafs and ORDER, *Diœcia Pentandria*: Male flowers containing five parts, and female flowers containing one part; upon distinct plants. There are two SPECIES: one of which, with due care, may be enured to this climate.

ZAN-

Zanthoxylum *Clava Herculis:* The Toothache-Tree; *a deciduous ſhrub;* native of Jamaica, Carolina, and Pennſylvania.

The Toothache-Tree (ſo called from its bark being ſaid to be efficacious in that complaint) will grow to the height of about twelve feet. The bark is rough, and armed with ſhort thick ſpines. The leaves are its greateſt ornament; for they are pinnated, are of a fine dark-green on their upper ſurface, and yellowiſh underneath, and grow without order on the branches. The folioles are ſpear-ſhaped, long; four or five pair are terminated by an odd one, and the whole leaf has much the reſemblance of thoſe of the Maſtich tree. The flowers come out in looſe panicles, from the ends of the branches; they are ſmall, and of little figure, having no petals, though the coloured ſegments of the calyx have been taken for petals. They are ſucceeded by roundiſh capſules, containing the ſeeds, which hardly ever ripen in England.

There is a *variety* of this genus, with leaves compoſed of oval, oblong folioles, which have prickly mid-ribs. This difference is permanent from ſeeds. They are numbered in the nurſeries as two diſtinct ſorts; the firſt is called the Lentiſcus-leaved Toothache-tree; the other the Aſh-leaved Toothache-tree.

Theſe trees are propagated from the ſeeds, which we receive from abroad; and theſe are ſeldom leſs than two, and often three or four, years before they come up. They muſt be ſown deep, in largiſh pots, filled with a good, light, ſandy compoſt; and after that, the pots may be plunged into ſome natural ſoil, in a ſhady place, and there left undiſturbed, except having conſtant weeding, during the next ſummer and winter. The ſpring following they may be taken up and plunged into an hotbed; and this will bring up many of the ſeeds. They muſt be next hardened by degrees; and afterwards plunged into their former ſtation, to remain there until autumn. In the enſuing winter they muſt be preſerved in the greenhouſe, or under a hotbed-frame; and in the ſpring they ſhould have a hotbed as before; and then you may expect to ſee the remainder of the whole crop. The ſame management muſt be repeated until the ſpring following, when
they

they muſt be all ſhaken out of the pots, and each be planted in a ſeparate pot. Watering ſhould be given them, to ſettle the mould to the roots ; and they ſhould be plunged into a hotbed as before. After this they muſt be hardened to the air, and ſet abroad in a ſhady place. The plants are now raiſed ; but they ſhould be treated as greenhouſe-plants for two or three years after ; when, in ſome ſpring, they may be turned out of their pots, with their mould, into the places where they are deſigned to remain. The places allotted them ſhould be naturally warm and well-ſheltered ; for although they are tolerably hardy when old, they require protection at firſt ; and with this, nothing but the ſevereſt winters can deſtroy them.

TIMBER.

T I M B E R.

TIMBER is the great and primary object of
planting. Ornament, abstracted from utility,
ought to be confined within narrow limits. Indeed,
in matters of planting, especially in the taller plantations, it were difficult to separate entirely the idea
of ornament from that of use. Trees in general
are capable of producing an ornamental effect;
and there is no tree which may not be said to be
more or less useful. But their difference in point
of value when arrived at maturity is incomparable;
and it would be the height of folly to plant a tree
whose characteristic is principally ornamental, when
another which is more useful and equally ornamental may be planted in its stead.

Therefore, previous to our entering at large upon
the business of planting, it will be proper to endeavour to specify the trees most useful to be planted. In attempting this we must look forward,
and endeavour to ascertain the species and proportional quantities of TIMBER which will hereafter
be wanted, when the trees now to be planted shall
have reached maturity. To do this with a degree
of certainty is impossible: Customs and fashions
alter as caprice and necessity dictate. All that appears capable of being done in a matter of this
nature is, to trace the great outlines, and, by observ-

ing

ing what has been permanently ufeful for ages paft,
judge what may, in all human probability, be alfo
ufeful in ages to come.

SHIPS, MACHINES, and
BUILDINGS, UTENSILS,

have been, are, and moft probably will continue to
be, the confumers of TIMBER in this country. We
will therefore endeavour to come at the principal
materials made ufe of in the conftruction of thefe
four great conveniencies of life. Indeed, whilft
mankind remain in their prefent ftate of civilization
and refinement, they are *neceffaries* of life which cannot be difpenfed with ; and are confequently objects
which the planter ought not to lofe fight of, as they
include in effect every thing that renders plantations ufeful ; FENCE-WOOD and FUEL excepted.

SHIPS are built chiefly of OAK : the keels, however, are now pretty generally laid with ELM ; and
part of the upper deck of men of war is of DEAL :
But thefe woods bear no proportion, in refpect of
the quantity ufed, to the Oak. The *timbers* of a fhip
are principally crooked, but the *planking* is cut out
of ftraight pieces. In a feventy-four gun fhip, the
crooked and ftraight pieces ufed are nearly equal,
but the *planking under water* is of FOREIGN OAK :
therefore, of ENGLISH OAK, the proportion of
crooked to ftraight pieces is almoft two to one.
Mafts and yards are of DEAL. The blockmakers
ufe Lignum Vitæ, Box, and other hard woods.
Upon the whole, it may be faid, that in the conftruction of a fhip, OAK is the only ENGLISH
WOOD made ufe of ; and that of this Englifh Oak,
nearly

nearly two-thirds is requifite to be more or lefs CROOKED.

BUILDINGS. In the metropolis, and towns in general, DEAL is the prevailing wood made ufe of by the *houfe-carpenter :* fome OAK is ufed for fafhes, alfo for window and door frames, and fome for wall-plates ; but in places fituated within the reach of water-carriage, DEAL is becoming every day more and more prevalent : Neverthelefs, there are many inland parts of the country where the houfe-carpen-ters ftill continue to work up great quantities of OAK and ELM. The *joiner* and the *carver* fcarcely ufe any other wood than DEAL, except in fome inland and well-wooded diftricts, where OAK is ftill in ufe for floors and ftair-cafes. Through the kingdom at large, perhaps three-fourths of the timber ufed in the conftruction of buildings is FOREIGN DEAL.

MACHINES. This clafs comprehends MILLS and other MACHINES OF MANUFACTORY, CAR-RIAGES of burden and pleafure, IMPLEMENTS OF HUSBRANDRY, with the other articles neceffary in rural affairs.

The *mill-wright's* chief material is OAK (As to the implements, utenfils, and machines of manu-factory, they are infinite, and various kinds of wood are worked up in making them).

The *waggon and cart-wright* ufes OAK for bodies, ASH for fhafts and axles, ELM for naves, and fometimes for fellies and linings.

The *plow-wright's* fheet anchor is ASH : in fome counties BEECH is fubftituted in its ftead, for every thing but plow-beams.

The

The *coachmakers* ufe Ash for poles, blocks, fplinter-bars, &c. Elm for naves; generally Ash for fpokes and fellies; and Rattan * for pannels.

Gates and Fences are made of Oak and Deal; fometimes of Ash, Elm, Maple, &c. but *pofts* are, or ought to be, univerfally of Oak.

Ladders, of Deal, Oak, &c.

Pumps and Water-Pipes, of Oak, Elm, Alder.

Wooden Bridges, *River Breaks*, and other *Water-works*, principally Oak; fome Elm and Alder under water.

UTENSILS. Under this head we clafs Furniture, Cooper's Ware, Mathematical Instruments, Trunks, Packing-Cases, Coffins, &c. &c.

The *cabinet-makers'* chief woods are Mahogany and Beech; next to thefe follow Dutch Oak (Wainfcot), Deal, Elm; and laftly Walnut-tree, Cherry-tree, Plum-tree, Box, Holly, Yew, and a variety of woods for inlaying and cabinets. In fome country places a confiderable quantity of English Oak is worked up into tables, chairs, drawers, and bedfteads; but in London, Beech is almoft the only Englifh wood made ufe of *at prefent* by the cabinet and chair-makers.

Coopers;—Oak (and fome Chesnut) for large cafks and veffels: Ash for dairy utenfils, butter-firkins, flour-barrels, &c. Oak for well-buckets and water-pails, and in fome places for milk-

* The Mahogany of the Bahama-Iflands.

pails

pails and other dairy veffels: BEECH for foap firkins, &c.

Turners ;—principally BEECH for large ware, if Beech is to be had ; if not, SYCAMORE, or other clean-grained wood : Box, HOLLY, &c. for fmaller utenfils.

Mathematical inftrument-makers ; — MAHOGANY, Box, HOLLY.

Trunk-makers ;—DEAL.

Packing-cafes ;—alfo DEAL.

Coffins ;—OAK, ELM, DEAL.

And finally, the *laft-makers*, who work up no inconfiderable quantity of wood, ufe BEECH for lafts ; ALDER and BIRCH for heels, patten-woods, &c.

We do not deliver the foregoing fketch as a perfectly correct account of the application of woods in this country : The attempt is new, and that which is new is difficult. We have not omitted to confult with profeffional men upon the fubject ; and we believe it to be fufficiently accurate for the purpofe of the planter. If we have committed any material error, we afk to be fet right. We do not wifh to defcend to minutiæ : it would be of little fignification to the planter, to be told what toys and toothpicks are made from : it is of much more importance to him to know, that, of ENGLISH WOODS, the OAK is moft in demand, perhaps three to one,—perhaps in a much greater proportion ; that the ASH, the ELM, the BEECH, and the Box, follow next ; and that the CHESNUT, the WALNUT, and the PRUNUS and PINUS tribes are principally

H h   valuable

valuable as fubftitutes for OAK and FOREIGN TIM-
BER. It likewife may not be improper in this place
to mention, that the OAK, though of flower growth
than the Afh, the Elm, the Beech, the Larch, the
Firs, and the Aquatics, is nearly of twice the value
of any of thefe woods at market; therefore, in a private and pecuniary point of view, the OAK is the
moft eligible tree to be planted : in a public
light, it rifes above comparifon.

HEDGES.

H E D G E S.

THE raifing of Live Hedges and Hedge-row Timber conftitutes no inconfiderable part of the bufinefs of planting. The value of good Hedges is known to every hufbandman ; and not-withftanding the complaints againft Hedge-row Timber as being liable to be knotty, &c. the *quality* of the timber itfelf is not queftioned : its faultinefs arifes wholly from an improper treatment of the tree, and not from the fituation of its growth. Indeed, we are clear in our opinion, that, under proper management, no fituation whatever is better adapted to the valuable purpofe of raifing Ship-timber than Hedges : The roots have free range in the adjoining inclofures, and the top is expofed to the exercife of the winds, with a fuffi-cient fpace to throw out lufty arms, and form, at a proper height, a fpreading head. Thus quicknefs of growth, with ftrength and crookedness of Timber, are at once obtained.

We are well aware of the injury refulting from woody Hedge-rows to arable inclofures ; but every man experienced in rural matters muft be con-vinced that it is not well-trained Timber-trees, but high Hedges and low Pollards, which are the bane of corn-fields. Thefe, forming a high and impervi-ous barrier, preclude the air and exercife fo effen-tial to the vegetable, as well as the animal, creation :

in

in Norfolk, lands thus encumbered are, with great strength and propriety of expreſſion, ſaid to be *wood-bound*. Beſides, Pollards and low-ſpreading trees are certain deſtruction to the Hedge-wood which grows under them.

Neither of theſe evils, however, reſult from tall Timber-oaks, and a Hedge trimmed down to four or five feet high : a circulation of air is, in this caſe, rather promoted than retarded ; and it is well known that a *trimmed* Hedge will thrive perfectly well under tall ſtemmed trees, Oaks more eſpecially. We will therefore venture to recommend, for arable incloſures, Hedges trimmed to four or five feet high, with Oak timbers of fifteen to twenty-five feet ſtem.

But for graſs lands, higher Hedges are more eligible. The graſſes affect warmth, which promotes their growth, and thereby increaſes their quantity, though their quality may be injured. Beſides, a tall fence affords ſhelter to cattle, provided it be thick and cloſe at the bottom ; otherwiſe, by admitting the air in currents, the blaſt is rendered ſtill more piercing. The ſhade of trees is equally friendly to cattle in ſummer, as thick Hedges are in the colder months ; therefore, in the Hedges of graſs incloſures, we wiſh to ſee the Oak wave its lofty ſpreading head, whilſt the Hedge itſelf is permitted to make its natural ſhoots : remembering, however, that the oftener it is cut down the more durable it will be as a fence, and the better ſhelter it will give to cattle ; more eſpecially if the ſides be trimmed the firſt and ſecond

years

years after cutting, in order to give it an upright tendency, and thicken it at the bottom.

Upon bleak hills, and in expofed fituations, it is well to have two or even three rows of Hedge-wood, about four feet apart from each other; the middle row being permitted to reach, and always remain at, its natural height ; whilft the fide-rows are cut down alternately to give perpetual fecurity to the bottom, and afford a conftant fupply of materials for Dead-hedges and other purpofes of Underwood.

Having thus given a general fketch of our ideas as to the different kinds of Hedges moft eligible to be raifed, we proceed to treat of the method of raifing them. In doing this it will be proper to confider,

1. The woods moft eligible for Hedges.
2. The time and manner of planting them.
3. The manner of defending the young plants.
4. The method of cleaning and training them.
5. The after-management.

1. The SPECIES OF HEDGE WOOD depends in fome meafure upon foil and fituation. That which is pro-per for a found foil in a temperate fituation, may not be eligible for a marfh or a mountain : and indeed a fence may be formed of any tree or ftrong fhrub, provided it will thrive in the given fituation. Neverthelefs, there are fome fpecies much more eligible than others; we particularize the following :

The Hawthorn.
The Crab-Tree.
The Aquatic Tribe.

The

The Holly.

The Furze.

The *Hawthorn* has been confidered, during time immemorial, as the wood moft proper for live fences. This pre-eminence probably arofe from the feedling-plants being readily collected in woods and waftes; the method of raifing them in feed-beds being formerly, and indeed in fome parts of the kingdom even to this day, but little practifed. The longevity of the Hawthorn, efpecially if it be frequently cropped, and its patience in cropping; its natural good qualities as a live fence, and its ufefulnefs as affording materials for dead hedges, are other reafons why it has been univerfally adopted. Another advantage of the Hawthorn—It will grow in almoft any foil, provided the fituation be tolerably dry and warm. However, if the foil hath not a degree of richnefs in itfelf, as well as a geniality of fituation, the Hawthorn will not thrive fufficiently, nor make a progrefs rapid enough, to recommend it, in preference, as a hedge-wood.

The *Crab-tree*, amongft the deciduous tribe, ftands next: indeed, taken all in all, it may be faid to rival, if not to exceed, the Hawthorn itfelf. Its growth is confiderably quicker, and it will thrive in poorer foils and in bleaker fituations; and altho' it may not be fo thorny and full of branches as the Hawthorn, yet it grows fufficiently rugged to make an admirable fence. Add to this, tho' its branches may not be preferable to thofe of the Hawthorn for *fhooting* dead-hedges, they undoubtedly afford a much greater quantity of *ftakes*; and no wood what-

whatever (the Yew perhaps only excepted) affords better ftakes than the Crab-tree. The feedling plants too are readily raifed, as the feeds of the Crab vegetate the firſt year. We do not mean, however, to force down the Crab-tree upon our Reader as being, in a general light, preferable to the Hawthorn : we wifh only to ftate, impartially, their comparative value ; leaving him to confult his own fituation and conveniency, and, having fo done, to judge for himfelf. Neverthelefs, from what has been adduced, we may venture to conclude, that upon a barren foil, and in a bleak fituation, the Crab-tree, as a hedge-wood, claims a preference to the Hawthorn.

The *Aquatics.* As the Crab excels the Hawthorn upon bleak barren hills, fo the Aquatics gain a preference in low fwampy grounds : for altho' the Hawthorn delights in a moift fituation, yet much ftagnant water about its roots is offenfive to it. Of the Aquatics, the *Alder* feems to claim a preference ; its growth is more forked and fhrubby than that of the Poplar or Willow ; and its leaves are particularly unfavory to cattle. In point of ornament, however, it is exceeded by the *Black Poplar,* which, if kept trimmed, will feather to the ground, and form a clofe and tolerably good fence.

The *Holly.* Much has been faid, and much has been written, of the excellency of Holly-hedges : neverthelefs, as fences to farm-inclofures, they ftill exift in books and theory only ; not having yet been introduced into general practice, we be-

lieve, in any part of the kingdom. Their fuperiority, however, whether in point of utility or ornament, is univerfally acknowledged. The perpetual
verdure they exhibit, the fuperior kind of fhelter
they afford during the winter months, and the
everlaftingnefs of their duration (an old decayed
Holly being an object rarely to be feen in nature),
all unite in eftablifhing their excellency. How
then are we to account for the fcarcity of Holly-
hedges? The difficulty of raifing them and the
flownefs of their growth have been held out as
obftacles; and fuch they are, in truth; but they
are obftacles arifing rather from a want of proper
management, than from any caufe inherent in the
Holly itfelf. Thoufands of young Hollies have
been deftroyed by being planted out improperly
in the fpring, at the time that the Hawthorn is
ufually planted : and the few which efcape total
deftruction by fuch injudicious removal, receive a
check which cripples their growth, probably for
feveral years.

We do not mean to intimate, that, by any treatment whatever, the progrefs of the Holly can be made
to keep pace with that of the Hawthorn, or Crab :
and we are of opinion that it ought, by reafon of
the comparative flownefs of its growth, to be raifed
under one or other of thefe two plants; more efpecially under the Crab, which, as has been obferved,
has a more upright tendency than the Hawthorn,
and confequently will afford more air, as well as
more room to the Holly rifing under it.

But

But whilft we thus venture to recommend raifing the Holly under the Crab, we are by no means of opinion that it is difficult to raife a hedge of Holly alone. The principal difadvantage arifing comparatively from this practice is, that the dead fence will be required to be kept up at leaft ten or twelve years inftead of fix or feven; in which time a Crab-tree hedge, properly managed, may be made a fence, and will remain fo without further expence until the Holly become impregnable; when the Crab may either be removed, or permitted to remain, as tafte, profit, or conveniency may point out.

The Holly will thrive upon almoft any foil; but thin-foiled ftony heights feem to be its natural fituation. We may venture to fay, that where corn will grow, Holly will thrive abundantly; and Holly hedges feem to be peculiarly well adapted to an arable country : for, being of flow growth, and its perfpiration being comparatively fmall, the Holly does not *fuck the land* (as the Countryman's phrafe is), and thereby rob the adjoining corn of its nourifhment, fo much as the Hawthorn; which, if fuffered to run up to that unpardonable height, and to ftruggle abroad to that fhameful width at which we frequently fee it, is not much lefs pernicious in its effects upon corn-land than the Afh itfelf.

The *Furze* is rather an affiftant Hedge-wood than a fhrub, which alone will make a fence. Upon light barren land, however, where no other wood will grow to advantage, tolerable fences may be made with Furze alone.

There

There is one material difadvantage of Furze as a live
Hedge-wood; the branches are liable to be killed
by fevere frofts, efpecially if the plants be fuffered
to grow tall, branchy, and thin at the bottom. It
follows, that the beft prefervative againft this
malady is to keep them cut or trimmed down low
and bufhy; indeed, they are of little ufe as a fence,
unlefs they be kept in that ftate.

In Norfolk it is a practice, which of late years
has become almoft univerfal, to fow Furze-feed
upon the top of the ditch-bank; efpecially when
a new Hedge is planted. In a few years the Furzes
get up, and become a fhelter and defence to the
young quick; and, affifted by the high ditch-bank
prevalent in that country, afford a comfortable
fhelter to cattle in winter; befides fupplying, at
every fall, a confiderable quantity of Farm-houfe
fuel.

2. The METHOD AND TIME OF PLANTING HEDGES
come next under confideration. The method varies
with the foil, and the time with the fpecies of wood
to be planted.

In a low level country, ditches become ufeful as
main-drains to the adjoining inclofures; but in a
dry upland fituation, drains are lefs wanted; and
here the Planter has it in choice whether he will
plant with or without a ditch.

The prevailing cuftom, taking the kingdom
throughout, is to plant with a fhallow ditch, laying
the plants in a leaning pofture againft the firft fpit
turned upfide-down, covering their roots with the

beft

beſt of the cultivated mould, and raiſing a bank over them with the remainder of the excavated earth of the ditch, without any regard being had to the wetneſs or dryneſs of the ſituation. It is a ſtriking faƈt, indeed, that in the vale of Glouceſter— where large plots of naturally rich land are chilled with ſurface water, and reduced to little value entirely thro' a want of proper ſewers and ditches— it is the cuſtom to plant Hedges with a paltry grip of twelve to fifteen inches deep; whilſt in Nor- folk—a dry ſandy country, where the natural abſor- bency of the ſub-ſtratum is ſeldom or ever ſatiat- ed—it is the univerſal praƈtice to raiſe Hedges with what is there called a " Six-foot Dyke;" and when freſh-made, they frequently run from ſix and a half to ſeven feet; meaſuring from the bottom of the ditch to the top of the bank.

What may appear equally extraordinary to the reſt of the kingdom, the Norfolk Huſbandmen, inſtead of planting the quick at the foot of the bank among the corn mould, lay it in, near the top of their wall-like bank, amongſt the crude earth taken out of the lower part of the ditch. It is no uncommon ſight, however, in that country, to ſee the face of the bank, which is uſually built very upright, waſhed down, quick and all, by beating rains, for rods together, into the bottom of the ditch. Nevertheleſs, if the plants eſcape this accident, it is aſtoniſhing to ſee the progreſs they will ſometimes make for a few firſt years after planting. But, as the roots enlarge, they become confined for want of room to range in; and the

bank

bank naturally mouldering down by time, they are left naked and expofed. It is common to fee fine young plants hanging with their heads downward againft the face of the bank; and the mould continuing to crumble away from their roots, they at length of courfe drop fingly into the ditch.

If we examine the unbroken flourifhing Hedges of that country, of fifteen, twenty, and thirty years ftanding (for many fuch there are, efpecially in the fleg hundreds); we fhall find them firmly rooted amongft the corn-mould at the foot of the bank, Neverthelefs, the Norfolk farmers in general are fo clofely wedded to the foregoing practice, that no arguments are fufficient to convince them of its impropriety.

We confefs ourfelves partial to the fuperior abilities of the Norfolk Hufbandmen, in their general management of rural affairs; and we hold eftablifhed practices in Hufbandry as things too refpectable to be wholly condemned without a full and candid examination : we will therefore endeavour, in as few words as poffible, to place the Norfolk practice of planting Hedges in its proper light.

There are not, generally fpeaking, any woodlands in Norfolk. The Hedges, it is true, efpecially of the eaftern part of the county, are full, much too full, of wood, chiefly Pollards. There are fome few Timber-groves fcattered here and there : but we find none of thofe extenfive tracts of Coppice or underwood in that county which we fee in other parts of the kingdom : confequently, the planter of Hedges experiences a fcarcity of materials for temporary

H E D G E S. 461

porary dead fences, having neither *flakes*, *edders*,
nor *rails*, to make them with. Fortunately for him,
however, the *foil* is of such a nature (a light sandy
loam of great depth, without a single stone to check
the spade), that by digging a deep trench, and raising
a mound with the soil, none of those materials are
wanted. The face of the bank being carried upright, and a little brush-wood set along the top of
it, a sufficient fence is formed ; whilst the depth of
the ditch prevents the cattle from browzing upon
the young plants. By this means Hedges are raised
in Norfolk at a trifling expence, compared with the
great cost bestowed upon them in some counties,
where two rows of posts and rails are used by way
of temporary fences. But the difficulty in raising a
Live-hedge in the Norfolk manner, arises from the
want of a proper place to plant the Quick in. If it
be put in towards the top of the bank, as is usually
done, the evil consequences above-mentioned follow :
if, on the contrary, it be laid in near the bottom,
the superincumbent pressure of the bank, and the
want of moisture in this part of it, render the progress of the young plants flow for the first three or
four years ; whilst those above, having loose *made
ground* for their fibres to strike amongst, and having
a sufficient supply of moisture collected from every
shower by the Brush-hedge, flourish apace ; until
the roots having grown too large for the bank, or
the upper part of the bank itself having been washed down or mouldered away, their career is stopt,
at a time when those below, having struggled thro'
the bank, and finding an ample supply of air, moisture, and rich cultivated soil, to work amongst, are

in

in their turn beginning to thrive ; whilft their main
roots being firmly fixed in the foil itfelf, there is no
fear of their afterwards receiving a check.

Thus it appears that the Norfolk method has its
advantage as being cheap, with its difadvantages
arifing from the want of a proper place to put the
plants in.

This is eafily cbviated by planting with an
OFFSET ; that is—inftead of continuing the face of
the bank with one unbroken flope—to fet it back a
few inches, fo as to form a break or fhelf, where the
Quick is planted ; for the obvious purpofe of giving
the young plants a fufficient fupply of moifture,
air, and pafturage, until their roots have had time
to extend themfelves to the adjoining inclofures.

This method of raifing a Hedge is not a mere
theoretical deduction, but has been practifed with
fuccefs in different parts of the kingdom ; and, in
a foil free from ftones and other obftructions of the
fpade, it is perhaps upon the whole the moft eli-
ble practice.

But the beft Live-hedges we remember to have feen
in any part of this kingdom, grow in the neighbourhood of Pickering, in the North Riding of Yorkfhire. Thefe Hedges ftand nearly upon level
ground, with little or no bank or ditch ; fo that the
plants have free range for pafturage on both fides ;
the fhallow trenches by which the Quick has been
planted being now grown up ; having, it is probable, never been fcoured out fince they were made.
Indeed, the affiftance of a ditch is not wanted, no
                                        temporary

temporary fence whatever being requifite to be made when the Hedge is topped : the ftems them-felves are a fufficient barrier, ftanding in rows like the heads of piles, and in fuch clofe order that not a fheep nor a hog, nor, in fome places, even a hare, can creep between them. In a few years thofe living piles throw out heads aftonifhingly luxuriant, and every fix or feven years afford an ample and profitable crop of Brufh-wood ; and this without any expence whatever except that of reaping it : whereas in Norfolk the renewal of the ditch and bank, when the Hedge is cut down, is nearly equal to the firft coft ; befides the difadvantage refulting from cutting off all communication with the inclo-fure on the ditch fide, and thereby robbing the Hedge of half its natural food.

Therefore, where a ditch is not neceffary as a drain, and where the nature of the fubftratum is fuch that it cannot be conveniently funk fufficiently deep to defend the young plants—the moft eligible method, in fuch a fituation, is to plant the Hedge upon the LEVEL GROUND, without either bank or ditch, in the manner hereafter defcribed ; which me-thod is now practifed in the neighbourhood above-mentioned with very promifing fuccefs.

Having thus endeavoured to deduce from actual practice what may be called the *theory* of raifing Hedges, we proceed to the application.

From what has been faid, it appears that there are three diftinct methods of raifing a Live-hedge :

1. With

1. With a ditch and plain bank.
2. With a ditch and offset.
3. Upon level ground.

The firſt has been aiready mentioned ; and being familiar to every countryman, it is needleſs to en-large upon it here.

The ſecond is to be praƈtiſed in wet ſituations where ſurface-drains are wanted, and when the ditch is neceſſary to be kept open ; and likewiſe in dry ſituations, where the ſubſoil is ſuch that a ditch can be conveniently ſunk deep enough to guard the young plants in front without an additional fence.

The manner of executing it is this : The ground may either be prepared by fallowing with the plow, or the work may be lined out upon the unbroken ground. In either caſe the plants ſhould be ſet *upon* the natural level of the ſoil, and at the diſtance of three to twelve inches from the brink of the ditch. This, in ordinary ſituations, ſhould be about four feet, ſay a quarter of a rod, wide at the top, and being brought to an angle at the bottom (or as near an angle as tools can bring it), its ſlope or ſides ſhould be about the ſame dimenſions ; the cavity of the ditch being made as nearly as may be an equi-lateral triangle. But if the ditch be wanted as a main-drain or common-fewer, its width ſhould be conſiderably greater ; for in this caſe it cannot be *pointed* at the bottom, and muſt therefore have a ſufficient width given it at the top to admit of its being made deep enough as a fence, and at the ſame time wide enough at the bottom to admit the given current of water. The bank ſhould riſe in front,

with

with a flope fimilar to that of the ditch; but, as
the back fhould be carried more upright at the foot,
fwelling out full towards the top, in order to admit
the infertion of a Brufh-hedge ; or rather, if it can
be conveniently had, a Dwarf Stake-and-Edder
Hedge, which will effectually compleat the fence
to the bank fide ; in either cafe, if any ftraggling
fpray overhang the young plants, it fhould be trimmed off with fome fharp inftrument, or be beaten
flat with the back of a fhovel, to prevent its drip from
injuring the tender fhoots.

The third method, namely, planting without a
ditch, is more particularly recommended for upland fhallow ftony foils. In executing this, the
ground muft be previoufly marked out, from four
to fix feet wide, be reduced to a fine tilth, and
made perfectly clean, either by a whole fummer's
fallow repeatedly ftirred with the plow, or by cultivating upon it, in a hufband-like manner, a crop
of Turnips, or, which is perhaps better, a crop of
Potatoes, efpecially if a little dung can be conveniently allowed them. At the approach of winter, the foil being fine and clean, and the crop, if
any, off, gather it up into a highifh round ridge or
land, and thus let it lie till the time of planting ;
when, opening a trench upon the ridge or middle of
the land, either with the fpade or the plow, infert
the plants upright, filling in the mould, and pref-
fing it gently to the roots in the common nurfery
manner.

The fame precautions fhould be obferved in
planting of Quick, which have been already recommended under the article TRAINING UP; namely, the

I i   plants

plants fhould be forted as to their fize, and fhould be either cut off within a few inches of the ground, or be trimmed up to fingle ftems.

The *diftance* fhould be regulated by the age and ftrength of the plants ; from four to fix inches is the ufual diftance; but if the plants have been previoufly tranfplanted from the feed-bed, as they ought ever to be, and have acquired four or five years of age and ftrength, as we would always wifh they fhould, from fix to nine inches is near enough.

The ufual *time of planting* is during the fpring months of February, March, and April ; and for the Hawthorn, the Crab, and the Aquatics, this is at leaft the moft convenient feafon ; but for the HOLLY, as has been already obferved under that article, Midfummer is the propereft time of planting.

Where much *ditching* is required, and hands fcarce, the foundations of the banks may be laid any time in winter, and left to fettle until the time of planting.

Thus far we have been fpeaking of raifing SINGLE HEDGES, whether of Hawthorn, Crab-tree or Holly ; we will now fay a word or two as to the method of raifing the HOLLY UNDER THE CRAB or Hawthorn. This may be done two ways; either by fowing the berries when the Quick is planted ; or by inferting the plants themfelves the enfuing Midfummer. The firft is by much the fimpleft, and perhaps upon the whole, the beft method. The feeds may either be fcattered among the roots of the deciduous plants, or be fown in a drill in front of them : and if plants of Holly be put in, they may either be planted be-

tween

tween thofe of the Crab, &c. or otherwife in front, in the Quincunx manner; the tablet of the offset, when a ditch is made ufe of, being left broad for that purpofe.

If the FURZE be made ufe of as an affiftant Hedge-wood, it is better to fow the feed on the *back* of the bank than upon the *top* of it; for in this cafe it is more apt to overhang the young plants in the face of the bank; whilft in the other it is better fituated to anfwer the purpofe intended; namely, guarding the back of the bank, as well as preventing its being torn down by cattle. The method of fowing the feed is this: Chop a drill with a fharp fpade about two-thirds of the way up the back of the bank, making the cleft gape as wide as may be, fo as not to break off the lip; and having the feed in a quart bottle, ftopt with a cork and goofe-quill, or with a perforated wooden ftopper, trickle it along the drill; covering it by means of a broom drawn gently above, and over, the mouth of the drill. This is better than clofing the drill entirely with the back of the fpade, the feeds being fufficiently covered without being fhut up too clofe, whilft the mouth of the drill is left open to receive the rain-water which falls on the top of the bank. One pound of feed will fow about forty ftatute rods. The French feed is the beft, as the plants from this feldom mature their feeds in this country, and confequently are lefs liable to fpread over the adjoining inclofure. It may be had at the feed-fhops in London for about fifteen-pence a-pound.

If a fence be required of Furze alone, a fimilar drill fhould be fown on the other fide of the bank;

and

and when the plants are grown up, the fides fhould be cut alternately.

Thus much as to *planting* the FENCE; we now proceed to the method of planting HEDGE-ROW TIMBER. It has already been given in opinion, that no fituation whatever is better adapted to the raifing of fhip-timber than Hedges; and we are clearly of opinion, that in thefe alone a fufficient fupply, of crooked timbers at leaft, might be raifed, to furnifh perpetually the Navy of Great Britain. It is a ftriking fact, that in Norfolk, where there is very little Oak, except what grows in the Hedges, and even in thefe, for one timber-tree there are ten pollards, yet the country experiences no want of oak-timber.

But whilft we recommend the Oak, as eligible to be planted in Hedges, we beg leave to condemn, as unfit for that purpofe, every other tree whatever (except perhaps the Aquatics in a marfh, the Beech and Pine-tribe upon a barren mountain, or the Elm where Oak has lately occupied the foil) and, more efpecially the Afh; not only as being the greateft enemy to the farmer, but becaufe the excellency of Afh-timber arifes from a length of ftem, and cleannefs of grain: groves, therefore, and not Hedges, are the natural fituation of the Afh; and, generally fpeaking, of every other timber-tree, the Oak only excepted.

The

The method of raifing the Oak in Hedges, may either be by fowing the acorns, or planting the feedlings, at the time of planting the fence-wood : we would wifh to recommend the practice of both; namely, to plant a well-rooted, thriving nurfery-plant (fuch as have previoufly been tapped and tranfplanted) at the diftance of every ftatute rod; and, at the fame time, to dibble round each plant three or four acorns, to guard againft a mifcarriage, and to give the judicious woodman a choice in the propereft plant to be trained.

This diftance may be objected to, as being too clofe; and fo it may in a deciduous Hedge; but, in a Holly-Hedge, we would not wifh to fee Oaks ftand at a greater diftance; for, fituated in a Hedge, they have unlimited room to fpread on either fide; and, by ftanding near each other, they are more likely to throw out main-branches, fit for fhip-timber, than they would if they had full head-room. For this reafon, it might not be amifs to plant at every half-rod, and, when the Hedge is perceived to begin to fuffer, to thin them in the manner moft conducive to the ends propofed, holding jointly in view the Fence and the Timber.

3. The Method of Defending the Young Plants. Little more remains to be faid upon this head. The ditch, bank, and dwarf-hedge have already been fully defcribed; and this is by much the cheapeft, and a very effectual, method where it can

be conveniently practised; but where the nature of the soil is such that a ditch sufficiently deep to defend the young plants cannot be sunk but at too great an expence, some other expedient must be sought for.

Posts and rails wound with bushes in the York-shire manner, are an effectual fence; but they are expensive in the extreme.

In Surry and Kent, the prevailing practice is to set a strong Stake-and-Edder-Hedge behind the Quick, and throw rough bushes into a shallow ditch in front: this in a Coppice-wood country may be done at a reasonable expence, but it is by no means effectual.

In some places Wattle-hedges are used, and in others Furze-faggots set in close order are found effectual for this purpose: in short, almost every country affords its own peculiar materials, and every judicious planter will endeavour to find out those which are most eligible.

4. The Method of Training. Much, very much indeed, depends upon this part of the business; nevertheless, it is the common idea of planters of Hedges every-where, that having performed the business of *planting*, and having *made* a fence sufficient to guard the plants at the time of planting, *their* part is *finished*; the rest is of course left to nature and chance.

The repairing of the fence,

The cleaning, &c. of the plants, and the

Trimming or pruning them, are not however less necessary operations, than the planting and fencing; for without proper attention to those, the expence
                                        bestowed

beftowed upon thefe is only fo much thrown away. A fingle gap, efpecially where fheep are to be fenced againft, may caufe to be undone in half an hour what has been doing for two or three years.

In this point of view a deep ditch-fence is preferable to one raifed upon the ground, provided the ditch be kept *pointed*; for without this precaution, a ditch, unlefs it be very deep indeed, muft not be depended upon as a fence either againft cattle or fheep : but neither the one nor the other will truft themfelves in a ditch without a bottom for them to ftand upon; nothing indeed is more terrible to them ; efpecially if part of the mould be formed into a fharp Banklet placed on the outer brink of the ditch.

Hares are great enemies to young Hedges : a ditch-fence is the beft prefervative againft them (paling or other clofe fences only excepted). An offset, however, is favourable to them ; they will run along it, and crop the plants from end to end : therefore, where hares are numerous, a tufted branch of Furze, Thorns, Holly, or other rough wood, fhould be ftuck here and there upon the platform, to prevent their running along it.

The next bufinefs is weeding either with the hoe or by hand; the former is more eligible where it can be ufed, as breaking the earth about the roots of the plants is of great fervice.

Fern is a great enemy to Quick ; it is difficult to be drawn by hand without endangering the plants, and being tough, it is equally difficult to be cut with the hoe ; and, if cut, will prefently fpring up again. The beft manner of getting rid of it, when got to a

head, is to give the ſtem a twiſt near the root, and
let the top remain on to wither and die by degrees :
this not only prevents its immediate ſpringing,
but to all appearance deſtroys the root.

Thiſtles, Docks, and other tall Weeds, are
equally injurious to the tender plants, in robbing
them of their nouriſhment and drawing them up
weak and ſlender, or ſmothering them out-right, if
not timely relieved by the foſtering hand of the
planter. Even the Graſſes are offenſive, and ſhould
be extirpated with all the care and attention neceſſary in a ſeed-bed or nurſery.

Nor is it enough to defend the young plants from
animal and vegetable intruders ; the plants themſelves muſt be taught how to grow, ſo as to beſt
anſwer the purpoſe for which they are intended.

The Hawthorn is naturally a ſhrubby plant,
throwing out ſtrong lateral ſhoots down cloſe to
the ground ; more eſpecially when planted by the
ſide of a ditch, which, by giving room, favours
this propenſity. Theſe horizontal branches, of
courſe, draw off their ſhare of nouriſhment from
the root ; which nouriſhment would be better ex-
pended upon the more upright ſhoots. They are
at the ſame time in the weeder's way, and by ſtraggle-
ing acroſs the ditch, become a temptation to cattle.
They ſhould therefore from time to time be ſtruck
off with a ſharp inſtrument, either of the hooked
or the ſabre kind.

In performing this, one rule muſt be obſerved
invariably ; that is, to leave the under-ſhoots the
longeſt, tapering the hedgeling upwards ; being very
careful however not to top the leading ſhoots ; for

by

by doing this, the upward tendency of the hedge will be checked; and, whilft its face is kept trimmed in the manner here defcribed, there is no fear of its becoming thin at the bottom.

Thus far we have been fpeaking of the method of training the SINGLE HEDGE, whether of Crab or Hawthorn. In raifing the HOLLY under either of thefe plants, a different kind of pruning is neceffary: for notwithftanding the Holly will ftruggle in a furprizing manner under the fhade and drip of other plants, yet the more air and head-room it is allowed, the greater progrefs it will make. In this cafe, therefore, the deciduous plants fhould be pruned to fingle ftems in the nurfery manner; for all that is required of thefe is ftrength and tallnefs, the Holly being a fufficient guard at the bottom.

This may be thought an endlefs bufinefs by thofe who have not practifed it; but is it not equally endlefs to prune the young plants of a nurfery? And we here beg leave to remind the young planter, that if he does not pay that care and attention to his hedgelings, in every ftage of the bufinefs, as he does to his nurfery rows, he is a ftranger to his own intereft. The advantage of obtaining a live-fence *upon a certainty* in feven or eight years, compared with that of *taking the chance* of one in fifteen or twenty, is fcarcely to be done away by any expence whatever beftowed upon planting and training it.

We are indeed fo fully impreffed with this idea, that we believe every gentleman would find his account in having even his fingle Hedges trained with fingle ftems, in order that they might the fooner
arrive

arrive at the defirable ftate above defcribed,— *a*
range of living piles. We beg to be underftood, however, that we throw this out as a hint to thofe who
wifh to excel in whatever they undertake, rather than
to recommend it as a practice to Hedge-planters
in general.

Neverthelefs, we recommend in general terms,
and in the ftrongeft manner, to keep the face of
a young Hedge *trimmed*; or, if the plants be
browzed by cattle, or otherwife become ftinted
and fhrubby, to cut them down within a few
inches of the ground; and by this and every other,
method promote as much as poffible their upward
growth. It is fome time before a young Hedge
becomes an abfolute fence againft refolute ftock;
and the fhorteft way of making it a *blind* is, by en-
couraging its upward growth, to raife it high enough
to prevent their looking over it; and, by trimming it
on the fides, to endeavour to render it thick enough
to prevent their feeing through it; giving it
thereby the appearance, at leaft, of a perfect
fence.

A *pruned* Hedge requires a different treatment
to perfect it as a fence. As foon as the ftems have
acquired a fufficient ftability, they fhould be cut
off Hedge-height; and, in order to give additional ftiffnefs, as well as to bring the live-ftakes
into drill, fome ftrong dead-ftakes fhould be driven
in here and there. This done, the whole fhould be
tightly eddered together, near the top.

As an adequate fence againft horned cattle, the
ftems are required to be of confiderable thicknefs;
but as a fufficient reftraint to fheep only, ftrong

plants

plants may be thus treated a few years after planting; efpecially thofe of Crab-tree. Upon a fheep-farm, pruning of plants would be eligible, were it only for the purpofe of getting their heads out of the way of thofe moft dangerous enemies.

5. THE AFTER MANAGEMENT. There is one general rule to be obferved in this bufinefs;—*cut often:* for the countryman's maxim is a good one; —" Cut thorns and have thorns."

The proper length of time *between the cuttings* depends upon the plant, the foil, and other circumftances: feven or eight years may be taken as the medium age at which the Hawthorn is cut in moft countries.

In Norfolk, however, the Hedges are feldom cut under twelve to fifteen years; and are fometimes fuffered to run twenty and even thirty years without cutting! The confequence is, the ftronger plants have by that time arrived at a tree-like fize, whilft the underlings are overgrown and fuffocated: the number of ftems are reduced in proportion, and at that age it is hazardous to fell the few which remain.

In Surry and Kent feven or eight years old is the ufual age at which the Farmers cut down their Quick Hedges: and in Yorkfhire they are frequently cut fo young as five or fix. This may be one reafon of the excellency of the Yorkfhire Hedges; for under this courfe of treatment every ftem, whether ftrong or weak, has a fair chance: the weak ones are enabled to withftand fo fhort a ftruggle, whilft the large ones are rather invigorated than checked by fuch timely cropping.

With

With refpect to the *firft cutting*, this alfo muft be guided by circumftances : a full-ftemmed thriving Hedge may ftand from twenty to thirty years between the planting and the firft fall ; but if the plants get moffy, or grow fhrubby and flat-topped, or put on any other appearance of being difeafed or ftinted ; or if they are unequal in ftrength, fo that the weaker are in danger of fuffering ; or if a young Hedge be much broken into gaps, or any other way rendered defective as a fence, the fooner it is cut down the better ; for time will not mend it, and tampering with it will make it worfe: whilft, on the contrary, cutting it down within a few inches of the ground, will give a falutary relief to the roots, and the frefh fhoots will furnifh a full fupply of ftems, without which no Hedge can be deemed perfect.

The ufual *time of cutting* is during the fpring months of February, March, April. The Hawthorn, however, may be cut any time in winter ; and it is obfervable, that the fhoots from the ftools of Hedges cut in May, when the leaves were breaking forth, have been equally as ftrong as thofe from Hedges felled early in the fpring. This late felling, however, is not recommended as a practice ; the brufh-wood cut out at that time, being of lefs value than that which is cut when the fap is down.

The *methods of cutting* are various. In Surry and Kent, the general practice is to fell to the ground, fcour out the ditch, fet a Stake-and-Edder Hedge behind or partially upon the ftubs, and throw fome rough thorns into the ditch.

In

In Norfolk there are two ways practifed : one, to cut within a few inches of the face of the bank, remake the ditch and bank, and fet a brufh-hedge as for the original planting : the other is called *Buck-ftalling*; which is to leave ftems about two feet long, without repairing the bank or fetting a Hedge; and only fhovelling out the beft of the mould of the ditch to form the bottoms of dung-hills with. This is a much cheaper way than the other, and where the Hedge ftands at the foot of the bank, and remains full ftocked with ftems, it is not ineligible; efpecially if a few of the flendereft of the old fhoots be layered in between the bank and the ftems, and kept there by a coping fod taken from the foot of the back of the bank : but when the roots lie high in the bank, and are of courfe more or lefs expofed, by the foil's mouldering away from them into the ditch, fuch treat-ment is deftructive to the Hedge; which, in this cafe, requires to be cut down within a few inches of the roots every eight or ten years, the ditch to be fcoured, and the bank to be faced and made fenci-ble by a Brufh-Hedge. This circumftance alone furnifhes fufficient argument againft planting high in the bank.

In Hertfordfhire, Gloucefterfhire, and fome parts of Yorkfhire, *plafhing* is much in ufe. This is done by cutting the larger ftems down to the ftub, and topping thofe of a middling fize Hedge-height by way of ftakes, between which the moft flender are interwoven, in the wattle-manner, to fill up the interftices and give an immediate live-fence. If

live

live ſtakes cannot be had, dead ones are uſually driven in their ſtead : and in order to keep the plaſhes in their places, as well as to bring the ſtakes into a line and ſtiffen the whole, it is cuſtomary in moſt places to edder ſuch Hedges.

If the ſtems alone are not ſufficient as a fence, this method of treatment may in ſome caſes be eligible, provided it be properly executed : much, however, depends upon the manner of doing it; many good Hedges have been ſpoiled by plaſhing. The plaſhers ſhould be numerous, and ſhould be trimmed to naked rods, in order that their ſpray may not incommode the tender ſhoots from the ſtools below : they ſhould be laid in an aſcending direction, ſo that they may be bent without nicking at the root, if poſſible : ſuch as will not ſtoop without danger of breaking, ſhould be nicked with an *upward* not with a *downward* ſtroke : *that,* if properly done, gives a *tongue* which conducts the rain-water from the wound ; *this,* a *mouth* to catch it.

However, in caſes where the *ſtems* ſtand regular, and are of themſelves ſtiff enough for a Fence, or where they can be readily made ſo by driving large ſtakes in the vacancies and weak places, plaſhing and every other expedient ought to be diſpenſed with : — where, upon examination, the ſtems are found inſufficient, it is generally the beſt practice to fell the whole to the ground, and train a ſet of new ones.

In caſe of gaps or vacancies too wide to be filled up by the natural branches of the contiguous ſtools, they ſhould be filled up by *layering* the neighbouring

ing young shoots, the first or second year after felling; being careful to weed and nurse up the
young layers until they be out of harm's way. If
such vacancies be numerous, it is best to keep the
whole Hedge, let its situation be what it may,
trimmed low, in order to give air and head-room to
the layers.

All renewed Hedges, whether layered or not,
should be trimmed on the sides the first and second
years after renewal ; at the same time weeding out
the brambles, thistles, dock, and every other *weed,*
whether herbaceous or ligneous; which, by crouding
the bottom, prevent the young branches from
uniting and interweaving with each other.

The proper time for performing this is when the
thistles are breaking into blow, before their seeds
have acquired a vegetative body. The large Spear-
Thistle *(Carduus lanceolatus)*, so mischievous in young
Hedges, and so conspicuously reproachful to
the Farmer when its seeds are suffered to be blown
about the country, is a biennial plant, which does
not blow till the second year ; when, having produced its seed, the root dies : it is therefore unpardonable to neglect taking this in the crisis; for by so
doing the whole race becomes at once extirpated.

The fittest instrument for the purpose of trimming
and weeding is a long hook, or rather a long straight
blade with a hooked point, which is convenient
for cutting out the brambles and weeds that grow
in the middle of the Hedge, as well as for other
purposes. We will venture to say, that whoever
puts this piece of husbandry in practice once will
not

not neglect doing it a fecond time; the ufes, as well as the neatnefs, refulting from it are numerous, and the expence of performing it little or nothing.

If the Hedge be intended to run up, either as a fource of ufeful materials, or as a fhelter in Grafs-land inclofures, the leading fhoots fhould not be touched; neverthelefs, it ought, in thefe two early trimmings, to be kept thin towards the top, leaving it to fwell out thicker towards the bottom : but, if it be intended to be kept down, as we have already faid it ought to be between arable inclofures, the leading fhoots fhould be cropped low both the firft and the fecond year, in order to check its upward tendency, and give it a dwarfifh habit; and the cropping muft be repeated from time to time, as occafion may require.

A Hedge under this treatment becomes a perpetual Fence, and its duration may be deemed everlafting. The age of the Hawthorn is probably unknown; but fuppofing that it will bear to be felled every ten years for two hundred years, during which time there will be twenty falls of wood (what a mountainous pile for one flip of land and one fet of roots to produce!) may we not be allowed to fuppofe that a fimilar hedge, kept in a dwarfifh ftate (in which ftate its produce, and confequently its exhauftion, could not be one-tenth fo much as that in the former fuppofition) would live to the age of three or four hundred years? Tenants have only a temporary property in the hedges of their refpective farms; and it is the bufinefs of landlords or their agents to fee that they are properly treated. The value of an eftate is
height-

heightened or depreciated by the good or bad ftate of its fences; which, it is well known, are expenfive to raife, and, when once let down, are difficult to get up again.

With refpeƈt to the rough and the worn-out Hedges, which conftitute a large majority of the Hedges of this country, it is not an eafy matter to lay down any precife rules of treatment. If the ground they grow in be fufficiently moift, they may be helped by felling and layering in the manner already defcribed, or by filling up the vacan-cies with young quicks, or with the cuttings of fallow, elder, &c. &c. firft clearing the ground from ivy and other encumbrances; but, in a dry bank which has been occupied by the roots of trees and fhrubs for ages, and which by its fituation throws off the rain-water that falls upon it, there can be little hope either of plants or cuttings taking to advantage.

The beft affiftance that can be given in this cafe is to drive ftakes into the vacancies, and in-terweave the neighbouring boughs between the ftakes, training them in the efpalier manner: or, if the vacancies be wide, to plafh tall boughs into them.

Thefe, however, are only temporary reliefs; for if the bodies of the plants themfelves be fuffered to run up and draw the nourifhment from the plafhers, the breaches will foon be opened again, and it will be found difficult to fill them up a fe-cond time: the only way by which to render this method of treatment in any degree lafting, is to keep the whole hedge trimmed as fnug and low as

K k the

the purpofe for which it is intended will permit;
weeding it with the fame care as a young Hedge.
By this means the vacancies in time will grow up,
and one regularly interwoven furface will be
formed.

After all, however, an old worn-out Hedge,
with all the care and attention that can be beftowed
upon it, cannot continue for any length of time;
and whenever it verges upon the laft ftage of decline, it is generally the beft management to grub
it up at once, and raife a new one in its place;
otherwife the occupier muft be driven, in the end,
to the humiliating and difgraceful employment of
patching with dead Hedge-work.

We are happy in having it in our power to fay,
that the practice of *replanting* Hedges has, of late
years, become prevalent in a county which has
long taken the lead in many important departments
of hufbandry; and although we have had occafion
to cenfure fome of its practices, with refpect to
Fences, we have great pleafure in giving to it
due praife in this particular; we fpeak of the
county of Norfolk. The beft way is to level the
old bank about Michaelmas, in order that the
mould may be thoroughly moiftened by the winter's rains, and tempered by the frofts. The
roots and old ftems will, in general, more than
repay the expence of grubbing and levelling, and
when the old ftools are numerous, and fuel is dear,
will, fometimes, go a good way towards raifing the
new Fence. One great advantage arifing from this

practice

practice, in an arable country, is the doing away the crookedness of old Hedges.

There is one general rule to be observed in renewing a Hedge in this manner, which is to plant a species of Hedge-wood different from that which formerly occupied the soil; and we know of no better change after the Hawthorn, than the Crab-tree and Holly.

Thus, having mentioned the several ways of raising and repairing Live Hedges, we now come to the training, and general treatment, of Hedge-row Timber: and, first, as to the young Oaks, which we recommended to be planted with the Hedge-wood.

The most eligible length of stem has been mentioned to be from fifteen to twenty-five feet; and, with due attention to their leading shoots, there will be little difficulty in training them to that or a greater height. If, by accident or disease, the head be lost, the stem should be taken off at the stub, and a fresh shoot trained. However, in this case, if the Hedge be got to any considerable height, it is best to let the stump stand until the first fall of the Hedge-wood; for then the young tree may be trained with less difficulty.

Next to the danger of being cropt by cattle, is that of the young trees being hurt by the Hedge-wood: first, from their being overhung and smothered amongst it; secondly, from their being drawn up too tall and slender; thirdly, from their being chafed against the boughs by the wind; and, lastly, from their stems getting locked in between

the

the branches, fo as to caufe an indenture in the
ftem, and thereby render it liable to be broken off
by the wind. The fimpleft way of guarding againft
thefe evils is to keep the Hedge-wood down to
fence-height; otherwife great care and attention
are requifite in training Hedge-timber. Even in
this cafe, the plants fhould be frequently looked
over, — to fee that the lower parts of them do not
interfere with the ftems of the Hedge-wood, — to
take off, as occafion may require, the lateral fhoots,—
and to give fimplicity and ftrength to the leaders,
until the plants have acquired a fufficient length
of ftem.

When this is obtained, it may not be amifs to
endeavour to throw the general tendency of the
head to one or the other fide of the Hedge, in
order to give air and head-room to the plants, and
crookednefs to the timber. In fhort, if trees in
Hedges are not treated with the fame attention as
thofe in Nurferies and Plantations, it were better
not to plant them; as they will become an encumbrance to the Hedge, without affording either
pleafure or profit to the planter or his fuccef-
fors.

What remain now to be confidered are, the
Grown Timbers, the Timber Stands, and the
Pollards, with which old Hedges are frequently
ftored.

There is not a more abfurd practice in the whole
circle of rural affairs, than that of making Falls
of Hedge-row Timber; which is neither more
nor lefs than for the woodman to begin at one end
of

of the Hedge, and hack down every timber-tree he comes at, whether full-grown, over-grown, or only half-grown, until he reaches the other. The impropriety is the fame, whether a young thriving tree be taken down before it has arrived at its full growth, or an old one be fuffered to remain ftanding after it has entered upon the ftage of decline.

A timbered eftate fhould frequently be gone over by fome perfon of judgement; who, let the price and demand for timber be what they may, ought to mark every tree which wears the fmalleft appearance of decay. If the demand be brifk and the price high, he ought to go two fteps farther, and mark not only fuch as are full grown, but fuch alfo as are near perfection; for the intereft of the money, the difencumbrance of the Hedge and the neighbouring young timbers, and the comparative advantages of a good market, are not to be bartered for any increafe of timber which can reafonably be expected from trees in the laft ftage of their growth.

There are men in this kingdom, who, from mifmanagement of their timber, are now lofing annually very handfome incomes. The lofs of price, which generally follows the refufal of a high offer, the certain lofs of intereft, the decay of timber, and the injuries arifing from the encumbrance of full-grown trees, are irretrievable loffes,

K k 3 which

which thofe who have the care and management of timber fhould ftudioufly endeavour to avoid.

But whilft we thus hold out the difadvantages of fuffering timber to ftand until it be overgrown, it is far from our intention to recommend, or even countenance, a premature felling,—of Hedge-row timbers more particularly: for, although in woods and clofe groves a fucceeding crop of faplings may repair in fome degree the lofs of growth in timber untimely fallen, yet it is not fo in Hedges, — where fapling ftands are liable to be fplit off from the ftool, as foon as they acquire any confiderable top; being expofed fingly, and on every fide, to the wind : and all that can be expected from the ftools of trees in Hedge-rows is a fufficiency of fhoots to fill up the breach in the Hedge.

With refpect to POLLARDS * in Hedges, fome general rules are obfervable. Pollards which are full-grown, but yet remain found, fhould be taken down before they become tainted at the heart;— for a good gate-poft is worth five fhillings; but a Firing Pollard, of the fame fize, is not worth one fhilling. Firing Pollards which, by reafon of their decay, or ftintednefs, will not, in the courfe of eighteen or twenty years, throw out tops equal in value to their prefent bodies, fhould alfo be taken down;—for the principal and intereft of the money

* Trees which have been *polled*, topt, or headed down to the ftem.

will

will be worth more at the end of that time than the body and top of the Pollard; befides the defirable riddance of fuch unfightly encumbrances. But in cafe a Pollard is already fo much tainted as to be rendered ufelefs as timber, yet found enough to all prefent appearances to throw out in the time abovementioned a top or tops of more than e-qual value to its prefent body;—it refts upon a variety of circumftances, whether, in ftrict propriety of management, fuch Pollard ought to ftand or fall.

We declare ourfelves enemies to Pollards; they are unfightly; they encumber and deftroy the Hedge they ftand in, (efpecially thofe whofe ftems are fhort) and occupy fpaces which might in general be better filled by timber-trees; and, at prefent, it feems to be the prevailing fafhion to clear them away: neverthelefs, in a country in which woodlands and coppices are fcarce, Hedge-pollards furnifh a valuable fupply of fuel, ftakes, &c.—and every man who clears away the clafs of Pollards laft-mentioned, without planting an adequate quantity of coppice-wood, commits a crime againft pofterity; more efpecially in a diftrict which depends wholly upon the fea for a fupply of coals. For, although Great Britain is at prefent miftrefs of her own coaft, what man is rafh enough to fay, that, amidft the revolutions in human affairs, fhe will always remain fo? She *once* was miftrefs of the fea at large!

With refpect to the Young Timbers which frequently abound in rough Hedge-rows, we venture to recommend the following management.

Upon eftates whofe Hedge-timber has been little attended to, (and, we are forry to fay, fuch are nine-tenths of the eftates in the kingdom) the firft ftep is to fet out the plants, and clear away the encumbrances.

After what has been faid, it may be needlefs to repeat here, that where the choice refts upon the fpecies of tree, the Oak fhould invariably be chofen; for every other fpecies we confider as a kind of encumbrance, which ought to be done away as foon as it can with any colour of propriety.

It is bad practice to permit Hedges to remain crouded with timber-ftands; they fhould, in general, be fet out fingle, and at diftances proportioned to their refpective fizes; fo that their tops be not fuffered to interfere with each other.

There is, however, one exception to this rule: where two trees, ftanding near each cther, have grown up in fuch a manner that their joint branches form, in appearance, but one top, they fhould both be permitted to ftand; for if one of them be removed, the other will not only take an unfightly outline, but will receive a check in its growth, which it will not overcome for feveral years. It is, neverthelefs, obfervable that twin trees, as well as thofe which are double-ftemmed, are dangerous to ftock: not only horned-cattle, but even horfes, have been known to be ftrangled by getting their heads locked in between them.

The

The method of training the young plants has already been defcribed; it now only remains to fay a few fords as to the PRUNING and SETTING-UP Hedge-row timbers.

Low-headed trees have been already condemned, as being injurious to the Hedge, as well as to the Corn which grows under them. To remove or alleviate thefe evils without injuring the tree it-felf, requires the beft fkill of the woodman. The ufual method is to hack off the offending bough; no matter how nor where; but, moft probably, a few inches from the body of the tree, with an axe; leaving the end of the ftump ragged, and full of clifts and fiffures, which, by receiving and retaining the wet that drips upon them, render the wound incurable. The mortification in a fhort time is communicated to the ftem, in which a recefs or hollow being once formed, fo as to receive and retain water, the decline of the tree, though otherwife in its prime, from that time muft be dated; and, if not prefently taken down, its properties as a timber-tree will, in a few years, be changed into thofe of fire-wood only. How many thoufand timber-trees ftand at this hour in the predicament here defcribed; merely through injudicious lopping. It is this vile treatment which has brought Hedge-row timber into a difrepute otherwife undeferved.

There is a wonderful fimilarity in the operations of Nature upon the Vegetable and the Animal

Cre-

Creation. A flight wound in the Animal Body foon heals up, and fkins over, whilft the wound fucceeding the amputation of a limb is with difficulty cicatrized. The effects are fimilar with refpect to the Vegetable Body : a twig may be taken off with fafety, whilft the amputation of a large bough will endanger the life of the tree. Again, pare off a fmall portion of the outer bark of a young thriving tree, the firft fummer's fap will heal up the wound : if a fmall twig had been taken off with this patch of bark, the effect would have been nearly the fame ; the wound would have been cicatrized, or barked over, in a fimilar manner ; and the body of the tree as fafely fecured from outward injury, as if no fuch amputation had taken place. Even a confiderable branch may be taken off in this manner with impunity, provided the furface of the wound be left fmooth and flufh with the *inner* bark of the Tree ; for, in a few years, it will be completely clofed up, and fecured from injury ; though an efchar may remain for fome years longer. But if a large bough be thus fevered, the wound is left fo wide, that it requires in moft trees a length of time to bark it over ; during which time the body of the tree having increafed in fize, the parts immediately round the wound become turgid, whilft the face of the wound itfelf is thrown back into a recefs ; and, whenever this becomes deep enough to hold water, from that time the wound is rendered incurable : Nature has, at leaft,

done

done her part ; and whether or not, in this cafe, affiftance may be given by opening the lower lip of the wound, remains yet (it is probable) to be tried by experiment : until that be afcertained, or fome other certain method of cure be known, it were the height of imprudence to rifk the welfare of a Tree on fuch hazardous treatment.

Further, although a branch of confiderable fize may be taken off clofe to the body of the Tree with fafety; yet if the fame branch be cut *a few inches* from it, the effect is not the fame; for, in this cafe, the ftump generally dies ; confequently the cicatrization cannot take place, until the ftem of the Tree has fwelled over the ftump, or the ftump has rotted away to the ftem ; and, either way, a mortification is the probable confequence. Even fuppofing the ftump to live, either by means of fome twig being left upon it, or from frefh fhoots thrown out, the cicatrization, even in this cafe, will be flow (depending entirely upon the feeble efforts of the bark of the ftump) ; and before it can be accomplifhed, the Tree itfelf may be in danger. But, had the amputation been made *at a diftance* from the ftem, and immediately *above a twig*, ftrong enough to draw up a fupply of fap, and keep the ftump alive upon a certainty, no rifque would have been incurred ; efpecially if the end of the ftump had been left fmooth, with the flope on the under-fide, fo that no water could hang, nor recefs be formed.

From

From what has been said, the following general rules with refpect to fetting up low-headed
trees may, we humbly conceive, be drawn with
fafety : *fmall boughs fhould be cut off clofe to the ftem ;
but large ones at a diftance from it, and above a lateral
branch large enough to keep the ftump alive.* Thus,
fuppofing the ftem of a tree in full growth to be
the fize of a man's waift, a bough the thicknefs of
his wrift may be taken off with fafety near the
ftem ; but one as thick as his thigh fhould be cut
at the diftance of at leaft two feet from it ; leaving a fide branch at leaft an inch in diameter,
with a top in proportion, and with air and head-
room enough to keep it in a flourifhing ftate. For
this purpofe, as well as for the general purpofe of throwing light into the head, the ftanding
boughs fhould be cleared from their lower
branches, particularly fuch as grow in a drooping
direction. In doing this no great caution is required ; for in taking a bough from a bough, let
their fizes be what they may, little rifque can be
thereby incurred upon the main body of the tree.

There is another general rule with regard to
pruning trees. The bough fhould be taken off
either by the *upward ftroke* of a fharp inftrument
(and, generally fpeaking, *at one blow)*, or with a
faw : in the latter cafe it fhould previoufly be
notched on the under-fide, to prevent its fplitting
off in the fall. If the bough to be taken off be
very large, the fafeft way (though fomewhat tedious) is firft to cut it off a few inches from the

ftem

ftem with an axe, and then to clear away the ftump
clofe and level with a faw, doing away the roughneffes left by the teeth of the faw with a plane, or
with the edge of a broad-mouthed axe, in order
to prevent the wet from hanging in the wound.
A faw for this purpofe fhould be fet very wide;
otherwife it will not make its way through the green
wood.

The fitteft opportunity for pruning and fetting
up young timbers, as well as for taking down Pollards and dotard timbers, and clearing away other
encumbrances, is when the Hedge itfelf is felled;
and it were well for landed individuals (as for the
nation at large) if no Hedge was fuffered to be cut
down without the whole bufinefs of the Hedge-
row being at the fame time properly executed.

Laftly, with refpeƈt to *filling up blanks* in
old Hedge-rows with young timber-trees, much
may be faid, tho' few general rules can be given.
Dry banks in general are unfriendly to young trees;
thofe more efpecially which are already fraught
with the roots of other trees or fhrubs, as the banks
of old Hedge-rows generally are.

There is another circumftance obfervable in filling up a vacancy caufed by felling a grown timber-tree: it is not enough to wait until fuch time
as the old roots are rotten; for, even then, it will
be in vain to plant a tree of the fame fpecies as that
taken down. It is a faƈt well known in the cyder
counties, that when once a piece of ground has
worn-out a fuit of fruit-trees, it is in vain to think
of making it an orchard a fecond time: the trees
may

may be made to *grow,* but never to *thrive* to advantage.

Another thing material in this bufinefs is the diftance to be obferved between the trees and the Hedge. A Hedge-tree fhould either ftand in the line of the fence, or at the diftance of three or four feet from it, fo that cattle may be able to walk between them without hurting themfelves or the fence. A tree which ftands near the fence, but not in it, is equally dangerous to the Hedge and to cattle, which coming to it, either by way of a rubbing-poft, or for the benefit of its fhade, feldom fail of tearing down the bank at leaft; and it frequently happens, that whilft one of them is ftanding with its head behind the tree, another takes it by furprize, and forces it over the Hedge.

Thefe particulars being premifed, it aptly follows, that before a Hedge-bank can be planted with timber-trees with any profpect of fuccefs, it is neceffary, in the firft place, that the part or parts where the trees are to be planted fhould be trenched two fpits deep, and at leaft fix feet wide, meafuring from the Hedge; in order to clear it from roots, and to meliorate the foil. Secondly, that, except the fituation be particularly moift, the foil ought to be thoroughly foaked with water at the time of planting, and as often afterwards as the droughtinefs of the feafons may require. Thirdly, that in the choice of plants proper for the occafion, the Planter fhould be guided by the fpecies which laft occupied the fpace now to be filled up. If the part to be planted has never been occupied, then the Oak
fhould

fhould be invariably chofen; or if it has been occupied by another tree (the Oak itfelf excepted), the choice ought to be the fame: if the Oak,
and that lately, it would be injudicious to throw
away labour and plants without a profpect of being repaid; therefore, in this cafe, and in this cafe
only, we recommend the Elm; provided the foil
be fufficiently ftrong: in a light upland foil, the
Beech, the Larch, or the Pine-Tribe, fhould be
fubftituted; or the foil fuffered to remain unoccu-
pied until time has rendered it fuitable to receive
the Oak again.

Laftly, the plants fhould be fet at about three
feet diftance from the Hedge, fhould be ftrong
healthy plants, at leaft ten or twelve feet high, and
fhould be effectually fenced from cattle. This
alone is enough to deter a man from an undertaking
of this nature; for if the young plants are not
effectually guarded againft cattle and other ftock,
all his labour is loft. We know but of one eligible way of getting over this difficulty: the work
fhould not be undertaken at any other time than
when the Hedge is felled; upon which occafion it
becomes neceffary that fome temporary fence
fhould be made to guard the young Hedge; and
if, inftead of placing this immediately behind the
Quick, it be fet at feven or eight feet from it, fo
as to include the fix-foot flip of trenched ground,
the young trees may be fecured at a reafonable
expence; whilft the flip inclofed may be planted
with

with Potatoes, Cabbages, Turnips, &c. until such
time as the plants be ftout enough to defend themfelves; — the face of the Hedge being guarded
with thorns placed in the ditch.

WOOD.

WOODLANDS;

OR,

USEFUL PLANTATIONS.

ALTHOUGH it may be a difficult taſk to diſtinguiſh nicely between *uſeful* and *ornamental* plantations *, yet the diſtinction between a rough coppice in a recluſe corner of a farm, and a flowering ſhrubbery under the windows of a manſion, is obvious : the one we view as an object of *pleaſure* and amuſement, whilſt the other is looked upon in the light of *profit* only. Upon theſe premiſes we ground our diſtinction. Under the preſent head we purpoſe to ſpeak of plantations whoſe leading features are of the more uſeful kind, and whoſe principal end is profit ; reſerving thoſe whoſe diſtinguiſhing characteriſtics are orna_mental, and whoſe primary object is pleaſure, for the article GROUNDS.

Perhaps, it will be expected, that before we begin to treat of the propagation of TIMBER, we ſhould previouſly prove an approach-

---

* See page 445.

L l

ing

ing fcarcity of that neceffary article in this country : for it may be argued, that every acre of
land applied to the purpofes of planting is loft to
thofe of agriculture; and as far as *culturable* land
goes, the argument is juft. To fpeak of this fubject generally as to the whole kingdom, and at
the fame time precifely, is perhaps what no man
is prepared for.

From an extenfive knowledge of the different parts
of the kingdom, we believe that the nation has not yet
experienced any real want of timber. We are happy
to find that in many parts of it there are great quantities now ftanding ; whilft in many other parts we
are forry to fee an almoft total nakednefs. With refpect to large well-grown OAK TIMBER, fuch as is
fit for the purpofes of SHIP-BUILDING, we believe
there is a growing fcarcity throughout the whole
kingdom.

We will explain ourfelves, by fpeaking particularly as to one diftrict — the Vale of Derwent,
in Yorkfhire. This diftrict for ages paft has fupplied in a great meafure the ports of Whitby and
Scarborough with fhip-timber. At prefent, notwithftanding the extenfive tracts of Woodlands ftill remaining, there is fcarcely a tree left ftanding with
a load of timber in it. Befides, the woods which
now exift have principally been raifed from the
ftools of timber-trees formerly taken down; the
faplings from which being numerous, they have
drawn each other up flender, in the grove manner ;

and

and confequently never will be fuitable to the more valuable purpofes of the fhip-builder.

When we confider the prodigious quantity of timber which is confumed in the conftruction of a large veffel, we feel a concern for the probable fituation of this country at fome future period. A feventy-four gun fhip (we fpeak from *good authority*) fwallows up nearly, or full, three thoufand loads of Oak timber. A load of timber is fifty cubical feet; a ton, forty feet; confequently, a feventy-four gun fhip takes 2,000 large well-grown timber-trees; namely, trees of nearly two tons each!

The diftance recommended by authors for planting trees in a *Wood* (a fubject we fhall fpeak to particularly in the courfe of this chapter) in which Underwood is alfo propagated, is thirty feet or upwards. Suppofing trees to ftand at two rods (33 feet, the diftance we recommend they fhould ftand at in fuch a plantation), each ftatute acre would contain 40 trees; confequently the building of a feventy-four gun fhip would clear, of fuch Woodland, the timber of 50 acres. Even fuppofing the trees to ftand at one rod apart (a fhort diftance for trees of the magnitude above-mentioned), fhe would clear twelve acres and an half; no inconfiderable plot of Woodland. When we confider the number of king's fhips that have been built during the late unfortunate war; and the Eaft Indiamen, merchants fhips, colliers, and fmall craft, that are launched daily in the different ports of the kingdom, we are ready to tremble for the confequences. Neverthelefs, there are men who.

treat

treat the idea of an approaching fcarcity as being chimerical; and, at prefent, we will *hope* that they have fome foundation for their opinion, and that the day of want is not near. At fome future opportunity we may endeavour to reduce to a degree of certainty, what at prefent is, in fome meafure, conjectural. The prefent ftate of this ifland with refpect to fhip-timber is, to the community, a fubject of the very firft importance.

However, in a work like the prefent, addreffed to individuals rather than to the nation at large, a true eftimate of the general plenty or fcarcity of timber is only important, as being inftrumental in afcertaining the local plenty or fcarcity which is likely to take place in the particular neighbourhood of the planter. This may be called a new doctrine in a Treatife upon Planting. It is fo, we believe, and we wifh to have it underftood, that we addrefs ourfelves to the private intereft, rather than to the public fpirit, of our readers; and we appeal to every man who has had extenfive dealings with mankind for the propriety of our conduct.

We are well aware that, fituated as this country appears to us to be at prefent, Planting ranks among the firft of public virtues; neverthelefs, we rather wifh to hold out that *lafting fame* which always falls to the fhare of the fuccefsful planter, and thofe *pecuniary advantages* which muft ever refult from plantations judicioufly fet about, and attentively executed, as being motives of a more *practical* nature.

We

We wifh, in the firft place, to do away a mif-taken notion, that when once a piece of ground is fet apart for a plantation, it becomes a dead-weight upon the eftate, or a blank in it at leaft. Nothing can be lefs true; for plantations, en-tered upon with judgment, and carried on with fpirit, accumulate in value like money at intereft upon intereft. If an eftate after a plantation has been made upon it is not worth more by the trouble and expence of making it, the under-taking was either ill-judged or badly executed.

An ozier-bed comes to profit the fecond or third year, and a coppice in fifteen or twenty, whilft an Oak may be a century before he reach the moft profitable ftate; but do they not in effeét all pay an annual income? Do not eftates fell at a price proportioned to the value of the timber which is upon them? and does not this value increafe an-nually? The fweets of a fall are well underftood, and the nearer we approach to this the more va-luable are the trees to be fallen.

We have fome knowledge of a gentleman now living, who during his life-time has made planta-tions, which, in all probability, will be worth to his fon as much as his whole eftate, handfome as it is. Suppofing that thofe plantations have been made fifty or fixty years, and that in the courfe of twenty or thirty more they will be worth 50,000l. may we not fay that at prefent they are worth fome twenty or thirty thoufand? What an incitement to planting!

Every thing, however, depends upon manage-ment. It is not fticking in a thoufand or ten thou-

fand

fand plants, as if for the fole purpofe of faying,
" I have done thofe things," without giving them
a fecond thought, that will ever bring in the profits of planting ; yet, how many gentlemen do we
fee fquandering their money, laying their lands
wafte, and rendering themfelves ridiculous, by fuch
management !

The firft ftep to be taken by a man who wifhes
to ferve his family and his country, and at the
fame time afford amufement and acquire credit
to himfelf, by planting, is to confider well his
own particular fituation.

Much depends upon LOCALITY, or relative
fituation, with refpect to water-carriage, and a
variety of other circumftances ; as contiguity to a
large town, or a manufacturing place, which generally enhances the value of land, and the price
of labour.

Much alfo depends upon the NATURAL FEA-
TURES, or pofitive fituation of his eftate : the
hang of a hill which is too fteep for the plow,
and a fwampy bottom too rotten to bear pafturing
ftock, and which cannot be rendered firm enough for
that purpofe but at too large an expence, may
in general be highly improved by planting.

Again, where the top-foil, or culturable ftratum,
is of an unproductive nature, whilft a bed of clay,
loam, or other good foil lies under it, planting
may fometimes be made greatly advantageous.
An inftance occurs in the Vale of Gloucefter, of
a coppice which pays at the rate of fourteen
or fifteen fhillings an acre annually, whilft the
land

land which furrounds it is not worth more than eight or ten fhillings. The foil is a *four* clay, and the fub-ftratum a calcareous loam. The valuable plantations above-mentioned afford a fimilar inftance; the top-foil is a light unproductive fand, under which lies a thick ftratum of ftrong clayey loam. Whereever we fee the Hawthorn flourifh upon *bad land*, we may venture to conclude, that, under ordinary cir cumftances, fuch land will pay for planting.

But with refpect to low lands which wear a profitable fward, and will bear the tread of cattle, or which by judicious draining can be rendered fuch at a reafonable expence; alfo up-lands which by proper management will throw out profitable crops of corn and other arable produce, more efpecially if the fub-ftratum is of a nature ungenial to the ligneous tribes; under thefe circumftances, we are of opinion, planting can feldom be carried on upon a large fcale with propriety. Neverthelefs, even under thefe circumftances Belts and other Skreen-plantations upon expofed heights; as well as Sheltering Groves, and ftripes or patches of planting to fill up the inconvenient crookedneffes of the borders of arable fields; may be productive of real and fub-ftantial improvement to an eftate.

The next ftep which a Gentleman ought to take before he fet about raifing plantations upon a large fcale, is to look round his neighbourhood, and make himfelf acquainted with its prefent ftate as to Woodlands, as well as with the comparative value which thefe bear to arable and grafs lands. He muft go ftill farther; he muft learn the natural confumption of the country; not only of timber in

general, but of the several species. Nor must he
stop here; he must endeavour to pry into futurity,
and form some judgment of the particular species,
whether it be Oak, Ash, Elm, Beech, the Aquatics, Pines, or Coppice Woods, which will be
wanted at the time his plantations arrive at maturity.

It is possible there may be situations in this island;
where, from a super-abundance of Woodlands, it
would be unprofitable to plant even hangs, bogs,
and bad top-soils: it is not probable, however,
that any such places are to be found; for in a country
situated near water-carriage, (and if the present
spirit of cutting canals continue to prevail, what part
of this island will, a century hence, be out of the
reach of water-carriage?) ship-timber will, in all
human probability, always find a market; and, in
situations remote from such cheap conveyance,
foreign timber will always bear a price proportionably high; consequently the timber raised in such
a country, in all probability, will find a market in
the neighbourhood of its growth.

Before we begin to speak of the several species of
plantations or woodlands, and the methods of raising
them, it will be proper to enumerate the different
species of trees which we conceive as most eligible
to be planted for the purpose of timber and underwood in this country.

Under the article TIMBER it appears that

The Oak,

The Ash,

The Elm, and

                              The

The Beech,

are the four principal *domeſtic* timbers now in uſe in this kingdom : To which muſt be added

The Pines, and

The Aquatics,

as ſubſtitutes for *foreign* timber at preſent imported in vaſt quantities into this iſland : And to thoſe muſt be added, as *underwoods,*

The Hazel,	The Ozier, and
The Sallow,	The Box.

There are four diſtinct ſpecies of WOODLANDS ;

Woods,

Timber-groves,

Coppices,

Woody Waſtes.

By *Woods* are meant a mixture of timber-trees and underwood ; by *Timber-groves,* a collection of timber-trees only, placed in cloſe order ; by *Coppices,* underwood alone, without an intermixture of timber-trees ; and by *Woody Waſtes,* graſs land over-run with rough woodineſſes ; or a mixture of Wood-land and graſſy patches ; which being thought an object of paſturage, the wood is kept under by being browſed upon by ſtock, whilſt the graſs in its turn is ſtinted by the trees, and rendered of an inferior quality by the want of a free admiſſion of ſun and air.

In practice, theſe Woody Waſtes ought firſt to be taken under conſideration ; for whilſt a Gentleman has an acre of ſuch land upon his eſtate, he ought not (generally ſpeaking) to think of ſetting about raiſing original plantations : for if graſſineſs prevail, and the ſoil be unkind for Wood, let this be clear-

ed

ed away, and the whole be converted to pasture or arable. But if, on the contrary, the woodiness prevails, fence out the stock, and fill up the vacancies in the manner *hereafter* described : for, in a systematic Treatise upon Planting, we think it most consistent with method to treat of Woodlands in the order already set down.

W O O D S.

OPEN Woods are adapted more particularly to the purpose of raising timber for ship-building, and perhaps for some few other purposes where *crookedness* is required. Where a *straightness* and length of stem and cleanness of grain are wanted, Close Woods or Groves are more eligible; and where Underwood is the principal object, Coppices unencumbered with timber-trees are most adviseable.

It follows, that no tree whatever but the Oak can be raised with propriety in open Woods, and this only when a supply of ship-timber is intended ; consequently open Woods are peculiarly adapted to places lying conveniently for water-carriage, or which may in all probability lie convenient for water-carriage a century or two hence.

Various opinions prevail with respect to the most eligible method of raising a Wood : some are warm advocates for *sowing*, others for *planting*; some again

again are partial to *rows*, whilft others prefer the *random* manner.

The difpute about fowing and planting may in fome meafure be reconciled in the following manner : Where the ftrength of the land lies in the fub-ftratum, whilft the top-foil is of an ungenial nature, *fow* in order that the roots may ftrike deep, and thereby reap the full advantage of the treafures below : but, on the contrary, when the top-foil is good, and the bottom of an oppofite quality, *plant*, and thereby give the roots the full enjoyment of the productive part of the foil ; or, under thefe laft circumftances, *fow*, and *tap* the young plants as they ftand (with a tapping inftrument), and thereby check their downward tendency, and ftrengthen their horizontal roots.

By *this* method of treating feedling plants, the peculiar advantage of planting is obtained. The difpute therefore feems to reft entirely upon this queftion : Which of the two methods is leaft expenfive ? To come at this, there are two things to be confidered—the *actual expence* of labour and other contingent matters, and the *lofs of time* in the land occupied. With refpect to the former, fowing is beyond comparifon the cheapeft method ; but in regard to the latter, planting may feem to gain a prefence ; for the feed-bed is fmall compared with the ground to be planted, and whilft that is rearing the feedling plants, this continues to be applied to the purpofes of hufbandry. However, if we confider the check which plants in general

receive

receive in tranfplantation *, and if (as we fhall
hereafter fhew) the interfpaces of an infant Wood
may for feveral years after fowing be ftill cultivated
to advantage, the preference we conceive is evidently and beyond all difpute on the fide of fowing.

With refpect to the arrangement of Wood Plants,—
the preference to be given to the *row*, or the *random*
manner, refts in fome meafure upon the nature
and fituation of the land to be ftocked with plants.
In boggy bottoms, and againft fteep hangs where the
plow cannot be conveniently ufed in cleaning and
cultivating the interfpaces during the infancy of
the Wood, either method may be adopted; and if
plants or cuttings are to be put in, the *quincunx*
manner will be found preferable to any. But in
found and more level fituations, we cannot allow
any liberty of choice: the *drill* manner is undoubtedly the moft eligible; and with this method

* We have known an inftance of tranfplanted Oaks remaining upon the ground fo long as eight years before they began to
move. And let us hear what MILLER fays upon this fubject; we have no reafon to doubt his fpeaking from his own
experience, though he does not particularize it.—" When
Oak-trees are cultivated with a view to profit, acorns fhould
be fown where the trees are defigned to grow; for thofe which
are tranfplanted will never arrive to the fize of thofe which
ftand where they are fown, nor will they laft near fo long.
For in fome places where thefe high trees have been tranfplanted with the greateft care, they have grown very faft for
feveral years after, yet are now decaying, when thofe which
remain in the places where they came up from the acorns, are
ftill very thriving, and have not the leaft fign of decay. Therefore, whoever defigns to cultivate thefe trees for timber, fhould
never think of tranfplanting them, but fow the acorns on the
fame ground where they are to grow; for timber of all
thofe trees which are tranfplanted is not near fo valuable as
that of the trees from acorns." (Art. QUERCUS.)

of

of raifing a Wood we will begin to give our directions.

But before we enter upon the immediate fubject, it will be proper to premife, that, previous to the commencement of any undertaking of this nature, it would be advifeable that the fpot or fpots intended to be converted into Woodland, fhould be determined upon, — the quantity of land afcertained, — and the whole (whether it be entire or in detached parts, and whether it be ten acres or a hundred) divided into *annual fowings*.

The exact number of thefe fowings fhould be regulated by the ufes for which the Underwood is intended. Thus, if, as in Surrey, ftakes, edders, and hoops are faleable, the fuite ought to confift of eight or ten fowings; or if, as in Kent, hop-poles are in demand, fourteen or fifteen fowings will be required; and if, as in Yorkfhire, rails be wanted, or, as in Gloucefterfhire, cordwood be moft marketable, eighteen or twenty fowings will be neceffary to produce a regular fucceffion of *annual falls*.

Many advantages will accrue from thus parcelling out the land into a fuite of fowings : the bufinefs, by being divided, will be rendered lefs burdenfome; a certain proportion being every year to be done, a regular fet of hands will, in proper feafon, be employed; and, by beginning upon a fmall fcale, the errors of the firft year will be corrected in the practice of the fecond, and thofe of the fecond in that of the third. The produce of the intervals will fall into regular courfe; and, when the whole is completed, the falls will follow

low each other in regular fucceffion. In fhort, the entire bufinefs, from beginning to ending, becomes methodical, and, to a contemplative mind, cannot fail of becoming pleafurable.

If it be found convenient to haften the bufinefs, two or three divifions may be fown in one year, the feparate falls being marked by the firft cutting. This, though by no means equal to regular fowings correfponding to the intended falls, is much better than hurrying over the whole bufinefs at once; — a piece of rafhnefs which no man who works upon an extenfive fcale fhould be guilty of.

The principal objections to raifing Woodlands in this progreffive manner, is the extra trouble in fencing. However, if the fowings lie detached from each other, the objection falls; if, on the contrary, they lie together, or in plots, let the entire plot be inclofed at once; and if it contain a number of fowings, fome fub-divifions will be neceffary, and the annual fowings of thefe fub-divifions may be fenced off with hurdles, or other temporary contrivance. If the adjoining land to be fown be kept under the plow, little temporary fencing will be wanted.

It may be further neceffary, before we enter upon the bufinefs of fowing, to give fome directions as to *fencing*; for, unlefs this be done effectually, that will be labour loft.

In raifing a Woodland from feeds, it is not only neceffary to fence againft cattle and fheep, but againft hares and rabbits alfo, efpecially if they be numerous. Nothing lefs than a clofe fence is adequate to this purpofe. Where the foil will admit of a ditch being funk, a ditch, bank, and

dwarf-

dwarf-paling, may be raifed, in the manner already defcribed under the article FENCES; except that, inftead of a ftake-and-edder-hedge, a clofe paling be fet upon the bank, in the following manner.

Before the bank be finifhed, the pofts, about five feet long, fhould be put down, their lower ends being firft *charred* (fuperficially burnt) to prevent their rotting. One rail is fufficient. To this the upper ends of the pales are nailed, their lower ends having been previoufly driven into the crown of the bank. The pales fhould be about three feet long, and ought to be made of Oak, or the bottom parts will foon rot off.

The fence is the ftronger, and more effectual, if the ditch be made on the outer-fide of it, and the paling fet fo as to lean outwards; but the Quick ftands a much better chance of being reared on the inner-fide of the paling, next to the feedling plants: therefore, the moft prudent method of making a fence of this kind, is to make the ditch on the outfide, without an off fet, leaning the paling over it, and planting the Quick at the foot of the bank, on the inner-fide: it then becomes what it ought always to be confidered, — a part of the *Nurfery*.

This, however, is an expenfive fence, and is better fuited to a fmall than a large fcale; and if, inftead of the dwarf-paling, a clofe rough ftake-and-edder hedge be fet upon the bank, it will (provided it be well made and carefully attended to from time to time, and the *muces*, if any be made, ftopt with a rough bufh, and a ftake driven through it),

continue

continue to be effectual againſt *hares* for a conſiderable time; and againſt *rabbits*, nothing leſs than death is effectual.

At length we come to treat particularly of the method of raiſing a Wood upon land ſufficiently found, and ſufficiently level to be cultivated conveniently with the plow.

The preparation of the ground.—If the ſoil be of a ſtiff clayey nature, it ſhould receive a whole year's fallow, as for wheat;—if light, a crop of turneps may be taken; at all events, it muſt be made perfectly clean before the tree-feeds be ſown, particularly from perennial root-weeds; for when once the feeds are ſown, all further opportunity of performing this neceſſary buſineſs is, in a great meaſure, loſt. If the ſituation be moiſt, the ſoil ſhould be gathered into wide lands; not high, but ſufficiently round to prevent ſurface-water from lodging upon them.

The time of ſowing is either autumn or ſpring. October may be called the fitteſt month for the autumnal ſowing, and March for the ſpring ſowing. A man of judgement, however, will attend to the ſeaſon, and to the ſtate of his ſoil, rather than to the Calendar.

The method of ſowing is this:—The land being in fine order, and the ſeaſon favourable, the whole ſurface ſhould be ſown with Corn or Pulſe, adapted to the ſeaſon of ſowing: if in autumn, Wheat or Rye may be choſen; in ſpring, Beans or Oats. Whichſoever of the three ſpecies of Corn is adopted, the quantity of ſeed ſhould be leſs than uſual, in
order

order to give a free admiffion of air, and prevent the crop from lodging.

The fowing of the grain being completed, that
of the Tree-feeds muft be immediately fet about.
Thefe muft be put in in lines, or drills, *acrofs* the
lands, and in the manner beft adapted to their
refpective natures : Acorns and Nuts fhould be dibbled in, whilft Keys and Berries fhould be fcattered in trenches or drills, drawn with the corner
of a hoe, in the manner the gardeners fow their
peas.

The diftance which we recommend to be obferved between the rows is a quarter of a ftatute-rod
(four feet and one and an half inch). This may
in theory, feem to be an unneceffary precifion ; but
in practice, there are many conveniencies accrue
from it. In fetting out the diftance between the
drills, a land-chain fhould be ufed, and not a line,
which is fubject to be fhortened or lengthened by
the weather. A chain is readily divided into rods,
and the quarters may be diftinguifhed by white
paint, or other obvious marks. Stakes being driven
at the ends of the drills, a line muft be ftretched, to
dibble or draw the trenches by *.

If the plot be extenfive, *glades*, for the purpofe
of roads, muft be left at convenient diftances.

* It may be unneceffary to obferve, that the drills fhould be
exactly perpendicular to the range of ftakes, otherwife the meafurement will be falfe. If the fowings or quarters could be fo
laid out, that the drills may be of fome determinate length, as
twenty rods for inftance, the bufinefs of meafuring would be
tendered ftill more eafy.

The

The species of underwood muſt be determined by
the conſumption, or demand, peculiar to the country in which it is intended to be raiſed. In
Surrey, where ſtakes, edders, and hoops, are in demand, the Oak, the Hazel, and the Aſh, are eſteemed
valuable as underwood. Upon the banks of the
Wye, and the country round about, in Hereford-
ſhire, Monmouthſhire, and Glouceſterſhire, where
great quantities of charcoal are made for the iron-
forges, Beech is the prevailing underwood; but
whether from choice or from its thriving well upon
thoſe bleak mountains, we cannot ſay. The Oak,
the Aſh, the Beech, the Birch, the Hazel, the Box,
may have their peculiar excellencies in different
countries, and the choice is left to the perſon
who has the care of the undertaking.

The species of timber has been already determined
upon; the *Oak* being the only tree admiſſible into
a *Wood*. The uſual ſpace allowed to timber-trees,
ſtanding amongſt underwood, is thirty feet: two
rods (thirty-three feet) will not be found, when
the trees have fully formed their heads, too wide a
ſpace. Therefore, every eighth drill at leaſt ſhould
be ſown with acorns, dibbled in about ſix inches
aſunder *.

The Oak and the Hazel, riſing the firſt year
after ſowing, their reſpective drills will be ſufficiently diſcriminable at harveſt; but the keys of the

* For the particulars reſpecting the propagation of the ſeveral
ſpecies under conſideration, ſee their reſpective genera in the
ALPHABET OF PLANTS.

Aſh

Afh lie two, and fometimes three, years in the ground before they vegetate; and it will be convenient to have fome diftinguifhing mark in the ftubble, in order to prevent their being difturbed in plowing the intervals after harveft. To this end, if Beans be the foftering crop, fcatter a few Oats along the drill among the keys, the ftubble of which will fhew itfelf plainly among that of the Beans; and, on the contrary, if Oats be the crop, a line of Bean-ftubble will have the fame beneficial effect.

At harveft the crop fhould be reaped, not mown, and be carried off with all convenient care. Be_tween harveft and winter, a pair of furrows fhould be laid back to back, in the middle of each interval, for the purpofes of meliorating the foil for the next year's crop, and of laying the feedling plants dry;—whilft the ftubble of the unplowed ground on each fide of the drills, will keep them warm during winter.

The next year's crop may be Potatoes, Cabbages, Turneps; or, if the firft was Corn, this may be Beans; or, if Beans, Wheat drilled in the Tullian manner.

All that the tree-drills will require this year, will be to be kept perfectly clean, by weeding and hand-hoeing.

In the fpring of the third year, the drills which rofe the firft year muft be looked over, and the vacancies filled up from the parts where the plants are fuperfluous: but thofe of the Afh fhould be deferred until the fourth year.

The

The whole fhould afterwards be looked over from time to time; and this, with cultivating the intervals, and keeping the drills free from weeds, will be all that will be neceffary until the tops of the plants begin to interfere.

However, if feedlings be wanted for the purpofe of laying into hedges, or if tranfplanted plants be faleable in the country, the fuperfluous feedlings may be drawn out of the drills, in the fpring of the third or fourth year, and tranfplanted into fome vacant ground.—None can be more proper, nor any fo convenient, as the contiguous intervals, in which they may remain two or three years without injury to the drills, and may afford a profitable crop; fubject, however, to this difadvantage, the fpade muft be made ufe of inftead of the plow, in cleaning the interfpaces. Neverthelefs, a ftock of plants of this kind are very valuable, not only as articles of fale, but for clumps, and fheltering plantations.

Further, with refpect to the crops.—Thefe may be continued fo long as the feedling plants remain within a narrow compafs; fo that the intervals can be cultivated to advantage, and without injury to the plants.

The crops fhould, of courfe, be varied, and the foil ought to receive its proportion, at leaft, of dung, which would be moft properly carried on for the Potatoe-crop.

If the foil be tolerably good, crops thus managed may anfwer for feveral years. If the produce pay
but

but the expence of plowing, and other contingent
charges, the advantage of ftirring and keeping
the intervals clean, will be confiderable; whilft in
an expofed fituation the crops will be a means of
faving the tree-plants from hares and rabbits.

Whenever the crops are difcontinued, the intervals muft ftill be kept ftirred; alternately throwing
the mould to the roots of the plants, and gathering
it into a ridge in the middle of the interval. The
moft complete manner of doing this, is to fplit the
intervals at the approach of winter, to preferve the
roots of the plants from the froft; gather in the
fpring, to check the weeds, and give a frefh fupply of air;—fplit at Midfummer, to keep the roots
from the drought; gather again in autumn, if neceffary; and fplit, as before, againft winter. The
fpring and the Midfummer plowings fhould be continued fo long as a plow can pafs between the
plants.

The firft cutting muft be timed by the plants
themfelves. Whenever the rows of Oaks intended
for timbers are in danger of being drawn up too
flender for their height, by reafon of their being
too much crowded by the interference of the rows,
the whole muft be cut down to within a hand-
breadth of the ground; except the Oaks intended
for ftands, which muft now be fet out at about
two rods diftance from each other, and as nearly
a quincunx as plants moft proper for the purpofe
will allow.

Strength,

Strength, cleannefs, and upward tendency, are the criterions by which the choice of thefe ought to be determined upon. If more than one plant of this defcription ftand near the point defired, it is advifeable not to take them down the firft fall (provided they do not interfere too clofely with each other), but to let them remain, in order to guard againft accidents, and to afford a future opportunity of making a fecond choice, when the plants are arrived at a more advanced ftate.

The young ftands will require to be more or lefs *pruned :* their leaders muft be particularly attended to, the lower fide-fhoots taken off, and their heads reduced in fuch a manner as to prevent their being rendered top-heavy.

However, if the firft fall of underwood be made in due time, their heads in general will want but little pruning ; for it is not in this cafe as in that of tranfplanting, where the roots have frefh fhoots to make, and a frefh fource of food to feek : here they are fully prepared to fend up the neceffary fupplies, and the more top there is to promote the afcent, the quicker progrefs the plants will be enabled to make.

It is therefore very imprudent to defer the firft fall until the plants be drawn up too flender to bear a well-fized top : We have known young Oaklings, raifed in a manner fimilar to that which is here defcribed, drawn up fo *weedy* by injudicious treatment, as not to be able to bear the fmalleft top without ftooping under the weight of their own

leaves ;

leaves; a fhower of fnow, falling without wind, would bow them to the ground.

The pruning being completed, the ftubs muft be cleared, and the fences repaired, and feduloufly attended to. The young timbers muft alfo be attended to from time to time, and the pruning repeated as often as may be found neceffary.

The fecond fall muft be timed according to the *ware* which the country calls for ; with this provifo, however, that the timber-ftands be not injured by being crouded amongft the underwood; for, rather than this fhould be the cafe, the fecond fall fhould take place, although the Coppice-wood may not have reached the moft profitable ftate.

After the fecond and every *fucceeding fall* of underwood, the timbers fhould be gone over, their leaders kept fingle, and their heads fet up, until the ftems have arrived at about twenty feet (more or lefs, as accidents, or their refpective tendencies, may happen to determine), when their heads fhould be permitted to fpread, and take their own natural form.

So foon as the branches are firmly eftablifhed (which may happen in ten, fifteen, or twenty years from the laft pruning, fooner or later, according to the foil, fituation, and other circumftances) *the heads fhould be pruned.*

In doing this, the leader muft be fhortened, to check the upward growth of the tree, and the main ftrength of the head thrown, as much as may be, into one principal arm, in order to obtain

with

with greater certainty the important end to which *Wood* timber is more peculiarly applicable

If the heads can be all trained in one direction, the interspaces will be more equally divided ; and if this direction can be pointed towards the sun, the quality of the timber will, by that means, be considerably improved.

Next, as to raising a Wood against a hang too steep to be cultivated conveniently with the plow, after the Wood-seeds are sown; but which may, nevertheless, be fallowed, and brought into proper tilth by the turn-wrest plow; namely, a plow which turns the furrows all one way, and which is in common use upon the hills of Kent and Surrey :

Under these circumstances, the planter has it in choice, whether he will sow seeds,—or put in seedlings,—or transplanted plants. If he adopt the first, the expence of cleaning by hand will fall heavy; and if the last, the labour of the Nursery will not be less burdensome. The middle path is therefore most adviseable.

The seedling plants may, in general, be permitted to remain until the third or fourth year in the seed-bed; by which time they will have acquired sufficient strength and stature to struggle with the lower order of weeds, whilst those of a more aspiring nature may be kept under, at a reasonable expence.

The arrangement of these plants may either be promiscuous, or in drills similar to those mentioned aforegoing. After the plants are in, a few acorns

may

may be dibbled in the interfpaces, that fuccefs may be rendered the more fecure.

The choice of underwood, and the after-management, under thefe circumftances, muft be fubject to the fame rules as under thofe already mentioned.

With refpect to hangs fo very fteep that even the turn-wreft plow cannot be ufed with fafety in preparing the foil, feedling plants may be put in without any previous preparations, except that of clearing away rubbifh, and burning off the weeds and rough grafs with which the furface may be encumbered. In this cafe, the number of plants, and the quantity of acorns, muft be greater than when the ground has been prepared by a fallow.

We omit faying any thing in this fection refpecting bogs which are too rotten to be cultivated with the plow ; becaufe, in fuch fituations, if the Oak would flourifh, fingle trees are liable to be blown up by the roots, by reafon of the inftability of the foil in which they are rooted, and which renders fuch fituations moft eligible for COPPICES.

It may be unneceffary to add, that, if the undertaking be large, and the fcite comprize the feveral defcriptions of fituation here treated of, the mode muft vary with the foil and fituation ; whilft glades muft be left at diftances moft commodious and in directions beft fuited to the general defign.

GROVES.

GROVES.

THE TIMBER-GROVE is the prevailing *planta-tion* of modern time. WOODS or COPPICES are fel-dom attempted; indeed, until of late years, clumps of Scotch Firs feem to have engaged, in a great meafure, the attention of the planter.

The Scotch Fir, however, is one of the laft trees that ought to engage the attention of the *Englifh* planter; and fhould be invariably excluded every foil and fituation in which any other timber-tree can be made to flourifh. The North afpect of bleak and barren heights is the only fituation in which it ought to be tolerated; and even there, the Nor-way Spruce, and the Larch, will fometimes out-brave it. In better foils and milder fituation, the wood of the Scotch Fir is worthlefs, and its growth fo licentious as to over-run every thing which grows in its immediate neighbourhood: this renders it wholly unfit to be affociated with other timber-trees: we fhall therefore confider it feparately at the clofe of this fection.

The fpecies of timber-trees which we beg leave to recommend to the planter's notice have been already mentioned at the opening of this chapter: They confift of

The Oak,	The Beech,
The Afh,	The Pines, and
The Elm,	The Aquatics.

Of

Of the two tribes laſt mentioned, we chiefly recommend

  The Larch,   The Poplar,
  The Spruce Fir, The Willow,
  The Weymouth Pine *, The Alder.

To this liſt may be added,
    The Cheſnut,
    The Walnut,
    The Cherry,

as ſubſtitutes for the Oak and the Beech ; and the
two latter as humble repreſentatives of the princely
Mahogany.

Reſpecting the *Elm* an error prevails : MILLER
and HANBURY tell us (ſpeaking more particularly
of the Fine-leaved ſort), that it will not flouriſh in
cloſe plantations. Experience, however, leads us
to be of a contrary opinion. How often do we ſee
two Elms ſtanding ſo cloſe together that a bird
could not fly through between them, yet both of
them equally well ſtemmed : indeed, the ſhoots of
the Elm will interweave with each other in a manner
we ſeldom ſee in any other ſpecies of tree. In
clumps and cloſe groves too we have ſeen them
thrive abundantly. It is obſervable, however, that
in theſe ſituations, their ſtems running up clean
and in a great meaſure free from ſide-ſhoots, the
timber takes a different nature from that which is

* There may be other individuals of the Pinus tribe, as the
Silver and Balm-of-Gilead Firs, &c. which may be equally
valuable with the three ſpecies here enumerated ; but we cannot ſpeak ſo particularly from our own knowledge concerning
them as we can of the Larch, the Spruce, and the Weymouth, whoſe woods we know to be of a good quality.

raiſed

raifed in more expofed places;—where the lateral
fhoots being numerous, and being lopped off from
time to time, the ftems become knotty ; by which
means the natural tenacity, in which the peculiar
excellency of the timber of the Elm entirely confifts, is confiderably increafed.

In a Grove, the *Afh* may be termed an *outfide* tree;
plow-beams, fhafts, fellies, and harrow-bulls requiring a curvature which generally takes place in
the outer-rows of a clofe plantation. The Afh, however, muft not be excluded a central fituation, as
a ftraightnefs of grain is frequently defirable.

The *Oak* (except for the purpofe of fhip-timber,
&c.) the *Beech*, the *Chefnut*, and the *Pinus-tribe*,
are *infide* trees ; the carpenter, the cooper, and
the turner, requiring a cleannefs of grain.

With refpect to *foil and fituation*, the Elm, the
Chefnut, the Walnut, and the Cherry, require a
good foil and mild fituation ; the Aquatics fhould
be confined to moift low grounds ; and the Beech
and the Pines to bleak mountainous places; whilft
the Oak and the Afh can accommodate themfelves
to almoft any foil or fituation.

We now come to the *method of raifing* the feveral
fpecies of Grove-timbers. The Oak, the Afh, the
coarfe-leaved Elm, the Beech, the Chefnut, the
Walnut, and the Cherry, may be raifed in drills in
the manner defcribed in the preceding fection,
without any variation, except in the method of
training. The Pines being of a hazardous nature
when in their infant ftate, it is advifeable to raife

them

them in feed-beds, and plant them out as feedling plants. The Fine-leaved Elm muft be raifed from layers; and the Aquatics from cuttings *.

The *method of training* Grove-timbers raifed in drills, is this: If feedling plants be wanted, the rows may be thinned, the third and fourth years, until the remaining plants ftand from twelve to eighteen inches apart. This done, nothing more will be requifite until fuch time as fome kind of *ware* can be cut out; as edders, hoops, ftakes, &c.

The plants having reached this ftage of their growth, the rows fhould be gone over every winter, and all the *underling* plants cut out within the ground (if practicable), which will in general kill the roots and fave the expence of grubbing. If the remaining plants are not already too much crouded, the *frugglers* ought to be left; in order to fupport and affift in drawing up with greater certainty the *conquering* plants.

This conduct fhould be obferved from the time of the firft cutting until the trees are fet out at diftances beft fuited to their refpective natures, and according to the accidental tendency which they happened to take in rifing. For, in thinning a timber-grove, little or no regard muft be had to a regularity of diftance at the root; an equal diftribution of head-room meriting a more particular attention.

* For the method of *planting* a Timber-Grove, fee the INTRODUCTION, p. 29.

The

The felection ought to be directed by the
ſtrength of the plants, and the uniformity of the
canopy, taken jointly : for a chaſm in what may
be called the foliage of a grove, is ſimilar to a vacancy in a coppice, or an unproductive patch in
a field of corn. The leaves are as labourers;
and every leaf deficient is a labourer loſt. The
pruner's eye ought therefore to be directed towards the tops, rather than to the roots, of his
plants.

There are other things obſervable in *thinning* a
grove. If it be thinned too faſt, its upward growth
will be checked, and the length of ſtem curtailed;
and if, on the other hand, the thinning be neglected, or be performed too leiſurely, the plants,
eſpecially in their taller ſtate, will be rendered too
ſlender and head-leſs, and thereby become liable to
laſh each other's tops with every blaſt of wind.
This evil is called *whipping of tops*, and many
fine groves, eſpecially of the pine-tribe, have been
very materially injured by it. Whenever two
trees are ſeen to be engaged in this ſort of con-
flict, one of them ſhould be taken down without
loſs of time; otherwiſe it will probably prove fatal to them both.

If the thinning be conducted with judgment,
little *pruning* will be neceſſary; ſome, however,
will be found requiſite : ſtrong maſter plants are
liable to throw out ſide-branches, to the annoy-
ance of their neighbours : thoſe ſhould be taken
off in time, and all dead branches ſhould be removed, eſpecially thoſe of the pines; otherwiſe
the

the heart of the timber will be rendered coarfe, knotty, and of a bad quality. The leaders fhould alfo have due attention paid to them; particularly if a group of foul-headed plants happen to fall together; for in this cafe, if nature be not affifted, a timber-tree will in the end be wanted.

This method of training holds good whether the grove be raifed from feeds, immediately, or from feedling, or other plants; and whether thefe be arranged in drills, or in the promifcuous manner; provided the body of the grove be formed of. one entire fpecies of timber-tree; for of the method of raifing that fpecies of grove we have hitherto been treating.

With regard to *mifcellaneous groves*, we have feen fo many evil effects arifing from injudicious mixtures of timber-trees, that we are inclined to condemn, *as unprofitable*, all mixtures whatever. It may be argued, however, that by affociating trees of different natures, the foil will be made the moft of; under an idea that each fpecies of plant has its own favourite food : and, indeed, it is well known that corn flourifhes after grafs, and grafs after corn; that the Afh will thrive after the Oak, and the Oak after the Afh, in a more profitable manner than any one of thefe plants would do if propagated repeatedly upon the fame fpot of foil.

This leads to an improvement in the method of raifing a Grove of Oaks ; and the fame method is applicable to any other fpecies of tree. Inftead of fowing every drill with acorns, let every fecond be fown with the feeds of a tree of a different nature ; and, under ordinary circumftances, with

those

thofe of the Afh : its feeds are eafily procured, and as underwood, no tree is applicable to fo many ufeful purpofes.

In this cafe, the method of training is nearly the fame as that already defcribed; except that, throughout, the Afh muft be made fubfervient to the Oak : if it rife too faft, it muft be cut down to the ftub as underwood : if afhen ftands be left to draw up the young Oaks, they muft be lopt, or taken down the moment they afpire to a fuperiority, or give the neighbouring plants an improper tendency.

When the Oaks have acquired a fufficient length of ftem, and have made good their canopy, the affiftance of the Afhes will be no longer wanted, nor will they be any longer valuable as underwood; they ought therefore to be entirely removed, and their ftubs grubbed up : and thus, the Oaks will receive at once a frefh fupply of air and pafturage.

In bleak fituations, a quicker-growing and better-feathered plant than the Afh, affords more valuable protection : the Scotch Fir, kept under due fubjection, is eligible in this cafe. The Furze is fometimes made ufe of for this purpofe : but the plant which we wifh to recommend in preference to either of thefe is the Broom ; as being lefs mifchievous, and at the fame time equally efficacious. Its feeds are readily procured ; its growth is rapid ; it will brave the bleakeft afpect ; and the natural foftnefs of its foliage renders it pleafant to work amongft, even in its talleft and moft crowded ftate.

His

His Grace the Duke of Portland,---whofe plantations, *alone*, are fufficient to immortalize his character, and hand down to fucceeding generations his Grace's public and private virtues, finds, that upon the bleak fandy fwells of Nottingham Foreft, the Birch affords the moft friendly protection to the Oak : and, when we confider the eafy manner in which this plant may be raifed, the quicknefs of its growth, the fhelter it gives, and its value in many places as an underwood, we muft allow great merit in the choice.

His Grace's plantations being carried on upon a fcale which is truly magnificent, and it being in the conducting of great undertakings that the human invention is raifed to the higheft pitch, it would be unpardonable, in a work of this nature, to omit inferting the following Letter from Mr. Speechly, his Grace's Gardener, to Dr. Hunter, Editor of a late edition of Evelyn's *Sylva*, defcribing the manner in which his Grace's plantations are conducted.

We introduce it in this place, as the ftyle of planting it defcribes is peculiarly adapted to raifing Groves againft Hangs, or acclivities of hills. The candour contained in the Letter itfelf precludes the neceffity of apprizing our readers, that it is not calculated for a strong level country, nor for raifing Woods in any foil or fituation.

N n

————" Few Noblemen plant more than his Grace the Duke of Portland ; and I think I may fay, without vanity, none with greater fuccefs. But as no man fhould think of planting in the very extenfive manner that we do, before he is provided with well-ftocked nurferies, it may not be amifs, before I proceed further, to give a fhort fketch of that neceffary bufinefs, as alfo to inform you of the foil and fituation of our feat of planting. The greateft part of our plantation is on that foil which in Nottinghamfhire is generally diftinguifhed by the name of Foreft-land. It is a continuation of hills and dales ; in fome places the hills are very fteep and high; but in general the afcents are gentle and eafy.

" The foil is compofed of a mixture of fand and gravel ; the hills abound moft with the latter, and the vallies with the former, as the fmaller particles are by the wind and rains brought, from time to time, from the high grounds to the lower. It is on the hilly grounds we make our plantations, which in time will make the vallies of much greater value, on account of the fhelter they will afford.

" After his Grace has fixed on fuch a part of this Foreft-land as he intends to have planted, fome well-fituated valley is chofen (as near the center of the intended plantations as may be) for the purpofe of a nurfery ; if this valley is furrounded with hills on all fides but the fouth, fo much the better. After having allotted a piece of ground, confifting of as many acres as is convenient for the purpofe, it is fenced about in fuch a man-ner as to keep out all obnoxious animals. At either end of the nurfery are large boarded gates, as alfo a walk down the middle, wide enough to admit carriages to go through, which we find exceedingly convenient when we remove the young trees from thence to the plantations. After the fence is completed, the whole is trenched (except the walk in the middle) about twenty inches deep, which work may be done for about three pounds ten fhillings, or four pounds, per acre, according as the land is more or lefs gravelly ; this work is beft done in the fpring when the planting feafon is over. If, after the trenching, two or three chaldrons of lime be laid on an acre, the land will pro-duce an excellent crop either of cabbages or turnips, which being eaten off by fheep in the autumn, will make the land in fine order for all forts of tree-feeds : but as the Oak is the fort of tree we cultivate in general, I fhall confine myfelf particularly to our prefent method of raifing and managing that moft va-luable fpecies. In the autumn, after the cabbage or turnips are eaten off, the ground will require nothing more than a common digging. So foon as the acorns fall, after being provided with a good quantity, we fow them in the following manner : Draw
drills

WOODLANDS.

drills with a hoe in the fame manner as is practifed for peafe, and fow the acorns therein fo thick as nearly to touch each other, and leave the fpace of one foot between row and row, and between every fifth row leave the fpace of two feet for the alleys. While the acorns are in the ground great care muft be taken to keep them from vermin, which would very often make great havock amongft the beds, if not timely prevented. Let this caution ferve for moft other forts of tree-feeds.

" After the acorns are come up, the beds will require only to be kept clean from weeds till they want thinning ; and as the plants frequently grow more in one wet feafon, where the foil is tolerably good, than in two dry ones, where the foil is but indifferent, the time for doing this is beft afcertained by obferving when the tops of the rows meet. Our rule is to thin them then, which we do by taking away one row on each fide the middle-moft, which leaves the remaining three rows the fame diftance apart as the breadth of the alleys. In taking up thefe rows we ought to be anxioufly careful neither to injure the roots of the plants removed, nor of thofe left on each fide. The reft of the young Oaks being now left in rows at two feet apart, we let them again ftand till their tops meet ; then take up every other row, and leave the reft in rows four feet afunder, till they arrive to the height of about five feet, which is full as large a fize as we ever wifh to plant. In taking up the two laft fizes, our method is to dig a trench at the end of each row full two feet deep, then undermine the plants, and let them fall into the trench with their roots entire.

" And here let me obferve, that much, very much, of their future fuccefs depends on this point of their being well taken up. I declare that I fhould form greater hopes from one hundred plants well taken up and planted, than from ten times that number taken up and planted in a random manner ; befides, the lofs of the plants makes the worft method the moft expenfive.

" But before I leave this account of our method of raifing Oaks, I fhall juft beg leave to obferve, that we are not very particular in the choice of acorns ; in my own opinion, it matters not from what tree the acorns are gathered, provided they are good ; for although there feems to be a variety of the Englifh Oak, in refpect to the form of the leaf and fruit, alfo their coming into leaf at different feafons, with fome other marks of diftinction, yet I am of opinion that they will all make good timber-trees if properly managed. It is natural to fuppofe that a tree will grow low and fpreading in a hedge-row ; on the contrary, it is very improbable that many fhould grow fo in a

thick

thick wood, where, in general, they draw one another up ſtraight and tall. And I have obſerved that the ſame diſtinctions hold good amongſt our large timber-trees in the woods, as in the low-ſpreading Oaks in the hedge-rows.

" Though I have not as yet taken notice of any other ſort of tree but the Oak, yet we have a great regard for, and raiſe great quantities of, Beech, Larch, Spaniſh Cheſnut, Weymouth Pine, and all ſorts of Firs, the Scotch excepted, as well as many other kinds, by way of thickening the plantations while young; among which the Birch has hitherto been in the greateſt eſti-mation, it being a quick growing tree, and taking the lead of moſt other ſorts on our poor foreſt hills ; and as we have an in-exhauſtible ſpring of them in the woods, where they riſe of themſelves in abundance from ſeed, we at all times plant them plentifully of different ſizes. As to the Elm and Aſh, we plant but few of them on the Foreſt, though we raiſe great quantities of both, but particularly the Aſh, which being an uſeful wood (but a bad neighbour amongſt the Oaks) we plant in places apart by itſelf. I ſhall diſmiſs this ſubject concerning the ma-nagement of our nurſeries, after ſaying a word or two relating to pruning : we go over the whole of the young trees in the nurſery every winter ; but in this we do little more than ſhorten the ſtrong ſide-ſhoots, and take off one of all ſuch as have double leads.

" Having thus pointed out the mode of forming and ma-naging our nurſeries, I ſhall now proceed to the plantations. The ſize of the plantations, at firſt beginning, muſt be in pro-portion to the ſtock of young trees in the nurſery ; for to un-dertake to plant more ground than we have young trees to go through with for thick plantations, would turn to poor account on our foreſt hills. We always plant thick, as well as ſow plen-tifully at the ſame time, provided it be a ſeaſon in which acorns can be had ; ſo that all our plantations anſwer in a few years as nurſeries to ſucceeding plantations.

" As to the form of the plantations, they are very irregular ; we ſometimes follow a chain of hills to a very great diſtance ; ſo that what we plant in one ſeaſon, which perhaps is ſixty, eighty, and ſometimes an hundred acres, is no more than a part of one great deſign.

" If the ground intended to be planted has not already been got into order for that purpoſe, it ſhould be fenced about at leaſt a twelvemonth before it is wanted to plant on, and imme-diately got into order for a crop of turnips ; two chaldrons of lime being laid on an acre will be of great ſervice, as it will not only be a means of procuring a better crop of turnips, but will bind

the

the land afterwards, and make it fall heavy, which is of great use when it comes to be planted, as some of the forest land is so exceedingly light as to be liable to be blown from the roots of the young trees after planting : therefore we find it to be in the best order for planting about two years after it has been plowed up from pasture, before the turf is too far gone to a state of decay. It will be neceffary to have a part of the turnips eaten off soon in the autumn, in order to get the ground into readiness for early planting ; for we find the forward planting generally succeeds the best.

" After the turnips are eaten off, we plow the ground with a double-furrow trenching plow made for that purpose, which, drawn by six horses, turns up the ground completely to the depth of twelve or thirteen inches : this deep plowing is of great service to the plants at the first, and also saves a great deal of trouble in making the holes. After the plowing is finished, we divide the ground into quarter for the planting by ridings. It will be a difficult matter to describe the laying out the ground for this purpose, efpecially where there is fuch a variety of land as we have on the Foreft ; much depends on the tafte of the perfon employed in this office. Between the hills, towards the outfides of the plantations, we frequently leave the ridings from fixty to an hundred yards in breadth, and contract them towards the middle of the woods, to the breadth of ten or twelve yards ; and on the tops of the hills where there are plains, we frequently leave lawns of an acre or two, which makes a pleafing variety.

" In fome of them we plant the Cedar of Libanus at good diftances, fo as to form irregular groves ; and this fort of tree feems to thrive to admiration on the foreft-land. On the out-fides of the woods, next to the ridings, we plant Evergreens, as Hollies, Laurels, Yews, Junipers, &c. and thefe we difpofe of in patches, fometimes the feveral forts entire, at other times we intermix them for variety ; but not fo as to make a regular fcreen or edging. Our defign in the diftribution of thefe plants, is to make the outfides of the woods appear as if fcalloped with Evergreens intermixed fometimes with rare trees, as the *Lirion-dendron Tulipifera*, or Virginian Tulip-tree, &c.

" After the ground is laid out into quarters for planting, we affign certain parts to Beech, Larch, Spanish Chefnuts &c. Thefe we plant in irregular patches here and there, throughout the plantations, which, when the trees are in leaf, has the moft pleafing effect, on account of the diverfity of fhades ; efpecially in fuch parts of the Foreft where four, five, and fometimes more of the large hill-points meet in the fame valley, and tend, as it were, to the fame center.

" After

" After thofe patches are planted, or marked out for that
purpofe, we then proceed to the planting in general. We always begin with planting the largeft young trees of every fort,
and end our work with thofe of the fmalleft fize : were we to
proceed otherwife, the making a hole for a larger-fized tree, after
the fmall ones are thick planted, would caufe the greateft con-
fufion.

" Birch is generally the fort of tree we make our beginning
with, which we find will bear to be removed with great fafety,
at the height of fix or feven feet though we commonly plant
rather under than at that fize. This fort of tree we are always
fupplied with from our plantations of five or fix years growth.
But before I proceed to the taking them up, it will be proper
to inform you, that in the planting feafon we divide our hands
into four claffes, which we term Takers-up, Pruners, Carriers,
and Planters: and here I fhall defcribe the feveral methods of
doing this work.

" Firft, in taking up we have the fame care to take up with
good roots in the plantations, as was recommended in the nurfery, though we cannot purfue the fame method ; but in both
places, fo foon as the plants are taken up we bed them in the
ground in the following manner : Dig a trench at leaft fifteen
inches deep, and fet the young trees therein with their tops
aflant, covering the roots well as we go along, and almoft half way
up the ftem of the plants, with the earth that comes out of a fecond trench, which we fill in the like manner, and fo proceed
on till we have a load more or lefs in a heap, as may be convenient to the place from whence they were taken. In our light
foil this trouble is but little, and we always have our plants fecure, both from their roots drying, and their fuffering by froft.
We have a low-wheeled waggon to carry them from the heaps,
where they are bedded, to the pruners, and generally take two
loads every other day. When they arrive, the planters, pruners,
&c. all affift to bed them there, in the fame manner as before
defcribed. We have a portable fhed for the pruners to work
under, which is alfo convenient for the reft of the work-people
to take fhelter under in ftormy weather. From the above heaps the
plants are taken only fo faft as they are wanted for pruning,
which work we thus perform : Cut off all the branches clofe to
the ftem to about half the height of the plant, fhortening the
reft of the top to a conical form in proportion to the fize of the
plant ; and in pruning of the roots, we only cut off the extreme
parts that have been bruifed by the taking up, or fuch as have
been damaged by accident, wifhing at all times to plant with as
much root as can be had.

" As

" As foon as they are pruned they are taken to the planters, by the carriers, who are generally a fet of boys, with fome of the worft of the labourers. The planters go in pairs ; one makes the holes, and the other fets and treads the plants faft, which work they commonly do by turns. In making of the holes we always take care to throw out all the bad foil that comes from the bottom : if the planting be on the fide of a hill, we lay the bad foil on the lower fide of the hole, fo as to form a kind of bafon ; for without this care our plants would lofe the advantage of fuch rains as fall haftily. We at all times make the holes fufficiently large, which is done with great eafe after our deep plowing.

" Before we fet the plant, we throw a few fpadefuls of the top foil into the hole, fetting the plant thereon with its top rather inclining to the weft ; then fill up the hole with the beft top foil, taking care that it clofes well with the roots, leaving no part hollow. When the hole is well filled up, one of the planters treads and faftens the tree firmly with his feet, while his partner proceeds to make the next hole.

" The faftening a tree well is a material article in planting; for if it once becomes loofe, the continual motion which the wind occafions, is fure to deftroy the fibres as faft as they are produced, which muft end in the deftruction of the plant, if not prevented. It is to guard againft this inconveniency that we take off fo much of the top, as has been defcribed in the article of pruning.

" We plant about three or four hundred Birches of the large fize on an acre, and nearly the fame number of the firft-fized Oaks ; we alfo plant here and there a Beech, Larch, Spanifh Chefnut, &c. exclufive of the patches of the faid forts of trees before planted. We then proceed to plant plentifully of the fecond and leffer-fized Oaks ; and laft of all a great number of the fmall Birches, which are procured from the woods at about three fhillings or three fhillings and fixpence per thoufand : thefe we remove to the fucceeding plantations after the term of five or fix years. Of the feveral fizes of the different kinds of trees, we generally plant upwards of two thoufand plants upon an acre of land, all in an irregular manner.

" After the planting is finifhed we then fow the acorns (provided it be a feafon that they can be had) all over the plantation, except amongft the Beech, Larch, &c. in the aforefaid patches. Great care fhould be taken to preferve the acorns intended for this purpofe, as they are very fubject to fprout, efpecially foon after gathering; the beft method is to lay them thin in a dry airy place, and give them frequent turnings. We fow thefe

acorns

acorns in fhort drills of about a foot in length, which work is
done very readily by two men, one with the acorns, the other
with a hoe for the purpofe of making the drills and covering
the feed.

" We are of cpinion that the plants produced from thefe
acorns will at laft make the beft trees ; however, I will not pretend to fay how that may be, as the Oaks, tranfplanted fmall,
grow equally well for a number of years : but it is probable that
a tree with its tap-root undifturbed may, in the end, grow to
a much larger fize.

" After the whole is finifhed to a convenient diftance round
the pruners, we then remove their fhed to a fecond ftation, and there proceed in the like manner ; and fo on till the
whole be finifhed.

" It would be well to get the planting done by the end of
February, efpecially for trees of the deciduous kind ; but from
the difappointments we meet with, occafioned by the weather,
we are fometimes detained to a later feafon.

" I have feveral times made trial of twelve or fourteen kinds
of American Oaks fent over to his Grace in great quantities. I
fowed them in the nurfery, and alfo in the beft and moft-fhelter-
ed parts of the plantations. In both places they come up very
plentifully ; but I now find that feveral of the forts will not
ftand the feverity of our winters, and thofe that do make fo
fmall a progrefs as to promife no other encouragement than to
be kept as curiofities.

" Towards the end of April, when the ground is moift, it
will be a great fervice to go over the whole plantations, and
faften all fuch trees as are become loofe fince their planting :
after this, nothing more will be required till the month of June,
when we again go over the whole with hoes, cutting off only
the tall-growing weeds ; for the fooner the ground gets covered
with grafs, in our light foil, fo much the better.

" I own there is fomething flovenly in the appearance of
this method, and on fome lands I would recommend keeping
the ground clean hoed for fome time at firft, as alfo planting in
rows, which in that cafe would be neceffary. More than once
I have tried this method on our foreft-hills, and always found,
after every hoeing, that the foil was taken away by the fucceeding winds into the valleys.

" Befides this inconvenience, the reflection of our fandy foil
is fo very great, that we find the plants ftand a dry feafon much
tetter in our prefent method, than in the former : and whoever
fancies that grafs will choak and deftroy feedling Oaks, will,
after a few years trial, find himfelf agreeably miftaken : I have

even

even recommended the fowing the poorer parts of the hills with furze or whin-feed, as foon as they are planted : we have fometimes permitted the furze to grow in the plantations by way of fhelter for the game, which though it-feems to choak and overgrow the Oaks for fome time, yet after a few years we commonly find the beft plants in the ftrongeft beds of whins. This fhews how acceptable fhelter is to the Oak whilft young ; and experience fhews us that the Oak would make but a flow progrefs on the foreft-hills for a number of years at the firft, were it not for fome kind nurfes ; and the Birch feems to anfwer that purpofe the beft, as I have already obferved.

" The feveral forts of Fir-trees, from appearance, feem to promife a greater. fhelter ; but on the foreft-land they do not grow fo faft as the former, and what is worfe, the Oak will not thrive under them, as they do immediately under the Birch.

" Where a plantation is on a plain, a fcreen of Firs for its boundary is of fingular ufe, but the fituation of the foreft-land denies us this advantage.

" We continue to cut down the tall-growing weeds two or three times the firft fummer, and perhaps once the next, or fecond feafon after planting ; which is all that we do in refpect to cleaning. The next winter after planting, we fill up the places with frefh plants where they have mifcarried ; after which there is little to be done till about the fourth or fifth year ; by which time the fmall-fized Birch, and feedling Oaks, will be grown to a proper fize for tranfplanting : in the thinning of thefe due care muft be had not to take too many away in one feafon, but, being properly managed, there will be a fupply of plants for at leaft half a dozen years to come.

" About the fame time that the leffer-fized Birch wants thinning, the large ones will require to have their lower branches taken off, fo as to keep them from injuring the Oaks ; and this is the firft profit of our plantations, the Birch-wood being readily bought up by the Broom-makers. This pruning we continue as often as required, till the Birches are grown to a fufficient fize to make rails for fencing ; we then cut them down to make room for their betters.

" By this time the Oaks will be grown to the height of twelve or fourteen feet, when they draw themfelves up exceedingly faft : each plant feems as it were in a ftate of ftrife with its neighbour, and in a ftrict fenfe they are fo, and on no other terms than life for life ; and he whofe fate it is to be once overtopped, is foon after compelled to give up the conteft for ever.

" After the Birches are cut down there is nothing more to be done but thinning the Oaks, from time to time, as may be required,

quired, and cutting off their dead branches as frequently as may
be neceffary.   We are very cautious in doing the former,
knowing well that if we can but once obtain length of timber,
time will bring it into thicknefs; therefore we let them grow
very clofe together for the firft fifty years.

"And here it may not be improper to obferve the progrefs
the Oak makes with us, by defcribing them in two of our plantations, one of twenty-eight, the other of fifty years growth.
In the former they are in general about twenty-five or twenty-
fix feet in height, and in girth about eighteen inches: the trees
in the latter, planted in 1725, are fomething more than fixty
feet in height, and in girth a little above three feet; and thefe
trees are in general about fifty feet in the bole, from which you
will eafily conceive the fmallnefs of their tops, even at this age.

"It would be a difficult matter to defcribe their farther progrefs with any degree of certainty, therefore let it fuffice to make
this laft obfervation on them in their mature ftate."

Welbeck, 16 *June,* 1775.

We fhall clofe this fection with the *Scotch Fir.*—
It has already been obferved, that the northern afpect
of bleak and barren hills is the only fituation in
which it can be propagated with any propriety in
this country.   The North of England abounds
with fituations bearing this defcription; and fo long
as they remain in a ftate of nakednefs and wafte,
whilft there is *any* tree with which they may be profitably covered, they lie, a reproach to their owners,
and as blots on the face of the country.

The late Sir CHARLES TURNER has made n attempt upon the heaths of Yorkfhire, but without any
great fhare of fuccefs. His method was to *dibble* in feed-
ling-trees, Larches, Oaks, and even Acorns, among
the long heath, without any previous preparations.
Some few of the Scotch Firs have got their heads
above the Ling, and are now (1783) doing well; efpecially under walls and in other fheltered fituations.

                                                    The

The Larches and the Oaks have fhared a worfe fate. The Spruce Fir has not been tried until within the laft two years ; the plants remain healthy, and promife fuccefs.

Sir CHARLES's want of fuccefs may not be owing to an improper method, more than to the want of a proper perfon to overlook his plantations : the perfon who has the care of plantations fhould refide upon the fpot.

Scotland is the country to which we muft apply for information refpecting the propagation of the Fir. It is there where it is cultivated to advantage. In the notes to HUNTER's EVELYN's *Sylva* we find a letter, written by JAMES FARQUHARSON, Efq. addreffed to the Editor of that compilation, defcribing the treatment of this tree in Scotland.

Marlee, June 22, 1775.

" ———————— In order to raife plantations of the Scotch Fir, let the cones be gathered in the month of February or March, from thriving young trees, as the old ones are not eafily acceffible, nor fo productive of feed. Thefe are to be expofed to the heat of the fun, thinly fpread on any kind of coarfe canvas, taking them under cover in the night-time, and only expofing them when the fun fhines. This foon makes the cones expand with a crackling noife. When any quantity of the feed is fhed, it muft be feparated from the cones by a fearch, otherwife the firft-dropped feeds would become too dry before the cones yielded their whole quantity, which often takes up a confiderable time ; fo that we are fometimes obliged to dry the cones in kilns, to make them give their contents in time for fowing, which ought to be done the end of April or beginning of May. The firft method of procuring the feed is certainly the moft eligible, tho' the other anfwers very well when attentively performed, fo as not to damage the feed by too much heat. A light loamy foil, trenched a foot and a half deep, and laid out in beds five feet broad, anfwers the beft for fowing. Let the feeds be fown very thick, and covered with a thick

fifting

fifting of mould from the alleys. Plants raifed in this manner will rife like a brufh. No kind of manure fhould be given to the beds, as productive of weeds; the drawing of which not only brings up many of the tender plants, but loofens the ground, and makes blanks that let in frofts in winter and drought in fummer. To give an idea of the fowing; I never confider my crop of plants good, unlefs I have above a thoufand in each foot long of the beds, that is, in five fquare feet. Upon their having two feafons growth, I plant them out irregularly from the feed bed, about three feet afunder, upon the mountainous grounds where they are to rife to perfection. I begin to plant the drieft ground in autumn, eighteen months after fowing, and perfift in this operation until the froft prevents me. I begin again in February, or rather as the weather admits, and continue this work fometimes to the end of April, fo as to plant out the product of the two-year-old feed-beds. I put the plants into the ground with two cuts of a fpade thus ⟩. I raife the point of the angle with what we call a Dibble, and laying the plant up to the neck, ftamp down the raifed fod with the foot. In this method two men may plant a thoufand in a day. When the ground is rocky, or very ftony, I ufe a Dibble fhod with iron, having a cleft at the extremity to lead down the root, putting the plants into the ground in the manner that cabbages are planted. One man will plant as many in this way as two in the other; yet the firft method is preferable where the ground admits of it, as I have always obferved fewer plants to fail. My reafon for planting from the feed-bed is, that it comes neareft to the operation of Nature. Plants removed from the feed-bed into the nurfery, muft have their roots pruned confiderably before they can be planted into the pits where they are to continue, which adds greatly to the expence. Befides, *nurfing* this hardy mountainous tree caufes a luxuriant growth, which fpoils its nature and robs it of longevity.

" It is generally believed that there are two kinds of Fir-trees, the produce of Scotland, viz. the red or refinous large trees, of a fine grain, and hard folid wood; the other, a white-wooded Fir, with a much fmaller proportion of refin in it, of a coarfer grain, and a foft fpongy nature, never comes to fuch a fize, and much more liable to decay. At firft appearance this would readily denote two diftinct fpecies, but I am convinced that all the trees in Scotland under the denomination of Scotch Fir, are the fame; and that the difference of the quality of the wood, and fize of the trees, is entirely owing to circumftances, fuch as the climate, fituation, and foil they grow in. The fineft Fir-trees appear in the moft mountainous parts of

the

the Highlands of Scotland, in glens or on sides of hills generally lying to a northerly aspect, and the soil of a hard gravelly
consistence, being the natural produce of these places; the
winged seeds are scattered in quantities by the wind, from the
cones of the adjacent trees, which expand in April and May
with the heat of the sun. These seedlings, when young, rise
extremely close together, which makes them grow straight, and
free from side-branches of any size, to the height of fifty or
sixty feet before they acquire the diameter of a foot: even in
this progress to height they are very slow, occasioned by the
poorness of the soil, and the numbers on a small surface, which
I may say makes them in a constant state of war for their scanty
nourishment, the stronger and tallest by degrees over-topping
the weaker, and when the winds blow they lash against one
another: this assists in beating off any horizontal branches that
might damage the timber with knots, as well as by degrees
crushes the over-topped trees. In such state of hostility they
continue struggling until the master-trees acquire some space
around them; then they begin to shoot out in a more bushy.
manner at the top, gradually losing their spiral form, increasing
afterwards more in size of body than height, some acquiring
four feet diameter, and about sixty feet of height to the branches,
fit for the finest deal board. The growth is still extremely slow,
as is plainly proved by the smallness of the grain of the wood,
which appears distinctly in circles from the center to the bark.
Upon cutting a tree over close at the root, I can venture to
point out the exact age, which, in these old Firs, comes to an
amazing number of years. I lately pitched upon a tree of two
feet and a half diameter, which is near the size of a planted Fir
of fifty years of age, and I counted exactly two hundred and
fourteen circles or coats, which makes this natural Fir above
four times the age of the planted one. Now, as to planted
Firs, these are raised first in dressed ground from the seed, where
they stand two seasons or more, then are planted out in the
ground they are to continue in at regular distances, have a clear
circumference round them for extending both roots and branches;
the one gives too quick nourishment to the tree, which shoots
out in luxuriant growth, and the other allows many of the
branches to spread horizontally, spoiling the timber with knots;
besides, this quick growth occasions these quick yearly circular
coats of wood, which form a coarse grain of a spongy soft nature. The juices never after ripen into a proportional quantity
their resinous preservative balm; so that the plantations decay
before the wood acquires age, or a valuable size; and the timber, when used in work, has neither strength, beauty, nor duration.

ration. I believe the climate has likewife a great fhare in forming the nature of the beft wood, which I account for in the following manner: The moft mountainous parts of the Highlands, particularly the northerly hanging fituations, where thefe fine Fir-trees are, have a much fhorter time of vegetation than a more foutherly expofure, or the lower open countries, being fhaded by high hills from the rays of the fun, even at mid-day, for months together; fo that, with regard to other vegetables, nature vifibly continues longer in a torpid ftate there than in any other places of the fame latitude. This dead ftate of nature for fo long a time yearly, appears to me neceffary to form the ftrength and health of this particular fpecies of timber. No doubt they may at firft fhew a gratefulnefs for better foil and more fun, by fhooting out fpontaneoufly; but if the plant or tree is fo altered by this luxury that it cannot attain any degree of perfection fit for the purpofes intended, the attempt certainly proves in vain.

" From what is faid above, it is not at all my intention to diffuade from *planting* Scotch Fir, but to encourage thofe that have the proper foil and fituation to do fo; being of opinion that where thefe circumftances agree, and there, planting not in lines, but irregularly and thicker than common, the trees will come to be of equal fize and value with the *natural* ones. In confidence of this, I have planted feveral millions on the fides of hills, out of the reach of feed from the natural Firs.

" As to the Larch, it grows in this country in great abundance, from the feed of our own plantations. I have found this beautiful and hardy tree to anfwer extremely well when planted out on barren grounds, from fix inches to fix feet high; and they are feldom known to fail, except where water has reached their roots. I have often remarked, with furprize, that when cattle or deer have broken off the main fhoots with their horns, another branch has taken the lead, and ftretched away at fuch a rate as to heal up the wound fo completely, that in a few years it was with difficulty I could difcover the traces of the injury. The amazing growth of the Larix far exceeds with me all the native as well as foreign trees, bearing the expofure and inclemency of the feafon better than any of them; and of late I have the pleafure to find that they naturalize themfelves by fowing. I wifh my experience could affift me in fpeaking with as much certainty with regard to the value and ufefulnefs of the timber; but in that I can give but little fatisfaction, as my oldeft trees are not thirty years from the feed. At Dunkeld I have feen a fmall fummer-houfe finifhed with Larix wood; the plants came from Loudon in earthen pots, about the year

1740, rather as a curiofity than from any expectation of their excellency. Though full of circular knots, the wood looked well, and did not feem to gall or warp fo much as Fir of the fame age and feafoning would have done. It will be neceffary to remark, that the heart or center of large trees is generally the knottieft part of the trunk, occafioned by the collateral branches, when young, fupporting the ftem to ftature, which, as the tree advances, die and fall off; and this is particularly evident in trees that grow in thickets. The furface foon heals over, and the body of the tree is annually increafed by circular rings of wood. I fhall fuppofe a tree to be a foot in diameter when the lower branches die and drop off. In courfe of time it acquires four feet in diameter, which gives a furrounding coat, one foot and a half in thicknefs, of clean timber, the center remaining knotty. The growth of the Larix, and manner of dropping its branches when clofe together, very much refembles the Fir; fo I am confident this fault of knottinefs, which feems to be the principal one, will amend by age. Yours, &c."

COPPICES.

AFTER what has been faid in the fection Woods, little remains to be added here concerning the method of raifing Coppices, excepting fo far as relates to bogs, and other rotten, fwampy fituations.

The fpecies of woods proper for this occafion are,

The Afh,	The Alder,
The Poplar,	The Ozier.

The *Afh* will thrive to great advantage in moift fituations, provided the roots have room enough to keep free from immediate contact with water.

The *Poplar* is feldom planted as a Coppice-wood;

for

for what reafon we know not : it will endure a great deal of moifture.

The wood which we moft commonly find propagated in fituations fimilar to thofe defcribed above, is the *Alder* ; a tree which feems to fet water at defiance, provided there be mould or mud enough to fheath its roots in. Its poles, however, are not equal to thofe of the Afh, nor is its fuel to be compared with it, though preferable to that of the Poplar.

But of all the Aquatics, the *Ozier* ftands firft as a Coppice-wood, whether it be cut annually for the ufe of the bafket-makers, or be fuffered to ftand three, four, five, or a greater number of years, for the purpofe of withs, hurdles, edders, ftakes, rake-handles, other implements of hufbandry, or poles of almoft any length or dimenfions.

The preparation of the foil confifts in digging trenches, at diftances proportioned to the fituation, fo as to lay the furface dry ; and the lower the water can be drawn beneath it, the fairer profpect there will be of fuccefs.

This work muft be done in autumn, when the foil, having had all the fummer to grow firm in, will ftand to the fpade ; and the fides of the trenches will be lefs liable to give way than they would in the fpring, when the foil is filled like a fponge with water ; which ouzing out from beneath the beds into the new-made trenches, their fides become undermined, and can never afterwards be

made

made to ftand properly; whereas, on the contrary, if the trenches be opened in autumn, and the mould which comes out of them be ufed in filling up the hollows, and laying the furface even and round, the winter's rains will not pafs through the foil, but will run off the furface, and will rather affift in eftablifhing the beds than in rendering them tender.

In March, the beds being firmly eftablifhed, and their furfaces in good working order, the foil fhould be thoroughly trenched with the fpade, and the fetts inferted.

The Afh will require rooted plants, but the Aquatics will grow from cuttings or from layers.

The fetts fhould be put in about two feet from each other, and a potatoe-plant dibbled into the center of each interfpace. During fummer the furface fhould be kept clean hoed, and the potatoes earthed up from time to time.

In autumn, after the potatoes are taken up, the foil ought to be drawn towards the roots of the plants, leaving channels between them to carry off the winter's rains. The enfuing fpring the plants muft be looked over, and fuch as have failed fhould be replaced with frefh ftrong fetts.

After this, little more will be neceffary than keeping under the taller weeds : if, however, in the courfe of three or four years the plants do not gain entire poffeffion of the foil, by overcoming the weeds and graffinefs, they muft be cut down to the ftub, the interfpaces dug, the rubbifh of the furface turned in, and the roots of the plants freed from incumbrances with the hoe : A fecond crop of potatoes may be táken, and the former treatment repeated.

Thus far as to the *Coppice :* we will conclude this fection with fome obfervations on the *Ozier-bed*.

Notwithftanding the Ozier is ufually planted near water, we have good reafon to believe it affects a dry fituation : this we know, that it will not flourifh in water ; that is, when water has a *conftant* communication with its roots. The places it moft delights in are drained moors and the banks of large rivers, both of which are peculiarly dry fituations : it has no diflike, however, to being flooded occafionally, but feems rather to be invigorated by fuch irrigation : therefore, the fandbanks which we frequently fee thrown up by the fides of rivers, and which fometimes lie for half a century before they become profitable, are peculiarly eligible to be converted into Ozier-grounds.

The method of planting an Ozier-ground is this : The foil being laid perfectly dry, and its furface made thoroughly clean, cuttings, of the fecond or third year's growth, and about twelve inches long, are planted in drills, about two feet and a half afunder, in the month of March. The cuttings ought to be thruft feven or eight inches deep, leaving four or five inches of head above ground.

The intervals muft be kept ftirred with a fmall plow ; or, the firft year a crop of potatoes may be taken ; the drills in either cafe muft be kept perfectly clean with the hand-hoe ; and at the approach

of

of winter the intervals muſt be ſplit, and the mould thrown to the roots of the young plants, in order to lay them dry and warm during winter.

In ſpring it will be well to trim off the firſt year's ſhoots (though not neceſſary), and replace the plants which have failed with freſh cuttings.

The ſecond ſummer the intervals muſt be kept ſtirred, the drills hoed, and the plants earthed up, as before, againſt winter.

The enſuing ſpring the ſtools muſt again be cleared; although the twigs as yet will be of little value. But the third cutting they will produce marketable ware, and will increaſe in quantity and value until the profits ariſing from them will be almoſt incredible. In ſituations which the Ozier affects, and in countries where the twigs are in demand, Ozier-grounds have been known to pay an annual rent of ten pounds an acre! Under or-dinary circumſtances, they will, if properly ma-naged, pay four or five.

In Yorkſhire the wands are ſold by the bundle; but in Glouceſterſhire, where Ozier-grounds abound upon the banks of the Severn, the grounds are lett under leaſe to the baſket-makers, who keep up the fences, and take upon themſelves the entire ma-nagement, during the term of the leaſe.

WOODY

W O O D Y   W A S T E S.

NO inconfiderable part of the face of this country, taken collectively, is disfigured by lands
bearing this defcription * ; the remedy, however,
is eafy, and the difgrace might foon be removed.

If the foil and fituation be favourable to grafs or
arable produce, grub up the bufhes, and clear
away the rubbifh ; but, on the contrary, if the
land, either from its own nature or from the proportion of woodinefs which has already got poffeffion of its furface, can be more profitably converted into Woodland, fill up the vacant patches
in the following manner.

The firft bufinefs is to fence it round, and the
next to cut down the underwood to the ftub, and
fet up the timbers. If the vacancies be fmall, they
ought to be trenched with the fpade ; if large, they
may be fallowed with the plow; or, in either cafe,
the plants may be put in without any other preparation than digging holes to receive them : however, with *this* kind of management fuccefs can
only be *hoped for*, whilft under *that* it may be
fecured.

The fpecies of wood and the mode of propagation depend upon locality and the fpecies of
plantation required. If underwood be an object,
the fmaller chafms may be filled up by layering;

* See the introductory part of this Chapter.

for

for which purpcfe young fhoots ought to be left, when the brufh-wood is felled, for layers: if timber alone be the defired object, feedling plants may be put in, and acorns or other feeds dibbled in the interfpaces : Whether the Wood, the Grove, or the Coppice, be intended, the large fpaces ought to be filled up in that way; or feeds only may be fown in drills, and treated as before directed; or they may be fcattered in the random manner, and the feedlings kept clean by weeding and hand-hoeing; or the foftering care may be left to nature alone : indeed, in *this* kind of way Wood and Timber Groves may be propagated.

We do not, however, mean to recommend to our readers practices dependent upon *chance*, after having been folicitous to point out thofe which may be purfued with *certainty*.

Gentlemen, when they fet about forming plantations or raifing Woodlands, ought to confider, that the labour, the fencing, the feeds or plants, the rent, and other contingent charges of the land, their own prefent credit, and their future fame, are *ftaked*. If after waiting eight or ten years a mifcarriage take place, the whole is *loft*. On the contrary, if, by judicious methods and careful management, no material failure happen, the prize is *won*; not only the principal but intereft is fecured : and this by a fmall additional expence; the trifling difference in labour beftowed upon the after-management only : for the labour in the firft

O o 3 inftance,

inſtance, rent, &c. &c. &c. are in both caſes ſimilar.

Much depends upon the perſon to whoſe care plantations are entruſted. If a Gentleman has not leiſure nor inclination to attend to them himſelf, he ought to appoint a man of experience; and, if poſ- ſible, one who is *ſettled* near the ſeat of planting; and who is likely to enjoy his appointment for ſome length of time. For he who plants ought to expeƈt to nurſe; and having planted he ought to nurſe, becauſe his own credit is at ſtake. On the contrary, a Gentleman who is continually changing his planter, muſt never expeƈt to ſee his planta- tions ſucceed; for the credit of the preſent riſes upon the miſcarriage of his predeceſſor: he has even an intereſt in negleƈting to nurſe; becauſe his own planting will be thereby ſet off to advantage. On the other hand, being without hopes of ſee- ing his own labours ſucceed, he loſes a neceſſary ſtimulus: indeed, he is no way intereſted; for he is ſure of an *excuſe* in the negleƈt of his ſucceſſor. This is not a theoretical idea; but is drawn from aƈtual obſervation.

CONCLUSION.

BEFORE we cloſe our diſcourſe upon WOOD- LANDS, it may be proper to ſay a few words reſpeƈt- ing the *ſale* and *felling* of timber and underwood.

In

In the fouthern counties timber is ufually fold by the ton of 40 feet, either naked, that is already fallen and divefted of its top, or ftanding; the buyer taking it down, and having the advantage of the top-wood, bark, &c. the *timber* (namely, every part exceeding fix inches *girt*, that is twenty-four inches in circumference) being meafured after it is fallen.

In Yorkfhire the prevailing practice is to fell the whole lot in the grofs, as it ftands; the buyer agreeing to take it down and pay a fum certain for timber, top-wood, and bark (we fpeak more particularly of Oak-timber), without any retrofpect as to quantity. This mode of felling faves fome trouble; and when timber ftands at a diftance from the refidence of the feller, the bufinefs is agreeably fhortened: befides, he is not liable to be impofed upon by falfe meafurement or other fraudulent practices. However, in other points of view, it is unpleafing and difadvantageous to the feller. The buyer being generally a profeffional man, and accuftomed to the valuing of ftanding-timber, can afcertain its worth with a great degree of accuracy; whilft the feller, unpractifed in the art of valuation, is under the neceffity of employing an agent in the bufinefs, at whofe mercy he lies, not only as to his judgement, but his honefty alfo.

In making a valuation of this kind, every tree is eftimated feparately, not only as to the quantity of timber it contains, but likewife as to its quality, or the ufe to which it is peculiarly adapted.

The

The method of *felling* depends upon two things :
If the ftubs or ftools be intended to be kept alive
for the purpofe of throwing out coppice-wood, as
in Kent and Surrey, or for raifing a fecond crop of
timber, as in Yorkfhire, the trees ought to be felled
high with a faw, leaving the ftools fix or eight
inches above ground ;—having great care not to
ftrip off nor bruife their bark : but, if the ground
be intended to be cleared, the trees ought either
to be grubbed up entirely by the roots, or *grub-
felled*, in the Norfolk manner ; which is to cut off
the roots clofe to the bottom of the ftem, leaving
this entire, with a conical piece of root annexed to
it ; fo that this method gives more timber even
than that of grubbing and afterwards cutting off
the butts with a faw.

If the wood be intended to be *fprung* again, the
ftools ought to be freed from the timber and top-
wood before the young fhoots make their appearance, or great mifchiefs muft neceffarily enfue.

Under the article FENCES we have declared againft
a *general fall* of Hedge-timber. But in a Wood,
where the trees ftand in contact with each other,
a different conduct ought to be obferved ; more
efpecially if it be defigned to be re-fprung : for, in
taking down the full-grown trees, the younger
ftands will, in all probability, be crufhed, or at leaft
maimed ; and if they efcape, they become injurious
to the rifing faplings.

Befides, it is obfervable, that fingle trees, left in
this manner, (diftinguifhed in Yorkfhire by the name
of *Wavers*) always receive a check, and generally
become

become *ftag-headed*; either from a change of atmofphere, which takes place upon the removal of their
wonted fhelter, or from a profufion of fide-fhoots;
which, under thefe circumftances, are ufually put
forth, and which confequently draw off a part of the
fap from the top; — or, perhaps, from the joint
operation of thefe two caufes.

Be this as it may, it is generally good management to clear away the whole to the ftub : the
coppice-wood, or fecond crop of faplings, will
generally rife to more advantage than the *Waver*, befides an incumbrance being removed.

The beft *time for taking down* timbers in general
is winter, when the fap is at reft : in this country,
however, the Oak may be confidered as an exception to this general rule; for, notwithftanding the
acknowledged fuperiority of the *timber* when taken
down at that time, the value of the bark, which
can only be ftripped during the rife of the fap, in
the Spring, or about Midfummer, is perhaps more
than a compenfation for the injury fuftained in the
quality of the timber.

The practice of barking trees, ftanding, and letting them remain upon their roots until the enfuing winter, has been ftrongly recommended; and
has been tried in different parts of the kingdom,
efpecially in Yorkfhire. There is, we believe, no
doubt as to the excellency of this method; but we
are much afraid it will not readily be brought into
general practice. The barking feafon is of fhort
duration, and is generally a time of hurry and
buftling;

buſtling: indeed, if the undertaking be large, diſpatch is neceſſary. The method of barking trees as they ſtand, is tedious in the firſt inſtance, beſides incurring what, we fear, will always be conſidered as *two troubles*; namely, the barking and the felling. Neverthelefs, the practice is highly recommendable to gentlemen who want to take down a few trees only for their own uſe; for although they are not, *in this caſe*, reſtricted by the general law relating to the bark, yet the value of the bark is too conſiderable to be thrown away.

In Norfolk it is ſold upon the tree at ſo much per load of timber, the tanners employing their own peelers, and taking all the trouble upon themſelves. The price of the bark generally runs from one-fifth to one-eighth part of the value of the timber; depending upon how the trees are *hung*, or furniſhed with branches.

In the ſouthern counties, bark is uſually ſold ready peeled, but in the rough; whilſt, in York-ſhire, it is generally *chopt* by the ſeller, and ſold to the tanner by the quarter.

With reſpect to Coppice and Underwood, the time of ſelling depends, in ſome meaſure, upon the ſpecies, and if that be Oak, upon the age at which it is felled. On the hills of Surrey, where it is uſually cut at ſeven or eight years old, for ſtakes and other purpoſes, it is generally felled in winter; but in places where it is ſuffered to ſtand fifteen or twenty years, for hop-poles, rails, &c: the Oak ſhould be taken down in the barking ſeaſon, not
only

only for the fake of the bark, but to guard againſt the worm.

Finally: with regard to the training of ſapling timbers, the care and management principally re- quiſite lie in keeping up the fences, and *weeding*— that is thinning—the young plants from time to time. The oftener this is repeated the more profit will generally ariſe. It is well to endeavour to train thoſe ſhoots which ſpring from the lower parts of the ſtools ; theſe being leſs liable to be ſplit off by the wind than thoſe which grow higher: in other reſpects, the conduct is much the ſame as that recommended in the ſection GROVES. The pro- greſs which thoſe ſaplings will make depends upon the ſoil they ſtand upon : in ordinary ſituations they will riſe to about forty feet high, and ſwell to *timber girt* (ſay from twenty to thirty inches in circum- ference) in forty years.

GROUNDS;

G R O U N D S:

O R,

ORNAMENTAL PLANTATIONS.

MANKIND no fooner find themfelves in faft poffeffion of the *neceffaries* of life, than they begin to feel a want of its *conveniencies*; and thefe obtained, feldom fail of indulging in one or more of its various *refinements*. Some men delight in the luxuries of the imagination; others in thofe of the fenfes. One man finds his wants fupplied in the delicacies of the table, whilft another flies to perfumes and effences for relief: few men are infenfible to the gratifications of the ear; and men in general are fufceptible of thofe of the eye. The imitative arts of painting and fculpture have been the ftudy and delight of civilized nations in all ages; but the art of embellifhing Nature herfelf has been referved for this age, and for this nation!

A fact the more aftonifhing, as ornamented Nature is as much fuperior to a Painting or a Statue, as a " Reality is to a Reprefentation;" — as the Man himfelf is to his Portrait. That the ftriking features—the beauties—of Nature, when-

ever

ever they have been *seen*, have always been *admired*
by men of fenfe and refinement, is undoubtedly
true; but why the good offices of art, in difclofing
thefe beauties, and fetting off thofe features to advantage, fhould have been fo long confined to the
human perfon alone, is, of all other facts in the
Hiftory of Arts and Sciences, the moft extraordinary.

The Tranflator of D'Ermenonville's Effay on
Landfcape has attempted to prove, in an introductory difcourfe, that the art is nothing *new*, for that
it was *known* to the Antients, though not *practifed*.
But the evidences he produces go no farther than to
fhew, that the Antients were *admirers of Nature in a
ftate of wildnefs*; for, whenever they attempted to
embellifh Nature, they appear to have been guided
by a kind of Otaheitean tafte; as the gardens of
the Greeks and Romans, like thofe of the modern
nations (until of late years in this country), convey
to us no other idea than that of *Nature tatoo'd* *.

Mr. Burgh, in a Note to his ingenious Commen-
tary upon Mr. Mafon's beautiful poem *The Englifh
Garden*, confirms us in thefe ideas; and, by a quo-
tation from the Younger Pliny, fhews the juft notions the Antients entertained of the powers of

---

* The inhabitants of Otaheitee, an ifland in the Southern
hemifphere, ornament their bodies by making punctures in the
fkin with a fharp-pointed inftrument, and call it *tatowing*. The
African Negroes are ftill groffer in their ideas of ornament,
gafhing their cheeks and temples in a manner fimilar to that
practifed by the Englifh Butcher in ornamenting a fhoulder of
mutton, or a Dutch gardener in embellifhing the environs of a
manfion.

human

human invention, in affociating and polifhing the rougher fcenes of Nature : for, after giving us a beautiful defcription of the natural fcenery round his Tufcan villa, upon the banks of the Tiber, he acknowledges " the view before him to refemble a " picture beautifully compofed, rather than a work " of Nature accidentally delivered."

We have been told that the Englifh Garden is but a copy of the Gardens of the Chinefe : this, however, is founded in Gallic envy rather than in truth ; for though their ftyle of Gardening may not admit of *tatooings* and *topiary works* *, it has as little to do with natural fcenery as the garden of an antient Roman, or a modern Frenchman :——THE ART OF *affifting* NATURE is, undoubtedly, all our own.

It cannot fail of proving highly interefting to our Readers, to trace the rife of this delightful Art.

Mr. Walpole, in his *Anecdotes of Painting in England,* has favoured the Public with *The Hiftory of modern Tafte in Gardening.* A pen guided by fo mafterly a hand muft ever be productive of information and entertainment when employed upon a fubject fo truly interefting as that which is now before us. Defirous of conveying to our Readers all the information which we can comprefs with propriety within the limits of our plan, we wifhed to have given the *fubftance* of this valuable paper ;

* Trees carved by a *Topiarius,* into the form of beafts, birds, &c.

but

but finding it already in the language of fimplicity, and being aware of the mifchiefs which generally enfue in *meddling* with the productions of genius, we had only one alternative; either wholly to tranfcribe, or wholly to reject. *This* we could not do, in ftrict juftice to our Readers; for, befides giving us, in detail, the advancement of the art, it throws confiderable light upon the art itfelf; and being only a fmall part of a work upon a different fubject, it is the lefs likely to fall into the hands of thofe to whom it cannot fail of proving highly interefting. We are, therefore, induced to exceed our intended limits in this refpect, by making a literal tranfcript; and hope, in the liberality of the Author, to be pardoned for fo doing.

GARDENING was probably one of the firft arts that fucceeded to that of building houfes, and naturally attended property and individual poffeffion. Culinary, and afterwards medicinal herbs were the objects of every head of a family: it became convenient to have them within reach, without feeking them at random in woods, in meadows, and on mountains, as often as they were wanted. When the earth ceafed to furnifh fpontaneoufly all thefe primitive luxuries, and culture became requifite, feparate inclofures for rearing herbs grew expedient. Fruits were in the fame predicament, and thofe moft in ufe or that demand attention, muft have entered into and extended the domeftic inclofure. The good man Noah, we are told, planted a vineyard, drank of the wine, and was drunken, and every body knows the confequences. Thus we acquired kitchen-gardens, orchards, and vineyards. I am apprized that the prototype of all thefe forts was the garden of Eden; but as that Paradife was a good deal larger than any we read of afterwards, being inclofed by the rivers Pifon, Gihon, Hiddekel, and Euphrates, as every tree that was pleafant to the fight and good for food grew in it, and as two other trees were likewife found there, of which not a flip or fucker remains, it does not belong to the prefent difcuffion. After the Fall, no man living
was

was fuffered to enter into the garden; and the poverty and neceffities of our firft anceftors hardly allowed them time to make
improvements in their eftates in imitation of it, fuppofing any
plan had been preferved. A cottage and a flip of ground for a
cabbage and a goofeberry-bufh, fuch as we fee by the fide of a
common, were in all probability the earlieft feats and gardens :
a well and bucket fucceeded to the Pifon and Euphrates. As
fettlements increafed, the orchard and the vineyard followed ;
and the earlieft princes of tribes poffeffed juft the neceffaries
of a modern farmer.

Matters, we may well believe, remained long in this fituation; and though the generality of mankind form their ideas
from the import of words in their own age, we have no reafon
to think that for many centuries the term Garden implied more
than a kitchen-garden or orchard. When a Frenchman reads
of the garden of Eden, I do not doubt but he concludes it was
fomething approaching to that of Verfailles, with clipt hedges,
berceaus, and trellis-work. If his devotion humbles him fo
far as to allow that, confidering who defigned it, there might
be a labyrinth full of Æfop's Fables, yet he does not conceive
that four of the largeft rivers in the world were half fo magnificent as an hundred fountains full of ftatues by Girardon. It
is thus that the word Garden has at all times paffed for whatever was underftood by that term in different countries. But
that it meant no more than a kitchen-garden or orchard for
feveral centuries, is evident from thofe few defcriptions that are
preferved of the moft famous gardens of antiquity.

That of Alcinous, in the Odyffey, is the moft renowned in
the heroic times. Is there an admirer of Homer who can read
his defcription without rapture; or who does not form to his
imagination a fcene of delights more picturefque than the landfcapes of Tinian or Juan Fernandez? Yet what was that
boafted Paradife with which

> the gods ordain'd
> To grace Alcinous and his happy land ? Pope.

Why, divefted of harmonious Greek and bewitching poetry, it
was a fmall orchard and vineyard, with fome beds of herbs
and two fountains that watered them, inclofed within a quick-
fet hedge. The whole compafs of this pompous garden inclofed—four acres.

> Four acres was th' allotted fpace of ground,
> Fenc'd with a green inclofure all around.

The trees were apples, figs, pomegranates, pears, olives, and
vines.

Tall

Tall thriving trees confefs'd the fruitful mold;
The redning apple ripens into gold.
Here the blue fig with lufcious juice o'erflows,
With deeper red the full pomegranate glows.
The branch here bends beneath the weighty pear,
And verdant olives flourifh round the year.

        *   *   *   *   *   *

Beds of all various herbs, for ever green,
In beauteous order terminate the fcene.

Alcinous's garden was planted by the poet, enriched by him
with the fairy gift of eternal fummer, and no doubt an effort
of imagination furpaffing any thing he had ever feen. As he
has beftowed on the fame happy prince a palace with brazen
walls and columns of filver, he certainly intended that the garden fhould be proportionably magnificent. We are fure therefore that as late as Homer's age, an inclofure of four acres,
comprehending orchard, vineyard and kitchen-garden, was a
ftretch of luxury the world at that time had never beheld.

The hanging gardens of Babylon were a ftill greater prodigy.
We are not acquainted with their difpofition or contents, but
as they are fuppofed to have been formed on terraffes and the
walls of the palace, whither foil was conveyed on purpofe, we
are very certain of what they were not; I mean they muft have
been trifling, of no extent, and a wanton inftance of expence
and labour. In other words, they were what fumptuous gardens have been in all ages till the prefent, unnatural, enriched
by art, pcffibly with fountains, ftatues, baluftrades, and fummer-
houfes, and were any thing but verdant and rural.

From the days of Homer to thofe of Pliny, we have no traces
to lead our guefs to what were the gardens of the intervening
ages. When Roman authors, whofe climate inftilled a wifh
for cool retreats, fpeak of their enjoyments in that kind, they
figh for grottos, caves, and the refrefhing hollows of mountains, near irriguous and fhady founts; or boaft of their porticos, walks of planes, canals, baths and breezes from the
fea. Their gardens are never mentioned as affording fhade and
fhelter from the rage of the dog-ftar. Pliny has left us defcriptions of two of his villas. As he ufed his Laurentine villa
for his winter retreat, it is not furprifing that the garden makes
no confiderable part of the account. All he fays of it is, that
the geftatio or place of exercife, which furrounded the garden
(the latter confequently not being very large) was bounded by
a hedge of box, and where that was perifhed, with rofemary;
that there was a walk of vines, and that moft of the trees were
fig and mulberry, the foil not being proper for any other forts.

On his Tufcan villa he is more diffufe; the garden makes a
confiderable part of the defcription:—and what was the principal

pal beauty of that pleafure-ground? Exactly what was the admiration of this country about threefcore years ago; box-trees cut into monfters, animals, letters, and the names of the mafter and the artificer. In an age when architecture difplayed all its grandeur, all its purity, and all its tafte; when arofe Vefpafian's amphitheatre, the temple of Peace, Trajan's fo-rum, Domitian's baths, and Adrian's villa, the ruins and vef-tiges of which ftill excite our aftonifhment and curiofity; a Ro-man conful, a polifhed emperor's friend, and a man of elegant literature and tafte, delighted in what the mob now fcarce ad-mire in a college-garden. All the ingredients of Pliny's cor-refponded exactly with thofe laid out by London and Wife on Dutch principles. He talks of flopes, terraffes, a wildernefs, fhrubs methodically trimmed, a marble bafon, * pipes fpout-ing water, a cafcade falling into the bafon, bay-trees, alter-nately planted with planes, and a ftrait walk, from whence if-fued others parted off by hedges of box, and apple-trees, with obelifks placed between every two. There wants nothing but the embroidery of a parterre, to make a garden in the reign of Trajan ferve for a defcription of one in that of king William†. In one paffage above Pliny feems to have conceived that natu-ral irregularity might be a beauty; *in opere urbaniffimo*, fays he, *fubita velut illati ruris imitatio*. Something like a rural view was contrived amidft fo much polifhed compofition. But the idea foon vanifhed, lineal walks immediately enveloped the flight fcene, and names and infcriptions in box again fucceeded to compenfate for the daring introduction of nature.

In the paintings found at Herculaneum are a few traces of gardens, as may be feen in the fecond volume of the prints. They are fmall fquare inclofures, formed by trellis-work, and efpaliers‡,

* The Englifh gardens defcribed by Hentzner in the reign of Elizabeth, are exact copies of thofe of Pliny. In that at Whitehall was a fun-dial and jet-d'eau, which on turning a cock fpurted out water and fprinkled the fpectators. In lord Burleigh's at Theobald's were obelifks, pyramids, and circular porticos, with cifterns of lead for bathing. At Hampton-court the garden walls were covered with rofemary, a cuftom, he fays, very common in England. At Theobald's was a labyrinth alfo, an ingenuity I fhall mention prefently to have been frequent in that age.

† Dr. Plot, in his Natural Hiftory of Oxfordfhire, p. 380, feems to have been a great admirer of trees carved into the moft heterogeneous forms, which he calls *topiary works*, and quotes one Laurembergius for faying that the Englifh are as expert as moft nations in that kind of fculpture, for which Hampton-court was particularly remarkable. The Doctor then names other gardens that flourifhed with animals and caftles, formed *arte topiaria*, and above all a wren's neft that was capacious enough to receive a man to fit on a feat made within it for that purpofe.

‡ At Warwick-caftle is an ancient fuit of arras, in which there is a garden exactly refembling thefe pictures of Herculaneum.

and

and regularly ornamented with vafes, fountains and eareatides, elegantly fymmetrical, and proper for the narrow fpaces allotted to the garden of a houfe in a capital city. From fuch I would not banifh thofe playful waters that refrefh a fultry manfion in town, nor the neat trellis, which preferves its wooden verdure better than natural greens expofed to duft. Thofe treillages in the gardens at Paris, particularly on the Boulevard, have a gay and delightful effect.—They form light corridores, and tranfpicuous arbours through which the fun-beams play and chequer the fhade, fet off the ftatues, vafes, and flowers, that marry with their gaudy hotels, and fuit the gallant and idle fociety who paint the walks between their parterres, and realize the fantaftic fcenes of Watteau and Durfè.

From what I have faid, it appears how naturally and infenfibly the idea of a kitchen-garden flid into that which has for fo mary ages been peculiarly termed a Garden, and by our anceftors in this country, diftinguifhed by the name of a Pleafure-garden. A fquare piece of ground was originally parted off in early ages for the ufe of the family :—to exclude cattle and afcertain the property it was feparated from the fields by a hedge. As pride and defire of privacy encreafed, the inclofure was dignified by walls ; and in climes where fruits were not lavifhed by the ripening glow of nature and foil, fruit-trees were affifted and fheltered from furrounding winds by the like expedient ; for the inundation of luxuries which have fwelled into general neceffities, have almoft all taken their fource from the fimple fountain of reafon.

When the cuftom of making fquare gardens inclofed with walls was thus eftablifhed, to the exclufion of nature and profpect *, pomp and folitude combined to call for fomething that might enrich and enliven the infipid and unanimated partition. Fountains, firft invented for ufe, which grandeur loves to difguife and throw out of the queftion, received embellifhments from coftly marbles, and at laft, to contradict utility, toffed their wafte of waters into air in fpouting columns. Art, in the hands of rude man, had at firft been made a fuccedaneum to nature ; in the hands of oftentatious wealth, it became the means of oppofing nature ; and the more it traverfed the march of the latter, the more nobility thought its power was demonftrated. Canals meafured by the line were introduced in lieu of meandering ftreams, and terraffes were hoifted aloft in oppo-

* It was not uncommon, after the circumadjacent country had been fhutout, to endeavour to recover it by raifing large mounts of earth to peep over the walls of the garden.

fition

fition to the facile flopes that imperceptibly unite the valley to the hill. Baluftrades defended thefe precipitate and dangerous elevations, and flights of fteps rejoined them to the fubjacent flat from which the terrafs had been dug. Vafes and fculpture were added to thefe unneceffary balconies, and ftatues furnifhed the lifelefs fpot with mimic reprefentations of the excluded fons of men. Thus difficulty and expence were the conftituent parts of thofe fumptuous and felfifh folitudes; and every improvement that was made, was but a ftep farther from nature. The tricks of water-works to wet the unwary, not to refrefh the panting fpectator, and parterres embroidered in patterns like a petticoat, were but the childifh endeavours of fafhion and novelty to reconcile greatnefs to what it had furfeited on. To crown thefe impotent difplays of falfe tafte, the fheers were applied to the lovely wildnefs of form with which nature has diftinguifhed each various fpecies of tree and fhrub. The venerable Oak, the romantic Beech, the ufeful Elm, even the afpiring circuit of the Lime, the regular round of the Chefnut, and the almoft moulded Orange-tree, were corrected by fuch fantaftic admirers of fymmetry. The compafs and fquare were of more ufe in plantations than the nurfery-man. The meafured walk, the quincunx, and the etoile, impofed their unfatisfying famenefs on every royal and noble garden. Trees were headed, and their fides pared away; many French groves feem green chefts fet upon poles. Seats of marble, arbours, and fummer-houfes, terminated every vifto; and fymmetry, even where the fpace was too large to permit its being remarked at one view, was fo effential, that, as Pope obferved,

———each alley has a brother,
And half the garden juft reflects the other.

Knots of flowers were more defenfibly fubjected to the fame regularity. Leifure, as Milton expreffed it,

in trim gardens took his pleafure.

In the garden of Marfhal de Biron at Paris, confifting of fourteen acres, every walk is buttoned on each fide by lines of flower-pots, which fucceed in their feafons. When I faw it, there were nine thoufand pots of Afters, or la Reine Marguerite. We do not precifely know what our anceftors meant by a bower, it was probably an arbour; fometimes it meant the whole frittered inclofure, and in one inftance it certainly included a labyrinth. Rofamond's bower was indifputably of that kind, though whether compofed of walls or hedges we

P p 3 cannot

cannot determine *. A fquare and a round labyrinth were fo capital ingredients of a garden formerly, that in Du Cerceau's architecture, who lived in the time of Charles IX. and Henry III. there is fcarce a ground-plot without one of each. The enchantment of antique appellations has confecrated a pleafing idea of a royal refidence, of which we now regret the extinction. Havering in the Bower, the jointure of many dowager queens, conveys to us the notion of a romantic fcene.

In Kip's Views of the Seats of our Nobility and Gentry, we fee the fame tirefome and returning uniformity. Every houfe is approached by two or three gardens, confifting perhaps of a gravel-walk and two grafs-plats, or borders of flowers. Each rifes above the other by two or three fteps, and as many walls and terraffes, and fo many iron gates, that we recollect thofe ancient romances, in which every entrance was guarded by nymphs or dragons. At Lady Orford's, at Piddletown, in Dorfetfhire, there was, when my brother married, a double inclofure of thirteen gardens, each I fuppofe not much above an hundred yards fquare, with an enfilade of correfpondent gates; and before you arrived at thefe, you paffed a narrow gut between two ftone terraffes, that rofe above your head, and which were crowned by a line of pyramidal yews. A bowling-green was all the lawn admitted in thofe times, a circular lake the extent of magnificence.

Yet though thefe and fuch prepofterous inconveniencies prevailed from age to age, good fenfe in this country had perceived the want of fomething at once more grand and more natural. Thefe reflections, and the bounds fet to the wafte made by royal fpoilers, gave origin to Parks. They were contracted forefts, and extended gardens. Hentzner fays, that, according to Rous of Warwick, the firft park was that at Woodftock. If fo, it might be the foundation of a legend that Henry II. fecured his miftrefs in a labyrinth : it was no doubt more difficult to find her in a park than in a palace, where the intricacy of the woods and various lodges buried in covert might conceal her actual habitation.

It is more extraordinary that having fo long ago ftumbled on the principle of modern gardening, we fhould have perfifted in retaining its reverfe, fymmetrical and unnatural gardens. That parks were rare in other countries, Hentzner, who travel-

---

* Drayton, in a note to his Epiftle of Rofamond, fays, her labyrinth was built of vaults under ground, arched and walled with brick and ftone ; but, as Mr. Gough obferves, he gives no authority for that affertion. V. pref. to 2d edit. of Britifh Topography, p. xxx. Such vaults might remain to Drayton's time, but did not prove that there had been no fuperftructure.

led over great part of Europe, leads us to suppose, by observing that they were common in England.   In France they retain the name, but nothing is more different both in compass and disposition.   Their parks are usually square or oblong inclosures, regularly planted with walks of chesnuts or limes, and generally every large town has one for its public recreation.   They are exactly like Burton's-court, at Chelsea-college, and rarely larger.

One man, one great man we had, on whom nor education nor custom could impose their prejudices ; who, " on evil days " though fallen, and with darkness and solitude compassed " round," judged that the mistaken and fantastic ornaments he had seen in gardens, were unworthy of the Almighty Hand that planted the delights of Paradise.   He seems, with the prophetic eye of taste [as I have heard taste well * defined], to have conceived, to have foreseen modern gardening; as Lord Bacon announced the discoveries since made by experimental philosophy.   The description of Eden is a warmer and more just picture of the present style than Claud Lorrain could have painted from Hagley or Stourhead.   The first lines I shall quote exhibit Stourhead on a more magnificent scale.

> Thro' Eden went a river large,
> Nor chang'd his course, but thro' the shaggy hil
> Pass'd underneath ingulph'd, for God had thrown
> That mountain as his garden-mound, high rais'd
> Upon the rapid current——

Hagley seems pictured in what follows :

> which thro' veins
> Of porous earth with kindly thirst updrawn,
> Rose a fresh fountain, and with many a rill
> Water'd the garden——

What colouring, what freedom of pencil, what landscape in these lines !

> ——from that saphire fount the crisped brooks,
> Rolling on orient pearl and sands of gold,
> With mazy error under pendent shades
> Ran nectar, visiting each plant, and fed
> Flow'rs worthy of Paradise, which not *nice art*
> In beds and curious knots, but *nature* boon
> Pour'd forth profuse on hill and dale and plain,

---

* By the great Lord Chatham, who had a good taste himself in modern gardening, as he shewed by his own villas in Enfield Chace and at Hayes.

Both where the morning fun firft warmly fmote
The *open field*, and where the unpierc'd fhade
Imbrown'd the noon-tide bow'rs.—*Thus was this place
A happy rural feat of various view.*

Read this tranfporting defcription, paint to your mind the
fcenes that follow, contraft them with the favage but refpectable
terror with which the poet guards the bounds of his Paradife,
fenced

————with the champaign head
Of a fteep wildernefs, whofe hairy fides
With thicket overgrown, grotefque and wild,
Accefs denied ; and over head upgrew
Infuperable height of loftieft fhade,
Cedar and pine, and fir, and branching palm,
A fylvan fcene, and, as the ranks afcend,
Shade above fhade, a woody theatre
Of ftatelieft view——

and then recollect that the author of this fublime vifion had never feen a glimpfe of any thing like what he has imagined, that
his favourite ancients had dropped not a hint of fuch divine
fcenery, and that the conceits in Italian gardens, and Theobalds
and Nonfuch, were the brighteft originals that his memory
could furnifh. His intellectual eye faw a nobler plan, fo little
did he fuffer by the lofs of fight. It fufficed him to have feen
the materials with which he could work. The vigour of a
boundlefs imagination told him how a plan might be difpofed,
that would embellifh nature, and reftore art to its proper office,
the juft improvement or imitation of it *.

It is neceffary that the concurrent teftimony of the age fhould
fwear to pofterity that the defcription above quoted was written
above half a century before the introduction of modern gardening,
or our incredulous defcendents will defraud the poet of half his
glory, by being perfuaded that he copied fome garden or gardens he had feen fo minutely do his ideas correfpond with the
prefent ftandard. But what fhall we fay for that intervening
half century who could read that plan and never attempt to put
it in execution ?

Now let us turn to an admired writer, pofterior to Milton, and
fee how cold, how infipid, how taftelefs is his account of what he
pronounced a perfect garden. I fpeak not of his ftyle, which it
was not neceffary for him to animate with the colouring and
glow of poetry. It is his want of ideas, of imagination, of tafte,

* Since the above was written, I have found Milton praifed and Sir William
Temple cenfured, on the fame foundations, in a poem called The Rife and
Progrefs of the prefent Tafte in Planting, printed in 1767.

that

that I cenfure, when he dictated on a fubject that is capable of all the graces that a knowledge of beautiful nature can beftow. Sir William Temple was an excellent man ; Milton, a genius of the firft order.

We cannot wonder that Sir William declares in favour of parterres, fountains, and ftatues, as neceffary to break the famenefs of large grafs-plats, which he thinks have an ill effect upon the eye, when he acknowledges that he difcovers fancy in the gardens of Alcinous. Milton ftudied the Antients with equal enthufiafm, but no bigotry, and had judgment to diftinguifh between the want of invention and the beauties of poetry. Compare his Paradife with Homer s Garden, both afcribed to a celeftial defign. For Sir William, it is juft to obferve, that his ideas centered in a fruit-garden. He had the honour of giving to his country many delicate fruits, and he thought of little elfe than difpofing them to the beft advantage. Here is the paffage I propofed to quote ; it is long, but I need not make an apology to the reader for entertaining him with any other words inftead of my own.

" The beft figure of a garden is either a fquare or an oblong, and either upon a flat or a defcent : they have all their beauties, but the beft I efteem an oblong upon a defcent. The beauty, the air, the view, makes amends for the expence, which is very great in finifhing and, fupporting the terras-walks, in levelling the parterres, and in the ftone-ftairs that are neceffary from one to the other.

" The perfecteft figure of a garden I ever faw, either at home or abroad, was that of Moor park in Hertfordfhire, when I knew it about thirty years ago. It was made by the Countefs of Bedford, efteemed among the greateft wits of her time, and celebrated by Doctor Donne ; and with very great care, excellent contrivance, and much coft ; but greater fums may be thrown away without effect or honour, if there want fenfe in proportion to money, or *if nature be not followed*, which I take to be the great rule in this, and perhaps in every thing elfe, as far as the conduct not only of our lives but our governments." [We fhall fee how *natural* that admired garden was.]

" Becaufe I take * the garden I have named to have been in all kinds the moft beautiful and perfect, at leaft in the figure and difpofition, that I have ever feen, I will defcribe it for a model to thofe that meet with fuch a fituation, and are above the regards of common expence. It lies on the fide of a hill,

---

* This garden feems to have been made after the plan laid down by Lord Bacon in his 46th Effay, to which, that I may not multiply quotations, I will refer the reader.

upon

upon which the houfe ftands, but not very fteep. The length
of the houfe, where the beft rooms and of moft ufe or pleafure
are, lies upon the breadth of the garden; the great parlour
opens into the middle of a terras gravel-walk that lies even
with it, and which may lie, as I remember; about three hundred
paces long, and broad in proportion ; the border fet with
ftandard laurels and at large diftances, which have the beauty of
orange-trees out of flower and fruit   From this walk are three
defcents by many ftone fteps, in the middle and at each end,
into a very large parterre.   This is divided into quarters by
gravel-walks, and adorned with two fountains and eight ftatues
in the feveral quarters.   At the end of the terras-walk are two
fummer-houfes, and the fides of the parterre are ranged with
two large cloifters open to the garden, upon arches of ftone, and
ending with two other fummer-houfes even with the cloifters,
which are paved with ftone, and defigned for walks of fhade,
there being none other in the whole parterre.   Over thefe two
cloifters are two terraffes covered with lead and fenced with
balufters ; and the paffage into thefe airy walks is out of the
two fummer-houfes at the end of the firft terras-walk.   The
cloifter facing the fouth is covered with vines, and would have
been proper for an orange-houfe, and the other for myrtles or
other more common greens, and had, I doubt not, been caft for
that purpofe, if this piece of gardening had been then in as
much vogue as it is now.

" From the middle of this parterre is a defcent by many
fteps flying on each fide of a grotto that lies between them,
covered with lead and flat, into the lower garden, which is all
fruit-trees ranged about the feveral quarters of a wildernefs
which is very fhady; the walks here 'are all green, the grotto
embellifhed with figures of fhell rock-work, fountains, and
water-works.   If the hill had not ended with the lower garden,
and the wall were not bounded by a common way that goes
through the park, they might have added a third quarter of all
greens; but this want is fupplied by a garden on the other fide
the houfe, which is all of that fort, very wild, fhady, and
adorned with rough rock-work and fountains.

" This was Moor-park when I was acquainted with it, and
the fweeteft place, I think, that I have feen in my life, either
before or fince, at home or abroad."—

I will make no farther remarks on this defcription.   Any
man might defign and *build* as fweet a garden, who had been
born in and never ftirred out of Holborn.   It was not peculiar
to Sir William Temple to think in that manner.   How many
Frenchmen are there who have feen *our* gardens, and ftill prefer
natural flights of fteps and fhady cloifters covered with lead !

Le

Le Nautre, the architect of the groves and grottos at Verfailles, came hither on a miffion to improve our tafte. He planted St. James's and Greenwich Parks—no great monuments of his invention.

To do farther juftice to Sir William Temple, I muft not omit what he adds. " What I have faid of the beft forms of gar‑ dens is meant only of fuch as are in fome fort regular ; for there may be other forms wholly irregular, that may, for aught I know, have more beauty than any of the others ; but they muft owe it to fome extraordinary difpofitions of nature in the feat, or fome great race of fancy or judgment in the contri‑ vance, which may reduce many difagreeing parts into fome figure, which fhall yet, upon the whole, be very agreeable. Something of this I have feen in fome places, but heard more of it from others, who have lived much among the Chinefes, a people whofe way of thinking feems to lie as wide of ours in Europe as their country does.   Their greateft reach of imagi‑ nation is employed in contriving figures, where the beauty fhall be great and ftrike the eye, but without any order or dif‑ pofition of parts, that fhall be commonly or eafily obferved. And though we have hardly any notion of this fort of beauty, yet they have a particular word to exprefs it; and where they find it hit their eye at firft fight, they fay the Sharawadgi is fine or is admirable, or any fuch expreffion of efteem :—but I fhould hardly advife any of thefe attempts in the figure of gardens among us ; they are adventures of too hard atchieve‑ ment for any common hands ; and though there may be more honour if they fucceed well, yet there is more difhonour if they fail, and it is twenty to one they will ; whereas in regular fi‑ gures it is hard to make any great and remarkable faults."

Fortunately Kent and a few others were not quite fo timid, or we might ftill be going up and down ftairs in the open air.

It is true, we have heard much lately, as Sir William Tem‑ ple did, of irregularity and imitations of nature in the gardens or grounds of the Chinefe.   The former is certainly true : they are as whimfically irregular as European gardens are formally uniform, and unvaried :—but with regard to nature, it feems as much avoided, as in the fquares and oblongs and ftrait lines of our anceftors.   An artificial perpendicular rock ftarting out of a flat plain, and connected with nothing, often pierced through in various places with oval hollows, has no more pretenfion to be deemed natural than a lineal terrafs or a parterre.   The late Mr. Jofeph Spence, who had both tafte and zeal for the prefent ftyle, was fo perfuaded of the Chinefe Emperor's pleafure‑ ground being laid out on principles refembling ours, that he

tranflated

tranflated and publifhed, under the name of Sir Harry Beau-
mont, a particular account of that inclofure from the Collection
of the Letters of the Jefuits. I have looked it over, and, except
a determined irregularity, can find nothing in it that gives me
any idea of attention being paid to nature. It is of vaft circumference, and contains 200 palaces, befides as many contiguous for the eunuchs, all gilt, painted, and varnifhed. There
are raifed hills from 20 to 60 feet high, ftreams and lakes, and
one of the latter five miles round. Thefe waters are paffed by
bridges :—but even their bridges muft not be ftrait—they fer-
pentize as much as the rivulets, and are fometimes fo lcng as
to be furnifhed with refting-places, and begin and end with
triumphal arches. Methinks a ftrait canal is as rational at
leaft as a meandering bridge. The colonades undulate in the
fame manner. In fhort, this pretty gaudy fcene is the work
of caprice and whim, and, when we reflect on their buildings,
prefents no image but that of unfubftantial tawdrinefs. Nor
is this all. Within this fantaftic Paradife is a fquare town,
each fide a mile long. Here the eunuchs of the court, to en-
tertain his Imperial Majefty with the buftle and bufinefs of the
capital in which he refides, but which it is not of his dignity
ever to fee, act merchants and all forts of trades, and even de-
fignedly exercife for his royal amufement every art of knavery
that is practifed under his aufpicious government. Methinks
this is the childifh folace and repofe of grandeur, not a retirement from affairs to the delights of rural life. Here too his
Majefty plays at agriculture: there is a quarter fet apart for
that purpofe; the eunuchs fow, reap, and carry in their harveft, in the imperial prefence; and his Majefty returns to Pekin,
perfuaded that he has been in the country *.

Having

---

* The French have of late years adopted our ftyle in gardens, but chufing
to be fundamentally obliged to more remote rivals, they deny us half the merit, or rather the originality of the invention, by afcribing the difcovery to the
Chinefe, and by calling our tafte in gardening Le Gout Anglo-Chinois. I
think I have fhewn that this is a blunder, and that the Chinefe have paffed
to one extremity of abfurdity, as the French and all antiquity had advanced to
the other, both being equally remote from nature; regular formality is the
oppofite point to faataftic Sharawadgis. The French, indeed, during the fafhionable paroxyfm of philofophy, have furpaffed us, at leaft in meditation on
the art. I have perufed a grave treatife of recent date, in which the author,
extending his views beyond mere luxury and amufement, has endeavoured to
infpire his countrymen, even in the gratification of their expenfive pleafures,
with benevolent projects. He propofes to them to combine gardening with
charity, and to make every ftep of their walks an act of generofity and a leffon
of morality. Inftead of adorning favourite points with a heathen temple, a
Chinefe pagoda, a Gothic tower, or fictitious bridge, he propofes to them at the

firft

GROUNDS. 573

Having thus cleared my way by afcertaining what have been
the ideas on gardening in all ages, as far as we have materials
to judge by, it remains to fhew to what degree Mr. Kent invented the new ftyle, and what hints he had received to fuggeft
and conduct his undertaking.

We have feen what Moor-park was, when pronounced a
ftandard. But as no fucceeding generation in an opulent and
luxurious country contents itfelf with the perfection eftablifhed
by its anceftors, more perfect perfection was ftill fought; and
improvements had gone on, till London and Wife had ftocked
our gardens with giants, animals, monfters*, coats of arms,
and mottos, in yew, box and holly. Abfurdity could go no
farther, and the tide turned. Bridgman, the next fafhionable
defigner of gardens, was far more chafte; and whether from
good fenfe, or that the nation had been ftruck and reformed
by the admirable paper in the Guardian, N° 173, he banifhed verdant fculpture, and did not even revert to the fquare
precifion of the foregoing age. He enlarged his plans, dif-
dained to make every divifion tally to its oppofite, and though
he ftill adhered much to ftrait walks with high clipped hedges,
they were only his great lines; the reft he diverfified by wil-

---

firft refting-place to erect a fchool, a little farther to found an academy, at a
third diftance a manufacture, and at the termination of the park to endow an
hofpital. Thus, fays he, the proprietor would be led to meditate, as he
faunters, on the different ftages of human life, and both his expence and
thoughts would march in a progreffion of patriotic acts and reflections.
When he was laying out fo magnificent, charitable, and philofophic an Uto-
pian villa, it would have coft no more to have added a foundling-hofpital, a
fenate-houfe, and a burying-ground.—If I fmile at fuch vifions, ftill one
muft be glad that in the whirl of fafhions, beneficence fhould have its turn in
vogue; and though the French treat the Virtues like every thing elfe, but as
an object of mode, it is to be hoped that they too will, every now and then,
come into fafhion again. The author I have been mentioning reminds me of
a French gentleman, who fome years ago made me a vifit at Strawberry-hill.
He was fo complaifant as to commend the place, and to approve our tafte in
gardens—but in the fame ftyle of thinking with the above-cited author, he
faid, "I do not like your imaginary temples and fictitious terminations of
views: I would have real points of view with moving objects; for inftance,
here I would have—(I forget what)—and there a watering-place." "That
is not fo eafy (I replied); one cannot oblige others to affemble at fuch or fuch
a fpot for one's amufement—however, I am glad you would like a watering-
place, for there happens to be one: in that creek of the Thames the inhabitants of the village do actually water their horfes; but I doubt whether, if it
were not convenient to them to do fo, they would frequent the fpot only to
enliven my profpect."—Such Gallo-Chinois gardens, I apprehend, will rarely
be executed.

† On the piers of a garden-gate not far from Paris I obferved two very
coquet fphinxes. Thefe lady monfters had ftraw hats gracefully fmart on one
fide of their heads, and filken cloaks half veiling their necks; all executed
in ftone.

dernefs,

derness, and with loose groves of oak, though still within surrounding hedges. I have observed in the garden * at Gubbins
in Hertfordshire many detached thoughts, that strongly indicate
the dawn of modern taste. As his reformation gained footing,
he ventured farther, and in the royal garden at Richmond
dared to introduce cultivated fields, and even morsels of a forest appearance, by the sides of those endless and tiresome
walks, that stretched out of one into another without inter-
mission. But this was not till other innovators had broke loose
too from rigid symmetry.

But the capital stroke, the leading step to all that has followed, was [I believe the first thought was Bridgman's] the
destruction of walls for boundaries, and the invention of fosses
—an attempt then deemed so astonishing, that the common
people called them Ha! Ha's! to express their surprize at
finding a sudden and unperceived check to their walk.

One of the first gardens planted in this simple though still
formal style, was my father's at Houghton. It was laid out by
Mr. Eyre, an imitator of Bridgman. It contains three-and-
twenty acres, then reckoned a considerable portion.

I call a sunk fence the leading step, for these reasons. No
sooner was this simple enchantment made, than levelling, mowing, and rolling, followed. The contiguous ground of the
park without the sunk fence was to be harmonized with the
lawn within; and the garden in its turn was to be set free
from its prim regularity, that it might assort with the wilder
country without. The sunk fence ascertained the specific garden, but that it might not draw too obvious a line of distinc-
tion between the neat and the rude, the contiguous out-lying
parts came to be included in a kind of general design; and
when nature was taken into the plan, under improvements,
every step that was made, pointed out new beauties and in-
spired new ideas. At that moment appeared Kent, painter
enough to taste the charms of landscape, bold and opinion-
ative enough to dare and to dictate, and born with a genius to
strike out a great system from the twilight of imperfect essays.
He leaped the fence, and saw that all nature was a garden.
He felt the delicious contrast of hill and valley changing imperceptibly into each other, tasted the beauty of the gentle
swell or concave scoop, and remarked how loose groves
crowned an easy eminence with happy ornament, and while

---

* The seat of the late Sir Jeremy Sambroke. It had formerly belonged
to Lady More, mother-in-law of Sir Thomas More, and had been tyrannically
wrenched from her by Henry VIII. on the execution of Sir Thomas, though
not her son, and though her jointure from a former husband.

they

they called in the diſtant view between their graceful ſtems, removed and extended the perſpective by deluſive compariſon. Thus the pencil of his imagination beſtowed all the arts of landſcape on the ſcenes he handled. The great principles on which he worked were perſpective, and light and ſhade. Groupes of trees broke too uniform or too extenſive a lawn; evergreens and woods were oppoſed to the glare of the champain; and where the view was leſs fortunate, or ſo much expoſed as to be beheld at once, he blotted out ſome parts by thick ſhades, to divide it into variety, or to make the richeſt ſcene more enchanting by reſerving it to a farther advance of the ſpectator's ſtep. Thus, ſelecting favourite objects, and veiling deformities by ſcreens of plantation; ſometimes allowing the rudeſt waſte to add its foil to the richeſt theatre; he realized the compoſitions of the greateſt maſters in painting. Where objects were wanting to animate his horizon, his taſte as an architect could beſtow immediate termination. His buildings, his ſeats, his temples, were more the works of his pencil than of his compaſſes. We owe the reſtoration of Greece and the diffuſion of architecture to his ſkill in landſcape.

But of all the beauties he added to the face of this beautiful country, none ſurpaſſed his management of water. Adieu to canals, circular baſons, and caſcades tumbling down marble ſteps, that laſt abſurd magnificence of Italian and French villas. The forced elevation of cataracts was no more. The gentle ſtream was taught to ſerpentize ſeemingly at its pleaſure, and where diſcontinued by different levels, its courſe appeared to be concealed by thickets properly interſperſed, and glittered again at a diſtance where it might be ſuppoſed naturally to arrive. Its borders were ſmoothed, but preſerved their waving irregularity. A few trees ſcattered here and there on its edges ſprinkled the tame bank that accompanied its meanders; and when it diſappeared among the hills, ſhades deſcending from the heights leaned towards its progreſs, and framed the diſtant point of light under which it was loſt, as it turned aſide to either hand of the blue horizon.

Thus, dealing in none but the colours of nature, and catching its moſt favourable features, men ſaw a new creation opening before their eyes. The living landſcape was chaſtened or poliſhed, not transformed. Freedom was given to the forms of trees; they extended their branches unreſtricted, and where any eminent Oak, or maſter Beech, had eſcaped maiming and ſurvived the foreſt, buſh and bramble was removed, and all its honours were reſtored to diſtinguiſh and ſhade the plain. Where the united plumage of an ancient wood extended wide its undulating canopy, and ſtood venerable in its darkneſs, Kent
thinned

thinned the foremoft ranks, and left but fo many detached and fcattered trees, as foftened the approach of gloom, and blended a chequered light with the thus lengthened fhadows of the remaining columns.

Succeeding artifts have added new mafter-ftrokes to thefe touches; perhaps improved or brought to perfection fome that I have named. The introduction of foreign trees and plants, which we owe principally to Archibald Duke of Argyle, contributed effentially to the richnefs of colouring fo peculiar to our modern landfcape. The mixture of various greens, the contraft of forms between our foreft-trees and the northern and Weft-Indian firs and pines, are improvements more recent than Kent, or but little known to him. The Weeping-willow and every florid fhrub, each tree of delicate or bold leaf, are new tints in the compofition of our gardens. The laft century was certainly acquainted with many of thofe rare plants we now admire. The Weymouth-pine has long been naturalized here; the patriarch plant ftill exifts at Longleat. The light and graceful Acacia was known as early; witnefs thofe ancient ftems in the court of Bedford houfe in Bloomfbury-fquare; and in the Bifhop of London's garden at Fulham are many exotics of very ancient date. I doubt therefore whether the difficulty of preferving them in a clime fo foreign to their nature did not convince our anceftors of their inutility in general; unlefs the fhapelinefs of the lime and horfe-chefnut, which accorded fo well with eftablifhed regularity, and which thence and from their novelty grew in fafhion, did not occafion the neglect of the more curious plants.

But juft as the encomiums are that I have beftowed on Kent's difcoveries, he was neither without affiftance or faults. Mr. Pope undoubtedly contributed to form his tafte. The defign of the Prince of Wales's garden at Carlton-houfe was evidently borrowed from the Poet's at Twickenham. There was a little of affected modefty in the latter, when he faid, of all his works he was moft proud of his garden. And yet it was a fingular effort of art and tafte to imprefs fo much variety and fcenery on a fpot of five acres. The paffing through the gloom from the grotto to the opening day, the retiring and again affembling fhades, the dufky groves, the larger lawn, and the folemnity of the termination at the cypreffes that lead up to his mother's tomb, are managed with exquifite judgment; and though Lord Peterborough affifted him

To form his quincunx and to rank his vines,

thofe were not the moft pleafing ingredients of his little perfpective.

I do

I do not know whether the difpofition of the garden at
Roufham, laid out for General Dormer, and in my opinion the
moft engaging of all Kent's works, was not planned on the
model of Mr. Pope's, at leaft in the opening and retiring
fhades of Venus's Vale. The whole is as elegant and antique
as if the Emperor Julian had felected the moft pleafing folitude
about Daphne to enjoy a philofophic retirement.

That Kent's ideas were but rarely great, was in fome meafure owing to the novelty of his art. It would have been difficult to have tranfported the ftyle of gardening at once from
a few acres to tumbling of forefts : and though new fafhions
like new religions, [which are new fafhions] often lead men
to the moft oppofite exceffes, it could not be the cafe in gardening, where the experiments would have been fo expenfive.
Yet it is true too that the features in Kent's landicapes were
feldom majeftic. His clumps were puny, he aimed at immediate effeft, and planted not for futurity. One fees no large
woods fketched out by his direftion. Nor are we yet entirely
rifen above a too great frequency of fmall clumps, efpecially in
the elbows of ferpentine rivers. How common to fee three or
four beeches, then as many larches, a third knot of cypreffes,
and a revolution of all three ! Kent's laft defigns were in a
higher ftyle, as his ideas opened on fuccefs. The north terras
at Claremont was much fuperior to the reft of the garden.

A return of fome particular thoughts was common to him
with other painters, and made his hand known. A fmall lake
edged by a winding bank with fcattered trees that led to a feat
at the head of the pond, was common to Claremont, Efher,
and others of his defigns. At Efher,

Where Kent and Nature vied for Pelham's love,

the profpects more than aided the Painter's genius—they
marked out the points where his art was neceffary or not ; but
thence left his judgment in poffeffion of all its glory.

Having routed profeffed art, for the modern gardener exerts
his talents to conceal his art, Kent, like other reformers, knew
not how to ftop at the juft limits. He had followed Nature,
and imitated her fo happily, that he began to think all her
works were equally proper for imitation. In Kenfington garden he planted dead trees, to give a greater air of truth to the
fcene—but he was foon laughed out of this excefs. His ruling
principle was, that Nature abhors a ftrait line. His mimics, for
every genius has his apes, feemed to think that fhe could love
nothing but what was crooked. Yet fo many men of tafte of
all ranks devoted themfelves to the new improvements, that it
is furprizing how much beauty has been ftruck out, with how

few abfurdities. Still in fome-lights the reformation feems to
me to have been pufhed too far.   Though an avenue croffing a
park or feparating a lawn, and intercepting views from the feat
to which it leads, are capital faults, yet a great avenue * cut
through woods, perhaps before entering a park, has a noble
air, and

> Like footmen running before coaches
> To tell the inn what lord approaches,

announces the habitation of fome man of diftinction.   In other
places the total banifhment of all particular neatnefs immediately about a houfe, which is frequently left gazing by itfelf
in the middle of a park, is a defect.   Sheltered and even clofe
walks in fo very uncertain a climate as ours, are comforts ill
exchanged for the few picturefque days that we enjoy : and
whenever a family can purloin a warm and even fomething of
an old-fafhioned garden from the landfcape defigned for them
by the undertaker in fafhion, without interfering with the picture, they will find fatisfactions on thofe days that do not in-
vite ftrangers to come and fee their improvements.

Fountains have with great reafon been banifhed from gardens as unnatural; but it furprifes me that they have not been
allotted to their proper pofitions, to cities, towns, and the
courts of great houfes, as proper accompaniments to architecture, and as works of grandeur in themfelves.   Their decora-
tions admit the utmoft invention, and when the waters are
thrown up to different ftages, and tumble over their border,
nothing has a more impofing or a more refrefhing found.   A
palace demands its external graces and attributes, as much as a
garden.   Fountains and cypreffes peculiarly become buildings,
and no man can have been at Rome, and feen the vaft bafons of
marble dafhed with perpetual cafcades in the area of St. Peter's,
without retaining an idea of tafte and fplendor.   Thofe in the
Piazza Navona are as ufeful as fublimely conceived.

Grottos in this climate are receffes only to be looked at
tranfiently.   When they are regularly compofed within of fymmetry and architecture, as in Italy, they are only fplendid im-
proprieties.   The moft judicioufly, indeed moft fortunately,

---

* Of this kind one of the moft noble is that of Stanftead, the feat of the
Earl of Halifax, traverfing an ancient wood for two miles and bounded by
the fea.   The very extenfive lawns at that feat, richly inclofed by venerable
beech woods, and chequered by fingle beeches of vaft fize, particularly when
you ftand in the portico of the temple and furvey the landfcape that waftes itfelf in rivers of broken fea, recall fuch exact pictures of Claud Lorrain, that
it is difficult to conceive that he did not paint them from this very fpot.

placed

placed grotto, is that at Stourhead, where the river burſts from the urn of its god, and paſſes on its courſe through the cave. But it is not my buſineſs to lay down rules for gardens, but to give the hiſtory of them. A ſyſtem of rules puſhed to a great degree of refinement, and collected from the beſt examples and practice, has been lately given in a book intituled, Obſervations on Modern Gardening. The work is very ingeniouſly and carefully executed, and in point of utility rather exceeds than omits any neceſſary directions. The author will excuſe me if I think it a little exceſs, when he examines that rude and unappropriated ſcene of Matlock-bath, and criticiſes Nature for having beſtowed on the rapid river Derwent too many caſcades. How can this cenſure be brought home to gardening? The management of rocks is a province can fall to few directors of gardens; ſtill in our diſtant provinces ſuch a guide may be neceſſary.

The author divides his ſubject into gardens, parks, farms, and ridings. I do not mean to find fault with this diviſion. Directions are requiſite to each kind, and each has its department at many of the great ſcenes from whence he drew his obſervations. In the hiſtoric light, I diſtinguiſh them into the garden that connects itſelf with a park; into the ornamented farm, and into the foreſt or ſavage garden. Kent, as I have ſhewn, invented or eſtabliſhed the firſt ſort. Mr. Philip Southcote founded the ſecond or *ferme ornée* *, of which is a very juſt deſcription in the author I have been quoting. The third I think he has not enough diſtinguiſhed. I mean that kind of alpine ſcene, compoſed almoſt wholly of pines and firs, a few birch, and ſuch trees as aſſimilate with a ſavage and mountainous country. Mr. Charles Hamilton, at Pain's hill, in my opinion has given a perfect example of this mode in the utmoſt boundary of his garden. All is great and foreign and rude; the walks ſeem not deſigned, but cut through the wood of pines; and the ſtyle of the whole is ſo grand, and conducted with ſo ſerious an air of wild and uncultivated extent, that when you look down on this ſeeming foreſt, you are amazed to find it contain a very few acres. In general, except as a ſcreen to conceal ſome deformity, or as a ſhelter in winter, I am not fond of total plantations of ever-greens. Firs in particular form a very ungraceful ſummit, all broken into angles.

Sir Henry Englefield was one of the firſt improvers on the new ſtyle, and ſelected with ſingular taſte that chief beauty of all gardens, proſpect and fortunate points of view: we tire of

all the painter's art when it wants thefe finifhing touches. The
faireft fcenes, that depend on themfelves alone, weary when
often feen. The Doric portico, the Palladian bridge, the Gothic ruin, the Chinefe pagoda, that furprife the ftranger, foon
lofe their charms to their furfeited mafter. The lake that
floats the valley is ftill more lifelefs, and its Lord feldom enjoys his expence but when he fhews it to a vifitor. But the ornament whofe merit fooneft fades, is the hermitage or fcene
adapted to contemplation. It is almoft comic to fet afide a
quarter of one's garden to be melancholy in. Profpect, animated profpect, is the theatre that will always be the moft frequented. Profpects formerly were facrificed to convenience
and warmth. Thus Burleigh ftands behind a hill, from the
top of which it would command Stamford. Our anceftors,
who refided the greateft part of the year at their feats, as
others did two years together or more, had an eye to comfort
firft, before expence. Their vaft manfions received and har-
boured all the younge branches, the dowagers and ancient
maiden aunts of the families, and other families vifited them
for a month together. The method of living is now totally
changed, and yet the fame fuperb palaces are ftill created, becoming a pompous folitude to the owner, and a tranfient entertainment to a few travellers.

If any incident abolifhes or reftrains the modern ftyle of gardening, it will be this circumftance of folitarinefs. The
greater the fcene, the more diftant it is probably from the capital, in the neighbourhood of which land is too dear to admit
confiderable extent of property. Mens tire of expence that is
obvious to few fpectators. Still there is a more imminent
danger that threatens the prefent, as it has ever done all tafte—
I mean the purfuit of variety. A modern French writer has
in a very affected phrafe given a juft account of this, I will
call it, diftemper. He fays, *l'ennui du beau amene le gout du
fingulier.* The noble fimplicity of the Auguftan age was driven
out by falfe tafte. The gigantic, the puerile, the quaint, and
at laft the barbarous and the monkifh, had each their fucceffive
admirers. Mufic has been improved, till it is a fcience of
tricks and flight of hand: the fober greatnefs of Titian is loft,
and painting fince Carlo Maratti has little more relief than
Indian paper. Borromini twifted and * curled architecture, as
if it was fubject to the change of fafhions like a head of hair.
If we once lofe fight of the propriety of landfcape in our gardens, we fhall wander into all the fantaftic Sharawadgis of the

---

* In particular, he inverted the volutes of the Ionic order.

Chinefe.

Chinefe. We have difcovered the point of perfection. We have given the true model of gardening to the world : let other countries mimic or corrupt our tafte ; but let it reign here on its verdant throne, original by its elega t fimplicity, and proud of no other art than that of foftening Nature's harſhneſſes, and copying her graceful touch.

The ingenious author of the Obfervations on Modern Gardening is, I think, too rigid when he condemns fome deceptions, becaufe they have been often ufed. If thofe deceptions, as a feigned fteeple of a diftant church, or an unreal bridge to difguife the termination of water, were intended only to furprife, they were indeed tricks that would not bear repetition ; but being intended to improve the landfcape, are no more to be condemned becaufe common, than they would be if employed by a painter in the compofition of a picture. Ought one man's garden to be deprived of a happy object, becaufe that object has been employed by another ? The more we exact novelty, the fooner our tafte will be vitiated. Situations are every where fo various, that there never can be a famenefs, while the difpofition of the ground is ftudied and followed, and every incident of view turned to advantage.

In the mean time how rich, how gay, how picturefque the face of the country ! The demolition of walls laying open each improvement, every journey is made through a fucceffion of pictures ; and even where tafte is wanting in the fpot improved, the general view is embellifhed by variety. If no relapfe to barbarifm, formality, and feclufion is made, what landfcapes will dignify every quarter of our ifland, when the daily plantations that are making have attained venerable maturity A fpecimen of what our gardens will be, may be feen at Pet worth, where the portion of the park neareft the houfe has been allotted to the modern ftyle. It is a garden of oaks two hundred years old. If there is a fault in fo auguft a fragment of improved nature, it is, that the fize of the trees are out of all proportion to the fhrubs and accompanyments. In truth, fhrubs fhould not only be referved for particular fpots and home de light, but are paffed their beauty in lefs than twenty years.

Enough has been done to eftablifh fuch a fchool of landfcape, as cannot be found on the reft of the globe. If we have the feeds of a Claud or a Gafpar amongft us, he muft come forth. If wood, water, groves, vallies, glades, can infpire or poet or painter, this is the country, this is the age to produce them. The flocks, the herds, that now are admitted into, now graze on the borders of our cultivated plains, are ready before the painter's eyes, and groupe themfelves to animate his picture.

One

One misfortune in truth there is that throws a difficulty on the artift. A principal beauty in our gardens is the lawn and fmoothnefs of turf: in a picture it becomes a dead and uniform fpot, incapable of *chiaro fcuro*, and to be broken infipidly by children, dogs, and other unmeaning figures.

Since we have been familiarized to the ftudy of landfcape, we hear lefs of what delighted our fportfmen-anceftors, *a fine open country*. Wiltfhire, Dorfetfhire, and fuch ocean-like extents, were formerly preferred to the rich blue profpects of Kent, to the Thames-watered views in Berkfhire, and to the magnificent fcale of Nature in Yorkfhire. An open country is but a canvaf on which a landfcape might be defigned.

It was fortunate for the country and Mr. Kent, that he was fucceeded by a very able mafter; and did living artifts come within my plan, I fhould be glad to do juftice to Mr. Brown; but he may be a gainer, by being referved for fome abler pen.

In general it is probably true, that the poffeffor, if he has any tafte, muft be the beft defigner of his own improvements. He fees his fituation in all feafons of the year, at all times of the day. He knows where beauty will not clafh with convenience, and obferves in his filent walks or accidental rides a thoufand hints that muft efcape a perfon who in a few days fketches out a pretty picture, but has not had leifure to examine the details and relations of every part.

Truth, which, after the oppofition given to moft revolutions, preponderates at laft, will probably not carry our ftyle of garden into general ufe on the continent. The expence is only fuited to the opulence of a free country, where emulation reigns among many independent particulars. The keeping of our grounds is an obftacle, as well as the coft of the firft formation. A flat country, like Holland, is incapable of landfcape. In France and Italy the nobility do not refide much, and make fmall expence, at their villas. I fhould think the little princes of Germany, who fpare no profufion on their palaces and country-houfes, moft likely to be our imitators; efpecially as their country and climate bears in many parts refemblance to ours. In France, and ftill lefs in Italy, they could with difficulty attain that verdure which the humidity of our clime beftows as the ground-work of our improvements. As great an obftacle in France is the embargo laid on the growth of their trees. As after a certain age, when they would rife to bulk, they are liable to be marked by the crown's furveyors as royal timber, it is a curiofity to fee an old tree. A landfcape and a crown-furveyor are incompatible.

GENERAL PRINCIPLES.

ARTS merely imitative have but one principle to work by, the *nature* or actual state of the thing to be imitated. In works of design and invention, another principle takes the lead, which is *taste*. And in every work in which mental gratification is not the only object, a third principle arises, *utility*, or the concomitant purpose for which the production is intended.

The art of *Gardening* is subject to these three principles: to nature, as being an imitative art; to utility, as being productive of objects which are useful as well as ornamental; and to taste, in the choice of fit objects to be imitated, and of fit purposes to be pursued, as also in the composition of the several objects and ends proposed, so as to produce the degree of gratification and use best suited to the *place* and to the *purpose* for which it is about to be ornamented: thus, a Hunting-Box and a Summer Villa,—an Ornamented Cottage and a Mansion, require a different *style* of ornament, a different *choice* of objects, a different *taste*. Nor can taste be confined to nature and utility,—the place and the purpose, alone; the object of the Polite Arts is the gratification of the human mind, and the state of refinement of the mind itself must be considered. Men's notions vary, not only in different ages, but individually in the same age: what would have gratified mankind a century ago in this country, will not please them now; whilst the country 'Squire and the Fine Gentleman of the

Qq 4 present

prefent day require a different kind of gratification:
neverthelefs, under thefe various circumftances,
every thing may be *natural*, and every thing adapted
to the *place* ; the *degree of refinement* conftituting
the principal difference.

We do not mean to enter into any argument
about whether a ftate of rufticity or a ftate of refinement, whether the foreft or the city be the
ftate for which the Author of Nature intended the
human fpecies : mankind are now found in every
ftate, and in every ftage of favagenefs, rufticity,
civilization, and refinement; and the particular
ftyle of ornament we wifh to recommend is, that
which is beft adapted to the ftate of refinement
that now prevails in this country; leaving individuals to vary it as their own peculiar taftes may
direct.

Before we proceed farther, it may be neceffary to
explain what it is we mean by *nature* and *natural*.
If in the idea of *natural ftate* we include *ground*,
water, and *wood*, no fpot in this ifland can be faid
to be in a *ftate of nature*. The *ground*, or the furface of the earth as left by Nature (or the con-
vulfions of Nature) remains, it is true, with but few
alterations; yet even here (efpecially among rocks
and fteep acclivities, the nobleft features in the face
of Nature) we frequently find the hand of Art has
been at work. Again, though rivers may ftill run
in the channels, or nearly in the channels, into
which Nature directed them ; yet *waters* taken
generally have been greatly controuled by human
art.

art. And with refpeft to *wood* we may venture to
fay, that there is not a tree, perhaps not a ftick,
now ftanding upon the face of the country which
owes its identical ftate of exiftence to Nature alone.
Wherever cultivation has fet its foot,—wherever the
plow and fpade have laid fallow the foil,—Nature is
become extinct; and it is in neglected or lefs cultivated places, in moffes and mountains, in forefts
and parochial waftes, we are to feek for any thing
near a ftate of Nature—we mean in this country :
and who would look for the ftandard of tafte, who
expeft to find the lovely mixture of wood and
lawn fo delightful to the human eye, in the endlefs
woods and the impenetrable roughneffes of America ? We may therefore conclude, that the objects of our imitation are not to be fought for in uncultivated Nature. The inhofpitable heaths of
Weftmoreland may aftonifh for the moment, may be
the pleafing amufement of a fummer's day, and
agreeable objects in their places; but are they
objects of imitation under the window of a drawing-room ? Rather let us turn our eyes to well-
foiled, well-wooded, well-cultivated fpots, where
Nature and Art are happily blended; leaving
thofe who are admirers of Art merely imitative
to contemplate Nature upon canvas ; and thofe
who wifh for Nature in a ftate of total neglect,
to take up their refidence in the woods of
America.

Far be it from us to rebel againft the laws of
Nature, or to queftion in any wife the perfection of
the Deity. A ftate of nature, in the eye of Omni-
fcience, is undoubtedly a ftate of perfection. But

in

in the littleneſs of human conception, ſomething is wanted to bring down natural objects to the level of our comprehenſion. What object in nature is in a ſtate of *human perfection?* Even in the fineſt woman a female critic will diſcover *faults:* and in the handſomeſt horſe a buyer will point out what in the human eye appear as *imperfections.* Did ever a landſcape painter find a ſcene, purely natural, which might not have been improved by the hand of Art, or which he did not actually improve by a ſtroke of his pencil? A ſtriking feature may ſometimes be caught where little addition is wanted; but in a rich picturesque view, which will bear to be placed repeatedly under the eye, a portion of *lawn* is requiſite *, and in the wilds of nature we know of no ſuch thing.

Therefore our idea of *natural* is not confined to *neglected* nature, but extends to *cultivated* nature—

to

* Mr. GRAY, whoſe letters to Dr. WARTON, deſcribing the natural ſcenery of the North of England, have been held out as models of their kind, corroborates our idea.

·" Juſt beyond this opens one of the ſweeteſt landſcapes that art ever attempted to imitate. The boſom of the mountain ſpreading here into a broad baſon, diſcovers in the midſt Graſmere Water: its margin is hollowed into ſmall bays, with bold eminences, ſome of rock, ſome of *ſoft turf*, that half conceal and vary the figure of the little lake they command: from the ſhore a low promontory puſhes itſelf far into the water, and on it ſtands a white village with the pariſh-church riſing in the midſt of it: hanging incloſures, corn-fields, and *meadows green as emerald*, with their trees, and hedges, and cattle, fill up the whole ſpace from the edge of the water: and juſt oppoſite to you is a large farm-houſe at the bottom of a ſteep *ſmooth lawn*, emboſomed in old woods which climb half way up the mountain-ſide, and diſcover above them a broken line of crags that crown the ſcene,

to nature *touched* by art, and rendered intelligible
to human perception: and we venture to recommend, as objects moſt worthy the ſtudy and imitation of the artiſts, ſuch *paſſages in nature* as give
the higheſt degree of gratification to cultivated
minds in general: paſſages like the following—
no matter whether produced by *accident* or *deſign*—
no matter whether it occur in a foreſt or a park—
or whether it occupy the corner of a common, or
fill up a conſpicuous quarter of an ornamental
ground :—a lofty wood hanging on a bold aſcent ;
its broken margin flowing negligently over the boſom of the valley, lying broad and bare beneath,
and falling gently to the brink of a river, winding
gracefully along the bottom.——We further beg
leave to add in this place, that if a paſſage like
this—eſpecially if the lawn be ſpread with cattle,
and the whole ſcene enlivened by the preſence of
the ſun, and animated by the fleeting ſhadows of
the clouds ſweeping its varied ſurface--is incapable of conveying a degree of gratification to the
mind of any of our Readers, we have no hope of
entertaining ſuch a mind in this part of our performance.

ſcene. Not a ſingle red tile, no flareing Gentleman's houſe, or
garden-walls, break in upon the repoſe of this little unſuſpected
paradiſe ; but all is peace, ruſticity, and happy poverty in its
neateſt, moſt becoming attire."
Gray's Letters to Dr. Warton, p. 181.

SITE.

S I T E.

BY *the Site* we mean, not only the *place* itfelf, but likewife fo much of the *furrounding ccuntry* as falls within the view.

If the place be already fuited to the furround-ing country, and to the particular purpofe for which it is intended, the affiftance of art is not wanted, the bufinefs of the Gardener is precluded. If the Site be *nearly* in this ftate, the *touchings* of art are only required. But if the place be greatly deficient, as places in general are, then it is the duty of the artift "to fupply its defects, to correct its faults, and to improve its beauties."

Every PLACE confifts either of *ground* alone, or of *ground* and *water*, or of *ground* and *wood*, or of *ground*, *water*, and *wood*.

G R O U N D.

BY Ground is meant that portion of naked fur-face, which is included within the place to be im-proved; whether that furface be *fwamp*, *lawn*, *roughet*, *broken-ground*, or *rock*; and whether it be a *hill*, a *valley*, a *plain*, or a compofition of *fwells*, *dips*, and *levels*.

Mr. GILPIN, in his excellent *Obfervations on the Wye*, *&c.* (page 62) gives us a fublime idea of what ground ought to be.—" Nothing, fays he,
gives

gives fo;juft an idea of the beautiful fwellings of
ground, as thofe of water, where it has fufficient
room to undulate and expand. In ground which
is compofed of very refractory materials, you are
prefented often with harfh lines, angular infertions,
and difagreeable abruptneffes. In water, whether
in gentle or in agitated motion, all is eafy, all is
foftened into itfelf; and the hills and the vallies
play into each other, in a variety of the moft beautiful forms. In agitated water, abruptneffes indeed
there are, but yet they are fuch abruptneffes
as, in fome part or other, unite properly with the
furface around them ; and are on the whole peculiarly harmonious. Now, if the ocean in any of
thefe fwellings and agitations could be arrefted and
fixed, it would produce that pleafing variety which
we admire in ground. Hence it is common to fetch
our images from water, and apply them to land :
we talk of an undulating line, a playing lawn, and
a billowy furface ; and give a much ftronger and
more adequate idea by fuch imagery, than plain
language could poffibly prefent."

The exertions of art, however, are here inade-
quate and the artift ought not to attempt to create
a *mountain*, a *valley*, or a *plain*; and fhould but
rarely meddle even with the fmaller inequalities of
grounds. The *rock* ftands equally above the reach
of human art, and to attempt to make or unmake
it is abfurd. *Roughets* and *broken-ground* may generally be reduced to lawn, or hid w th wood ; and
                                                    a *fwamp*

a *swamp* may be drained, or covered with water; whilst *lawn* may be variegated at pleasure by wood, and sometimes by water.

WATER.

THIS is either *sea, lake, pool, river, rivulet,* or *rill.*

A broad *lake* and a copious *river* are too great for human art to cope with : nevertheless, the margin and the bank may be ornamented, and the surface of the water disclosed to advantage. *Rivulets* are often in themselves delightful, and, where broad waters are wanted, may be turned to great advantage by art. * Stowe affords a proof of what may be accomplished even with a *rill.* If the base of the valley be broad, a lake may be made ; if narrow, a river.

In countries where natural waters abound, art may improve, but should not attempt to create : but in places naturally dry, the artist may frequently call forth the creative powers with success. In any situation, however, art must miscarry, if Nature has not furnished a sufficient supply of materials : *stagnant pools* are always disgusting : *stews,* indeed, may often be necessary ; but, like the kitchen-garden, they ought not to be *seen.*

* The seat of Lord Temple, near Buckingham.

WOOD.

W O O D.

OVER this element of the rural art the power
of the artist is absolute ; he can increase or diminish
at pleasure : if the place be over-wooded, he can
lighten it with lawn, or with water: if too naked,
he can supply the deficiency by PLANTING.

In forming ORNAMENTAL PLANTATIONS, two
things are to be considered, the *species of plantation*, and the *species of tree*.

The different species of plantation are the *Wood*,
the *Grove*, the *Coppice* or *Thicket*, the *Shrubbery-
Quarter*, the *Border*, the *Clump*, the *Group*, and
the *Single-Tree*.

Woods, Groves, and extensive Thickets, are
more particularly adapted to the sides of hills and
elevated situations : the larger Clumps, Groups and
Single-Trees, to the lower grounds. A naked hill
gives an idea of bleakness; as a valley filled with
wood does that of dankness. The Shrubbery depends more upon artificial accompaniments than
natural situation.

Much depends upon the disposition of the several
distinct woodinesses (whether accidental or designed) with respect to each other; and much also
upon the respective outlines, particularly those of
the larger kind. The Atmosphere and the Earth are
equally bountiful in affording the rural artist fit
subjects for study. The margins of seas and lakes
give us, in their bays and promontories, an ample
choice

choice of outline; whilft the blue expanfe fcattered with fummer's clouds, difcovers infinite variety both of figure and difpofition.

In the choice of trees four things are obfervable:
the *height*, the *form*, the *colour*, and the *ufe*. *This*
is more effential to a good choice than may appear
at firft fight: nothing heightens the idea of ornament, efpecially in the eye of the owner, more than
utility; nor on the contrary does any thing flatten
it, or throw a damp upon the gratification, more
than the worthleffnefs of the object before us. Immediately under the eye, the gaudy Shrub, and the
ornamental though ufelefs Exotic, may be admitted; but for more diftant objects, and in lefs embellifhed fituations, the Timber tree ought to prevail. We fhould endeavour to make fuch a choice
as will gratify the prefent age and benefit the future.

In mixing trees there is in refpect of *height*, a
general rule: the talleft fhould be made to occupy
the central parts, defcending gradually to the
margin: but with refpect to *colour* all precepts
would be vague; the tints ought to be as wild and
various as the evening fky tinged by the fetting
fun.

NATURAL ACCOMPANIMENTS.

THE moft judicious admixture of wood and
lawn appears flat and unmeaning, unlefs it be enlivened

livened by animated nature. What fprightlinefs and elegance are added to the plain, in the playful attitudes and racings of the horfe;—and how much additional grandeur the vale receives in the fcattered herd!—How ftrikingly beautiful the bofom of a hill enlivened by the pafturing flock; and how peculiarly delightful the fequeftered lawn, while the hare is prefent! Even the fquirrel gives a chearfulnefs to the grove; whilft the plumy tribes difperfe an agreeable animation through the whole fcene.

FACTITIOUS ACCOMPANIMENTS.

UNDER this head we arrange · *Fences*, *Walks*, *Roads*, *Bridges*, *Seats*, and *Buildings*.

The FENCE, where the place is large, becomes neceffary; yet the eye diflikes conftraint. Our ideas of liberty carry us beyond our own fpecies: the imagination feels a diflike in feeing even the brute creation in a ftate of confinement. The birds wafting themfelves from wood to grove are objects of delight; and the hare appears to enjoy a degree of happinefs unknown to the barriered flock. Befides, a tall fence frequently hides from the fight objects the moft pleafing; not only the flocks and herds themfelves, but the furface they graze upon. Thefe confiderations have brought the *unfeen fence* into general ufe.

This fpecies of barrier it muft be allowed incurs a degree of deception, which can fcarcely be warranted upon any other occafion. In this inftance, however, it is a fpecies of fraud which we obferve in nature's practice: how often have we feen two diftinct herds feeding, to appearance, in the fame extended meadow; until coming abruptly upon a deep-funk rivulet, or an unfordable river, we difco-ver the deception.

Befides the *funk* fence, another fort of unfeen barrier may be made, though by no means equal to *that*; efpecially if near the eye. This is conftructed of paling, painted of the *invifible green*. If the colour of the back-ground were permanent, and that of the paint made exactly to correfpond with it, the deception would, at a diftance, be complete; but back-grounds, in general, changing with the feafon, this kind of fence is the lefs eligible.

Clumps and patches of woodinefs fcattered pro-mifcuoufly on either fide of an unfeen winding fence, affift very much in doing away the idea of conftraint. For by this means

'I he wand'ring flocks that broufe between the fhades,
Seem oft to pafs their bounds, the dubious eye
Decides not if they crop the mead or lawn.

MASON.

The WALK, in extenfive grounds, is as neceffary as the Fence. The beauties of the place are dif-clofed that they may be feen; and it is the office of the walk to lead the eye from view to view; in order that, whilft the tone of health is pre-ferved by the favourite exercife of nature, the mind may be thrown into unifon by the harmony of the furrounding objects.

The

The direction of the walk muft be guided by the points of view to which it leads, and the nature of the ground it paffes over : it ought to be made fubfervient to the natural impediments — the Ground, Wood, and Water — which fall in its way, without appearing to have any direction of its own. It can feldom run, with propriety, any diftance in a ftrait line ; a thing which rarely occurs in a *natural walk*. The paths of the Negroes, and the Indians, are always crooked; and thofe of the brute creation are very fimilar. Mr. Mafon's defcription of this *Path of Nature* is happily conceived.

> The peafant driving through each fhadowy lane
> His team, that bends beneath th' incumbent weight
> Of laughing CERFS, marks it with his wheel ;
> At night and morn, the milk-maid's carelefs ftep
> Has, thro' yon pafture green, from ftile to ftile
> Impreft a kindred curve ; the fcudding hare
> Draws to her dew-fprent feat, o'er thymy heaths,
> A path as gently waving.——— *Eng. Gard.* v. 60.

THE ROAD may be a thing of neceffity, as an *approach* to the manfion, or a matter of amufement only, as a *drive* or a *ride* from which the grounds, and the furrounding country, may be feen to advantage. It fhould be the ftudy of the artift to make the fame road anfwer, as far as may be, the two-fold purpofe

The Road and the Walk are fubject to the fame rule of *Nature* and *Ufe*. The direction ought to be natural and eafy, and adapted to the purpofe intended. A Road of neceffity ought to be ftraighter

than one of mere conveniency : in this, recreation is
the predominant idea; in that, utility. But, even
in this, the direct line may be difpenfed with. The
natural roads upon heaths and open downs, and the
graffy glades and green roads acrofs forefts and extenfive waftes, are proper fubjects to be ftudied.

THE BRIDGE fhould never be feen where it is not
wanted: a ufelefs bridge is a deception; deceptions are frauds; and fraud is always hateful; unlefs when practifed to avert fome greater evil. A
bridge without water is an abfurdity; and half
an one ftuck up as an eye-trap is a paltry trick,
which, though it may ftrike the ftranger, cannot
fail of difgufting when the fraud is found out.

In low fituations, and wherever water abounds,
bridges become *ufeful,* and are therefore *pleafing
objects :* they are looked for, and ought to appear;
not as objects of ornament only, but likewife as
matters of utility. The walk or the road, therefore, ought to be directed in fuch a manner as to
crofs the water at the point in which the bridge
will appear to the greateft advantage.

In the conftruction of bridges, alfo, regard muft
be had to ornament and utility. A bridge is an artificial production, and as fuch it ought to appear.
It ranks among the nobleft of human inventions :
the fhip and the fortrefs alone excel it. Simplicity
and firmnefs are the leading principles in its conftruction. Mr. Wheatley's obfervation is juft when
he fays, " The fingle wooden arch, now much in
fafhion, feems to me generally mifapplied. Elevated
without

without occafion fo much above, it is totally de-tached from the river; it is often feen ftraddling in the air, without a glimpfe of water to account for it; and the oftentation of it as an ornamental object, diverts all that train of ideas, which its ufe, as a communication, might fuggeft." (*Obf. on Mod. Gard.* 73.) But we beg leave to differ from this ingenious Writer when he tells us, that it is " fpoiled, if adorned; it is disfigured, if only painted of any other than a dufky colour." In a ruftic fcene, where Nature wears her own coarfc garb, " the vulgar foot-bridge of planks only, guarded on one hand by a common rail, and fup-ported by a few ordinary piles," may be in cha-racter; but amidft a difplay of ornamented Nature, a contrivance of that kind would appear mean and pal-try; and would be an affectation of fimplicity, rather than the lovely attribute itfelf. In cultivated fcenes, the bridge ought to receive the ornaments which the laws of architectural tafte allow; and the more po-lifhed the fituation, the higher fhould be the ftyle and finifhings.

SEATS have a two-fold ufe; they are ufeful as places of reft and converfation, and as guides to the *points of view,* in which the beauties of the furround-ing fcene are difclofed. Every point of view fhould be marked with a feat, and, fpeaking generally, no feat ought to appear but in fome favourable point of view. This rule may not be invariable, but it ought feldom to be deviated from.

In the ruder ſcenes of neglected Nature, the ſimple trunk, rough from the woodman's hands, and the butts or ſtools of rooted trees, without any other marks of tools upon them than thoſe of the ſaw which ſevered them from their ſtems, are ſeats in character; and in romantic or recluſe ſituations, the cave or the grotto are admiſſible. But where-ever human deſign has been executed upon the na-tural objects of the place, the ſeat and every other artificial accompaniment ought to be in uniſon; and, whether the bench or the alcove be choſen, it ought to be formed and finiſhed in ſuch a manner as to unite with the wood, the lawn, and the walk, which lie round it.

The colour of ſeats ſhould likewiſe be ſuited to ſituations : where uncultivated Nature prevails, the natural brown of the wood itſelf ought not to be altered : but where the rural art preſides, white, or ſtone-colour, has a much better effect.

BUILDINGS may be admitted into ornamented Nature; provided they be at once uſeful and orna-mental. Mere ornament without uſe, and mere uſe without ornament, are equally inadmiſſible. Nor ſhould their uſes be diſguiſed; a barn dreſſed up in the habit of a country church, or a farm-houſe figuring away in the fierceneſs of a caſtle, are ridiculous deceptions. A landſcape daubed upon a board, and a wooden ſteeple ſtuck up in a wood, are beneath cenſure.

There is another ſpecies of uſeleſs ornament ſtill more offenſive, becauſe more coſtly, than thoſe com-

paratively

paratively innocent eye-traps ; we mean Temples *.
Whether they be dedicated to Bacchus, Venus, Pri-
apus, or any other demon of debauchery, they are,
in this age, enlightened with regard to theological
and fcientific knowledge, equally abfurd.

We are far, however, from wifhing to exclude
architecture from ornamented Nature. We wifh to
fee it exercifed in all its beauty and fublimity upon
a chapel †, a maufoleum ‡, a monument ||,— fcat-
tered judicioufly among the natural ornaments :
not too open or confpicuous, to give them tho
air of principals; nor too reclufe, to lofe their
full effect as fubordinate parts of the one great
whole,

---

* Notwithftanding thoufands, or tens of thoufands, have, in one
inftance, been facrificed to Vanity and falfe Wit (O ! Temple !
" how delightful are thy Temples !"), we flatter ourfelves that as
few men's names can apologize for committing fo great an act of
folly, the example will not be copied.

† The late Sir William Harbord, whofe tafte and judgement,
upon every occafion, difcovered a goodnefs of heart and a greatnefs of character, has given us a model of this kind, at Gunton,
in Norfolk. The parifh-church ftanding in his park, and being
an old unfightly building, he had it taken down, and a *beautiful
temple*, under the direction of the Adams', erected upon its
fite.

‡ The maufoleum at Caftle-Howard, in Yorkfhire, the feat
of the Earl of Carlifle, is a noble building.

|| The *temple* of Concord and Victory at Stowe, erected to the
memory of the great Lord Chatham and *his* glorious war, is a
beautiful *monumental* building, fuited to the greatnefs of the
occafion.

R r 4   In

In ſcenes leſs ornamented, buildings of an œco-
nomical nature may appear with good effect. Sir
George Warren, at his ſeat near Fetcham in Surrey,
has turned a *temple* into a wind-mill with great ſuc-
ceſs. What was before a uſeleſs, *lifeleſs* fabrick,
now ſtands an emblem of activity and induſtry.
Under the heads of large artificial lakes, water-mills
may generally be erected, and with good effect.
A mill is not only a ſtriking inſtance of the power
of the human invention, but is frequently a great
relief to the poor in its neighbourhood. Subſtan-
tial farm-houſes, and neat comfortable cottages,
ſcattered at a proper diſtance, are always pleaſing
objects. The banquetting-houſe and the porter's
lodge, being more ſuſceptible of ornament, may be
permitted nearer the eye.

GENERAL APPLICATION.

HAVING thus enumerated the elements, and
ſet forth the leading principles, we now proceed to
the execution.

We beg leave to preface this part of our performance with apprizing our Readers, that all that
can be written upon this delightful art muſt be
more or leſs general. — All that *ſcience* can do is
to give *a comprehenſive view of the ſubject*; and all
that

that *precept* should attempt, is to lay down *general rules* of practice. The nature of the place itself--- and the purpose for which it is about to be improved, must ever determine the particular application. It follows, that a gentleman who, from long residence, is fully acquainted with the former, and whose will is a rule to the latter, is the properest person to improve his own place ; — provided he be intimately acquainted with the *Art* — as well as with the *place* and the *purpose :* the three are equally and essentially necessary to be understood. It would be as great an impropriety in a gentleman to set about the execution of a work of this nature upon a large scale, before he had acquired a comprehensive knowledge of the subject, studied its leading principle from Nature, made ample observation upon places already ornamented, and had established his theory by some actual practice, at least upon a small scale;—as it would be in a professional artist to hazard his own reputation, and risque the property of his employer, before he had studied maturely the nature of the place, and been made fully sensible of the intentions of its owner.

The nature and style of improvement, — *the purpose,* — depends entirely upon the intention and taste of the proprietor, and is consequently as various as the nature of places themselves : nevertheless, improvements in general may be classed under the following heads :

The

The Hunting-Box, The Villa, and
The Ornamented Cottage, The Principal Refi-
       dence.

But, before we enter upon the detail, it will be
proper to make fome general obfervations.

It is unneceffary to repeat, that wherever Nature
or accident has already adapted the place to the
intended purpofe, the affiftance of Art is precluded :
but wherever Nature is improveable, Art has an
undoubted right to ftep in, and make the requifite
improvement. The diamond, in its natural ftate,
is highly improveable by art.

In the lower claffes of rural improvements, Art
fhould be feen as little as may be ; and in the more
negligent fcenes of Nature, every thing ought to
appear as if it had been done by the general laws of
Nature, or had grown out of a feries of fortuitous
circumftances. But, in the higher departments, Art
cannot be hid ; and the *appearance* of defign ought
not to be excluded. A human production cannot
be made perfectly natural ; and, held out as fuch,
it becomes an impofition. Our art lies in endeavouring to adapt the productions of Nature to human tafte and perceptions ; and, if much art be ufed,
do not *attempt* to hide it. Who confiders an accomplifhed well-dreffed woman as in a ftate of Nature ?
and who, feeing a beautiful ground adorned with
wood and lawn, with water, bridges, and buildings, believes it to be a natural production ? Art
           feldom

feldom fails to pleafe when executed in a mafterly manner: nay, it is frequently the defign and execution, more than the production itfelf, that ftrikes us. It is the *artifice*, not the *defign*, which ought to be avoided. It is the *labour*, and not the *art*, which ought to be concealed. A well-written poem would be read with lefs pleafure, if we *knew* the painful exertions it gave rife to in the compofition; and the rural artift ought, upon every occafion, to endeavour to avoid labour; or, if indifpenfibly neceffary, to conceal it. No trace fhould be left to lead back the mind to the *expenfive toil*. A mound raifed, a mountain levelled, or a ufelefs temple built, convey to the mind feelings equally difgufting.

But though the aids of Art are as effential to gardening, as education is to manners; yet Art may do too much: fhe ought to be confidered as the handmaid, not as the miftrefs, of Nature: and whether fhe be employed in carving a tree into the figure of an animal, or in fhaping a view into the form of a *picture*, fhe is equally culpable. The nature of the place is facred. Should this tend to *landfcape*, from fome principal point of view, affift Nature, and perfect it; provided this can be done without injuring the views from other points. But do not disfigure the natural features of the place:—do not facrifice its native beauties, to the arbitrary laws of landfcape painting.

Great Nature fcorns controul ; fhe will not bear
One beauty foreign to the fpot or foil
She gives thee to adorn : 'Tis thine alone
To mend, not change her features.   MASON.

In a picture bounded by its frame, a perfect
landfcape is looked for : it is of itfelf a *whole*, and
the frame muſt be filled. But it is not fo in ornamented Nature : for, if a fide-fcreen be wanting,
the eye is not offended with the frame, or the wainfcot ; but has always fome natural and pleafing object to receive it. Suppofe a room to be hung with
one continued rural reprefentation, — would *pretty
pictures* be expected ? would correct landfcapes be
looked for ? Nature fcarcely knows the thing man.
kind call a *landfcape*. The landfcape-painter feldom, if ever, finds it perfected to his hands ; —
fome addition or alteration is almoſt always wanted.
Every man who has made his obfervations upon
natural fcenery, knows that the Mifletoe of the Oak
occurs almoſt as often as a perfect natural landfcape ;
and to attempt to make up artificial landfcape, upon
every occafion, is unnatural and abfurd.

It is far from our intention to intimate any thing
the leaſt difrefpectful to *landfcape painting* : let the
ingenious artiſt cull from Nature her choiceſt beauties, and let him affociate them in the manner beſt
fuited to his own fingle and permanent point of
view : but do not let us carry his production back
again to Nature, and contract her unbounded beauties within the limits of a picture-frame. If, indeed,

the

the eye were fixed in one point, the trees could be
raifed to their full height at command, and the fun
be made to ftand ftill,—the rural artift might work
by the rules of *light and fhade*, and compofe his
landfcape by the painter's law. But, whilft the fun
continues to pour forth its light impartially, and the
trees to rife with flow progreffion, it would be ridiculous to attempt it. Let him rather feek out, imitate, and affociate, fuch STRIKING PASSAGES IN
NATURE as are immediately applicable to the place
to be improved, without regard to rules of landfcape, merely human ;—and let him,

———————————. in this and all
Be various, wild, and free, as Nature's felf." MASON.

Inftead of facrificing the natural beauties of the
place to one formal landfcape, let every ftep difclofe
frefh charms uniought-for. How ftrikingly beau
tiful the changes formed by the iflands, and their
refpective mountains, in failing through the Weft-
Indies! The eye does not catch the fame view
twice: the fcene is ever changing,—ever beautiful.

We fhould not have offered our fentiments fo
freely upon landfcape, had not a French writer of
fome eminence *, in a work lately publifhed, laid

* The Marquis D'Ermenonville, friend of the celebrated
Roufleau, who died at his houfe, and whofe remains were depo-
fited in his grounds, at Ermenonville.

it

it down as an invariable rule, that all ornamental grounds fhould have a complete landfcape, to be feen from fome part of the houfe; and to be made from a perfpective drawing, previoufly taken from the window of the faloon, or the top of the manfion. The work, in other refpects, has, neverthelefs, great merit, and is in fact an ingenious *Effay on Englifh Gardening*. The Frenchman's vanity, however, will not fuffer him to make this acknowledgement: no, it is neither Antients, nor Moderns, nor Englifh, nor Chinefe; and there is fome reafon to fufpect, that the Marquis holds out landfcape for no other purpofe, than to endeavour to give his work the air of originality; for, in other refpects, it contains, in effect, what Wheatley and Mafon, Kent and Brown, have previoufly taught and practifed.

Notwithftanding, however, the nature of the place ought not to be facrificed to the manfion; — the houfe muft ever be allowed to be a principal in the compofition. It ought to be confidered as the center of the fyftem; and the rays of art, like thofe of the fun, fhould grow fainter as they recede from the center. The houfe itfelf being entirely a work of art, its immediate environs fhould be highly finifhed; but as the diftance increafes, the appearance of defign fhould gradually diminifh, until Nature and fortuitoufnefs have full poffeffion of the fcene.

In general, the approach fhould be to the back-front, which, in fuitable fituations, ought to lie

open

open to the pafture-grounds. On the fides more
highly ornamented, a well-kept gravel-walk may
embrace the walls; to this the fhaven lawn and
fhrubbery fucceed; next, the grounds clofely paftured; and, laftly, the furrounding country, which
ought not to be confidered as out of the artift's
reach: for his art confifts not more in decorating
particular fpots, than in endeavouring to render the
whole face of Nature delightful.

Another reafon for this mode of arrangement is,
objects immediately under the eye are feen more
diftinctly than thofe at a diftance, and ought to be
fuch as are pleafing in the detail. The beauties of
a flower can be difcerned on a near view only;
whilft, at a diftance, a roughet of coppice-wood,
and the moft elegant arrangement of flowering-
fhrubs, have the fame effect. The moft rational
entertainment the human mind is capable of receiving, is that of obferving the operations of Nature.
The foliation of a leaf, the blowing of flowers, and
the maturation of fruit, are among the moft delightful fubjects that a contemplative mind can be
employed in. Thefe proceffes of Nature are flow,
and except the object fall fpontaneoufly under the
eye of the obferver, the inconveniencies of vifiting
it in a remote part, fo far interfere with the more
important employments of life, as to blunt, if not
deftroy, the enjoyment. This is a ftrong argu
ment in favour of fhrubs and flowers being planted
under or near our windows, efpecially thofe from
whence

whence they may be viewed during the hours of leifure and tranquillity.

Further, the vegetable creation being fubject to the animal, the fhrub may be cropt, or the flower trodden down, in its day of beauty. If, therefore, we wifh to converfe with Nature in private, intruders muft be kept off, — the fhrubbery be fevered from the ground;— yet not in fuch a manner as to drive away the pafturing ftock from our fight. For this reafon, the fhaven lawn ought not to be too extenfive, and the fence which inclofes it fhould be fuch as will not interrupt the view: But whether it be *feen* or *unfeen*, *fufpected* or *unfufpected*, is a matter of no great import: its utility in protecting the fhrubs and flowers,—in keeping the horns of the cattle from the window, and the feet of the fheep from the gravel and broken ground,—in preferving that neatnefs on the outfide, which ought to correfpond w th the finifhings and furniture within,— render it of fufficient importance to become even a part of the ornament.

Be ore any ftep can be taken towards the execution of the defign, be it large or fmall, a map or plan of the place, exactly as it lies in its unimproved ftate, fhould be made ; with a correfponding fketch, to mark the intended improvements upon. Not a hovel nor a twig fhould be touched, until the artift has ftudied maturely the natural abilities of the place, and has decidedly fixed in his mind, and finally fettled on his plan, the propofed alterations :

ations: and even then, let him " dare with caution."

There is a ſtriking ſimilarity between a neglected ſcene in Nature, and a neglected cottage-beauty; and the mode of improvement is, in either inſtance, ſimilar. If the face unwaſhed, and uncombed hair, be conſidered as ornamental,—Art is not wanted. If ruſtic bloom and native ſimplicity be deemed more deſirable,—waſh the face, and comb the hair in flowing ringlets, and ſuch ornament will be had in its higheſt perfection. If that elegance of carriage, and gracefulneſs of deportment, which flow from education and a refined underſtanding, be thought requiſite, Art may be employed in giving this grace and elegance; for thus far ſhe may go with propriety. But, if ſhe attempt to go farther, if ſhe preſume to *cut and carve*, or diſguiſe the native beautifulneſs of features with *paint and patches*, or to hide the lovelineſs of form with *fantaſtic or formal dreſſes*,— ſhe does too much.

It would be needleſs to add, that Art may be employed in concealing, or in doing away the deformities of Nature. But, even in this, ſhe ought to be cautiouſly circumſpect: for, throughout, there is more danger of doing too much than too little; and nothing ſhould ever be attempted which cannot be performed in a maſterly manner.

H U N T I N G - B O X.

HERE Art has little to do. Hunting may be called the amufement of Nature; and the place appropriated to it ought to be no farther altered from its natural ftate than decency and conveniency require:---With men who live in the prefent age of refinement, " a want of decency is a want of fenfe."

The ftyle throughout fhould be *mafculine.* If fhrubs be required, they fhould be of the hardier forts; the Box, the Holly, the Lauruftinus. The trees fhould be the Oak and the Beech, which give in Autumn an agreeable variety of foliage, and anticipate, as it were, the feafon of diverfion. A fuite of paddocks fhould be feen from the houfe; and if a view of diftant covers can be caught, the background will be compleat. The ftable, the kennel, and the leaping-bar, are the factitious accompaniments; in the conftruction of which fimplicity, fubftantialnefs, and conveniency, fhould prevail.

O R N A M E N T E D   C O T T A G E.

NEATNESS and fimplicity ought to mark the ftyle of this rational retreat. Oftentation and fhow fhould be cautioufly avoided; even elegance fhould not be *attempted*; though it may not be *hid*, if it offer itfelf fpontaneoufly.

Nothing,

Nothing, however, fhould appear vulgar, nor fhould fimplicity be pared down to baldnefs; every thing whimfical or expenfive ought to be ftudioufly avoided;—chaftenefs and frugality fhould appear in every part.

Near the houfe a *ftudied neatnefs* may take place; but, at a diftance, *negligence* fhould rather be the characteriftic.

If a tafte for botany lead to a collection of native fhrubs and flowers, a fhrubbery will be requifite; but, in this, every thing fhould be native. A gaudy exotic ought not to be admitted; nor fhould the lawn be kept clofe fhaven; its flowers fhould be permitted to blow; and the herbage, when mown, ought to be carried off, and applied to fome ufeful purpofe.

In the artificial accompaniments, ornament muft be fubordinate; utility muft prefide. The buildings, if any appear, fhould be thofe in actual ufe in rural economics. If the hovel be wanted, let it appear; and, as a fide-fcreen, the barn and rickyard are admiffible; whilft the dove-houfe and poultry-yard may enter more freely into the compofition.

In fine, the ORNAMENTED COTTAGE ought to exhibit cultivated Nature in the firft ftage of refinement. It ranks next above the farm-houfe. The plain garb of rufticity may be fet off to advantage; but the ftudied drefs of the artift ought not to appear. That becoming neatnefs, and thofe domeftic conveniencies, which render the rural life agreeable to a cultivated mind, are all that fhould be aimed at.

THE VILLA.

HERE, a ſtyle very different from the preceding ought to prevail : It ought to be *elegant*, *rich*, or *grand*, according to the ſtyle of the houſe itſelf, and the ſtate of the ſurrounding country ; the principal buſineſs of the artiſt being to connect theſe two in ſuch a manner, that the one ſhall not appear naked or flareing, nor the other deſolate and inhoſpitable.

If the houſe be ſtately, and the adjacent country rich and highly cultivated, a ſhrubbery may intervene, in which Art may ſhew her utmoſt ſkill. Here, the artiſt may even be permitted to *play at landſcape :* for a place of this kind being ſuppoſed to be ſmall, the purpoſe principally ornamental, and the point of view probably confined ſimply to the houſe, ſide-ſcreens may be formed, and a foreground laid out ſuitable to the beſt diſtance that can be caught.

If buildings or other artificial ornaments abound in the offſcape, ſo as to mark it ſtrongly, they ought alſo to appear more or leſs in the fore-ground : if the diſtance abound with wood, the fore-ground ſhould be thickened, leſt baldneſs ſhould offend ; if open and naked, elegance rather than richneſs ought to be ſtudied, leſt heavineſs ſhould appear.

It is far from being any part of our plan to cavil unneceſſarily at artiſts, whether living or dead ; we cannot, however, refrain from expreſſing a concern for the almoſt total neglect of the principles
here

here laid down, in the prevailing practice of a late celebrated artist, in ornamenting the vicinages of villas. We mention it the rather, as Mr. Brown seems to have *set the fashion*, and we are sorry to find it copied by the inferior artists of the day. Without any regard to uniting the house with the adjacent country, and, indeed, seemingly without any regard whatever to the offscape, one invariable plan of embellishment prevails; namely, that of stripping the fore-ground entirely naked, or nearly so, and surrounding it with a wavy border of shrubs and a gravel walk; leaving the area, whether large or small, one naked sheet of greensward.

In small confined spots, this plan may be eligible. We dislike those bolstered flower-beds which abound in the suburbs of the metropolis, where the broken-ground sometimes exceeds the lawn: nevertheless, to our apprehension, a simple border round a large unbroken lawn only serves to shew what more is wanted. Simplicity in general is pleasing; but even simplicity may be carried to an extreme, so as to convey no other idea than that of poverty and baldness. Besides, how often do we see in natural scenery, the holly and the foxglove flourishing at the foot of an oak, and the primrose and the campion adding charms to the hawthorn scattered over the pastured lawn? And we conceive that single trees footed with evergreens and native flowers, and clumps as well as borders of shrubs, are admissible in *ornamental* as well as in *natural* scenery.

Sf 3   The

The fpecies of fhrub will vary with the purpoft. If the principal intention be a winter retreat, ever-greens and the early-blowing fhrubs fhould predo-minate; but in a place to be frequented in fummer and autumn, the deciduous tribes ought chiefly to be planted.

PRINCIPAL RESIDENCE.

HERE the whole art centers. The artift has here full fcope for a difplay of tafte and genius. He has an extent of country under his eye, and will endeavour to make the moft of what nature and accident have fpread before him.

Round a Principal Refidence, a gentleman may be fuppofed to have fome confiderable eftate, and it is not a fhrubbery and a ground only, which fall under the confideration of the artift: he ought to endeavour to difclofe to the view, either from the houfe or fome other point, as much as he conveni-ently can of the adjacent eftate. The love of poffeffion is deeply planted in every man's breaft; and places fhould bow to the gratification of their owners. To curtail the view by an artificial fide-fcreen, or any other unnatural machinery, fo as to deprive a man of the fatisfaction of over-looking his own eftate, is an abfurdity which no artift ought to be permitted to be guilty of. It is very dif-ferent, however, where the property of another in-

trudes

trudes upon the eye: Here the view may; with some colour of propriety, be bounded by a woody screen.

After what has been said under the head GENE-RAL APPLICATION, little remains to be added here. Indeed, it would be in vain to attempt to lay down particular rules: different places are marked by sets of features as different from each other as are those in men's faces. Much muft be left to the skill and taste of the artift; and let those be what they may, nothing but mature study of the natural abilities of the particular place to be improved, can render him equal to the execution, so as to make the moft of the materials that are placed before him.

Some few general rules may neverthelefs be laid down. The approach ought to be conducted in such a manner, that the striking features of the place shall burst upon the view at once: no trick however should be made use of: all should appear to fall in naturally. In leading towards the house its direction should not be fully in front, nor exact-ly at an angle, but should pafs obliquely upon the house and its accompaniments; so that their position with refpect to each other, as well as the perspec-tive appearance of the house itfelf, may vary at every step: and, having shewn the front and the prin-cipal wing, or other accompaniment, to advantage, the approach should wind to the back-front, which,

as

as has been already obferved, ought to lie open to the park or paftured grounds.

The improvements and the rooms from which they are to be feen fhould be in *unifon*. Thus, the view from the drawing-room fhould be highly embellifhed, to correfpond with the beauty and elegance within : every thing here fhould be *feminine*— elegant—beautiful—fuch as attunes the mind to politenefs and lively converfation The breakfaft-ing-room fhould have more mafculine objects in view : wood, water, and an extended country for the eye to roam over ; fuch as allures us imperceptibly to the ride or the chace. The eating and banquetting rooms need no *exterior* allurements.

There is a harmony in tafte as in mufic : variety, and even wildnefs upon fome occafions, may be admitted ; but difcord cannot be allowed. If, therefore, a place be fo circumftanced as to confift of properties totally irreconcileable, the parts ought, if poffible, to be feparated in fuch a manner, that, like the air and the recitative, the adagio and the allegro, in mufic, they may fet off each other's charms by the contraft. We will endeavour to illuftrate our meaning, and clofe our prefent performance, with a defcription and propofed improvement of *Persfield*, the feat of Mr. Morris, near Chepftow in Monmouthfhire ; a place upon which Nature has been peculiarly lavifh of her favours ; and which has been fpoken of by Mr. Wheatley, Mr. Gilpin,

GILPIN, and other writers in the moſt flattering terms.

Persfield is ſituated upon the banks of the river Wye, which divides Glouceſterſhire and Monmouth ſhire, and which was formerly the boundary between England and Wales. The general tendency of the river is from north to ſouth, but about Persfield it deſcribes by its winding courſe the letter S, ſome-what compreſſed, ſo as to reduce it in length and increaſe its width. The grounds of Persfield are lifted high above the bed of the river, ſhelving, and form the brink of a lofty and ſteep precipice, towards the ſouth weſt.

The lower limb of the letter is filled with *Perſe-wood*, which makes a part of Perſefield; but is at preſent an impenetrable thicket of coppice-wood. This dips to the ſouth-eaſt down to the water's edge; and, ſeen from the top of the oppo-ſite rock, has a good effect.

The upper limb receives the farms of *Llan-cot*: rich and highly cultivated: broken into in-cloſures, and ſcattered with groups and ſingle trees: two well looking farm-houſes in the center, and a neat white chapel on one ſide: altogether a lovely little paradiſaical ſpot. The lowlineſs of its ſitua-tion ſtamps it with an air of meekneſs and humi-lity; and the natural barriers which ſurround it adds that of peacefulneſs and ſecurity. Theſe pictureſque farms do not form a low flat bottom, ſubject to be overflowed by the river; but take the form of a gorget, riſing fulleſt in the middle, and

fall-

falling on every fide gently to the brink of the Wye;
except on the eaft-fide, where the top of the gorget
leans in an eafy manner againft a range of perpendicular rock ; as if to fhew its difk with advantage
to the walks of Perfefield.

This rock ftretches acrofs what may be called
the Ifthmus, leaving only a narrow pafs down into
the fields of Llancot, and joins the principal range
of rocks at the lower bend of the river.

To the north, at the head of the letter, ftands
an immenfe rock (or rather a pile of immenfe rocks
heaped one above another) called Windcliff; the
top of which is elevated as much above the grounds
of Perfefield as thofe are above the fields of
Llancot.

Thefe feveral rocks, with the wooded precipices
on the fide of Perfefield, form a circular inclofure,
about a mile in diameter, including Perfe-wood.
Llancot, the Wye, and a fmall meadow lying at
the foot of Windcliff.

The grounds are divided into the upper and
lower lawn *, by the approach to the houfe: a
fmall irregular building ; ftanding near the brink of
the precipice ; but facing down the lower lawn : a

---

* Mr. Wheatley fays, the park contains about three hundred
acres : but we think the two *lawns* cannot contain fo much ;
and if the hanging-wood at the bottom of the lower lawn, with
the face of the Precipice and Perfewood be added, they contain
a great deal more.

beau-

beautiful ground, falling " precipitately every way
into a valley which fhelves down in the middle ;"
and is fcattered with groups and fingle trees in an
excellent ftyle.

The view from the houfe is foft, rich, and
beautifully picturefque :—the lawn and woods of
Perfefield and the oppofite banks of the river :—
the Wye, near its mouth, winding thro' " meadows green as emerald," in a manner peculiarly
graceful :—the Severn, here very broad, backed by
the wooded and highly cultivated hills of Gloucefterfhire, Wiltfhire, and Somerfetfhire. Not one
rock enters into the compofition :—The whole view
confifts of an elegant arrangement of lawn, wood,
and water.

The upper lawn is a lefs beautiful ground,
and the view from it, though it command the
" cultivated hills and rich vallies of Monmouthfhire," bounded by the Severn and backed by the
Mendip-hills, is much inferior to that from the
houfe.

To give variety to the views from Perfefield, to
difclofe the native grandeur which furrounds it, and
to fet off its more ftriking features to advantage,
walks have been cut through the woods and on the
face of the precipice which border the grounds to
the fouth and eaft. The viewer enters thefe walks
at the lower corner of the lower lawn.

The

The firſt point of view is marked by an alcove, from which are ſeen the bridge and the town of Chepſtow, with its caſtle ſituated in a remarkable manner on the very brink of a perpendicular rock, waſhed by the Wye: and beyond theſe the Severn ſhews a ſmall portion of its ſilvery ſurface.

Proceeding a little farther along the walk, a view is caught which the painter might call a complete landſcape: The caſtle with the ſerpentine part of the Wye below Chepſtow, *intermixed* in a peculiar manner with the broad waters of the Severn, form the fore-ground; which is backed by diſtant hills: the rocks, crowned with wood, lying between the alcove and the caſtle, to the right; and Caſtle-hill farm, elevated upon the oppoſite banks of the river, to the left—form the two ſide-ſcreens. This point is not marked, and muſt frequently be loſt to the ſtranger.

The grotto, ſituated at the head of Perſe-wood, commands a near view of the oppoſite rocks:— magnificent beyond deſcription! The littleneſs of human art was never placed in a more humiliating point of view:—the caſtle of Chepſtow, a *noble fortreſs*, is, compared with theſe natural bulwarks, a mere *houſe of cards*.

Above the grotto, upon the iſthmus of the Perſe-field ſide, is a ſhrubbery:—ſtrangely miſplaced! an unpardonable intruſion upon the native grandeur of this ſcene. Mr. GILPIN's obſervations upon this —as upon every other occaſion---are very juſt.

He

He fays, " It is pity the ingenious Embellifher of thefe fcenes could not have been fatisfied with the great beauties of nature which he commanded. The fhrubberies he has introduced in this part of his improvements I fear will rather be efteemed paltry."——" It is not the fhrub which offends : it is the *formal introduction* of it. Wild underwood may be an appendage of the grandeft fcene : it is a beautiful appendage. A bed of violets or of lillies may enamel the ground with propriety at the foot of an oak ; but if you introduce them artificially in a border, you introduce a trifling formality, and difgrace the noble object you wifh to adorn." (GILPIN *on the Wye*, p. 42.)

The walk now leaves the wood and opens upon the lower lawn, until coming near the houfe it enters the alarming precipice facing Llancot; winding along the face of it in a manner which does great honour to the artift. Sometimes the fragments of rock which fall in its way are avoided, at other times partially removed, fo as to conduct the path along a ledge carved out of the rock ; and in one inftance, a huge fragment, of a fomewhat conical fhape and many yards high, is perforated ; the path leading through its bafe. This is a thought which will hand down to future times the greatnefs of Mr. MORRIS's tafte : the defign and the execution are equally great : not a mark of a tool to be feen ; all appears perfectly natural. The arch-way is made winding, fo that on the approach

proach it appears to be the mouth of a cave; and, on a nearer view, the idea is ftrengthened by an allowable deception; a black dark hole on the fide next the cliff, which, feen from the entrance before the perforation is difcovered, appears to be the darkfome inlet into the body of the cave.

From this point, that vaft inclofure of rocks and precipices which marks the peculiar magnificence of Perfefield, is feen to advantage. The area, containing in this point of view the fields of Llancot and the lower margin of Perfe-wood, is broken in a manner peculiarly picturefque by the graceful winding of the Wye; here wafhing a low graffy fhore, and there fweeping at the feet of the rocks, which rife in fome places perpendicular from the water : but in general they have a wooded offfett at the bafe; above which they rife to one, two, or perhaps three or four hundred feet high; expofing one full face, filvered by age, and bearded with ivy, growing out of the wrinkle-like feams and fiffures. If one might be allowed to compare the paltry performances of art with the magnificent works of nature, we fhould fay, that this inclofure refembles a prodigious fortrefs which has lain long in ruins. It is in reality one of nature's ftrong-holds ; and as fuch has probably been frequently made ufe of — Acrofs the ifthmus on the Gloucefterfhire-fide there are the remains of a deep intrenchment, called to this day the Bulwark ; and tradition ftill teems with the extraordinary warlike feats that have been performed among this romantic fcenery.

From

From the perforated rock, the walk leads down
to the cold-bath (a complete place), feated about
the mid-way of the precipice, in this part lefs fteep:
and from the cold-bath a rough path winds down
to the meadow, by the fide of the Wye, from
whence the precipice on the Perfefield-fide is feen
with every advantage: the giant fragments, hung
with fhrubs and ivy, rife in a ghaftly manner from
amongft the underwood, and fhew themfelves in
all their native favagenefs *.

From the cold-bath upward, a coach-road (very
fteep and difficult) leads to the top of the cliff, at
the upper corner of the upper lawn. Near the top
of the road is a point which commands one of the
moft pleafing views of Perfefield : The Wye fweeping through a graffy vale which opens to the left :—
Llancot backed by its rocks, with the Severn immediately behind them ; and, feen in this point of
view, feems to be divided from the Wye by only a
fharp ridge of rock, with a precipice on either fide ;
and behind the Severn the vale and wooded hills of
Gloucefterfhire.

From this place a road leads to the top of Windcliff—aftonifhing fight ! The face of nature probably affords not a more magnificent fcene ! Llancot in all its grandeur ; the grounds of Perfefield ;
the caftle and town of Chepftow ; the graceful

---

* There is another way down into this meadow: a kind of
winding ftair-cafe, furrowed out of the face of the precipice,
behind the houfe, and leading down into a walk made on the
fide of the river ; but being at prefent out of repair, the defcent
this way is rendered very difficult, and fomewhat dangerous.

windings of the Wye below, and its conflux with
the Severn: to the left, the foreft of Dean:
to the right, the rich marfhes and picturefque
mountains of South Wales: a broad view of
the Severn, opening its fea-like mouth: the conflux of the Avon, with merchant fhips at anchor in
King-road, and veffels of different defcriptions under fail: Auft-Cliff, and the whole vale of Berke-
ley, backed by the wooded fwells of Gloucefterfhire; the view terminating in clouds of diftant
hills, rifing one behind another, until the eye becomes unable to diftinguifh the earth's billowy
furface from the clouds themfelves *.

The leading principle of the propofed improvement would be to feparate the *fublime* from the
beautiful; fo that in viewing the one, the eye might
not fo much as fufpect that the other was near.

Let the *hanging-walk* be conducted entirely along
the precipices, or through the thickets, fo as to
difclofe the natural fcenery, without once difcover-
ing the lawn or any other acquired foftnefs. Let
the path be as rude as if trodden only by wild beafts
and favages, and the refting-places, if any, as ruftic as poffible.

Erafe entirely the prefent fhrubbery, and lay out
another as elegant as nature and art could render
it before the houfe, fwelling it out into the lawn,

---

o The waters of the Severn and Wye, being principals in
thefe views, and being fubject to the ebbings and flowings of the
tide, which at the bridge of Chepftow rifes to the almoft incre-
dable height of forty or fifty feet; it follows, that the time of fpring-
tide and high water is the propereft time for going over Perfefield.

towards

towards the ſtables; between which and the
kitchen-garden make a narrow winding entrance.

Convert the upper lawn into a deer-paddock,
ſuffering it to run as wild, rough, and foreſt-like
as total negligence would render it.

The viewer would then be thus conducted: He
would enter the *hanging-walk* by a ſequeſtered path,
at the lower corner of the lawn *, purſuing it thro'
the wood to beneath the grotto; and round the
head-land, or winding through Perſe-wood, to the
perforated rock and the cold-bath, without once
conceiving an idea (if poſſible) that art, or at
leaſt that much art, had been made uſe of in
diſcloſing the natural grandeur of the ſurrounding objects; which ought to appear as if they pre-
ſented themſelves to his view, or, at moſt, as if
nothing was wanted but his own penetration and
judgement to find them out. The walk ſhould
therefore be conducted in ſuch a manner, that the
breaks might be quite natural; yet the points of
view obvious, or requiring nothing but a block or
a ſtone to mark them. A ſtranger at leaſt wants no
ſeat here; he is too eager, in the early part of his
walk, to think of lounging upon a bench.

From the cold-bath he would aſcend the ſteep,
near the top of which a commodious bench or
benches might be placed: the fatigue of aſcending
the hill would require a reſting-place; and there are

* A young plantation below the entrance into the lower lawn
has been placed as it were for that purpoſe.

T t   few

few points which afford a more pleafing view than
this; it is grand, without being too broad and
glaring.

From thefe benches he would enter the *foreft*
part. Here the idea of Nature in her primitive
ftate would be ftrengthened : the roughneffes and
deer to the right, and the rocks in all their native
wildnefs to the left. Even Llancot might be fhut
out from the view by the natural fhrubbery of the
cliff. The Lover's-Leap, however (a tremendous
peep), might remain; but no benches nor other
work of art fhould here be feen. A natural path,
deviating near the brink of the precipice, would
bring the viewer down to the lower corner of the
park; where benches fhould be placed in a happy
point, fo as to give a full view of the rocks and
native wildneffes, and, at the fame time, hide the
farm-houfes, fields, and other acquired beauties of
Llancot.

Having fatiated himfelf with this favage fcene,
he would be led, by a ftill ruftic path, through the
labyrinth—when the fhrubbery, the lawn, with all
its appendages, the graceful Wye and the broad
filver Severn, would break upon the eye with every
advantage of ornamented nature: the tranfition
could not fail to ftrike.

From this foft fcene he would be fhewn to the
top of Windcliff, where, in one vaft view, he
would unite the fublime and beautiful of Perfefield.

F I N I S.

ENGLISH NAMES,

AND OTHER

NON-LINNEAN TERMS,

IN THE

ALPHABET OF PLANTS.

A.

ALMOND,	See	*Amygdalus.*
Angelica Tree,		*Aralia.*
Andrachne,		*Arbutus.*
Alder,		*Betula.*
Allfpice Tree,		*Calycanthus.*
Alder, American,		*Clethra.*
Azarole,		*Cratægus.*
Aria,		*Gratægus.*
Afh,		*Fraxinus.*
Acacia, three-thorned,		*Gleditfia.*
Althea frutex,		*Hibifcus.*
Amelanchier,		*Mefpilus.*
Afpen,		*Populus.*
Apricot,		*Prunus.*
Apple,		*Pyrus.*

Alder,

Bullace,

Bullace, - - - -	*Prunus.*
Buckthorn, - - -	*Rhamnus.*
Briar, sweet, - - -	*Rosa.*
Bramble, — —	*Rubus.*
Broom, butcher's, ——	*Ruscus.*
Bindweed, — —	*Smilax.*
Bittersweet, — —	*Solanum.*
Broom, — —	*Spartium.*
Bladder Nut, ——	*Staphylea.*
Briony, black, — —	*Tamus.*

C.

Custard Apple, - -	*Annona.*
Catalpa, - - -	*Bignonia.*
Climber, Virginia, - -	*Bignonia.*
Climbers, Class of, - -	*Clematis.*
Cherry, Cornelian, - -	*Cornus.*
Colutea, jointed, - -	*Coronilla.*
Cypress, - -	*Cupressus*
Cneorum, - - -	*Daphne.*
Chamelæa, - -	*Daphne.*
Chesnut, - - -	*Fagus.*
Chinquepin, - - -	*Fagus.*
Cytisus of Montpelier, - -	*Genista,*
Creeper, Virginia, - -	*Hedera.*
Cedars, Class of, - -	*Juniperus.*
Savin, - - -	*Juniperus.*
Crab, - -	*Pyrus.*
Cork-Tree, - -	*Quercus.*
Christi Thorn, - -	*Rhamnus.*

T t 3   Chamæ

Chamæ-ciftus,	- -	*Rhododendron.*
Coccygria,	- - -	*Rhus.*
Caragana,	- - -	*Robinia.*
Cat-Whin,	- - -	*Rofa.*
Cæfius,	- - -	*Rubus.*
Cafine,	- - - -	*Viburnum.*
Caffioberry,	- - -	*Viburnum.*
Chafte Tree,	- -	*Vitex.*
Cytifus, prickly,	- -	*Spartium.*
Cherry, common,	- -	*Prunus.*
Cherry, dwarf,	- -	*Lonicera.*
Cucumber Tree,	- -	*Philadelphus.*
Cembro-Pine,	- -	*Pinus.*
Cedar of Lebanon,	- -	*Pinus.*

D.

Date Plum,	- -	*Diofpyros.*
Diervilla,	- -	*Lonicera.*
Dog's Bane,	- -	*Periploca.*
Dewberry,	- - -	*Rubus.*
Dogwoods, Clafs of,	- -	*Cornus.*

E.

Efculus,	- -	*Æfculus.*
Evonymus, baftard,	-	*Celaftrus.*
Evonymus, Clafs of,	-	*Euonymus.*
Elder,	- - -	*Sambucus.*
Elm,	- -	*Ulmus.*
Eglantine,	- ‹	*Rofa.*
Elder, marfh,	- -	*Viburnum.*

Fringe

F.

Fringe Tree,	- -	*Chionanthus.*
Filbert,	- - -	*Corylus.*
Firs,	- - -	*Pinus.*
Frangula,	- -	*Rhamnus.*
Furze,	- - -	*Ulex.*
Flammula,	- - -	*Clematis.*

G.

Gale, spleenwort-leaved,	-	*Liquidamber.*
Gale, sweet,	- -	*Myrica.*
Gelder Rose Spiræa,	- -	*Spiræa.*
Gelder Rose Marsh Elder,		*Viburnum.*
Grape,	- -	*Vitis.*
Gorse,	- -	*Ulex.*
Glafswort,	-	*Salsola.*
Glaftonbury Thorn,	-	*Cratægus.*

H.

Horse Chesnut,	- -	*Æsculus.*
Honeysuckle, upright,	-	*Azalea.*
Hartwort, Ethiopian,	-	*Bupleurum.*
Hornbeam,	- -	*Carpinus.*
Hazel,	- -	*Corylus.*
Hawthorn,	- -	*Cratægus.*
Horse-tail,	- -	*Ephedra.*
Hazel, dwarf,	- -	*Hamamelis.*
Holly,	- - -	*Ilex.*
Hickery Nut,	- - -	*Juglans.*

T t 4   Honey.

Honeyfuckles, Clafs of,	-	*Lonicera,*
Hypogloffum,	-	*Rufcus.*
Hypericum frutex,	-	*Spiræa.*
Haw, black,	- -	*Viburnum.*
Hep-Tree,	- - ˉ	*Rofa.*

I.

Indigo, baftard,	- -	*Amorpha*
Jafmine, Virginia,	-	*Bignonia.*
Judas Tree,	- - -	*Cercis.*
Ivy,	- - -	*Hedera.*
Jafmines, Clafs of,	- -	*Jafminus.*
Jeffamy,	—— —	*Jafminus.*
Ilex,	- - -	*Quercus.*
Juniper,	- -	*Juniperus.*

K.

Kidneybean plant,	- -	*Glycine.*
Kermes,	- - -	*Quercus.*

L.

Laburnum,	- - -	*Cytifus.*
Lucerne Tree	- -	*Medicago.*
Larch,	— -	*Pinus.*
Larix,	- -	*Pinus.*
Laurel, Common and Portugal,		*Prunus.*
Laurel, American Mountain,		*Rhododendron.*
Laurel, Alexandrian,	-	*Rufcus.*
Lilac,	- - -	*Syringa.*

Lime,

Lime,	- - -	*Tilia.*
Linden,	- -	*Tilia.*
Lauruſtinus,	- -	*Viburnum.*
Lote-Tree,	- - -	*Celtis.*

M.

Mezereon,	- - -	*Daphne.*
Milkvetch,	- - -	*Glyene..*
Mallow, Cyrian,	-	*Hibiſcus.*
Mallow-Tree,	- -	*Lavatera.*
Medick-Tree,	- -	*Medicago.*
Moon Trefoil,	- -	*Medicago.*
Medlar,	- -	*Meſpilus.*
Mulberry,	- -	*Morus.*
Myrtle, Candleberry,	-	*Myrica.*
Myric,	- -	*Myrica.*
Mock Orange,	-	*Philadelphus.*
Mock Privet,	-	*Phillyrea.*
Mahaleb,	- -	*Prunus.*
Mountain-Aſh,	-	*Sorbus.*
Meally Tree,	- -	*Viburnum.*
Marſh Elder,	-	*Viburnum.*
Miſletoe,	- -	*Viſcum.*
Maples, Claſs of,	- -	*Acer.*
Moon-Seed,	- -	*Moniſpermum.*
Myrtle, Dutch,	-	*Myrica.*

Nettle

N.

Nettle-Tree,	—	*Celtis.*
Nickar-Tree,	—	*Guilandina.*
Nightſhade, woody,	—	*Solanum.*

O.

Old Man's Beard,	—	*Clematis,*
Oleaſter,	—	*Eleagnus.*
Olive, wild,	—	*Elæagnus.*
Orange, mock,	—	*Philadelphus,*
Oak,	—	*Quercus.*
Oak, poiſon,	—	*Rhus.*
Ozier,	—	*Salix.*

P.

Peach,	—	*Amygdalus,*
Papaw,	—	*Anuona,*
Periploca,	——	*Cynanchum.*
Piſhamin Plum,	—	*Dioſpyros.*
Privet,	——	*Liguſtrum.*
Pyracantha,	—	*Meſpilus.*
Paſſion-Flower,	—	*Paſſiflora.*
Privet, mock,	—	*Phillyrea.*
Pine,	——	*Pinus.*
Plane,	——	*Platanus.*
Poplar,	——	*Populus.*
Padus,	——	*Prunus.*
Plum,	——	*Prunus.*
Pear,	——	*Pyrus.*
Paliurus,	——	*Rhamnus.*
Perriwinkle,	——	*Vinca.*

Per-

Pervinca,	——	*Vinca.*
Pepper Tree,	—	*Vitis.*
Pinaster,	——	*Pinus.*
Petty Whin,	—	*Genista.*

Q.

Quince, dwarf,	—	*Mespilus.*
———, bastard,	—	———.
——, common,	- -	*Pyrus.*
Quick Beam,	—	*Sorbus.*
Quicken Tree,	—	———.
Quick,	——	*Cratægus.*

R.

Red Twig,	——	*Ceanothus.*
Red Bud,	—	*Cercis.*
Rock Rose,	——	*Cistus.*
Rest Harrow, climbing,		*Glycine.*
Rest Harrow, shrubby,		*Ononis.*
Rosebay, dwarf,	—	*Rhododendron.*
Roses, Class of,	—	*Rosa.*
Raspberry,	—	*Rubus.*
Roan-Tree,	—	*Sorbus.*

S.

Sycamore,		*Acer.*
Strawberry-tree,	—	*Arbutus.*
Sea-Purslain-Tree,	—	*Atriplex.*
Staff-Tree,	—	*Celastrus.*
Snow-drop-Tree,	—	*Chionanthus.*
Sumach, myrtle-leaved,	-	*Coriaria.*
Scorpion, Sena,	—	*Coronilla.*
Service, wild,	—	*Cratægus.*

Scam-

Thymelæa,	- -	*Daphne.*
Tarton-raire,	- -	*Daphne.*
Tulip-Tree, Virginia,		*Liriodendron.*
Tulip-Tree, bay-leaved,		*Magnolia.*
Thorn, evergreen,	-	*Mespilus.*
Tupelo-Tree,	- -	*Nyssa.*
Turpentine-Tree,	- -	*Piſtacia.*
Thorn, black,	- -	*Prunus.*
Trefoil Shrub,	- -	*Ptelea.*
Toxicodendron,	- -	*Rhus.*
Tamariſk,	- -	*Tamarix.*
Thea, South-Sea,	-	*Viburnum.*
Toothache Tree,	-	*Zanthoxylum.*
Thorn of Chriſt,	-	*Rhamnus.*

V.

Virginia Jaſmine,	- -	*Bignonia.*
Virginia Climber,	- -	*Bignonia.*
Virgin's Bower,	- -	*Clematis.*
Varniſh Tree,	- -	*Rhus.*
Vine,	- -	*Vitis.*

U.

Umbrella-Tree,	- -	*Magnolia.*

W.

Wormwood-Tree,	-	*Artemiſia.*
Widow-wail,	- -	*Cneorum.*
White-Thorn,	- -	*Cratægus.*
White-Beam,	- -	———.
White-Leaf,	- -	———.
Wild Olive,	- -	*Elæagnus.*

Wood-

Printed in the United States
By Bookmasters